International
REVIEW OF
Neurobiology
Volume 70

International
REVIEW OF
Neurobiology

Volume 70

SERIES EDITORS

Auditory Spectral Processing

EDITED BY

MANUEL S. MALMIERCA
Laboratory for the Neurobiology of Hearing
The Institute of Neuroscience of 'Castilla y Leon' (INCyL)
Faculty of Medicine, University of Salamanca
Campus 'Miguel de Unamuno'
37007 Salamanca, Spain

DEXTER R. F. IRVINE
Monash University
Department of Psychology
School of Psychology, Psychiatry, and Psychological Medicine
Faculty of Medicine, Nursing, and Health Sciences
Vic 3800, Australia

ELSEVIER

AMSTERDAM • BOSTON • HEIDELBERG • LONDON
NEW YORK • OXFORD • PARIS • SAN DIEGO
SAN FRANCISCO • SINGAPORE • SYDNEY • TOKYO
Academic Press is an imprint of Elsevier

Elsevier Academic Press
525 B Street, Suite 1900, San Diego, California 92101-4495, USA
84 Theobald's Road, London WC1X 8RR, UK

This book is printed on acid-free paper. ♾

For all information on all Elsevier Academic Press publications visit our Web site at www.books.elsevier.com

ISBN-13: 978-0-1236-6871-4
ISBN-10: 0-12-366871-9

Printed and bound by CPI Group (UK) Ltd, Croydon, CR0 4YY

Transferred to Digital Print 2012

CONTENTS

Across-Channel Spectral Processing

JOHN H. GROSE, JOSEPH W. HALL III, AND EMILY BUSS

Speech and Music Have Different Requirements for Spectral Resolution

ROBERT V. SHANNON

Non-Linearities and the Representation of Auditory Spectra

ERIC D. YOUNG, JANE J. YU, AND LINA A. J. REISS

Spectral Processing in the Inferior Colliculus

KEVIN A. DAVIS

Neural Mechanisms for Spectral Analysis in the Auditory Midbrain, Thalamus, and Cortex

MONTY A. ESCABÍ AND HEATHER L. READ

Spectral Processing in the Auditory Cortex

MITCHELL L. SUTTER

Processing of Dynamic Spectral Properties of Sounds

Adrian Rees and Manuel S. Malmierca

Representations of Spectral Coding in the Human Brain

Deborah A. Hall

Spectral Processing and Sound Source Determination

Donal G. Sinex

Spectral Information in Sound Localization

Simon Carlile, Russell Martin, and Ken McAnally

Plasticity of Spectral Processing

DEXTER R. F. IRVINE AND BEVERLY A. WRIGHT

Spectral Processing in Cochlear Implants

COLETTE M. MCKAY

CONTRIBUTORS

Numbers in parentheses indicate the pages on which the authors' contributions begin.

Emily Buss (87), Department of Otolaryngology/Head & Neck Surgery, University of North Carolina at Chapel Hill, Chapel Hill, North Carolina 27599, USA

Simon Carlile (399), Auditory Neuroscience Laboratory, School of Medical Sciences, University of Sydney, Sydney NSW 2006, Australia

Kevin A. Davis (169), Departments of Biomedical Engineering & Neurobiology and Anatomy & Center for Navigation and Communication Sciences, University of Rochester, Rochester, New York 14642, USA

Monty A. Escabí (207), Department of Electrical Engineering, University of Connecticut, Storrs, Connecticut 06269, USA

John H. Grose (87), Department of Otolaryngology/Head & Neck Surgery, University of North Carolina at Chapel Hill, Chapel Hill, North Carolina 27599, USA

Deborah A. Hall (331), MRC Institute of Hearing Research, Nottingham NG7 2RD, United Kingdom

Joseph W. Hall III (87), Department of Otolaryngology/Head & Neck Surgery, University of North Carolina at Chapel Hill, Chapel Hill, North Carolina 27599, USA

Dexter R. F. Irvine (1, 435), Department of Psychology, Faculty of Medicine, Nursing, and Health Sciences, Monash University, Victoria 3800, Australia

Enrique A. Lopez-Poveda (7), Instituto de Neurociencias de Castilla y León, Universidad de Salamanca, Salamanca 37007, Spain

Manuel S. Malmierca (1, 299), Laboratory for the Neurobiology of Hearing, The Institute of Neuroscience of 'Castilla y Leon' and Faculty of Medicine, University of Salamanca, Salamanca 37007, Spain

Russell Martin (399), Defense Science and Technology Organisation, Fisherman's Bend, Victoria 3207, Australia

Ken McAnally (399), Defense Science and Technology Organisation, Fisherman's Bend, Victoria 3207, Australia

Colette M. McKay (473), School of Life and Health Sciences, Aston University, Birmingham B4 7ET, United Kingdom

Brian C. J. Moore (49), Department of Experimental Psychology, University of Cambridge, Cambridge CB2 3EB, United Kingdom

Heather L. Read (207), Department of Psychology, The University of Connecticut, Storrs, Connecticut 06269, USA

Adrian Rees (299), School of Neurology, Neurobiology, and Psychiatry, University of Newcastle, Newcastle upon Tyne NE2 4HH, United Kingdom

Lina A. J. Reiss (135), Department of Speech Pathology and Audiology, University of Iowa, Iowa City, Iowa 52242, USA

Robert V. Shannon (121), House Ear Institute, Los Angeles, California 90057, USA

Donal G. Sinex (371), Department of Psychology, Utah State University, Logan, Utah 84322, USA

Mitchell L. Sutter (253), Center for Neuroscience and Section of Neurobiology, Physiology, and Behavior, University of California Davis, Davis, California 95616, USA

Beverly A. Wright (435), Department of Communication Sciences and Disorders, Northwestern University, Evanston, Illinois 60208, USA

Eric D. Young (135), Department of Biomedical Engineering, Johns Hopkins University, Baltimore, Maryland 21205, USA

Jane J. Yu (135), Department of Biomedical Engineering, Center for Hearing Sciences, Johns Hopkins University, Baltimore, Maryland 21205, USA

AUDITORY SPECTRAL PROCESSING: AN OVERVIEW

Dexter R. F. Irvine* and Manuel S. Malmierca[†]

*Department of Psychology, Faculty of Medicine, Nursing, and Health Sciences
Monash University, Victoria 3800, Australia
[†]Laboratory for the Neurobiology of Hearing, The Institute of Neuroscience of 'Castilla y Leon'
and Faculty of Medicine, University of Salamanca,
Salamanca 37007, Spain

Almost all natural sounds and sounds produced by man-made devices (such as musical instruments) are composed of multiple frequencies and their frequency composition varies over short time periods, i.e., they have complex and time-varying frequency spectra. Although a great deal of information can be derived from auditory signals in which the spectral information is considerably degraded (such as those transmitted by most telephones), it is nevertheless the case that a full perceptual appreciation of most complex sounds depends on the availability of a non-degraded representation of the spectral detail. Examples of such perceptual experiences are the appreciation of music and the extraction of so-called indexical information from speech, i.e., information about the speaker's age, gender, mood, and so on (e.g., Nygaard and Pisoni, 1995). Detailed spectral information has also been shown to be critically important in the processes of identifying the discrete auditory objects comprising complex auditory scenes, in which the sound waves generated by multiple sources sum together to form a single complex pressure wave at each eardrum, and of sound localization. The importance of detailed spectral information in these important aspects of human and animal hearing indicates that this information must be encoded in the peripheral auditory system and then made available to and processed in the central auditory brain pathways and associated structures that ultimately give rise to our auditory perceptual experience. It is the aim of this volume to provide a comprehensive overview of the neural mechanisms of auditory spectral processing and of the ways in which spectral information is used by human listeners to generate their auditory perception of the world around them.

Each point on the basilar membrane (BM) in the cochlea has a characteristic frequency (CF) at which it responds maximally, and the BM carries out a real-time spectral analysis of the pressure wave delivered to the inner ear by the outer and middle ears. The filtering properties of the BM depend partly on its passive mechanical properties, first described in the pioneering studies of von Békésy (1960), and partly on an active process of amplification (produced by forces generated by the outer hair cells [OHCs]) which feeds energy back into the cochlea at sound pressure levels (SPLs) close to threshold (Robles and Ruggero,

INTERNATIONAL REVIEW OF
NEUROBIOLOGY, VOL. 70
DOI: 10.1016/S0074-7742(05)70000-3

2001). As a consequence of these processes, each inner hair cell (IHC) is sharply tuned to the CF of the point at the BM at which it is located, and each of the auditory nerve (AN) fibers that innervate that hair cell has the same CF and sharp frequency tuning as the hair cell from which its input is derived (Patuzzi and Robertson, 1988). One way in which spectral information is conveyed to the central nervous system (CNS), therefore, is in the distribution of activity across the AN fiber array. At frequencies below 3 to 4 kHz, the discharge of AN fibers is phase-locked to the stimulus waveform, with the consequence that information about spectral components in this frequency range is also encoded in the temporal pattern of the discharge of AN fibers. These peripheral mechanisms, and models of the processes involved, are discussed in detail by Lopez-Poveda in Chapter 1.

The mechanical and neural filtering that occurs in the peripheral auditory system underlies the frequency resolution ability exhibited by human and animal listeners. Psychophysical evidence on the frequency resolving ability of human listeners is reviewed by Moore in Chapter 2. As previously noted, however, most natural sounds have complex frequency spectra, and our perception of those sounds involves the integration of information in different frequency bands. This integration begins in the cochlear nucleus, where all of the AN fibers carrying information to the brain terminate, and is continued in the ascending auditory pathways that extend from the cochlear nucleus to the auditory cortex. In Chapter 3, Grose and colleagues discuss the major auditory phenomena that reflect across-channel frequency processing and consider the ways in which this processing can be understood. A number of the aspects of across-channel processing discussed in this chapter relate to the analysis of complex auditory scenes and the segregation of the separate auditory "streams" produced by different sound sources. The role of spectral information in this process is examined explicitly in Chapter 11 by Sinex, who also presents physiological evidence from the auditory midbrain nucleus, the inferior colliculus (IC) relating to some of the neural mechanisms involved. This chapter depends in part on some of the material presented in the chapters on the processing of spectral information at different levels of the auditory neuraxis and, therefore, it follows those chapters. The use of spectral information for sound localization similarly involves integration across frequency bands, and psychophysical and neurophysiological evidence on these processes is reviewed in Chapter 12 by Carlile and his colleagues.

The capacity of human listeners to perform complex perceptual tasks despite degraded spectral information was alluded to in the previous text, and is examined explicitly by Shannon in Chapter 4, in which he contrasts the limited amount of spectral information needed for the perception of speech with the detailed information required for the recognition and appreciation of music. The robustness of speech perception in the face of degraded spectral information reflects the fact that our auditory perceptual experience depends not only on the processing of spectral and other information contained in the acoustic signal but also the interpretation of this information in terms of stored information based on our auditory experience of the world (i.e., on what have been termed "bottom-up" and "top-down" processes). Although the emphasis of this book is on the former, all of the chapters dealing with the more complex aspects of auditory spectral processing point to the overwhelming importance of top-down processing.

Bottom-up processing is the focus of Chapters 5 to 9, in which the central processing of spectral information at successively "higher" levels of the ascending auditory pathway is examined: in the cochlear nucleus by Young and his colleagues (Chapter 5); in the midbrain by Davis (Chapter 6), in the thalamocortical system by Escabí and Read (Chapter 7), and in auditory cortical fields by Sutter (Chapter 8). The psychophysical evidence for integration of information across frequency bands reviewed by Grose *et al.* in Chapter 3 is complemented by neurophysiological evidence for such integration at all levels of the central auditory pathway. Much of this integration is highly non-linear, and in Chapter 5, Young and his colleagues examine in detail the non-linear integration in the circuitry in the dorsal cochlear nucleus, and the difficulties raised by this non-linearity for the task of measuring spectral sensitivity and predicting neural responses to complex stimuli. These issues are also examined in the context of the thalamocortical levels of the system in Chapters 7 and 8.

The spectral selectivity of individual auditory CNS neurons in animal preparations has been measured in a number of ways over the more than 60-year history of auditory single-unit neurophysiology. In what appears to have been the first measurement of neuronal frequency selectivity, Galambos and Davis (1943) defined what they called the "response areas" of AN fibers by determining the range of frequencies by which the fiber was excited at each of a range of SPLs. Similar procedures remain in wide use today: response amplitude (and other response characteristics) are measured over a wide range of frequency-SPL combinations, usually presented in pseudo-random order under computer control, to generate what are commonly called frequency response areas (FRAs). An alternative procedure is simply to measure threshold as a function of frequency (sometimes using an automated threshold-tracking algorithm) to generate a frequency tuning curve (FTC), which corresponds to the boundaries of the FRA. In the absence of spontaneous activity, neither of these methods is able to reveal inhibitory components of neuronal FRAs, a problem solved by the use of more complex two-tone stimuli. More recently, a range of methods using spectrally complex stimuli and reverse-correlation analytic techniques have been employed to generate spectro-temporal receptive fields (STRFs), which reveal the manner in which excitatory and inhibitory components of the response area vary over time. These methods and their applications, strengths, and weaknesses are discussed by Young and his colleagues in Chapter 5, by Escabí and Read in Chapter 7, and by Sutter in Chapter 8.

All of these methods are based on measurement of the action potentials generated by the neurons, and thus do not provide information on excitatory synaptic inputs that are below the threshold for action potential generation. Intracellular recordings from auditory cortical neurons have shown that, like neurons in other sensory cortices, they receive such subthreshold inputs in response to a range of stimuli (in this case, frequency-SPL combinations) outside the "classical" response areas defined in terms of the neurons' spike activity (e.g., Kaur *et al.*, 2004).

In Chapter 9, Rees and Malmierca review the ways in which neurons at different levels of the auditory pathway respond to frequency modulations. Dynamic changes in frequency, whether as modulations of single tones or clusters of harmonics, are important information bearing elements of natural sounds, and acoustically distinguishing attributes in the vocal repertoire of many species. Specificity for parameters like the direction of frequency change is apparent at all levels above the auditory nerve. At more central locations, there is evidence for increased specificity, topographic mapping of frequency modulation parameters, and differences in response properties between cortical subdivisions.

The single unit recording techniques which have provided such rich information about spectral processing mechanisms in animal models are not applicable to the human auditory CNS, except under exceptional conditions, such as when such recordings are made in association with surgery and electrical recording to identify epileptogenic foci (e.g., Brugge *et al.*, 2005). However, the study of human auditory cortical activity, and specifically of spectral processing in human auditory cortex, has been revolutionized in recent years by the development of a range of functional imaging techniques, culminating in functional magnetic resonance imaging (fMRI). These methods, and the data on spectral processing in human auditory cortex that they have yielded, are examined by Hall in Chapter 10. In contrast to the dependence of most animal electrophysiological recording techniques on neural spiking activity, the signals generated by imaging techniques such as fMRI reflect predominantly population synaptic activity.

The processing of spectral information occurs in auditory circuitry that until relatively recently was thought to be modifiable by experience early in development, but stable in adults. It has now been realized that sensory cortical and subcortical structures retain substantial capacity for plasticity in adult animals as a result of altered input consequent on either injury (e.g., Kaas and Florence, 2001) or various types of learning (e.g., Weinberger, 2004). Much of the evidence for such plasticity in the auditory system has been derived from psychophysical and electrophysiological studies of the processing of spectral information, and this evidence is reviewed by Irvine and Wright in Chapter 13.

Almost certainly the most interesting application of our knowledge of spectral processing is provided by the cochlear implant, which has now been used to restore functional hearing to tens of thousands of profoundly deaf people. The design of cochlear implants and the processing strategies utilized in them have been based on our knowledge of peripheral spectral processing, i.e., understanding of cochlear place and temporal coding mechanisms was applied to the task of electrically stimulating surviving auditory nerve fibers in the manner that was most effective in evoking normal auditory percepts. But the flow of information has been bi-directional: the perceptual experiences of implant users have also provided valuable information about basic mechanisms of spectral processing. Furthermore, the improvement in performance of implant users over the post-implantation period stands as perhaps the most dramatic practical example of plasticity in auditory processing. These issues are reviewed in detail by McKay in Chapter 14, and are also examined by Shannon in Chapter 4.

A final comment is required concerning the relationship between the chapters comprising this volume. It is our hope that the chapters together make up an integrated whole which gives a comprehensive overview of auditory spectral processing, although as previously noted more attention is given to bottom-up than to top-down processes. It is realized, however, that the interest of many readers will be focused on particular issues in the field. Therefore, each chapter has been designed to stand alone, even though this might involve some overlap in the content of various chapters, or even disagreement between authors as to the interpretation of particular effects or the value of certain techniques. The overlap is an unavoidable consequence of making the chapters stand-alone; the disagreement is a characteristic of any dynamic research area.

References

Brugge, J. F., Volkov, I. O., Reale, R. A., Garell, P. C., Kawasaki, H., Oya, H., Li, Q., and Howard, M. A. (2005). The posterolateral superior temporal auditory field in humans: Functional organization and connectivity. *In* "Auditory Cortex—A Synthesis of Human and Animal Research" (R. König, P. Heil, E. Budinger, and H. Scheich, Eds.), pp. 145–162. Erlbaum, Mahwah, N.J.

Galambos, R., and Davis, H. (1943). The response of single auditory-nerve fibers to acoustic stimulation. *J Neurophysiol* **6,** 39–58.

Kaas, J. H., and Florence, S. L. (2001). Reorganization of sensory and motor systems in adult mammals after injury. *In* "The Mutable Brain" (J. H. Kaas, Ed.), pp. 165–242. Harwood Academic Publishers, Amsterdam.

Kaur, S., Lazar, R., and Metherate, R. (2004). Intracortical pathways determine breadth of subthreshold frequency receptive fields in primary auditory cortex. *J Neurophysiol* **91,** 2551–2567.

Nygaard, L. C., and Pisoni, D. B. (1995). Speech perception: New directions in research and theory. *In* "Speech, Language, and Communication" (J. L. Miller and P. D. Eimas, Eds.). Academic, San Diego.

Patuzzi, R., and Robertson, D. (1988). Tuning in the mammalian cochlea. *Physiol Rev* **68,** 1009–1082.

Robles, L., and Ruggero, M. A. (2001). Mechanics of the mammalian cochlea. *Physiolo Rev* **81,** 1305–1352.

von Békésy, G. (1960). *In* "Experiments in Hearing" (E. G. Wever, Ed. and transl.). McGraw Hill, N.Y.

Weinberger, N. M. (2004). Experience-dependent response plasticity in the auditory cortex: Issues, characteristics, mechanisms, and functions. *In* "Plasticity of the Auditory System" (T. N. Parks, E. W. Rubel, A. N. Popper, and R. R. Fay, Eds.), pp. 173–227. Springer, N.Y.

SPECTRAL PROCESSING BY THE PERIPHERAL AUDITORY SYSTEM: FACTS AND MODELS

Enrique A. Lopez-Poveda

Instituto de Neurociencias de Castilla y León, Universidad de Salamanca
Salamanca 37007, Spain

The chapter provides an updated review of the basic characteristics of the response of individual auditory nerve (AN) fibers to sounds with simple spectra (one and two tones). Specifically, it summarizes current knowledge on fiber types, dynamic range, adaptation, synchronization, frequency selectivity, level-dependent rate and phase responses, suppression, and distortion. Also reviewed is our understanding on how these characteristics, particularly cochlear-related non-linearities, determine the AN representation of sounds with complex spectra (e.g., speech). Finally, eight representative phenomenological non-linear AN models are reviewed and compared with regard to their ability to account for

INTERNATIONAL REVIEW OF
NEUROBIOLOGY, VOL. 70
DOI: 10.1016/S0074-7742(05)70001-5

0074-7742/05 $35.00

the experimental phenomena described. It is concluded that, despite the reasonable performance of some models, none of them actually reproduces *all* physiological characteristics of the AN response. Furthermore, the analysis reveals that different models not only use different approaches but also are validated using different criteria. The suggestion is made that comparative studies are timely aimed at contrasting models' performance against identical, consented AN phenomena and datasets.

I. Introduction

Psychophysiological models of basic auditory percepts commonly use signal processing analogs to simulate the function of the peripheral auditory system. The objective of this chapter is to review the latter, as the former will be reviewed in other chapters of this book and have already been reviewed elsewhere (e.g., Delgutte, 1996). Specific aims of this chapter are to summarize the basic properties of the auditory nerve (AN) response that underlie the coding of the stimulus spectrum and to review state-of-the-art computational models of signal processing by the peripheral auditory system. Given the topic of this book, special attention is paid to discussing the ability of current models to reproduce the main physiological results pertaining to spectral processing. The discussion focuses on the physiological properties of the (healthy) mammalian AN response. Rather than providing a historical account, the chapter describes the current knowledge and discusses future directions. Readers interested in an in-depth analysis of the seminal work are referred to the relevant papers.

II. Basic Response Properties of Individual Auditory Nerve Fibers

The response properties of single AN fibers have been the subject of exhaustive study over the last four decades. The phenomena that have been described and the associated experimental data (reviewed in the following text and also by Javel, 1986; Ruggero, 1992; and Møller, 2000) constitute invaluable information with which to develop and test computational AN models. On the other hand, a consensus is lacking on a set of properties (or on specific datasets) that all AN models should be able to simulate. Indeed, different models usually focus on specific phenomena and employ different datasets as their validation criteria. The aim of this section is to provide a succinct updated review of the basic properties of the AN response to sounds with simple spectra (one and two tones). The idea is to set a common framework that will be used below to compare the virtues and performance of a representative, although admittedly subjective, selection of phenomenological AN models.

A. ANATOMY

The AN is comprised of about 30,000 fibers in man and 50,000 in cats (Harrison and Howe, 1974). Of them, between 85% and 90% are afferent fibers that innervate inner hair cells (IHCs) and send information from the peripheral to the central auditory system. The remaining 10% to 15% are efferent fibers that innervate outer hair cells (OHCs) (reviewed by Geisler, 1998; Møller, 2000; Pickles, 1988). About 20 fibers innervate each IHC whereas about 6 fibers innervate each OHC. The efferent fibers modulate the functioning of the peripheral system (reviewed by Guinan, 1996).

B. SPONTANEOUS RATE, RATE THRESHOLD, AND DYNAMIC RANGE

On the basis of their spontaneous rate (SR), AN fibers have been classified in either two (Evans and Palmer, 1980; Kiang *et al.*, 1965) or three (Liberman, 1978; Winter *et al.*, 1990) types. Fibers with high spontaneous rates (HSR) (>15 spikes/s) comprise approximately 60% of the population. In response to pure tones at the characteristic frequency (CF; the term characteristic frequency is defined in Section E in the following text), these fibers have low thresholds (<10 dB sound pressure levels [SPLs]) and dynamic ranges of approximately 30 to 40 dB (Evans and Palmer, 1980; Sachs and Abbas, 1974). The remaining 40% of the fibers are of a medium spontaneous rate (MSR) or low spontaneous rate (LSR) type. These have higher thresholds (>15 dB SPL) and wider dynamic ranges (~50 to 60 dB) (Evans and Palmer, 1980; Sachs and Abbas, 1974). The dynamic range of AN fibers is wider in response to narrow- or broad-band noise than in response to tones (Ruggero, 1973; see also Fig. 13 of Rhode and Greenberg, 1994), possibly as a result of suppression (Rhode and Greenberg, 1994; Schalk and Sachs, 1980; see Section I in the following text).

The rate-level functions of HSR fibers for on-CF tones are approximately sigmoidal in shape and symmetrical with respect to a mid-range sound level (Fig. 1A). By contrast, the rate-level functions of MSR fibers (Fig. 1C) are described to be "sloping-saturation" in form (Sachs and Abbas, 1974). This term refers to the fact that the slope of the rate-level function decreases above a certain sound level which is still below the fiber's saturation threshold. It has been shown that such level corresponds to the compression threshold of basilar membrane (BM) responses and that the reduction in slope reflects BM compression (Sachs and Abbas, 1974; Yates, 1990; Yates *et al.*, 1990).

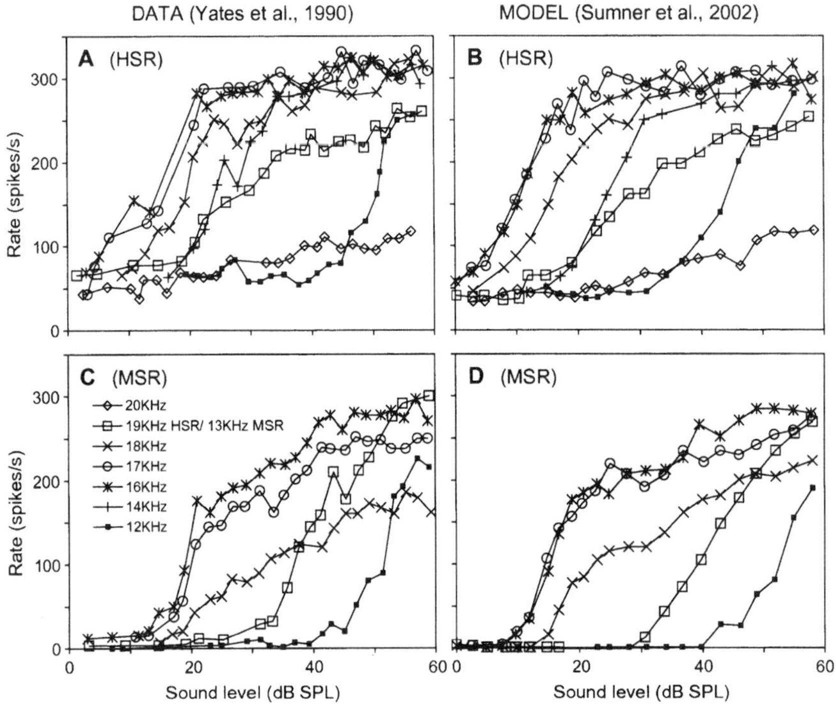

FIG. 1. Panels A and C illustrate rate-level functions for HSR and MSR auditory nerve fibers, respectively (data adapted from Yates *et al.*, 1990, Copyright © 1990 with permission from Elsevier). Panels B and D illustrate corresponding rate-level functions simulated with the auditory nerve model of Sumner *et al.* (2002) (adapted with permission Copyright © 2002 American Institute of Physics). The characteristic frequency of both experimental fibers is approximately 16 kHz. Different symbols illustrate rate-level functions for different stimulation frequencies (as indicated by the inset in panel C). The on-CF rate-level function of the MSR fiber (asterisks) is of the sloping saturation type. The decrease in slope occurs for a level of approximately 20 dB SPL. Note the close correspondence between the experimental and simulated functions.

C. Adaptation

The AN afferent fibers show adaptation effects. The term adaptation encompasses a number of properties. On the one hand, it refers to the observation that the fibers' discharge rate is highest at the stimulus onset and then decays over time until it becomes stable, when it is said to be "adapted" (Westerman, 1985; Westerman and Smith, 1984). The decay is faster over approximately the first 10 to 20 ms of the stimulus and then becomes slower. The rapid and slow decays can be reasonably described as exponential with level-independent time constants of around 3 and 60 ms, respectively (Westerman, 1985). On the other hand, adaptation also refers to the recovery of AN responses from prior stimulation. Immediately after the stimulus termination, the fiber's discharge rate falls below spontaneous activity to its lowest value from which it gradually increases with time until it reaches a value equal to the fiber's spontaneous rate (Westerman, 1985). The onset decrement in discharge rate is assumed to be related to the adapted response of the fiber, which in turn is level dependent. Finally, adaptation also refers to the phenomenon by which an increment/decrement in discharge rate occurs when the sound level suddenly increases/decreases. These discharge rate increments/decrements occur even when the response rate of the fiber prior to the level change is at saturation or at spontaneous values (Smith and Zwislocki, 1975). A short review of adaptation is provided by Sumner *et al.* (2003a).

Adaptation is thought to reflect the dynamics of neurotransmitter release at the synapse between the IHC and the corresponding AN fiber (e.g., Meddis, 1986; Sumner *et al.*, 2003a; Westerman, 1988).

D. Phase-Locking and Synchronization

In the absence of acoustic stimulation, action potentials occur at random (Kiang *et al.*, 1965). However, in response to sinusoidal stimulation the timing of spike occurrences depends on the frequency of the stimulus. For low-frequency sinusoids, spikes tend to occur during the positive half-cycle of the stimulus waveform. This property was termed phase-locking (Rose *et al.*, 1967) and applies to AN fibers of all vertebrate classes (reviewed in the Introduction of Palmer and Russell, 1986), regardless of their CF (defined in the following text). The degree of phase-locking is usually expressed in terms of the synchronization index (for a definition see Anderson, 1973; and Johnson, 1980). The synchronization index assumes a value between 0 and 1. A value of 0 means that the spikes occur at times evenly spread over a full stimulus cycle, i.e., the spike histogram for one cycle of the stimulus is flat. A synchronization index of unity means that all spikes occur in one bin.

The synchronization index, hence the degree of phase-locking, decreases with increasing stimulus frequency (Johnson, 1980). Palmer and Russell (1986) showed that this reflects the low-pass filtering characteristics of IHCs. The frequency for which synchronization ceases to occur depends on the species (a useful review is provided in the Introduction of Palmer and Russell, 1986).

E. Frequency Selectivity and Cochleotopy

The properties described so far apply to all AN afferent fibers, regardless of the cochlear region they innervate. However, other properties do depend on the cochlear region of innervation. Most of them directly reflect the response properties of the BM (see the following text).

Fibers respond to pure tones in a frequency-selective manner. The SPL required to raise their rate responses to values just above their spontaneous rates[1] depends on the stimulus frequency. For a given fiber, the contour plot of the sound levels required to produce a certain discharge rate as a function of the stimulus frequency is termed the "frequency tuning curve" (Fig. 2). For a discharge rate criterion near the fiber's rate threshold, the tip of the tuning curve indicates the frequency at which the fiber is most sensitive and is termed the characteristic frequency or CF. For a given species, the range of CFs extends over approximately the frequency range of hearing. Fibers with low and high CFs innervate IHCs located in the apical and basal regions of the cochlea, respectively. A map can be established between the fibers' CFs and the distance between the oval window and the IHCs they innervate (reviewed by Greenwood, 1990; and Chapter 12 of Geisler, 1998).

The degree of a fiber's frequency selectivity is usually measured by the bandwidth of its tuning curve at some fixed level above the lowest threshold or by the slopes of its tuning curve above and below the CF (reviewed by Pickles, 1988). The bandwidth of AN fibers rises with increasing CF in a predictable manner (Fig. 3A) (reviewed by Greenwood, 1990; and Chapter 12 of Geisler, 1998). Additionally, the tuning curves of high CF fibers are asymmetric with high-frequency slopes steeper than low-frequency slopes (Figs. 2 and 3B). However, the tuning curves of low CF fibers are more symmetric (Fig. 2 and 3B; Evans, 1972).

For stimulus levels near threshold, there is a close correspondence between the tuning characteristics of auditory nerve fibers and the mechanical response of the cochlear partition they innervate (Narayan *et al.*, 1998; Ruggero *et al.*, 1986). Furthermore, the CF-cochlear position map of auditory nerve fibers is understood to reflect closely the BM cochleotopy (Greenwood, 1990). This, among other evidence (reviewed by Ruggero *et al.*, 2000; and Robles and Ruggero, 2001), suggests that the near-threshold tuning characteristics of auditory nerve fibers mostly reflect basilar-membrane (BM) tuning.

F. Level-Dependent Tuning and Best Frequency Shifts

Iso-intensity curves (Fig. 4A) illustrate the discharge rate of AN fibers as a function of the stimulus frequency with sound level as a parameter. Their width increases as the sound level increases, regardless of the fiber's CF (e.g., Rose *et al.*, 1971). This occurs as a result of: (1) the saturation of the fiber's discharge rate at high levels; and (2) the gradual broadening of the BM response with increasing

[1]Typically, a fiber is considered to respond just above threshold when its discharge rate exceeds the fiber's spontaneous rate by 20%.

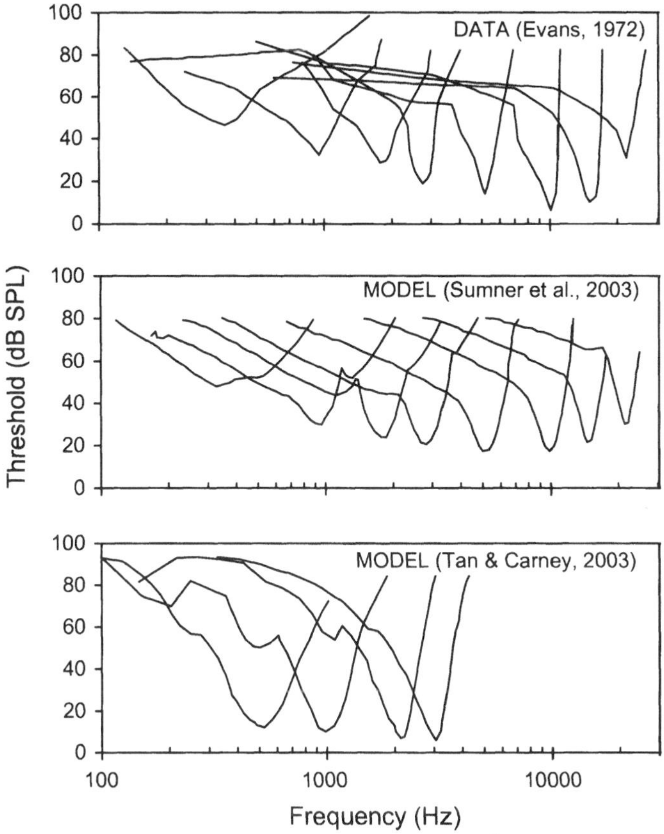

FIG. 2. Tuning curves of auditory nerve fibers of various characteristic frequencies. (Top panel) Data reproduced from Evans (1972) (Copyright © 1972 The Physiological Society). (Middle panel) Simulations for the guinea pig filter bank model of Sumner *et al.* (2003b) (adapted with permission Copyright © 2003 American Institute of Physics). (Bottom panel) Simulations for the model of Tan and Carney (2003) (adapted with permission Copyright © 2003 American Institute of Physics). Note that the tips of the tuning curves for this latter model occur at approximately the same level regardless of the fiber's CF. This is unrealistic (i.e., compare with top panel), but could be solved by including a middle ear in the model (c.f. Tan and Carney, 2003).

level (e.g., Ruggero *et al.*, 1997). The effects of level on frequency tuning have been reviewed by Møller (2000).

As the level increases, not only the degree of frequency selectivity decreases, but also the best frequency[2] (BF) of the fiber shifts to a value different from the CF

[2]The term best frequency (BF) refers to the frequency to which the fiber responds most strongly at any given sound level. Therefore, at a level near threshold, the CF is the best frequency.

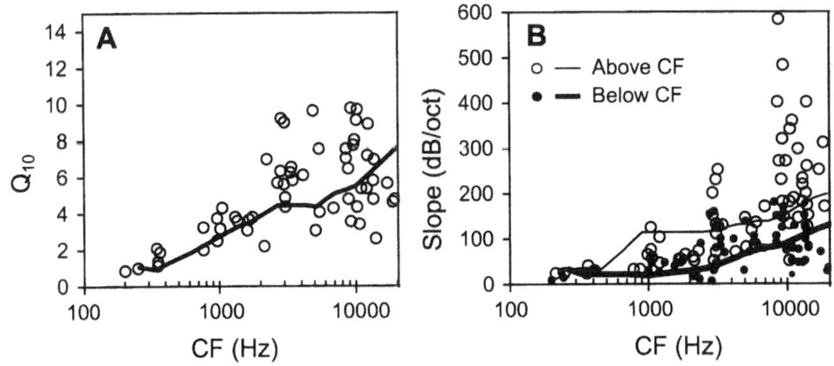

FIG. 3. (A) The bandwidth (expressed in terms of the quality factor, Q10) of auditory nerve fiber tuning curves as a function of the fibers' CF. Symbols illustrate experimental data reproduced from Evans (1972) (Copyright © 1972 The Physiological Society). The line illustrates the results for the guinea pig filter bank model of Sumner *et al.* (2003b) (adapted with permission Copyright © 2003 American Institute of Physics). (B) Slopes of guinea pig auditory nerve fiber tuning curves for frequencies below (filled symbols) and above (open circles) the characteristic frequency of the fibers (data reproduced from Evans, 1972, Copyright © 1972 The Physiological Society). Lines illustrate the results for the Sumner *et al.* (2003b) auditory nerve filter bank model. The thin and thick lines correspond to the model's slopes for frequencies above and below the fiber's CF, respectively. Note that the slope values for high CF fibers are greater for frequencies above the CF than for frequencies below it, both in the data and in the model. For low CFs, however, both slopes are comparable.

(Fig. 4A). The shift occurs in different directions depending on the CF of the fiber (e.g., Carney, 1999; Carney *et al.*, 1999; Rose *et al.*, 1971). For fibers with CFs above approximately 1.5 kHz, the shift occurs from the CF to lower frequencies as the sound level increases. For fibers with CFs below approximately 1 kHz, the shift is rather idiosyncratic although it generally occurs from the CF toward higher frequencies with increasing level. Finally, BF shifts are hardly observed for fibers with CFs of approximately 1.0 to 1.5 kHz. It is believed (e.g., Carney, 1999) that the BF shifts observed in the AN responses directly reflect the response properties of the BM (see the following text).

G. Level-Dependent Phase

The AN spikes tend to occur synchronously for stimulus frequencies within the range of phase-locking. However, the time at which they occur differs from the time of the peaks in the stimulus waveform (Anderson *et al.*, 1971). This phase shift increases linearly with increasing frequency. Furthermore, the phase angle also depends on the stimulus level. For frequencies below CF, spikes tend to occur progressively later in the cycle as the sound level increases. The opposite trend

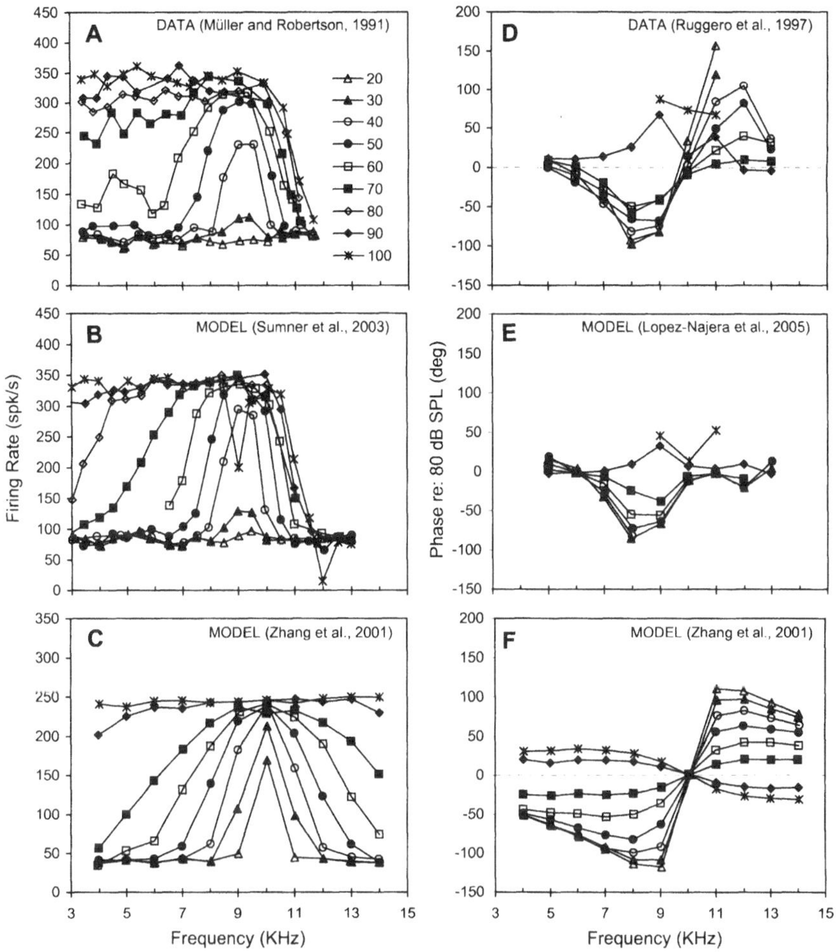

FIG. 4. (A to C) Auditory nerve iso-intensity rate responses for auditory nerve HSR fibers with a CF ~9 kHz. (D to F) Basilar membrane phase responses for a cochlear site with approximately the same CF. Each symbol illustrates the response for a different sound level, as indicated by the inset in panel A. Phases are expressed relative to the phase of the response for an 80-dB SPL tone and plotted using the convention of Zhang *et al.* (2001). (A) Experimental data for guinea pig (reproduced from Müller and Robertson, 1991, Copyright © 1991 with permission from Elsevier). (D) Experimental data for chinchilla (reproduced with permission from Ruggero *et al.*, 1997, Copyright © 1997 American Institute of Physics). (B) Simulations with the model of Sumner *et al.* (2003b) (adapted with permission Copyright © 2003 American Institute of Physics). (E) Simulations with the DRNL filter, a component of the Sumner *et al.* model, using optimum parameters provided by Lopez-Najera *et al.* (2005). (C and F) Simulations with the model of Zhang *et al.* (2001) (adapted with permission Copyright © 2001 American Institute of Physics). Note that the model of Sumner *et al.* (2003b) is better at reproducing the asymmetric aspect of the iso-intensity rate responses but the model of Zhang *et al.* (2001) accounts better for the level-dependent aspect of the phase of the response, particularly for frequencies above the CF.

occurs for frequencies above CF. At or near CF, the phase angle is approximately independent of sound level (Anderson *et al.*, 1971).

Similar types of level-dependent phase shifts have been reported in recordings of BM motion, originally by Rhode and Robles (1974) and more clearly later by Ruggero *et al.* (1997) (Fig. 4D). Therefore, it is likely that the AN level-dependent phase shifts are, at least in part, a reflection of the phase angles generated by the BM (Geisler and Rhode, 1982; Ruggero *et al.*, 2000). Geisler and Rhode (1982) suggested that these shifts relate to the level-dependent bandwidth of the BM magnitude response.

Another level-dependent aspect of AN phase responses is related to what is known as "Nelson's notch." Nelson Kiang and colleagues (e.g., Kiang *et al.*, 1986) reported that the discharge rate of AN fibers decreased drastically from saturation values to a value near the spontaneous rate and back again to saturation rates all within a range of levels between 90 to 100 dB SPL. This result occurs only for certain combinations of CF and stimulus frequency and always coincides with a rapid change in phase of approximately 180° over a 10-dB range. This result has a correlate with observations of IHC receptor potential (Cheatham and Dallos, 1998) but not with BM responses (reviewed by Ruggero *et al.*, 2000). Therefore, it is unlikely to reflect a characteristic of the BM response. Evidence provided by Cheatham and Dallos (1998) supports that the result reflects what is known as "two-factor cancellation" (Liberman and Kiang, 1984; Lin and Goldstein, 1995; Lopez-Poveda and Meddis, 2001). That is, it occurs when two excitatory inputs to the IHC have similar amplitudes and are 180° out of phase. The origin of such inputs is unclear.

H. Suppression

Suppression is the phenomenon by which the rate response of an AN fiber to a given sound is suppressed (reduced) by a second sound, termed the "suppressor." Suppression is usually illustrated by stimulation with pairs of tones (Sachs and Kiang, 1968). In such a case, the response of the fiber to one tone (the excitor) is reduced by the simultaneous presentation of a second tone (the suppressor) at an appropriate frequency and level (Fig. 5A). Suppression may occur even for suppressor tones that do not excite the fiber by themselves.

A distinction needs to be made between the two ways in which suppression effects have been measured: *rate* suppression (e.g., Delgutte, 1990; Javel *et al.*, 1978; Sachs and Kiang, 1968) refers to the decrease in the *average* discharge rate of an AN fiber that occurs during the presentation of the suppressor, while *synchrony* (or *phasic*) suppression (e.g., Javel, 1981; Javel *et al.*, 1983) refers to the discharge rate decrease of the Fourier component phase-locked to the excitor tone that occurs during presentation of the suppressor tone. Synchrony suppression

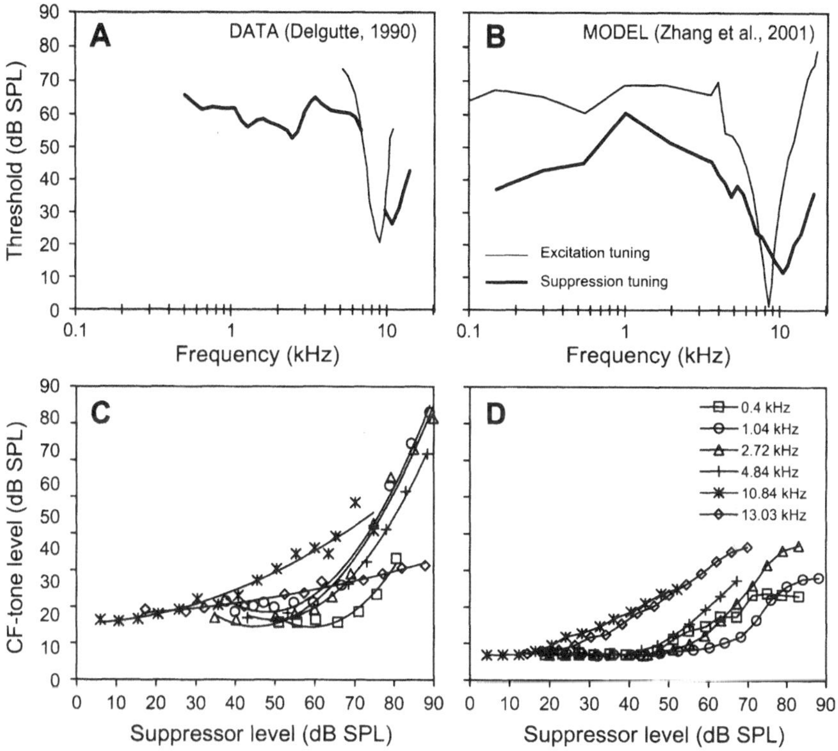

FIG. 5. Auditory nerve two-tone suppression effects (adapted from Zhang *et al.*, 2001, Copyright © 2001 Acoustical Society of America). The left panels illustrate experimental data from Delgutte (1990) (Copyright © 1990, adapted with permission from Elsevier). The right panels illustrate simulations with the model of Zhang *et al.* (2001) (adapted with permission Copyright © 2001 American Institute of Physics). The top panels illustrate the tuning of excitation (thin lines) and suppression (thick lines). The bottom panels illustrate the asymmetric growth of suppression for low- and high-side suppressors. The experiment consists in measuring the level of a CF tone required for the fiber to produce a constant rate response for suppressor tones of different levels (abscissa) and frequencies (indicated by the inset in panel D). Note that for a given increase in suppressor level, the required level increase for the CF tone is larger for suppressor frequencies below the fiber's CF (~7.9 kHz) than for frequencies above it. Note that the model accounts for the main features of suppression, although it produces less suppression than required for low-side suppressors.

can be measured only for excitor tones with frequencies within the range of phase-locking, and it has the advantage that it can be used to estimate the amount of suppression even when the two-tone stimulus elicits a saturated rate response (Javel, 1981).

As a first approximation, the suppressor tone has the effect of shifting the rate versus level (e.g., Javel *et al.*, 1978) and the synchrony strength versus level (e.g., Javel *et al.*, 1983) functions horizontally. For this reason, the amount of suppression is generally, although not always (e.g., Abbas and Sachs, 1976; Cai and Geisler, 1996a), expressed as the level increase (in dB) of the excitor tone required to maintain a certain discharge rate (Delgutte, 1990; Javel *et al.*, 1978) or synchrony strength (Javel *et al.*, 1983) criterion while the suppressor is being played.

The amount of suppression depends both on the frequency and on the SPL of the suppressor simultaneously. *Low-side* and *high-side* suppression are said to occur when the frequency of the suppressor tone is lower or higher, respectively, than that of the excitor tone, which is commonly assumed to be at the CF. Suppression is restricted to frequencies and intensities flanking the excitatory tuning curves of AN fibers (Sachs and Kiang, 1968) (Fig. 5A). The lowest suppressor level that produces suppression (i.e., the suppression threshold) is lower for high-side than for low-side suppressors (Sachs and Kiang, 1968) (Fig. 5A). As the suppressor sound level increases, the amount of suppression generally increases (e.g., Javel *et al.*, 1978, 1983). On the other hand, the amount of suppression grows faster with increasing the suppressor level for low-side than for high-side suppressors (e.g., Delgutte, 1990; Javel *et al.*, 1983) (Fig. 5C). Low-side suppression extends to suppressor frequencies well below the CF (e.g., Delgutte, 1990). Indeed, suppressor-tone frequencies as low as 1 kHz can reduce the activity of AN fibers with CFs as high as 21 kHz if they are intense enough (e.g., Delgutte, 1990). The rate of growth of suppression produced by any given low-frequency suppressor increases as a function of the CF (Delgutte, 1990). By contrast, the range of suppressor frequencies that produces high-side suppression depends on the CF: for fibers with CFs >1.5 kHz, it is restricted to frequencies within one octave of the CF, but it is wider for fibers with CFs <1.5 kHz (Javel *et al.*, 1978). Delgutte (1990) reported a similar result, but he set the CF cut-off at 2 kHz.

The amount of suppression also depends on the spontaneous rate of the fibers (Cai and Geisler, 1996a). Low-side suppressors affect fibers with low-spontaneous rates more than fibers with high-spontaneous rates (Cai and Geisler, 1996a; Fahey and Allen, 1985; reviewed by Geisler, 1998).

Stimulation of the efferent system shifts AN rate level functions horizontally (e.g., Guinan and Gifford, 1988; Guinan, 1996) in a way that resembles the suppression phenomena previously described. However, the latency of this efferent effect is of the order of a few hundred milliseconds (reviewed by Guinnan, 1996), whereas the latency of the suppression phenomena previously described is insignificant (Arthur *et al.*, 1971). Furthermore, with the exception of its dependence on the fiber's spontaneous rate (see the previous text), AN suppression shares many of the characteristics of BM (reviewed by Robles and Ruggero, 2001; and Cooper, 2004) and IHC (e.g., Cheatham and Dallos, 1992; Nuttall and Dolan, 1993b) suppression. Also, suppression is associated with BM nonlinear compression (Ruggero and Rich, 1991). For these reasons, it is widely accepted that AN suppression reflects mostly BM suppression (evidence reviewed by Geisler, 1998; and Cooper, 2004).

I. Distortion

The AN fibers may respond to pairs of tones even when both tones have frequencies (f_1 and f_2, $f_2 > f_1$) outside the frequency response area of the fiber and hence do not evoke a response by themselves (Goldstein and Kiang, 1968). Furthermore, a fiber with a CF equal to $2f_1 - f_2$ responds to the two-tone stimulus in a manner that is indistinguishable, both in magnitude and phase, from its response to a single tone at a CF of an appropriate level (Goldstein and Kiang, 1968; Buunen and Rhode, 1978). Responses may be evoked by primary tones with levels as low as 34 dB SPL (Kim *et al.*, 1980) and the magnitude of the response decreases rapidly with increasing frequency separation of the primaries (Buunen and Rhode, 1978). It is as though, for certain stimulus conditions, the AN responds to "new" tones that are not present in the acoustic stimulus. These "new" tones are referred to as two-tone distortion products or combination tones.

Combination tones can be actually heard and were first described psychophysically (e.g., Goldstein, 1967). They are perceived at frequencies which are combinations of the primary frequencies: $f_2 - f_1$, $(n + 1)f_1 - nf_2$, or $(n + 1)f_2 - nf_1$ ($n = 1, 2, 3, \ldots$). Of them, the cubic tone ($2f_1 - f_2$) and the difference tone ($f_2 - f_1$) are the most prominently heard and, for this reason, have been most extensively studied. It is possible to hear combination tones whose frequencies are lower (Goldstein, 1967) and higher (Zurek and Sachs, 1979) than the frequency of the primaries, although the latter are heard with more difficulty (Goldstein, 1967; Zurek and Sachs, 1979). This suggests that two-tone stimuli must evoke a response in AN fibers with CFs lower and higher than the primary frequencies.

Combination tones have also been recorded in the BM (Robles *et al.*, 1991, 1997) and IHC (Nuttall and Dolan, 1990, 1993a) responses. The characteristics of BM and AN responses to combination tones are similar (Robles *et al.*, 1997). For this reason, it is accepted that combination tones originate at the BM non-linearity. Furthermore, they originate at the place of the BM best tuned to the frequency of the primaries and *propagate* to the location with a CF equal to the frequency of the combination tone (Kim *et al.*, 1980; Robles *et al.*, 1997).

Another form of distortion, referred to as "rectifier distortion," affects the response properties of AN fibers to low-frequency tones or vowel formants (e.g., Sachs and Young, 1980; Young and Sachs, 1979). The Fourier transform of a fiber's period histogram in response to these types of stimuli reveals a large frequency component at the frequency of the stimulating tone (or formant), but also significant energy at integer multiples (harmonics) of that frequency. These harmonics arise, in part, from the rectification inherent to the fact that auditory neurons cannot discharge with negative pulses and, therefore, the inhibitory half cycles of a period histogram are clipped at a zero rate (Sachs and Young, 1979).

III. Auditory Nerve Representation of the Spectra of Complex Sounds

The mechanisms by which the stimulus spectrum is represented at the AN have been investigated using mostly synthetic speech (i.e., artificially generated speech-like sounds) and head-related transfer function (HRTF) stimuli (for a definition of the HRTF, see the chapter entitled Spectral Information in Sound Localization by Carlile *et al.*). Studies on the encoding of speech are reviewed by Delgutte (1997) while those on the encoding of HRTF stimuli are reviewed by Lopez-Poveda (1996).

The stimulus spectrum may be represented in the (average) discharge rate (Sachs and Young, 1979, 1980) and in the (instantaneous) synchronized rate response (Palmer, 1990; Palmer *et al.*, 1986; Young and Sachs, 1979) of a population of AN fibers with CFs spanning the spectral band of the stimulus.[3] Both representations, however, are largely determined by the fundamental response properties of individual fibers, particularly by the peripheral nonlinearities described in the previous section (e.g., Holmes *et al.*, 2004; Lopez-Poveda, 1996; May, 2003; Sachs and Young, 1980; Tan and Carney, 2005). For instance, Sachs and Young (1979, 1980) concluded that the rate-place representation of vowel spectra is appropriate at low levels but is negatively affected by stimulus level as a result of the narrow dynamic range of most fibers (rate saturation) and two-tone *rate* suppression. By contrast, the temporal-place representation is more robust at high levels because it benefits from *synchrony* suppression (Sachs and Young, 1980; Young and Sachs, 1979), although it is negatively affected by two-tone and rectification-related distortion (for a full discussion see Sachs and Young, 1980).

Temporal- and rate-place codes are *not* mutually exclusive and possibly act simultaneously to encode spectral features with frequencies *below* the cut-off of phase-locking (e.g., mostly vowel information). However, temporal codes are of limited use to represent high-frequency spectral information,[3] which is crucial for

[3]It is important to bear in mind that phase-locking occurs for all AN fibers, regardless of their CF. For this reason, the low-frequency information can be represented simultaneously in at least two ways: 1) in the timing of the response of all fibers; and 2) in the response rate of fibers with CFs spanning approximately the spectral band of the stimulus. This implies, for instance, that high CF fibers can encode the envelope of complex sounds, which is low-frequency information, in the timing of their response. High-frequency information, by contrast, can be represented only in the rate response of fibers with CFs spanning approximately the spectral band of the stimulus.

speech (i.e., consonant sounds) and music perception as well as for sound localization. These spectral features are most likely encoded by means of a rate-place code and this, as has already been explained, degrades at high levels. A number of mechanisms have been proposed to account for the encoding of high-frequency spectral features at high levels (reviewed by Alves-Pinto and Lopez-Poveda, submitted 2005). Delgutte and Kiang (1984a, 1984b), for instance, suggested that at high levels, when the adapted response of many of the fibers is saturated, the stimulus spectrum may still be reasonably well-represented in the onset rate profiles. Their suggestion is based on the fact that AN fibers have a wider dynamic range over the first few (5 to 20) milliseconds of their response (Smith and Brachman, 1980). Alternatively, several investigators (Delgutte and Kiang, 1984a; Rice *et al.*, 1995; Sachs and Young, 1979) have suggested that, since LSR fibers have higher thresholds and wider dynamic ranges, high-frequency spectral features could be encoded in the rate profile of HSR and LSR fibers at low and high levels, respectively. Nevertheless, the apparent quality[4] of the rate profiles of LSR fibers' also deteriorates at high levels (Delgutte and Kiang, 1984a,b; Rice *et al.*, 1995; Sachs and Young, 1979), possibly as a result of the broadening of the fibers frequency response with level (Rose *et al.*, 1971), the saturation of their discharge rate[5] (Sachs and Young, 1979), and the fiber-to-fiber variation in rate (Rice *et al.*, 1995). A more recent review on the topic (May, 2003) concludes that, although some deterioration of their rate profile may occur, LSR fibers, are capable of rate-place encoding the spectrum of vowels due to the effects of rate suppression. Rice *et al.* (1995), for their part, proposed that high-frequency spectral features may be better encoded in the difference between the rate profiles in response to two (successive) stimuli with different spectra. This code would eliminate the fiber-to-fiber variation in rate but would be highly affected by adaptation phenomena, as explained in the following text. Finally, it is noteworthy that recent modeling studies (Heinz *et al.*, 2001b,c) have suggested that, despite the very low synchronization coefficients for fibers at high frequencies, psychophysical frequency discrimination is still based on what little temporal (rather than rate) information may be available for frequencies up to 10 kHz.

Adaptation also affects the AN representation of stimuli with time-varying spectra (e.g., Delgutte and Kiang, 1984b), as is the case for most natural sounds. For these sounds, the activity of the AN at any instant in time depends in part on

[4]In this context, the word "quality" must be understood to refer to the degree of similarity between the stimulus spectrum and the auditory nerve rate-profile representation.

[5]Some studies (e.g., Schalk and Sachs, 1980) report LSR fibers that saturate at high levels, while others report LSR fibers with straight, non-saturating, rate-level functions (Winter *et al.*, 1990). Whether or not LSR fibers saturate probably depends on the species. The issue is still controversial (see Winter *et al.*, 1990).

previous stimulation. As explained by Delgutte and Kiang (1984b, p. 904), "... adaptation tends to increase the contrast between successive speech segments separated by an abrupt change in spectral characteristics." This occurs for spectral features regardless of their frequency.

It is clear from the previous discussion that further research is still needed to elucidate the codes by which the spectra of complex sounds are represented in the AN response. Holmes *et al.* (2004, p. 3534) provide a succinct description of the problem: "(...) it is difficult to infer the processing mechanisms from the measured auditory responses alone. Furthermore, the non-linear nature of the system makes it difficult to predict the responses to new complex stimuli, often requiring more data collection. Animal experiments are complex and time consuming, even using efficient measurement techniques (...). A non-linear computational model able to simulate the physiological data would be very useful in assessing how this coding would apply to a much wider range of (...) stimuli."

It is also clear from the previous discussion that the AN representation of sounds with complex spectra must be based on population responses which are ultimately determined by the response characteristics of individual AN fibers. For this reason, the goal must be to design computational models capable of reproducing, in a realistic manner, the response characteristics of populations (of an arbitrary size) of AN fibers with different spontaneous rates and CFs, each of which shows adaptation, phase-locking, level-dependent tuning and phase responses, rate and synchrony suppression, and distortion. Furthermore, the models should be able to process acoustic stimuli in the time domain, much as it is done by the peripheral auditory system. The underlying assumption is that any model that meets these requirements will be valid for exploring the codes used by the peripheral auditory system to convey simple and complex sounds to the central auditory system. The following sections are dedicated to reviewing the extent to which this goal has been achieved by a selection of *non-linear* models of signal processing by the peripheral auditory system.

IV. Phenomenological Models of Signal Processing by the Peripheral Auditory System

This section reviews a representative sample of *phenomenological* models of peripheral auditory processing. These models are concerned with producing the right output for a given input without paying much attention to the actual biophysical processes underlying a given physiological result. Therefore, they employ any powerful description of the system that is available, regardless of its nature. Models of a different kind are reviewed elsewhere: a succinct review of transmission line models is provided by Duifhuis (2004); mechanical cochlear models are reviewed by de Boer (1996) and Hubbard and Mountain (1996);

physical models of the external and middle ear function are reviewed by Lopez-Poveda (1996), Lopez-Poveda and Meddis (1996), and Rosowski (1996); electrical circuit descriptions of hair cell function are reviewed by Mountain and Hubbard (1996); and finally, physiological models of basic auditory percepts are reviewed by Delgutte (1996) and Moore (2003).

Given the importance of the cochlear nonlinearities in determining the representation of complex spectra at the AN (see Section II), the review focuses on *non-linear* models intended to reproduce a wide range of physiological AN phenomena. Linear models, models of psychophysical data, and non-linear models of specific physiological phenomena (e.g., two-tone suppression) are not reviewed in depth, although they will be commented on when appropriate. The discussion focuses on the ability of models to reproduce the response characteristics of *individual* AN fibers to stimuli with simple spectra (as enumerated in the previous text), as well as that of *population* responses to stimuli with complex spectra.

A. Approach Guidelines and Assumptions Common to All Phenomenological Models

As previously explained, many of the AN response characteristics reflect directly those of previous processing stages of the peripheral auditory system. For instance, AN non-linear frequency selectivity is determined (at least in mammals) by the non-linear frequency selective properties of the BM response. The same applies to two-tone suppression and rate distortion. Rectifier distortion and phase-locking are thought to reflect the half-wave rectification and low-pass filtering characteristics of the IHC response. As for adaptation phenomena, they are thought to occur at the synapse between IHCs and AN fibers. For that reason, many models of the auditory periphery are structured as a cascade of processing stages (Fig. 6), each of which simulates the function of a corresponding physiological stage. The assumption is that a reasonable AN model can be produced by realistic modeling of the response properties of each of the main previous processing stages. In such an arrangement, the output of one stage serves as the input to the next stage and it is assumed that no feedback occurs between "physiological" stages. Models of this sort are referred to as "composite models."

Unlike older composite cochlear models, whose design was based on transmission-line electrical analogues and which were suitable to deal with continuous-time signals, current models are designed and tested with digital signals. That is, they operate on a digital time-varying input signal to produce a digital time-varying output signal. This imposes two important restrictions on their algorithms: (1) they must have a corresponding digital version, which must

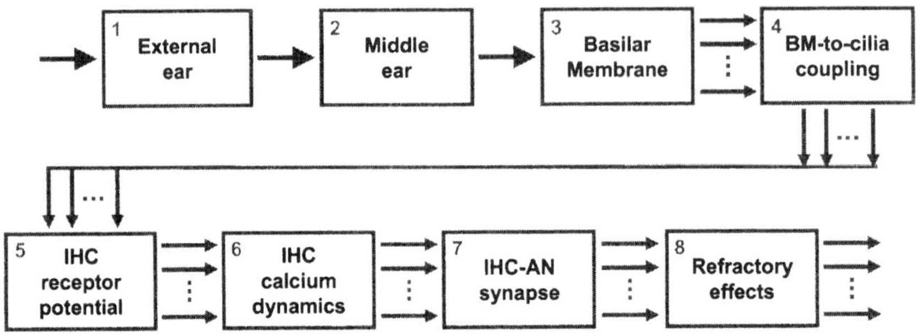

FIG. 6. General structure of phenomenological composite models of the auditory nerve response. Not all the models include the stages as illustrated.

be stable in the digital (z-domain) sense; and (2) their computation times must be reasonably short.

Phenomenological models always have a number of parameters whose nature depends on the nature of the algorithm. The ideal is to reproduce the largest possible number of phenomena with the fewest possible parameters.

B. The Model of Deng and Geisler (1987b)

The design of this model was motivated mainly to account for "synchrony capture" (Deng and Geisler, 1987a; Deng *et al.*, 1987), a phenomenon by which "one formant in the speech syllable (or the probe component in a multi-tone stimulus) produces more response synchrony to itself than is predicted by linear methods from the fiber's threshold frequency tuning curve" (Deng and Geisler, 1987b, p. 2001).

The model consists of four stages. The first stage is a linear middle ear filter fitted to experimental data. The second stage reproduces the response of the whole cochlear partition by a set of difference equations (Viergever, 1980) describing the behavior of the one-dimensional transmission line model of Wickesberg and Geisler (1986). Its parameters are the damping and stiffness along the BM and they were tuned to account for AN threshold tuning curves with different CFs. Both linear and non-linear versions of this stage were provided. To achieve the desired non-linear behavior, the damping parameter was made to vary as a function of the output partition displacement. The third stage represents the IHC function by means of a simple sigmoid-type of memory-less non-linearity followed by a low-pass filter. It transforms partition velocity into IHC receptor potential. The fourth stage represents the synaptic effects and transforms the receptor potential into probability of firing of AN fibers. It is based on the Oono and Sujaku (1974) reservoir-type of synapse used by Geisler and Greenberg (1986).

The model is unstable for long signals due to accumulated round-off errors. It reproduces the response of HSR fibers with level-dependent tuning. Although it has built-in capabilities to model suppression and adaptation phenomena, a detailed account of the model's ability to reproduce related physiological data is lacking. Instead, the model responses to speech sounds were compared with experimental data for single fibers and populations of fibers. The main conclusions were that the cochlear non-linearity (responsible for level-dependent tuning) and the synaptic non-linearity (responsible for adaptation) were crucial for the model to produce a realistic AN representation of speech. Specifically, the model reproduces synchrony capture due to the broadening of its frequency response at high levels. The hair cell non-linearity, by contrast, had little effect on achieving a realistic output.

C. The MBPNL Model of Goldstein (1990)

The multiple bandpass non-linear (MBPNL) model of Goldstein (1988, 1990, 1993, 1995) was developed in an attempt to provide a unified account of complex non-linear phenomena characteristic of BM and AN responses, such as compression, suppression, distortion, and simple-tone interference (defined in the following text). Actually, the MBPNL model reproduces the response of the BM rather than that of the AN. It simulates the filtering function of a given cochlear partition (a given CF) by cascading a narrowly tuned bandpass filter followed by a compressive memory-less non-linear gain, followed by another more broadly tuned bandpass filter (Fig. 7B). This structure is similar to the bandpass non-linear (BPNL) filter proposed by Pfeiffer (1970) and Duifhuis (1976). The narrow and broad filters represent the "tip" and the "tail" filters described by Kiang and Moxon (1974). By carefully choosing their shapes and the compressive gain, the model reproduces level-dependent tuning and best-frequency shifts. This confers on the model the ability to reproduce "synchrony capture" (see the previous text) (Goldstein, 1990).

The model was specifically designed to reproduce the non-linear cyclic interactions between a moderate-level tone at CF and another highly-intense tone with a very low frequency, a phenomenon usually referred to as "simple-tone interaction" (or simple-tone interference) (Patuzzi *et al.*, 1984). This required incorporating an expanding non-linearity (inverse in form to the compressing non-linearity) whose role in the model is to enhance the low frequencies before they interact with on-CF tones at the compressive stage (Fig. 7B). With this expanding non-linearity, the model reproduces low- and high-side suppression. Most interestingly, it accounts for the fact that the amount of suppression grows

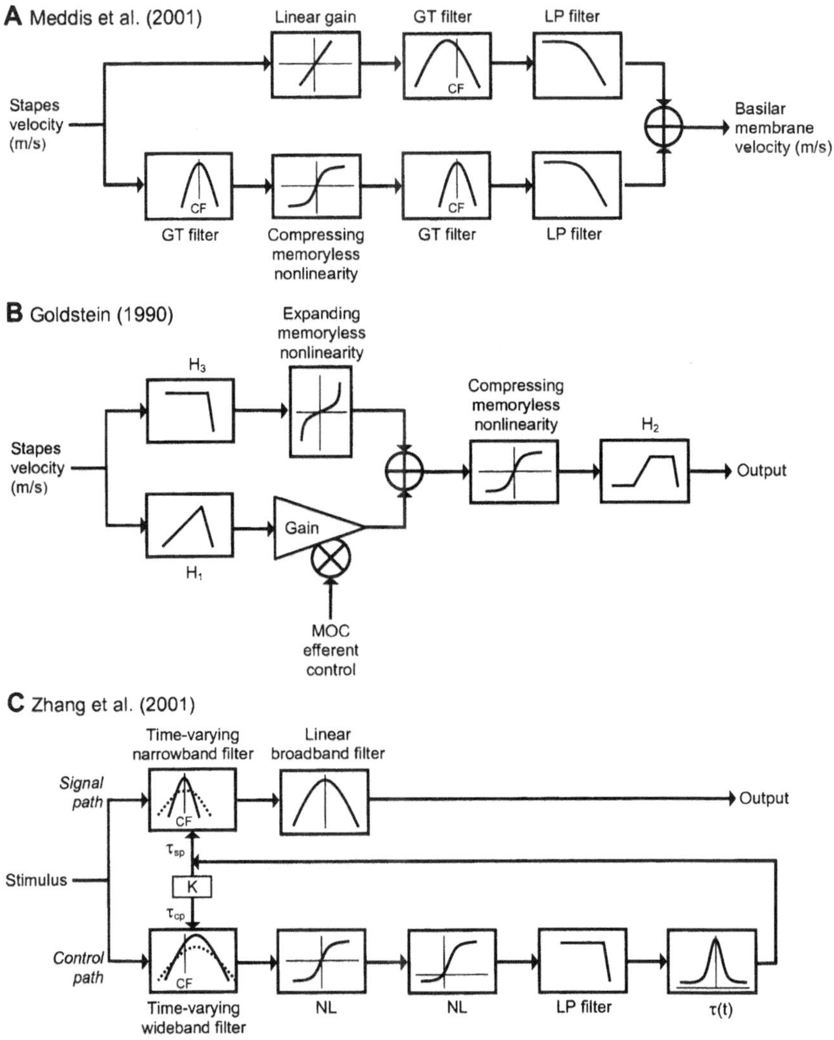

FIG. 7. Comparative architecture of three phenomenological non-linear basilar membrane models. (A) The dual-resonance non-linear filter of Meddis *et al.* (adapted from Lopez-Poveda and Meddis, 2001. Copyright © 2001 American Institute of Physics). (B) The multiple bandpass non-linear filter of Goldstein (adapted from Goldstein, 1990, Copyright © 1990 with permission from Elsevier). H_1, H_2, and H_3 represent three different filters. (C) The model of Zhang *et al.* (adapted with permission from Zhang *et al.*, 2001, Copyright © 2001 American Institute of Physics). See text for details. GT = gamma-tone; LP= low-pass; NL = non-linearity; MOC = medio-olivocochlear.

faster with increasing suppressor level for low-side than for high-side suppressors (Goldstein, 1995). Moreover, the model reproduces the $2f_2 - f_1$ combination tone correctly. However, propagation of combination tones (Kim *et al.*, 1980; Robles *et al.*, 1997) is lacking in the model, although it appears necessary to account for the experimental data regarding the $2f_1 - f_2$ combination tone (Goldstein, 1995).

The MBPNL model was further developed into a version capable of reproducing the response of the whole cochlear partition by means of a bank of *interacting* MBPNL filters (Goldstein, 1993). This newer version gave the model the ability to account for propagating combination tones. However, systematic harness tests[6] have not been reported to date on this MBPNL filter bank.

As previously detailed, the model accounts for many non-linear aspects of the BM response that have a direct AN correlate. However, since it is essentially a cochlear model, it does *not* reproduce AN rate-level functions, different types of fibers, phase-locking, or adaptation.

D. The Model of Jenison *et al.* (1991)

This model was developed to explore the importance of level-dependent auditory frequency selectivity for the representation of speech in the AN. Specifically, the authors aimed at reproducing the data of Young and Sachs (1979), particularly "synchrony suppression."

Unlike other models whose structure resembles more or less closely the processing stages of the peripheral auditory system in the right order, this model reproduces a number of physiological properties with only three stages. Stage 1 consists of a bank of AN filters. The filters are of the finite impulse response (FIR) type and are derived for each signal level by applying an inverse Fourier transform to the iso-intensity response functions of a large number of fibers with CFs spanning the range from 0.375 to 13 kHz. This approach is based on the assumption that the iso-intensity response functions of single HSR fibers reflect the summed filtering mechanisms of the cochlea, IHC transduction, synaptic effects, and spike generation. Stage 2 provides the model with appropriate level-dependent gains by applying sigmoidal functions that represent rate-level functions that map a particular input level to the resulting instantaneous rate of each frequency channel. Finally, stage 3 gives the model realistic frequency-dependent synchronization by applying a simple low-pass filter.

The model reproduces the level-dependent tuning properties of HSR fibers only (LSR are omitted). It also reproduces frequency-dependent synchronization and rate-level functions for on-CF tones. However, it does not reproduce adaptation, suppression, or rate-level functions for tones with frequencies different from the CF. By contrast to linear models, this model does reproduce the spread of formant-related synchrony to high-CF channels reported by Young and Sachs (1979). This is achieved due to the broadening of the response of individual fibers as a function of level (Jenison *et al.*, 1991).

E. The Model of Giguère and Woodland (1994a,b)

This model consists of an analogue electrical network that reproduces the functional properties of the peripheral auditory system up to the AN response. Its design is based on the analogy that exists between mechanoacoustical and electrical systems (c.f. Merhaut, 1981), a common approach to designing peripheral auditory models (c.f. Rosowski, 1996). The authors also provide a digital version of the model that allows computation of digital time-varying signals.

The model consists of five stages. The first stage (adapted from Killion and Clemis, 1981) simulates the diffraction of a free-field sound by the torso, the head, and the pinna flange. Stage 2 (adapted from Gardner and Hawley, 1972) reproduces the acoustic resonances at the concha and the ear canal cavities. The third stage (adapted from Lutman and Martin, 1979) simulates the transmission of pressure through the middle ear. It explicitly allows modeling the effect of the acoustic middle ear reflex. The fourth stage reproduces non-linear BM motion by means of a classical transmission line cochlear model (Zwicker, 1986). The role of OHCs is explicitly modeled. In the model, they provide instant feedback whose net effect is not to add energy to the simulated traveling wave, but rather to reduce the damping of BM displacement. Finally, stage 5 simulates IHC transduction (with adaptation and phase-locking properties) by means of an electrical network equivalent to the model of Meddis (1988). The input to the IHC is assumed to be proportional to the viscous drag of the surrounding fluid, which itself is proportional to the velocity of BM motion.

This is one of the few models (other examples are the models of Ghitza, 1988 or Kates, 1991) that simulate the effects of OHC efferent innervation (Giguère and Woodland, 1994b). The assumption in this model is that the efferent system regulates the AN average firing rate by modulating the input signal to the IHC indirectly through modulating the velocity of the fluid surrounding IHCs.

The model allows access to the output signal from each of the intermediate processing stages. That is, it allows monitoring of the pressure at the eardrum, the stapes velocity, BM displacement or velocity, and AN discharge rate for arbitrary input signals. It has built-in capabilities to reproduce a wide range of phenomena including level-dependent tuning and phase, best-frequency shifts with level, phase-locking, adaptation phenomena, and rate-level functions for the three types of fibers, distortion, and, to some unspecified extent, suppression. Unfortunately, to date no attention has been paid to testing the model performance with regard to AN phenomena other than non-linear tuning and the rate-level function of an MSR fiber.

Giguère and Woodland (1994b) investigated the effects of BM level-dependent tuning as well as those of the descending paths (middle ear reflex and efferent control on OHC function) on the cochleogram representation of speech. Their analysis was based on purely visual (thus, subjective) judgments of the cochleogram quality. Their main conclusions were that: (1) level-dependent tuning contributes to the representation of consonant sounds through spectral sharpening; and (2) the cochlear efferent system contributes to the representation of stop consonants and fricatives.

F. The Compressive Gamma-Chirp Filter of Irino and Patterson (1997, 2001)

It is not possible to understand this model (or the models described in the following sections) without first understanding the characteristics of its predecessor: the gamma-*tone* filter. The gamma-tone filter was developed to simulate the impulse response of AN fibers as estimated by reverse correlation techniques (Aersten and Johannesma, 1980; de Boer, 1975; de Boer and de Jongh, 1978; Flanagan, 1960). The impulse response of the gamma-tone filter basically consists of the product of two components: a carrier tone of a frequency equal to the best frequency of the fiber and a statistical gamma-distribution function that determines the shape of the impulse response envelope. One of the advantages of the gamma-tone filter is that its digital, time-domain implementation is relatively simple and computationally efficient (Slaney, 1993), and for this reason it has been largely used to model both physiological and psychophysical data pertaining to auditory frequency selectivity. It has also been used to simulate the excitation pattern of the whole cochlear partition by approximating the functioning of the BM to that of a bank of parallel gamma-tone filters with overlapping passbands, a *filter bank* (e.g., Patterson *et al.*, 1992).

On the other hand, the gamma-tone filter is linear, thus level-independent, and has a symmetric frequency response. Therefore, it is inadequate to model the asymmetric, non-linear frequency selective characteristics of both BM and AN responses that have been previously described. Several attempts have been made to design more physiological versions of the gamma-one filter. For instance, Lyon (1997) proposed an *all*-pole digital version of the gamma-tone filter with an asymmetric frequency response. The all-pole gamma-tone filter also has the advantage of being simpler than the conventional gamma-tone filter in terms of parameters, as its gain at center frequency and its bandwidth are both controlled by a single parameter, namely, the quality factor (Q) of the filter (N.B.: the quality factor of a filter is defined as the ratio of the filter center frequency to the filter bandwidth measured at a certain number of decibels below the maximum gain, $Q = CF/BW$; c.f. Hartmann, 1998).

The gamma-*chirp* filter of Irino and Patterson (1997), like the all-pole gamma-tone filter, was designed to produce an asymmetric gamma-tone–like filter. This was achieved by making the carrier-tone term of the analytic impulse response of the gamma-tone filter modulated in frequency, thus the suffix *chirp*. This property was inspired by the fact that the impulse response of AN fibers is also frequency modulated (Carney *et al.*, 1999). However, in its original form, the gamma-chirp filter was still level-*in*dependent, hence inadequate to simulate the non-linear, compressive growth of BM response with level. Furthermore, it showed a level-dependent chirp inconsistent with the level-*in*dependent character of the chirp of BM (de Boer and Nuttall, 1997; Recio *et al.*, 1998) and AN (Carney *et al.*, 1999) impulse responses.

Further refinements of the filter led to a *compressive* gamma-chirp filter with a level-*in*dependent chirp (Irino and Patterson, 2001), hence more consistent with the physiology. The compressive gamma-chirp filter can be viewed as a cascade of three fundamental filter elements: a gamma-tone filter followed by a low-pass filter, followed by a high-pass filter with a *level-dependent* corner frequency. Combined, the first two filters produce an asymmetric gamma-tone–like filter, which can be approximated to represent the "passive" response of the BM. Because of its asymmetric frequency response, the associated impulse response of this "passive" filter shows a chirp.

The third element in the cascade, the high-pass filter, is responsible for the level-dependent gain and tuning characteristics of the compressive gamma-chirp filter. It is designed to affect only those frequencies near the center frequency of the gamma-tone filter in a level-dependent manner. At low levels, its corner frequency is configured to compensate for the effect of the low-pass filter, thus making the frequency response of the global gamma-chirp filter symmetric. At high levels, by contrast, its corner frequency is set so that the frequency response of the "passive" filter be almost unaffected and thus asymmetric. The chirping properties of the gamma-chirp filter are largely determined by those of its "passive" asymmetric filter at *all* levels, and have been shown to fit well those of AN fibers (Irino and Patterson, 2001).

The compressive gamma-chirp filter by itself is *not* a model of the AN. It does *not* reproduce physiological properties of the AN response such as half-wave rectification, phase-locking, adaptation, or rate-level functions. Instead, it can be better understood as a non-linear filter of auditory frequency selectivity inspired by some physiological properties of the BM and the AN response (level-dependent tuning, compression, and level-independent chirps). As such, it has proved adequate to design filter banks that reproduce psychophysically estimated human auditory filters over a wide range of center frequencies and levels (Patterson *et al.*, 2003). It could probably be used to simulate physiological BM iso-intensity responses directly, although no studies have been reported to date aimed at testing the filter in this regard. Its best frequency shifts with level as do BM and AN iso-intensity curves, but the trends shown by Irino and Patterson (2001) are not consistent with the physiological data (see the Introduction of Tan and Carney, 2003). More importantly, we still lack detailed studies aimed at examining the ability of this filter to account for other non-linear phenomena such as level-dependent phase responses, combination tones, or two-tone suppression. Some authors have suggested that it can not reproduce two-tone suppression because it is not a "true" non-linear filter, but rather a "quasi-linear" filter whose shape changes with level (Plack *et al.*, 2002).

Should it prove useful to account for BM data, the compressive gamma-chirp filter could then be used in conjunction with models of IHC transduction and of the synapse between the IHC and the AN to produce a full AN model. Similar arrangements have been very successful for the gamma-tone filter (e.g., Meddis and Hewitt, 1991; Patterson *et al.*, 1992).

G. The Model of Robert and Eriksson (1999)

This model simulates the response of a population of AN fibers with arbitrary CFs in response to arbitrary sounds. It consists of four stages. Stage 1 simulates the frequency response of the outer and middle ears using simple, linear bandpass filters. Stage 2, the cochlea, consists of a set of parallel filters, a *filter bank*, each representing a specific location along the BM. In stage 3, the output from each cochlear filter is fed to the Meddis (1986, 1988) computational model of the synapse between the IHC and the AN. This model allows for the reproduction of the AN phase-locking and adaptation. It reproduces, and provides the probability of firing for any single AN fiber. Finally, stage 4 translates the probability of firing into actual spike occurrences, considering absolute and relative refractory periods.

The main original aspect of this model is, perhaps, its cochlear filterbank, as it shows important physiological-like properties absent in other filter bank models described in this chapter. Each filter in the bank consists of the cascade of a passive filter, with time-independent parameters, followed by an "active" filter whose parameters vary dynamically in time. Both the active and the passive filters are implemented as digital all-pole gamma-tone filters (Lyon, 1997; see the previous text). Inspired by an idea of Carney (1993), a feed*back* process controls the quality factor, Q, of the active filter, thus its tuning and gain. Peculiar to this model is the fact that the feedback control signal of any given filter depends not only on the output signal from that filter, but also on the output signals from neighboring filters. Furthermore, the feedback is designed to enhance the suppressive effects of frequencies lower than the CF of the corresponding filter (i.e., of low-side suppressors). Finally, this filter bank also considers that the output from each filter serves as the input to the next, as happens in the classical "traveling wave" interpretation of basilar membrane motion (e.g., de Boer, 1996). This also contrasts with several other filter bank models reviewed here (e.g., Irino and Patterson, 1997, 2001; Lopez-Poveda *et al.*, 2001; Meddis, 2001; Patterson *et al.*, 2003; Sumner *et al.*, 2003b; Zhang *et al.*, 2001) that assume that all filters in the bank share a common input signal.

The model was shown to realistically reproduce an important number of characteristics of BM responses, namely level-dependent tuning, low- and high-side suppression, and responses to notch noises. Low-side suppression is particularly well represented in this model thanks to the distributed form of feedback, although the asymmetry in suppression growth for low- and high-side suppressors has not been addressed to date with this model. The model also reproduces important characteristics of the AN response, such as rate-intensity functions of LSR and HSR fibers, adaptation, and responses to pure tones in background noise. However, this model has not yet accounted for the dependence of the synchronization index on level and frequency, or the level dependence of the phase of phase-locked components. Finally, this model may be particularly useful to investigate the AN representation of broad-band signals (e.g., speech). However, harness tests have not been run to date to convincingly demonstrate the model performance regarding this task.

H. The Composite AN Model of Carney and Colleagues

Carney and colleagues (Heinz *et al.*, 2001a; Zhang *et al.*, 2001) have proposed an improved version of Carney's (1993) composite phenomenological model of the AN response that reproduces a large number of non-linear AN response characteristics, including level-dependent phase responses and two-tone suppression. A newer version of this model (Tan and Carney, 2003) also reproduces level-independent frequency glides (N.B.: the term "frequency glide" is synonymous with the term "chirp" and both refer to the frequency-modulated character of BM and AN impulse responses).

In its more general form, the model consists of five stages. Stage 1 is a linear bandpass filter that models the frequency response properties of the middle ear (Tan and Carney, 2003). This stage was omitted in the model version of Zhang *et al.* (2001) but is important to account for the downward frequency glide observed for fibers with low CFs (Carney *et al.*, 1999). Stage 2 (Fig. 7C) reproduces roughly the response of a single BM site and is key to the model. In essence, it consists of a gamma-tone filter whose gain and bandwidth vary dynamically in time depending on the level of the input signal (this filter is referred to in the original reports as "the signal path"). For a gamma-tone filter, both of these properties, gain and bandwidth, depend on the filter's time constant, τ (see equation 2 of Zhang *et al.*, 2001). In the model, the value of this time constant varies dynamically in time depending on the amplitude of the output signal from a feed-*forward* control path, which itself depends on the level of the input signal. As the level of the input signal to the control path increases, then the value of τ decreases, thus increasing the filter's bandwidth and decreasing its gain.

The structure of the control path is carefully designed to reflect the "active" cochlear process of the corresponding local basilar-membrane site as well as that of neighboring sites. It consists of a cascade of a wideband filter followed by a saturating non-linearity. This saturating non-linearity can be understood to represent the transduction properties of OHCs and is responsible for the compressive character of the model input/output response. Finally, the bandwidth of the control-path filter also varies dynamically with time, but it is always set to a value greater than that of the signal-path filter. This is necessary to account for two-tone suppression, as it allows for frequency components outside the passband of the signal-path filter to reduce its gain and thus the net output amplitude from stage 2. Also, the center frequency of the control-path filter is set higher than that of the signal-path filter (Fig. 7C) to account for the asymmetry in the suppression thresholds for low- and high-side suppressors (Fig. 5B).

Stage 3 of the model simulates the IHC transduction. Essentially, it consists of a cascade of a half-wave rectifier, logarithmic (thus compressive) function, followed by a low-pass filter. This low-pass filter is required to account for the roll-off of phase-locking with increasing stimulus frequency. Stage 4 of the model simulates the synapse between the IHC and the AN fiber. It consists of a simplified implementation of the diffusion model of Westerman and Smith (1988), whose output is discharge rate prior to the inclusion of refractory effects. Finally, stage 5 makes discharge rate statistics akin to a non-homogeneous Poisson process (Winter and Palmer, 1991) and introduces refractory effects.

This model is very comprehensive in so far as it allows inspection of the output signal from intermediate physiological processing stages, such as BM excitation or the IHC receptor potential, and also because it has been specifically tested and proved adequate to account for a large number of AN response characteristics (Table 1). Particularly interesting is that it accounts for certain physiological phenomena that other models fail to reproduce such as level-dependent phase responses (Fig. 4F) (although this may be at the expense of being unable to reproduce asymmetric iso-intensity responses) (Fig. 4C), suppression tuning curves (Fig. 5B), asymmetric suppression thresholds and suppression growth for low- and high-side suppressors (Fig. 5D), and realistic responses to noise (Zhang *et al.*, 2001). A version exists that simulates the response of fibers with different spontaneous rates (Heinz *et al.*, 2001a).

The model of Zhang *et al.* (2001) uses symmetric gamma-tone filters and, therefore, does *not* produce asymmetric iso-intensity curves (see Fig. 5) or click

TABLE I

SUMMARY OF THE CHARACTERISTICS OF EACH MODEL AND OF THE PHENOMENA THEY ACCOUNT FOR. SYMBOLS SHOULD BE INTERPRETED AS FOLLOWS: √: THE MODEL REPRODUCES OR INCLUDES THIS FEATURE CORRECTLY; ●: THE MODEL REPRODUCES THIS FEATURE PARTIALLY RIGHT; ○: BUILT-IN FEATURE, BUT NOT SPECIFICALLY TESTED FOR WITH THE MODEL. BLANK SPACES INDICATE THAT THE MODEL DOES NOT INCLUDE OR REPRODUCE THE CORRESPONDING FEATURE CORRECTLY; §: INDICATES A CHARACTERISTIC WHICH IS DEMONSTRATED IN A PUBLICATION (OR A MODEL VERSION) DIFFERENT FROM THE ONE INDICATED IN THE TOP ROW (SEE MAIN TEXT FOR FURTHER DETAILS)

	Deng & Geisler (1987)	Goldstein (1990)	Jenison *et al.* (1991)	Giguère & Woodland (1994)	Robert & Eriksson (1999)	Irinno & Patterson (2001)	Zhang *et al.* (2001)	Sumner *et al.* (2002)
Stages of the model								
Outer- & middle-ear	√	√		√	√	√	√	√
Nonlinear basilar membrane	√	√			√	√	√	√
Basilar-membrane to IHC-stereocilia coupling								√
OHC control				√	√		√	●
Middle-ear reflex			√					
IHC receptor potential	√						√	√
IHC calcium dynamics								√
IHC-auditory nerve synapse	√			√	√		√	√
Refractory effects				√	√		√	√
Auditory-nerve response characteristics reproduced by the model								
Fibers with different SR	○			√	√		√ §	√
Rate-level functions	○		●	●	√		√	√
Level-dependent tuning	√	√	√	√	√	√	●	√
BF-shifts with level		√	√	●	○		● §	√
Level-dependent phase	○	○	○	○	○	○	√	● §
Two-tone suppression		√		○	√		√	● §
Single-tone interaction		√						
Filterbank (or equivalent)	√	○ §	√	√	√	√ §	√	√ §
Phase-locking	○		√	√	○		√	√
Response to steady-state speech	●		√	●			√ §	√ §
Response to HRTF sounds								√ §
Adaptation	○			√	○		●	√
Two-tone distortion	○	●	○	○	○		●	●
Level-independent chirp						√	√ §	● §

responses showing frequency glides. These shortcomings are accounted for by an improved version of the model proposed by Tan and Carney (2003). This new version uses *a*symmetrical digital filters that are designed in the complex plane (i.e., by positioning their poles and zeros) to have the appropriate glide (or "chirp"). Furthermore, by making the *relative* position of these poles and zeros in the complex plane independent of level, the model can also account for level-*in*dependent frequency glides, consistent with the physiology (Carney *et al.*, 1999; de Boer and Nuttall, 1997; Recio *et al.*, 1998). This new model version has not been so thoroughly tested as its predecessor, the version of Zhang *et al.* (2001). Indeed, this new version has been tested only for fibers with CFs below 4 kHz (Fig. 2) and for a narrower range of phenomena.

An important shortcoming of both model versions is that they do *not* account for the tails of AN fiber tuning curves (Fig. 2). This is because its signal-path filter consists of a single exciting resonance, while some properties of the AN response, like Nelson's notches previously described, as well as recent studies (Lin and Guinan, 2000, 2004) indicate that the AN must be driven by multiple exciting resonances. Furthermore, the ability of the model to account for level-dependent adaptation effects (Westerman, 1985) has not been fully explored.

Versions of this model have been described (Heinz *et al.*, 2001a) that may be useful for predicting the AN representation of complex stimuli, such as speech, in both normal and impaired ANs. An example application of this kind was investigated by Bruce *et al.* (2003). Based on the model response, they showed that damage to both OHCs and IHCs affect the synchronization of the AN response to vowel formants, and hence contribute to the degradation of the tonotopic representation of vowels in the AN. Heinz *et al.* (2001b,c) used the model to investigate the physiological mechanisms underlying auditory performance limits in frequency and level discrimination tasks. The model version of Tan and Carney (2003) has been used by Tan and Carney (2005) to investigate the cues used by the auditory system in (vowel) formant-frequency discrimination tasks.

I. The Composite AN Model of Meddis and Colleagues

Meddis and colleagues (Lopez-Poveda, 1996; Lopez-Poveda and Meddis, 2005; Lopez-Poveda *et al.*, 1998; Sumner *et al.*, 2002, 2003a,b) have proposed a comprehensive model of signal processing by the peripheral auditory system that reproduces an important number of the characteristics of both BM and AN responses.

The model contains five stages. Stage 1 consists of either one or two linear bandpass filters adequately arranged in parallel to reproduce the combined frequency response of the outer and middle ears as measured in terms of stapes velocity of vibration. Stage 2 simulates the velocity of vibration of a given site on the BM by means of a dual-resonance non-linear (DRNL) filter (Fig. 7A) (Lopez-Poveda and Meddis, 2001; Meddis *et al.*, 2001). The DRNL filter is inspired by Goldstein's MBPNL model and its predecessors (see Sec. II.3 above IV.C), although the structure of the DRNL filter is itself unique. The input signal to the filter is processed through two asymmetric bandpass filters arranged in parallel: one linear and broadly tuned, and one non-linear and narrowly tuned. Gamma-tone filters are employed that are made asymmetric by filtering their output through a low-pass filter. A compressing memory-less (i.e., instantaneous) gain is applied to the narrow filter that produces linear responses at low levels but compressive responses for moderate levels. The output from the DRNL filter is the sum of the output signals from both paths. Level-dependent tuning is achieved by setting the relative gain of the two filter paths so that the output from the narrow and broad filters dominate the total filter response at low and high levels, respectively. Level-dependent best-frequency shifts are accounted for by setting the center frequency of the broad filter to be different from that of the narrow filter.

The model reproduces suppression because the narrow non-linear path is actually a cascade of a gamma-tone filter followed by the compressive non-linearity, followed by another gamma-tone filter (Fig. 7A). For a two-tone suppression stimulus, the first gamma-tone filter passes both the suppressor and the probe tone, which are then compressed together by the non-linear gain. Because the probe tone is compressed with the suppressor, its level at the output of the second filter is less than it would be if it were presented alone. This means that the form of the suppression contour (Fig. 8) describes approximately the tuning properties of the first gamma-tone filter (Plack *et al.*, 2002). Some versions of the DRNL filter assume that the two gamma-tone filters in this pathway are identical (Lopez-Poveda and Meddis, 2001; Meddis *et al.*, 2001; Sumner *et al.*, 2002), while others (e.g., Plack *et al.*, 2002) allow for the two filters to have different center frequencies and bandwidths to account for the data more realistically (specifically, it can be assumed that the first filter is broader and has a higher center frequency than the second filter). On the other hand, the characteristics of the first gamma-tone filter in this non-linear pathway determine the range of primary frequencies for which combination tones occur, while the second gamma-tone filter determines the amplitude of the generated combination tones.

The DRNL filter has proved adequate to reproduce frequency- and level-dependent BM amplitude responses for a wide range of CFs (Lopez-Najera, 2005; Meddis *et al.*, 2001). It also reproduces local combination tones (i.e., combination tones that originate at BM regions near the measurement site) and some aspects of two-tone suppression (Meddis *et al.*, 2001; Plack *et al.*, 2002). However, the model does *not* account for the different suppression growth rates of low- and high-side suppressors (Plack *et al.*, 2002). Its impulse response resembles that of the BM and it shows frequency glides (Lopez-Najera *et al.*, 2005;

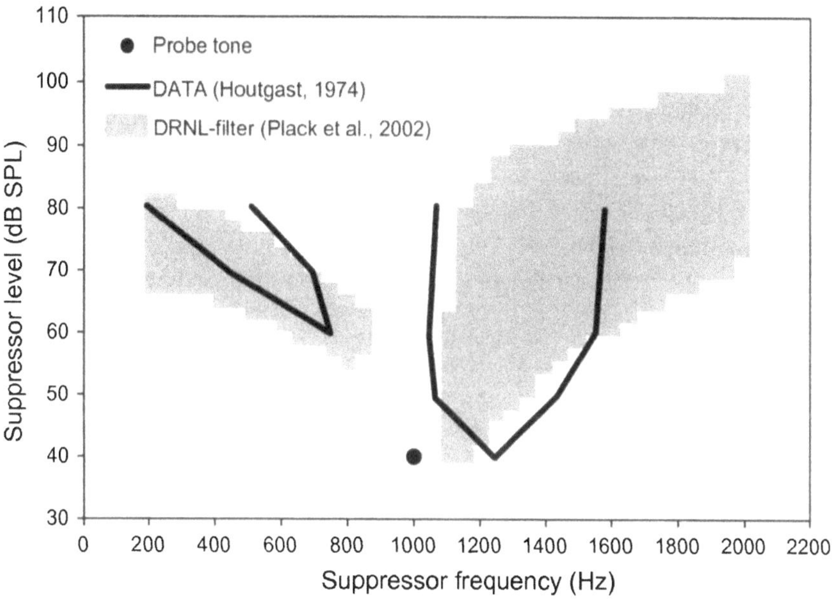

FIG. 8. Two-tone suppression as reproduced by the DRNL filter (adapted from Plack *et al.*, 2002, Copyright © 2002 European Acoustics Association). The solid lines illustrate the boundaries of psychophysical suppression (Houtgast, 1974). A suppressor tone with a level and frequency within the boundary regions produced at least a 3-dB reduction in the effective level of the probe tone (denoted by a circle). The shaded region corresponds to the suppression region determined with the DRNL filter for similar experimental conditions (see Plack *et al.*, 2002 for details).

Meddis *et al.*, 2001). The filter's phase response and the frequency glide of its impulse response are emergent properties of the filter that vary with sound level depending on the filter's bandwidth and on the asymmetry of its frequency response, respectively. These characteristics are very sensitive to the values of the model parameters, particularly to the total order of the filters in both paths and to the frequency response of the middle ear filter used in the model (Lopez-Najera, 2005; Lopez-Najera *et al.*, 2005). With carefully selected parameter values, the frequency glides of the model can be made approximately independent of level (see Fig. 4 of Lopez-Najera *et al.*, 2005) and its phase close to the experimental phase, although only for frequencies below the CF (Fig. 4E) (Lopez-Najera, 2005; Lopez-Najera *et al.*, 2005). The filter also accounts for Nelson's notches (Meddis *et al.*, 2001; Lopez-Poveda and Meddis, 2001).

Filter bank versions of the DRNL filter have been proposed for humans (Lopez-Poveda and Meddis, 2001), guinea pigs (Sumner *et al.*, 2003b) (Fig. 2), and chinchillas (Lopez-Najera, 2005) based on corresponding experimental data. These filter banks do not consider interaction between neighboring filters or propagation of combination tones. The parameters of the DRNL filter may be simply adjusted to model BM responses in cochleae with OHC loss (Lopez-Poveda and Meddis, 2000, 2001). Finally, despite its non-linear character, an approximate transfer function exists for the filter that allows rapid evaluations of both the magnitude and phase of its frequency response at different sound levels (Lopez-Poveda, 2003, 2004).

The third stage of the AN composite model simulates IHC transduction. It is based on a biophysical model of the IHC proposed by Shamma *et al.* (1986). It accounts for the fact that the motion of IHC stereocilia follows BM velocity at low frequencies but displacement at high frequencies (reviewed by Dallos, 1996). It realistically simulates the receptor potential of the IHCs in response to sound, including half-wave rectification and low-pass filtering (Lopez-Poveda, 1996; Shamma, 1986). Stage 4 of the model accounts for the role played by calcium concentration on the probability of release of neurotransmitter to the synaptic cleft between the IHC and the AN fiber (Sumner *et al.*, 2002). It is crucial for the model to account for the response characteristics of fibers with different spontaneous rates. Finally, stage 5 consists of an improved version of the Meddis (1986, 1988) model of synaptic adaptation proposed by Lopez-Poveda (1996) and Lopez-Poveda *et al.* (1998). It considers that the release of the transmitter into the synaptic cleft is quantal and stochastic. This stage simulates the functional effects of adaptation, which are assumed to be due to pre-synaptic transmitter depletion.

This composite model accounts for the non-linear characteristics of the AN response in so far as they are accounted for by its DRNL filter stage. Additionally, it realistically reproduces rate-level functions for fibers with low, medium, and high spontaneous rates (Fig. 1) for stimuli of different frequencies (Sumner *et al.*, 2002). It correctly simulates frequency- and level-dependent phase-locking effects (Sumner *et al.*, 2002). It also correctly simulates level-dependent adaptation effects for fibers with different spontaneous rates (Sumner *et al.*, 2003a). Finally, a filter bank version exists for simulating the activity of a population of guinea pig AN fibers in response to arbitrary sounds (Sumner *et al.*, 2003b).

The model has been successfully employed for predicting the AN representation of stimuli with complex spectra, such as HRTF (Lopez-Poveda, 1996), speech (Holmes *et al.*, 2004), harmonic complexes (Gockel *et al.*, 2003; Wiegrebe and Meddis, 2004), or amplitude modulated stimuli (Meddis *et al.*, 2002). It has also been used to account for data pertaining to the recovery of AN responses from prior stimulation (Meddis and O'Mard, 2005), a phenomenon that depends substantially on the adaptation properties of the AN response. The model has also been used to drive models of brainstem units (e.g., Wiegrebe and Meddis, 2004). It has also been used as the basis to build a biologically inspired speech processor for cochlear implants (Wilson *et al.*, 2005a,b). With this new processor, the intelligibility of speech in noise improves in certain tests (Wilson *et al.*, 2005b).

J. Other Phenomenological Non-Linear Models

The phenomenological models previously reviewed amount to a reduced, although representative, selection of different approaches to simulate a wide range of characteristics of the AN response. Other models of this sort have been proposed to account for specific physiological phenomena. For example, Cai and Geisler (1996b) and Cooper (1998) have proposed models that account for AN low-side suppression and BM harmonic distortion, respectively.

Other models of this sort have been produced in an attempt to design better cochlear implants (e.g., Geurts and Wouters, 1999) or speech recognition systems (e.g., Tchorz and Kollmeier, 1999), rather than to reproduce physiological data. The idea here is to take advantage of the properties of the auditory periphery as a speech signal processor (reviewed by Goldstein, 1994; Wilson *et al.*, 2003; Lopez-Poveda and Meddis, 2005) and apply them to the design of those devices. Of course, the models reviewed in the previous text can also be used for these applications.

K. Discussion and Summary: Comparative Account and Recommendations

The approaches to designing phenomenological models of signal processing by the peripheral auditory system have been diverse. As a result, the user must generally compromise between the complexity of a model and its ability to account for a wide range of physiological phenomena. The decision is sometimes difficult, as different models are designed to account for different phenomena. Comparative studies on the ability of different *non-linear* AN models to account for a consented dataset are currently lacking. As models become more useful tools for speech recognition systems, artificial intelligence, speech processors for auditory prostheses, or basic research, such comparative studies should be encouraged. As a first attempt, Table 1 provides a digest of the merits of the models reviewed herein based on the information available thus far. The Table reveals that none of the models reviewed account for all of the phenomena.

It seems clear that, for most purposes, that there are two key stages in producing a reasonable auditory model: (1) the stage that simulates BM excitation; and (2) the stage that models AN adaptation effects. Comparative studies that will help decide on the "most appropriate" phenomenological non-linear model of BM responses are lacking. However, existing physiological evidence may be helpful in this respect. A recent analysis of AN and BM click responses indicates that these can be better accounted for by the interaction of two (or more) excitation drives with different latencies and frequency contents (Lin and Guinan, 2000, 2004). This supports phenomenological models that incorporate multiple resonances, like the MBPNL filter of Goldstein (1990) or the DRNL filter of Meddis *et al.* (2001). As for the type of compressive non-linearity, existing physiological evidence suggests that for any given level of excitation, a linear system is sufficient to describe basilar membrane responses to clicks (e.g., Recio *et al.*, 1998) or noise-evoked otoacoustic emissions (Maat *et al.*, 2000), while at the same time basilar membrane vibration grows compressively with level and frequency and shows distortion and suppression. Such behavior appears more consistent with the type of level-dependent non-linearity considered by the model of Zhang *et al.* (2001) than with the compressive memory-less type of non-linearity considered by the MBPNL or the DRNL filters (see also Harte *et al.*, 2005). An optimum solution might be to combine the best of both approaches in a way that is not yet clear.

As for models of AN adaptation, the only a comparative study was reported by Hewitt and Meddis (1991), who evaluated eight computer models on the same subset of well-reported AN properties. Although no single model completely replicated the data considered, the authors favored the Meddis (1986, 1988) model based both on its good agreement with physiological data and its computational efficiency. It should be noted, however, that none of the models tested included a non-linear basilar membrane model, which is now common in most composite AN models. Therefore, a new comparative study of current models appears timely.

Acknowledgments

The ideas presented herein were obtained by useful discussions over the years with Laurel Carney, Roy Patterson, Toshio Irino, Xuedong Zhang, Julius Goldstein, Hendrikus Duifhuis, Chris Plack, Lowel O'Mard, Chris Sumner, Alberto López Nájera, and Ray Meddis. The author is indebted to Laurel Carney, Ray Meddis, and Ana Alves-Pinto for their comments on an earlier version of this chapter, and to Almudena Eustaquio Martín for producing the chapter figures. The author is supported by the Spanish *Fondo de Investigaciones Sanitarias* (refs. PI020343 and G02/203).

References

Abbas, P. J., and Sachs, M. B. (1976). Two-tone suppression in auditory-nerve fibers: Extension of stimulus-response relationship. *J. Acoust. Soc. Am.* **59,** 112–122.

Aertsen, A. M., and Johannesma, P. I. (1980). Spectro-temporal receptive fields of auditory neurons in the grassfrog. I. Characterization of tonal and natural stimuli. *Biol. Cybern.* **38,** 223–234.

Alves-Pinto, A., and Lopez-Poveda, E. A. (2005). Detection of high-frequency spectral notches as a function of level. *J. Accoust. Soc. Am.* **118,** 2458–2469.

Anderson, D. J. (1973). Quantitative model for the effects of stimulus frequency on synchronization of auditory nerve discharges. *J. Acoust. Soc. Am.* **54,** 361–364.

Anderson, D. J., Rose, J. E., Hind, J. E., and Brugge, J. F. (1971). Temporal position of discharges in single auditory nerve fibers within the cycle of a sine-wave stimulus: Frequency and intensity effects. *J. Acoust. Soc. Am.* **49,** 1131–1139.

Arthur, R. M., Pfeiffer, R. R., and Suga, N. (1971). Properties of "two-tone inhibition" in primary auditory neurons. *J. Physiol. (London)* **212,** 593–609.

Bruce, I. C., Sachs, M. B., and Young, E. D. (2003). An auditory-periphery model of the effects of acoustic trauma on auditory nerve responses. *J. Acoust. Soc. Am.* **113,** 369–388.

Buunen, T. J. F., and Rhode, W. S. (1978). Responses of fibers in the cat's auditory nerve to the cubic difference tone. *J. Acoust. Soc. Am.* **64,** 772–781.

Cai, Y., and Geisler, C. D. (1996a). Suppression in auditory-nerve fibers of cats using low-side suppressors. II. Effect of spontaneous rates. *Hearing Res.* **96,** 113–125.

Cai, Y., and Geisler, C. D. (1996b). Suppression in auditory-nerve fibers of cats using low-side suppressors. III. Model results. *Hearing Res.* **96,** 126–140.

Carney, L. H. (1993). A model for the responses of low-frequency auditory nerve fibers in cat. *J. Acoust. Soc. Am.* **93,** 401–417.

Carney, L. H. (1999). Temporal response properties of neurons in the auditory pathway. *Curr. Opin. Neurobiol.* **9,** 442–446. *J. Acoust. Soc. Am.* **105,** 356–369.

Carney, L. H., and Yin, T. C. (1988). Temporal coding of resonances by low-frequency auditory nerve fibers: Single-fiber responses and a population model. *J. Neurophysiol.* **60,** 1653–1677.

Carney, L. H., McDuffy, M. J., and Shekhter, I. (1999). Frequency glides in the impulse response of auditory-nerve fibers. *J. Acoust. Soc. Am.* **105,** 2384–2391.

Cheatham, M. A., and Dallos, P. (1992). Two-tone suppression in inner hair cell responses: Correlates of rate suppression in the auditory nerve. *Hearing Res.* **60,** 1–12.

Cheatham, M. A., and Dallos, P. (1998). The level dependence of response phase: Observations from cochlear hair cells. *J. Acoust. Soc. Am.* **104,** 356–369.

Cooper, N. P. (1998). Harmonic distortion on the basilar membrane in the basal turn of the guinea pig. *J. Neurophysiol.* **509,** 277–288.

Cooper, N. P. (2004). Compression in the peripheral auditory system. *In* "Compression. From Cochlea to Cochlear Implants" (S. P. Bacon, R. R. Fay, and A. N. Popper, Eds.), pp. 18–61. Springer-Verlag, New York.

Dallos, P. (1996). Overview: Cochlear neurobiology. *In* "The Cochlea" (P. Dallos, A. N. Popper, and R. R. Fay, Eds.), Chap. 1, pp. 1–43. Springer-Verlag, New York.

de Boer, E. (1975). Synthetic whole-nerve action potentials for the cat. *J. Acoust. Soc. Am.* **58,** 1030–1045.

de Boer, E. (1996). Mechanics of the cochlea. *In* "The Cochlea" (P. Dallos, A. N. Popper, and R. R. Fay, Eds.), Chap. 5, pp. 258–317. Springer-Verlag, New York.

de Boer, E., and de Jongh, H. R. (1978). On cochlear encoding: Potentialities and limitations of the reverse correlation technique. *J. Acoust. Soc. Am.* **63,** 115–135.

de Boer, E., and Nuttall, A. L. (1997). The mechanical waveform of the basilar membrane. I: Frequency modulation ("glides") in impulse responses and across-correlation functions. *J. Acoust. Soc. Am.* **101,** 3583–3592.

Delgutte, B. (1990). Two-tone rate suppression in auditory-nerve fibers: Dependence on suppressor frequency and level. *Hearing Res.* **49,** 225–246.

Delgutte, B. (1996). Physiological models for basic auditory percepts. *In* "Auditory Computation" (H. L. Hawkins, T. A. McMullen, A. N. Popper, and R. R. Fay, Eds.), 6, Chap. 5, pp. 157–220. Springer-Verlag, New York.

Delgutte, B. (1997). Physiological models for basic auditory precepts. *In* "Auditory Computation" (H. L. Hawkins, T. A. McMullen, A. N. Popper, and R. R. Fay, Eds.), **6**, Chap. 5, pp. 157–220. Springer-Verlag, New York.

Delgutte, B. (1997). Auditory neural processing of speech. *In* "The Handbook of Phonetic Sciences" (W. J. Hardcastle and J. Laver, Eds.), Chap. 16. Blackwell Publishers, Cambridge.

Delgutte, B., and Kiang, N. S.-Y. (1984a). Speech coding in the auditory nerve: III. Voiceless fricative consonants. *J. Acoust. Soc. Am.* **75,** 887–896.

Delgutte, B., and Kiang, N. S.-Y. (1984b). Speech coding in the auditory nerve: IV. Sounds with consonant-like dynamic characteristics. *J. Acoust. Soc. Am.* **75,** 897–907.

Deng, L., and Geisler, C. D. (1987a). Responses of auditory-nerve fibers to nasal consonant-vowel syllables. *J. Acoust. Soc. Am.* **82,** 1977–1988.

Deng, L., and Geisler, C. D. (1987b). A composite auditory model for processing speech sounds. *J. Acoust. Soc. Am.* **82,** 2001–2012.

Deng, L., Geisler, C. D., and Greenberg, S. (1987). Responses of auditory-nerve fibers to multiple tone complexes. *J. Acoust. Soc. Am.* **82,** 1989–2000.

Duifhuis, H. (1976). Cochlear nonlinearity and second filter: Possible mechanism and implications. *J. Acoust. Soc. Am.* **59,** 408–423.

Duifhuis, H. (2004). Comment on "An approximate transfer function for the dual-resonance nonlinear filter model of auditory frequency selectivity" [*J. Acoust. Soc. Am.* **114,** 2112–2117] (L). *J. Acoust. Soc. Am.* **115,** 1889–1890.

Evans, E. F. (1972). The frequency response and other properties of single fibers in the guinea-pig cochlear nerve. *J. Physiol. London* **226,** 263–287.

Evans, E. F., and Palmer, A. R. (1980). Relationship between the dynamic range of cochlear nerve fibers and their spontaneous activity. *Exp. Brain Res.* **40,** 115–118.

Fahey, P. F., and Allen, J. B. (1985). Nonlinear phenomena as observed in the ear canal and at the auditory nerve. *J. Acoust. Soc. Am.* **77,** 599–612.

Flanagan, J. L. (1960). Models for approximating basilar membrane displacement. *Bell Sys. Tech. J.* **39,** 1163–1191.

Gardner, M. B., and Hawley, M. S. (1972). Network representations of the external ear. *J. Acoust. Soc. Am.* **52,** 1620–1628.

Geisler, C. D. (1998). From sound to synapse: Physiology of the mammalian ear. Oxford University Press, New York.

Geisler, C. D., and Greenberg, S. G. (1986). A two-stage nonlinear cochlear model possesses automatic gain control. *J. Acoust. Soc. Am.* **80,** 1359–1363.

Geisler, C. D., and Rhode, W. (1982). The phases of basilar membrane vibration. *J. Acoust. Soc. Am.* **71,** 1201–1203.

Geurts, L., and Wouters, J. (1999). Enhancing the speech envelope of continuous interleaved sampling processors for cochlear implants. *J. Acoust. Soc. Am.* **105,** 2476–2484.

Ghitza, O. (1988). Auditory neural feedback as a basis for speech processing. *Proc. IEEE Int. Conf. Acoust. Speech Signal Process (New York)*, 91–94.

Giguère, C, and Woodland, P. C. (1994a). A computational model of the auditory periphery for speech and hearing research. I. Ascending path. *J. Acoust. Soc. Am.* **95,** 331–342.

Giguère, C, and Woodland, P. C. (1994b). A computational model of the auditory periphery for speech and hearing research. II. Descending paths. *J. Acoust. Soc. Am.* **95,** 343–349.

Gockel, H., Moore, B. C. J., Patterson, R. D., and Meddis, R. (2003). Louder sounds can produce less forward masking effects: Effects of component phase in complex tones. *J. Acoust. Soc. Am.* **114,** 978–990.

Goldstein, J. L. (1967). Auditory nonlinearity. *J. Acoust. Soc. Am.* **41,** 676–689.

Goldstein, J. L. (1988). Updating cochlear driven models of auditory perception: A new model for nonlinear auditory frequency analysing filters. *In* "Working Models of Human Perception" (B. A. G. Elsendoorn and H. Bouma, Eds.), pp. 19–58. Academic Press, London.

Goldstein, J. L. (1990). Modeling rapid waveform compression on the basilar membrane as multiple-bandpass-nonlinearity filtering. *Hearing Res.* **49,** 39–60.

Goldstein, J. L. (1993). Exploring new principles of cochlear operation: Bandpass filtering by the organ of Corti and additive amplification by the basilar membrane. *In* "Biophysics of Hair Cell Sensory Systems" (H. Duifhuis, J. W. Horst, P. van Dijk, and S. M. van Netten, Eds.), pp. 315–322. World Scientific, Singapore.

Goldstein, J. L. (1995). Relations among compression, suppression, and combination tones in mechanical responses of the basilar membrane: Data and MBPNL model. *Hearing Res.* **89,** 52–68.

Goldstein, J. L., and Kiang, N. Y.-S. (1968). Neural correlates of the aural combination tone $2f_1 - f_2$. *Proc. IEEE* **56,** 981–992.

Goldstein, M. H. (1994). Auditory periphery as speech signal processor. *IEEE Eng. Med. Biol.* April/May, 186–196.

Greenwood, D. D. (1990). A cochlear frequency-position function for several species-29 years later. *J. Acoust. Soc. Am.* **87,** 2592–2605.

Guinan, J. J., Jr. (1996). Physiology of olivocochlear efferents. *In* "The Cochlea" (P. Dallos, A. N. Popper, and R. R. Fay, Eds.), Chap. 1, pp. 1–43. Springer-Verlag, New York.

Guinan, J. J., Jr., and Gifford, M. L. (1988). Effects of electrical stimulation of efferent olivocochlear neurons on cat auditory-nerve fibers. I. Rate-level functions. *Hearing Res.* **33,** 97–114.

Harrison, J. M., and Howe, M. E. (1974). Anatomy of the afferent auditory nervous system of mammals. *In* "Handbook of Sensory Physiology" (W. D. Keidel and W. D. Neff, Eds.), Vol. 5/1, pp. 283–336. Springer, Berlin.

Harte, J. M., Elliott, S. J., and Rice, H. J. (2005). A comparison of various nonlinear models of cochlear compression. *J. Acoust. Soc. Am.* **117,** 3777–3786.

Hartmann, W. M. (1998). "Signals, Sound, and Sensation." AIP Press, New York.

Heinz, M. G., Colburn, H. S., and Carney, L. H. (2001b). Evaluating auditory performance limits: I. One-parameter discrimination using a computational model for the auditory nerve. *Neural Computation* **13,** 2273–2316.

Heinz, M. G., Colburn, H. S., and Carney, L. H. (2001c). Evaluating auditory performance limits: II. One-parameter discrimination with random-level variation. *Neural Computation* **13,** 2317–2339.

Heinz, M. G., Zhang, X., Bruce, I. C., and Carney, L. H. (2001a). Auditory nerve model for predicting performance limits of normal and impaired listeners. *Acoust. Res. Lett. Online* **2,** 91–96.

Hewitt, M. J., and Meddis, R. (1991). An evaluation of eight computer models of mammalian inner hair-cell function. *J. Acoust. Soc. Am.* **90,** 904–917.

Holmes, S. D., Sumner, C. J., O'Mard, L. P. O., and Meddis, R. (2004). The temporal representation of speech in a nonlinear model of the guinea pig cochlea. *J. Acoust. Soc. Am.* **116,** 3534–3545.

Houtgast, T. (1974). "Lateral Suppression in Hearing." Free University of Amsterdam.

Hubbard, A. E., and Mountain, D. C. (1996). Analysis and synthesis of cochlear mechanical function using models. *In* "Auditory Computation" (H. L. Hawkins, T. A. McMullen, A. N. Popper, and R. R. Fay, Eds.), Chap. 3, pp. 62–120. Springer-Verlag, New York.

Irino, T., and Patterson, R. D. (1997). A time-domain, level-dependent auditory filter: The gammachirp. *J. Acoust. Soc. Am.* **101,** 412–419.

Irino, T., and Patterson, R. D. (2001). A compressive gammachirp auditory filter for both physiological and psychophysical data. *J. Acoust. Soc. Am.* **109,** 2008–2022.

Javel, E. (1981). Suppression of auditory nerve responses. II. Temporal analysis, intensity effects and suppression contours. *J. Acoust. Soc. Am.* **69,** 1735–1745.

Javel, E. (1986). Basic response properties of auditory nerve fibers. *In* "Neurobiology of Hearing: The Cochlea" (R. A. Altschuler, D. W. Hoffman, and R. P. Bobbin, Eds.), pp. 213–245. Raven, New York.

Javel, E., Geisler, C. D., and Ravindran, A. (1978). Two-tone suppression in auditory nerve of the cat: Rate-intensity and temporal analysis. *J. Acoust. Soc. Am.* **63,** 1093–1104.

Javel, E., McGee, J., Walsh, E. J., Farley, G. R., and Gorga, M. P. (1983). Suppression of auditory nerve responses. II. Suppression threshold and growth, iso-suppression contours. *J. Acoust. Soc. Am.* **74,** 801–813.

Jenison, R. L., Greenberg, S., Kluender, K. R., and Rhode, W. S. (1991). A composite model of the auditory periphery for the processing of speech based on the filter response functions of single auditory-nerve fibers. *J. Acoust. Soc. Am.* **90,** 773–786.

Johnson, D. H. (1980). The relationship between spike rate and synchrony in responses of auditory-nerve fibers to single tones. *J. Acoust. Soc. Am.* **68,** 1115–1122.

Kates, J. M. (1991). A time-domain digital cochlear model. *IEEE Trans. Signal Process* **39,** 2573–2592.

Kiang, N. Y.-S., and Moxon, E. C. (1974). Tails of tuning curves of auditory-nerve fibers. *J. Acoust. Soc. Am.* **55,** 620–630.

Kiang, N. Y.-S., Liberman, M. C., Sewell, W. F., and Guinnan, J. J. (1986). Single units clues to cochlear mechanics. *Hearing Res.* **22,** 171–182.

Kiang, N. Y.-S., Watanabe, T., Thomas, E. C., and Clark, L. F. (1965). "Discharge Patterns of Single Fibers in the Cat's Auditory Nerve." MIT Press, Cambridge, Massachusetts.

Killion, M. C., and Clemis, J. D. (1981). An engineering view of middle ear surgery. *J. Acoust. Soc. Am* **69**(Suppl. 1), S44.

Kim, D. O., Molnar, C. E., and Matthews, J. W. (1980). Cochlear mechanics: Nonlinear behaviour in two-tone responses as reflected in cochlear-nerve-fiber responses and in ear-canal sound pressure. *J. Acoust. Soc. Am.* **67,** 1704–1721.

Liberman, M. C. (1978). Auditory-nerve response from cats raised in a low-noise chamber. *J. Acoust. Soc. Am.* **63,** 442–455.

Liberman, M. C., and Kiang, N. Y.-S. (1984). Single-neuron labeling and chronic cochlear pathology. IV. Stereocilia damage and alterations in rate- and level-functions. *Hearing Res.* **16,** 75–90.

Lin, T., and Goldstein, J. L. (1995). Quantifying 2-factor phase relations in nonlinear responses from low characteristic-frequency auditory-nerve fibers. *Hearing Res.* **90,** 126–138.

Lin, T., and Guinan, J. J., Jr. (2000). Auditory-nerve-fiber responses to high-level clicks: Interference patterns indicate that excitation is due to the combination of multiple drives. *J. Acoust. Soc. Am.* **107,** 2615–2630.

Lin, T., and Guinan, J. J., Jr. (2004). Time-frequency analysis of auditory-nerve fiber and basilar-membrane click responses reveal glide irregularities and non-characteristic frequency skirts. *J. Acoust. Soc. Am.* **116,** 405–416.

Lopez-Najera, A. (2005). "Simulación Computacional de la Respuesta de la Membrana Basilar." Ph.D. thesis, University of Salamanca, Spain.

Lopez-Najera, A., Meddis, R., and Lopez-Poveda, E. A. (2005). A computational algorithm for computing non-linear auditory frequency selectivity: Further studies. *In* "Auditory Signal Processing: Physiology, Psychoacoustics, and Models" (D. Pressnitzer, A. de Cheveigné, S. McAdams, and L. Collet, Eds.), pp. 14–20. Springer, New York.

Lopez-Poveda, E. A. (1996). "The Physical Origin and Physiological Coding of Pinna-based Spectral Cues." Ph. D. thesis, Loughborough University, UK.

Lopez-Poveda, E. A. (2003). An approximate transfer function for the dual-resonance nonlinear filter model of auditory frequency selectivity. *J. Acoust. Soc. Am.* **114,** 2112–2117.

Lopez-Poveda, E. A. (2004). Reply to comment on "An approximate transfer function for the dual resonance nonlinear filter model of auditory frequency selectivity" (L). *J. Acoust. Soc. Am.* **115,** 1891.

Lopez-Poveda, E. A., and Meddis, R. (1996). A physical model of sound diffraction and reflections in the human concha. *J. Acoust. Soc. Am.* **100,** 3248–3259.

Lopez-Poveda, E. A., and Meddis, R. (2000). A computational model for simulating basilar membrane nonlinearity in subjects with normal and impaired hearing. *1st International Hearing Aid Research Conference*, PA9, Lake Tahoe, CA.
Lopez-Poveda, E. A., and Meddis, R. (2001). A human nonlinear cochlear filterbank. *J. Acoust. Soc. Am.* **110,** 3107–3118.
Lopez-Poveda, E. A., and Meddis, R. (2005). Oídos artificiales. *Mente y Cerebro* **10,** 24–32.
Lopez-Poveda, E. A., O'Mard, L. P. O., and Meddis, R. (1998). A revised computational inner hair cell model. *In* "Psychophysical and Physiological Advances in Hearing" (A. R. Palmer, A. Rees, Q. Summerfield, and R. Meddis, Eds.), pp. 112–121. Whurr Publishers, London.
Lutman, M. E., and Martin, A. M. (1979). Development of an electroacoustic analogue model of the middle ear and acoustic reflex. *J. Sound Vib.* **64,** 133–157.
Lyon, R. F. (1997). All-pole models of auditory filtering. *In* "Diversity in Auditory Mechanics" (Lewis, *et al.*, Eds.), pp. 205–211. World Scientific Publishing, Singapore.
Maat, B., Wit, H. P., and van Dijk, P. (2000). Noise-evoked ototacoustic emissions in humans. *J. Acoust. Soc. Am.* **108,** 2272–2280.
May, B. J. (2003). Physiological and psychophysical assessments of the dynamic range of vowel representations in the auditory periphery. *Speech Communication* **41,** 49–57.
Meddis, R., O'Mard, L. P. O., and Lopez-Poveda, E. A. (2001). A computational algorithm for computing non-linear auditory frequency selectivity. *J. Acoust. Soc. Am.* **109,** 2852–2861.
Meddis, R. (1986). Simulation of mechanical to neural transduction in the auditory receptor. *J. Acoust. Soc. Am.* **79,** 702–711.
Meddis, R. (1988). Simulation of auditory-neural transduction: Further studies. *J. Acoust. Soc. Am.* **83,** 1056–1063.
Meddis, R., and Hewitt, M. J. (1991). Virtual pitch and phase sensitivity of a computer model of the auditory periphery. I: Pitch identification. *J. Acoust. Soc. Am.* **89,** 2866–2882.
Meddis, R., and O'Mard, L. P. O. (2005). A computer model of the auditory-nerve response to forward-masking stimuli. *J. Acoust. Soc. Am.* **117,** 3787–3798.
Meddis, R., Delahaye, R., O'Mard, L. P. O., Sumner, C., Fantini, D. A., Winter, I., and Pressnitzer, D. (2002). A model of signal processing in the cochlear nucleus: Comodulation masking release. *Acta Acustica United with Acustica* **88,** 387–398.
Merhaut, J. (1981). "Theory of Electroacoustics." McGraw-Hill, New York.
Møller, A. R. (2000). "Hearing: Its Physiology and Pathophysiology." Academic Press, San Diego.
Moore, B. C. J. (2003). "An Introduction to the Psychology of Hearing," 5th ed. Academic Press, London.
Mountain, D. C., and Hubbard, A. E. (1996). Computational analysis of hair cell and auditory nerve processes. *In* "Auditory Computation" (H. L. Hawkins, T. A. McMullen, A. N. Popper, and R. R. Fay, Eds.), Chap. 4, pp. 121–156. Springer-Verlag, New York.
Müller, M., and Robertson, D. (1991). Relationship between tone burst discharge pattern and spontaneous firing rate of auditory nerve fibers in the guinea-pig. *Hearing Res.* **57,** 63–70.
Narayan, S. S., Temchin, A. N., Recio, A., and Ruggero, M. A. (1998). Frequency tuning of basilar membrane and auditory nerve fibers in the same cochleae. *Science* **282,** 1882–1884.
Nuttall, A. L., and Dolan, D. F. (1990). Inner hair cell responses to the $2f_1 - f_2$ intermodulation distortion product. *J. Acoust. Soc. Am.* **87,** 782–790.
Nuttall, A. L., and Dolan, D. F. (1993a). Intermodulation distortion ($f_2 - f_1$) in inner hair cell and basilar membrane responses. *J. Acoust. Soc. Am.* **93,** 2061–2068.
Nuttall, A. L., and Dolan, D. F. (1993b). Two-tone suppression of inner hair cell and basilar membrane responses in the guinea pig. *J. Acoust. Soc. Am.* **93,** 390–400.
Oono, Y., and Sujaku, Y. (1974). A probabilistic model for discharge patterns of auditory nerve fibers. *Trans. Inst. Elect. Comm. Eng. (Japan)* **57,** 35–36.

Palmer, A. R. (1990). The representation of the spectra and fundamental frequencies of steady-state single- and double-vowel sounds in the temporal discharge patterns of guinea pig cochlear-nerve fibers. *J. Acoust. Soc. Am.* **88,** 1412–1426.

Palmer, A. R., and Russell, I. J. (1986). Phase-locking in the cochlear nerve of the guinea-pig and its relation to the receptor potential of inner hair-cells. *Hearing Res.* **24,** 1–15.

Palmer, A. R., Winter, I. M., and Darwin, C. J. (1986). The representation of steady-state vowel sounds in the temporal discharge patterns of the guinea pig cochlear nerve and primary-like cochlear nucleus neurons. *J. Acoust. Soc. Am.* **79,** 100–113.

Patterson, R. D., Robinson, K., Holdsworth, J., McKeown, D., Zhang, C., and Allerhand, M. (1992). Complex sounds and auditory images. *In* "Auditory Physiology and Perception" (Y. Cazals, K. Horner, and L. Demany, Eds.), pp. 429–443. Pergamon Press, Oxford.

Patterson, R. D., Unoki, M., and Irino, T. (2003). Extending the domain of center frequencies for the compressive gammachirp auditory filter. *J. Acoust. Soc. Am.* **114,** 1529–1542.

Patuzzi, R., Sellick, P. M., and Johnstone, B. M. (1984). The modulation of the sensitivity of the mammalian cochlea by low frequency tones III. Basilar membrane motion. *Hearing Res.* **13,** 19–27.

Pfeiffer, R. R. (1970). A model for two-tone inhibition of single cochlear-nerve fibers. *J. Acoust. Soc. Am.* **48,** 1373–1378.

Pickles, J. O. (1988). "An Introduction to the Physiology of Hearing," 2nd ed. Academic Press, London.

Plack, C. J., Oxenham, A. J., and Drga, V. (2002). Linear and nonlinear processes in temporal masking. *Acta Acustica United with Acustica* **88,** 348–358.

Recio, A. R., Rich, N. C., Narayan, S. S., and Ruggero, M. A. (1998). Basilar-membrane responses to clicks at the base of the chinchilla cochlea. *J. Acoust. Soc. Am.* **103,** 1972–1989.

Rhode, W. S., and Greenberg, S. (1994). Lateral suppression and inhibition in the cochlear nucleus of the cat. *J. Neurophysiol.* **71,** 493–514.

Rhode, W. S., and Robles, L. (1974). Evidence from Mössbauer experiments for nonlinear vibration in the cochlea. *J. Acoust. Soc. Am.* **55,** 588–596.

Rice, J. J., Young, E. D., and Spirou, G. A. (1995). Auditory-nerve encoding of pinna-based spectral cues: Rate representation of high-frequency stimuli. *J. Acoust. Soc. Am.* **97,** 1764–1776.

Robert, A., and Eriksson, J. L. (1999). A composite model of the auditory periphery for simulating responses to complex sounds. *J. Acoust. Soc. Am.* **106,** 1852–1864.

Robles, L., and Ruggero, M. A. (2001). Mechanics of the mammalian cochlea. *Physiol. Rev.* **81,** 1305–1352.

Robles, L., Ruggero, M. A., and Rich, N. C. (1991). Two-tone distortion in the basilar membrane of the cochlea. *Nature* **349,** 413–414.

Robles, L., Ruggero, M. A., and Rich, N. C. (1997). Two-tone distortion on the basilar membrane of the chinchilla cochlea. *J. Neurophysiol.* **77,** 2385–2399.

Rose, J. E., Brugge, J. F., Anderson, D. J., and Hind, J. E. (1967). Phase-locked response to low-frequency tones in single auditory nerve fibers of squirrel monkey. *J. Neurophysiol.* **30,** 769–793.

Rose, J. E., Hind, J. E., Anderson, D. J., and Brugge, J. F. (1971). Some effects of stimulus intensity on response of auditory nerve fibers in the squirrel monkey. *J. Neurophysiol.* **34,** 685–699.

Rosowski, J. J. (1996). Models of external- and middle-ear function. *In* "Auditory Computation" (H. L. Hawkins, T. A. McMullen, A. N. Popper, and R. R. Fay, Eds.), Chap. 1, pp. 15–61. Springer-Verlag, New York.

Ruggero, M. A. (1973). Response to noise of auditory nerve fibers in the squirrel monkey. *J. Neurophysiol.* **36,** 569–587.

Ruggero, M. A. (1992). Physiology and coding of sound in the auditory nerve. *In* "The Mammalian Auditory Pathway: Neurophysiology" (A. N. Popper and R. Fay, Eds.), pp. 34–93. Springer-Verlag, New York.

Ruggero, M. A., Narayan, S. S., Temchin, A. N., and Recio, A. (2000). Mechanical bases of frequency tuning and neural excitation at the base of the cochlea: Comparison of basilar-membrane vibrations and auditory-nerve-fiber responses in chinchilla. *Proc. Nat. Acad. Sci.* **97,** 11744–11750.

Ruggero, M. A., and Rich, N. C. (1991). Furosemide alters organ of Corti mechanics: Evidence for feedback of outer hair cells upon the basilar membrane. *J. Neurosci.* **11,** 1057–1067.

Ruggero, M. A., Rich, N. C., Recio, A., Narayan, S. S., and Robles, L. (1997). Basilar-membrane responses to tones at the base of the chinchilla cochlea. *J. Acoust. Soc. Am.* **101,** 2151–2163.

Ruggero, M. A., Robles, L., Rich, N. C., and Costalupes, J. A. (1986). Basilar membrane motion and spike initiation in the cochlear nerve. *In* "Auditory Frequency Selectivity" (B. C. J. Moore and R. Patterson, Eds.), pp. 189–198. Plenum Pub. Corp.

Sachs, M. B., and Abbas, P. J. (1974). Rate versus level functions for auditory-nerve fibers in cats: Tone-burst stimuli. *J. Acoust. Soc. Am.* **56,** 1835–1847.

Sachs, M. B., and Kiang, N. Y.-S. (1968). Two-tone inhibition in auditory-nerve fibers. *J. Acoust. Soc. Am.* **43,** 1120–1128.

Sachs, M. B., and Young, E. D. (1979). Encoding of steady-state vowels in the auditory nerve: Representation in terms of discharge rate. *J. Acoust. Soc. Am.* **66,** 470–479.

Sachs, M. B., and Young, E. D. (1980). Effects of nonlinearities on speech encoding in the auditory nerve. *J. Acoust. Soc. Am.* **68,** 858–875.

Schalk, T. B., and Sachs, M. B. (1980). Nonlinearities in auditory-nerve fiber responses to bandlimited noise. *J. Acoust. Soc. Am.* **67,** 903–913.

Shamma, S. A., Chadwick, R. S., Wilbur, W. J., Morrish, K. A., and Rinzel, J. (1986). A biophysical model of the cochlear processing: Intensity dependence of pure tone responses. *J. Acoust. Soc. Am.* **80,** 133–145.

Slaney, M. (1993). An efficient implementation of the Patterson-Holdsworth auditory filter bank. *Apple Computer Technical Report* #35, Apple Computer, Inc.

Smith, R. L., and Brachman, M. L. (1980). Operating range and maximum response of single auditory nerve fibers. *Brain Res.* **184,** 499–505.

Smith, R. L., and Zwislocki, J. J. (1975). Short-term adaptation and incremental responses in single auditory-nerve fibers. *Biol. Cybernetics* **17,** 169–182.

Sumner, C., Lopez-Poveda, E. A., O'Mard, L. P. O., and Meddis, R. (2002). A revised model of the inner hair cell and auditory nerve complex. *J. Acoust. Soc. Am.* **111,** 2178–2188.

Sumner, C., Lopez-Poveda, E. A., O'Mard, L. P. O., and Meddis, R. (2003a). Adaptation in a revised model of the inner-hair cell. *J. Acoust. Soc. Am.* **113,** 893–901.

Sumner, C., O'Mard, L. P. O., Lopez-Poveda, E. A., and Meddis, R. (2003b). A non-linear filter-bank model of the guinea-pig cochlear nerve. *J. Acoust. Soc. Am.* **113,** 3264–3274.

Tan, Q., and Carney, L. H. (2003). A phenomenological model for the responses of auditory-nerve fibers. II. Nonlinear tuning with a frequency glide. *J. Acoust. Soc. Am.* **114,** 2007–2020.

Tan, Q., and Carney, L. H. (2005). Encoding of vowel-like sounds in the auditory nerve: Model predictions of discrimination performance. *J. Acoust. Soc. Am.* **117,** 1210–1222.

Tchorz, J., and Kollmeier, B. (1999). A model of auditory perception as front end for automatic speech recognition. *J. Acoust. Soc. Am.* **106,** 2040–2050.

Viergever, M. A. (1980). "Mechanics of the Inner Ear-A Mathematical Approach." Ph. D. thesis, Delft U. P., Delft, the Netherlands.

Westerman, L. A. (1985). "Adaptation and Recovery of Auditory Nerve Responses," Special report ISR-S-24, Syracuse University.

Westerman, L. A., and Smith, R. L. (1984). Rapid and short-term adaptation in auditory nerve responses. *Hearing Res.* **15,** 249–260.

Westerman, L. A., and Smith, R. L. (1988). A diffusion model of the transient response of the cochlear inner hair cell synapse. *J. Acoust. Soc. Am.* **83,** 2266–2276.

Wickesberg, R. E., and Geisler, C. D. (1986). Longitudinal stiffness coupling in a 1-dimensional model of the peripheral ear. *In* "Peripheral Auditory Mechanics" (J. B. Allen, J. L. Hall, A. Hubbard, S. T. Neely, and A. Tubis, Eds.), pp. 113–120. Springer, Berlin.

Wiegrebe, L., and Meddis, R. (2004). The representation of periodic sounds in simulated sustained chopper units of the ventral cochlear nucleus. *J. Acoust. Soc. Am.* **115,** 1207–1218.

Wilson, B. S., Lawson, D. T., Muller, J. M., Tyler, R. S., and Kiefer, J. (2003). Cochlear implants: Some likely next steps. *Annu. Rev. Biomed. Eng.* **5,** 207–249.

Wilson, B. S., Schatzer, R., and Lopez-Poveda, E. A. (2005b). Possibilities for a closer mimicking of normal auditory functions with cochlear implants. *In* "Cochlear Implants" (S. B. Waltzman and J. T. Roland, Eds.), 2nd ed. Thieme Medical Publishers, New York, scheduled for publication in 2005.

Wilson, B. S., Schatzer, R., Lopez-Poveda, E. A., Sun, X., Lawson, D.T, and Wolford, R. D. (2005a). Two new directions in speech processor design for cochlear implants. *Ear* & *Hear* (scheduled for publication in a supplement to the July/August 2005 issue.) **26,** 73S–81S.

Winter, I. M., and Palmer, A. R. (1991). Intensity coding in low-frequency auditory-nerve fibers of the guinea pig. *J. Acoust. Soc. Am.* **90,** 1958–1967.

Winter, I. M., Robertson, D., and Yates, G. K. (1990). Diversity of characteristic frequency rate-intensity functions in guinea pig auditory nerve fibers. *Hearing Res.* **45,** 191–202.

Yates, G. K. (1990). Basilar membrane nonlinearity and its influence on auditory nerve rate-intensity functions. *Hearing Res.* **50,** 145–162.

Yates, G. K., Winter, I. M., and Robertson, D. (1990). Basilar membrane nonlinearity determines auditory nerve rate-intensity functions and cochlear dynamic range. *Hearing Res.* **45,** 203–220.

Young, E. D., and Sachs, M. B. (1979). Representation of steady-state vowels in the temporal aspects of the discharge patterns of populations of auditory nerve fibers. *J. Acoust. Soc. Am.* **66,** 1381–1403.

Zhang, X., Heinz, M. G., Bruce, I. C., and Carney, L. H. (2001). A phenomenological model for the responses of auditory-nerve fibers: I. Nonlinear tuning with compression and suppression. *J. Acoust. Soc. Am.* **109,** 648–670.

Zurek, P. M., and Sachs, R. M. (1979). Combination tones at frequencies greater than the primary tones. *Science* **205,** 600–602.

Zwicker, E. (1986). A hardware cochlear nonlinear preprocessing model with active feedback. *J. Acoust. Soc. Am.* **80,** 163–176.

BASIC PSYCHOPHYSICS OF HUMAN SPECTRAL PROCESSING

Brian C. J. Moore

Department of Experimental Psychology, University of Cambridge
Cambridge CB2 3EB, United Kingdom

The peripheral auditory system contains a bank of bandpass filters, the auditory filters, with center frequencies spanning the range from about 50 to 15,000 Hz. The mechanical properties of the basilar membrane (BM) and surrounding structures appear to provide the initial basis of the filtering process, giving the passive linear "tail" filter. An additional more sharply tuned "tip" filter is produced by the active mechanism in the cochlea, which appears to depend strongly on the operation of the outer hair cells (OHCs). The gain of the tip filter is level dependent. The auditory filter can be thought of as a weighting function which characterizes frequency selectivity at a particular center frequency.

INTERNATIONAL REVIEW OF
NEUROBIOLOGY, VOL. 70
DOI: 10.1016/S0074-7742(05)70002-7

0074-7742/05 $35.00

The shape of the auditory filter at a given center frequency can be estimated using the notched-noise masking technique as well as the assumptions of the power-spectrum model. Its bandwidth for frequencies above 1 kHz is about 10% to 17% of the center frequency. At moderate sound levels, the auditory filter is roughly symmetric on a linear frequency scale. At high sound levels the low-frequency side of the filter becomes less steep than the high-frequency side.

The excitation pattern of a given sound represents the distribution of activity evoked by that sound as a function of the characteristic frequency (CF) of the neurons stimulated. In psychophysical terms, the excitation pattern can be defined as the output of the auditory filters as a function of center frequency. The shapes of the excitation patterns for sinusoids or narrowband noises are similar to the masking patterns of narrowband noises.

Non-simultaneous masking reveals suppression effects similar to the suppression observed on the BM and in primary auditory neurons. This suppression is not revealed in simultaneous masking. The frequency selectivity measured in forward masking is greater than that measured in simultaneous masking. It is likely that auditory filter shapes estimated using data from simultaneous masking data are less sharp than the underlying physiological filters.

People with hearing loss of cochlear origin show reduced frequency selectivity, probably because of damage to the active mechanism in the cochlea. This contributes to the difficulties experienced by hearing-impaired people when trying to understand speech in background noise.

I. Introduction

Frequency selectivity refers to the ability of the auditory system to separate or resolve (to a certain extent) the sinusoidal components in a complex sound. This ability is also known as frequency resolution and frequency analysis; these terms will be used interchangeably in this chapter. Frequency resolution is most often demonstrated and quantified by studying masking. Masking may be regarded as reflecting the limits of frequency resolution. If a sound of a given frequency is masked by another sound with a different frequency, then the auditory system has failed to resolve the two sounds. Hence, by measuring when one sound is just masked by another, it is possible to characterize the frequency resolution of the auditory system.

It is widely believed that the frequency selectivity measured behaviorally in masking experiments is largely determined by the filtering that occurs on the basilar membrane (BM) (Evans *et al.*, 1989; Fletcher, 1940; Moore, 1986). The tuning of a specific point of the BM is determined by two components: a passive component reflecting the hydromechanical properties of the BM and surrounding fluids and structures, and an active component reflecting the influence of the outer hair cells (OHCs) on BM motion (see Chapter 5). The passive component appears to be roughly linear and does not vary with level, while the active component is highly non-linear (Robles and Ruggero, 2001). The non-linearity greatly complicates the analysis and interpretation of masking experiments. However, this chapter begins with a discussion of a simple model of masking which ignores these complications.

II. The Power-Spectrum Model and the Concept of the Critical Band

Fletcher (1940) measured the threshold for detecting a sinusoidal signal presented in a bandpass noise masker, as a function of the bandwidth of the masker. The noise was always centered at the signal frequency, and the noise power density was held constant. Thus, the total noise power increased as the bandwidth increased. This experiment has been repeated several times since then, with similar results (Bernstein and Raab, 1990; Greenwood, 1961b; Hamilton, 1957). The threshold of the signal increases at first as the noise bandwidth increases, but then flattens off; further increases in noise bandwidth do not change the signal threshold significantly.

To account for this pattern of results, Fletcher (1940) suggested that the peripheral auditory system behaves as if it contains a bank of bandpass filters, with overlapping passbands. These filters are now called the "auditory filters." The center frequencies of the filters range from about 50 to 15,000 Hz (Moore *et al.*, 1997). Fletcher suggested that the signal was detected by using the output of the auditory filter centered on the signal frequency. Increases in noise bandwidth result in more noise passing through that filter, as long as the noise bandwidth is less than the filter bandwidth. However, once the noise bandwidth exceeds the filter bandwidth, further increases in noise bandwidth do not increase the noise passing through the filter. Fletcher called the bandwidth at which the signal threshold ceased to increase the "critical bandwidth" (CB). It is usually assumed that this bandwidth is closely related to the bandwidth of the auditory filter at the same center frequency. However, Fletcher's band-widening experiment does not provide a precise way of estimating the bandwidth of the auditory filter, as small errors of measurement can lead to rather large errors in the estimated CB (Patterson and Moore, 1986).

Fletcher's experiment led to a model of masking known as the power-spectrum model, which is based on the following assumptions:

1. The peripheral auditory system contains an array of linear overlapping bandpass filters.
2. When trying to detect a signal in a noise background, the listener is assumed to make use of just one filter with a center frequency close to that of

the signal. Usually, it is assumed that the filter used is the one that has the highest signal-to-masker ratio at its output.

3. Only the components in the noise which pass through the filter have any effect in masking the signal.

4. The threshold for detecting the signal is determined by the amount of noise passing through the auditory filter; specifically, threshold is assumed to correspond to a certain signal-to-noise ratio, K, at the output of the filter. The stimuli are represented by their long-term power spectra, i.e., the relative phases of the components and the short-term fluctuations in the masker are ignored.

It is now known that none of these assumptions is strictly correct: the filters are not linear, but are level-dependent (Glasberg and Moore, 2000; Lutfi and Patterson, 1984; Moore and Glasberg, 1987b; Patterson *et al.*, 2003; Rosen *et al.*, 1998; Weber, 1977); simultaneous masking can be influenced by non-linear effects such as suppression (Delgutte, 1988, 1990; Moore and Vickers, 1997); listeners can combine information from more than one filter to enhance signal detection (Buus *et al.*, 1986; Moore *et al.*, 1997; Spiegel, 1981); noise falling outside the passband of the auditory filter centered at the signal frequency can affect the detection of that signal (Hall *et al.*, 1984, 1995), as described in Chapter 3; the threshold may not correspond to a constant signal-to-masker ratio at the output of the auditory filter, because the salience of the detection cues used by the subject can change when the masker spectrum is altered (Kluk and Moore, 2004; Moore *et al.*, 1998); and fluctuations in the masker and the phase spectrum of the masker can play a strong role (Alcántara *et al.*, 2003; Gockel *et al.*, 2002; Kohlrausch and Sander, 1995; Kohlrausch *et al.*, 1997; Moore, 1988; Patterson and Henning, 1977; Summers and Leek, 1998). These failures of the assumptions of the model do not mean that the basic concept of the auditory filter is wrong. Indeed, the concept is widely accepted and has proved to be very useful. Nevertheless, it should be remembered that simplifying assumptions are often made in attempts to characterize and model the auditory filter.

Fletcher (1940) pointed out that the value of the CB could be estimated indirectly, by measuring the power of a sinusoidal signal (P_s) required for the signal to be detected in broadband white noise, given the assumptions of the power-spectrum model. For a white noise with power density N_0, the total noise power falling within the CB is approximately $N_0 \times$ CB. According to assumption (4) above,

$$P_s/(\mathrm{CB} \times N_0) = K \tag{1}$$

where K is a constant representing the "detection efficiency" of the listener (Patterson and Moore, 1986). Hence,

$$\mathrm{CB} = P_s/(K \times N_0). \tag{2}$$

By measuring P_s and N_0 and by estimating K, the value of the CB can be evaluated.

Fletcher estimated that K was equal to 1, indicating that the value of the CB should be equal to P_s/N_0, which is called the "critical ratio." However, Fletcher's estimate of K was found not to be accurate. More recent experiments show that K is typically about 0.4 (Scharf, 1970). Thus, at most frequencies the critical ratio is about 0.4 times the value of the CB estimated by more direct methods, such as the band-widening experiment. Also, K varies with center frequency, increasing markedly at low frequencies, so the critical ratio does not give a correct indication of how the CB varies with center frequency (Moore *et al.*, 1997; Patterson and Moore, 1986).

III. Estimating the Shape of the Auditory Filter

Most methods for estimating the shape of the auditory filter at a given center frequency are based on the assumptions of the power-spectrum model of masking. If the masker is represented by its long-term power spectrum, $N(f)$, and the weighting function or shape of the auditory filter is $W(f)$, then the power-spectrum model is expressed by:

$$P_s = K \int_{0}^{\infty} W(f)N(f)df \tag{3}$$

where P_s is the power of the signal at threshold, and K is the same as previously described. By manipulating the masker spectrum, $N(f)$, and measuring the corresponding changes in P_s, it is possible to derive the filter shape, $W(f)$.

The masker chosen to measure the auditory filter shape should be such that the assumptions of the power-spectrum model are not strongly violated. A number of factors affect this choice. If the masker is composed of one or more sinusoids, beats between the signal and masker may provide a cue to the presence of the signal (Alcántara *et al.*, 2000; Egan and Hake, 1950; Kluk and Moore, 2004; Moore *et al.*, 1998). This makes sinusoids unsuitable as maskers for estimating the auditory filter shape, since the salience of beats changes as the masker frequency is altered; this violates the assumption of the power-spectrum model that threshold corresponds to a constant signal-to-masker ratio at the output of the auditory filter.

In general, noise maskers are more suitable than sinusoids for estimating the auditory filter shape, since noises have inherent amplitude fluctuations which make beats much less effective as a cue. However, for narrowband noises, which have relatively slow fluctuations, temporal interactions between the signal and masker may still be audible (Kluk and Moore, 2004; Moore *et al.*, 1998). In addition, the slow fluctuations may strongly influence the detectability of the signal in a way which depends on the difference between the center frequency of the masker and the frequency of the signal (Buus, 1985; Moore and Glasberg, 1987a; Nelson and Schroder, 1996). For these reasons, the assumptions of the power-spectrum model are best satisfied using reasonably broadband noise maskers.

A second important consideration in choosing a noise masker for measuring auditory filter shapes is that the filter giving the highest signal-to-masker ratio is not necessarily centered at the signal frequency. For example, if the signal has a frequency of 1 kHz, and the masker spectrum consists entirely of frequencies above 1 kHz, the highest signal-to-masker ratio may occur for a filter centered below 1 kHz. The process of detecting the signal through a filter which is not centered at the signal frequency is called "off-frequency listening." In this context, it is the center frequency of the filter that is "off frequency." Furthermore, if the masker spectrum is concentrated primarily above or below the signal frequency, there may be a range of filter center frequencies over which the signal-to-masker ratio is sufficiently high enough to provide useful information. Under these conditions, the observer may combine information over several auditory filters, rather than using the output of a single filter as assumed by the power-spectrum model.

A. Psychophysical Tuning Curves (PTCs)

The measurement of PTCs involves a procedure which is analogous in many ways to the determination of a neural tuning curve (Chistovich, 1957; Small, 1959). The signal is fixed in level, usually at a very low level, e.g., 10 dB Sensation Level (SL). The masker can be either a sinusoid or a narrow band of noise, but a noise is generally preferred, for the reasons previously listed.

For each of several masker center frequencies, the level of the masker needed just to mask the signal is determined. Because the signal is at a low level it is assumed that it will produce activity primarily at the output of one auditory filter. It is further assumed that, at the threshold, the masker produces a constant output from that filter, in order to mask the fixed signal. Thus, the PTC indicates the masker level required to produce a fixed output from the auditory filter as a function of frequency. Normally, a filter characteristic is determined by plotting the output from the filter for an input varying in frequency and fixed in level. However, if the filter is linear the same result can be obtained by plotting the input required to give a fixed output. Thus, if linearity is assumed, the shape of the auditory filter can be obtained simply by inverting the PTC. Examples of some PTCs are provided in Fig. 1; the data are taken from Vogten (1974). These data were obtained using a sinusoidal masker, but the signal duration was short (50 ms) so as to reduce the influence of beats as a cue.

Thus far, it has been assumed that only one auditory filter is involved in the determination of a PTC. However, there is now good evidence that off-frequency listening can influence PTCs. When the masker frequency is above the signal

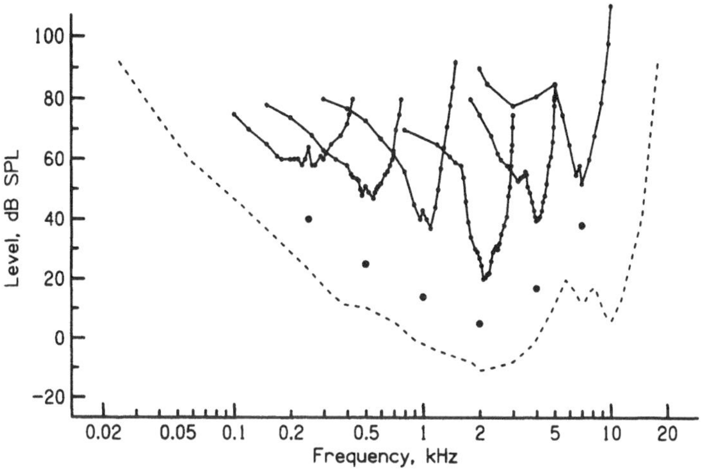

FIG. 1. PTCs determined in simultaneous masking, using sinusoidal signals at 10 dB SL. For each curve, the solid circle below it indicates the frequency and level of the signal. The masker was a sinusoid which had a fixed starting phase relationship to the brief 50-ms signal. The masker level required for threshold is plotted as a function of masker frequency, on a logarithmic scale. The dashed line shows the absolute threshold for the signal. Data from Vogten (1974).

frequency, the highest signal-to-masker ratio occurs for a filter centered below the signal frequency. Conversely, when the masker frequency is below the signal frequency, the highest signal-to-masker ratio occurs for a filter centered above the signal frequency. In both these cases, the masker level required for threshold is higher than would be the case if off-frequency listening did not occur. When the masker frequency equals the signal frequency, the signal-to-masker ratio is similar for all auditory filters that are excited, and off-frequency listening is not advantageous. The overall effect of off-frequency listening is that the PTC has a sharper tip than would be obtained if only one auditory filter were involved (Johnson-Davies and Patterson, 1979; O'Loughlin and Moore, 1981a,b).

One way to limit off-frequency listening is to add to the masker a fixed, low-level noise with a spectral notch centered at the signal frequency (Moore *et al.*, 1984a; O'Loughlin and Moore, 1981b). Such a masker should make it disadvantageous to use an auditory filter whose center frequency is shifted much from the signal frequency. The effect of using such a noise, in addition to the variable narrowband masker, is to broaden the tip of the PTC; the slopes of the skirts are relatively unaffected.

The detection of beats can influence PTCs even when the masker is a narrowband noise (Kluk and Moore, 2004). Kluk and Moore (2004) found that the noise bandwidth needed to exceed 160 Hz to eliminate the effects of beat detection. However, the use of a noise bandwidth above 160 Hz is problematic when the signal frequency is below approximately 1500 Hz, since the noise bandwidth in such cases is greater than the auditory filter bandwidth, and the tip of the PTC becomes "smeared" (Sek *et al.*, 2005).

B. The Notched-Noise Method

To satisfy the assumptions of the power-spectrum model, it is necessary to use a masker which limits the amount by which the center frequency of the filter can be shifted (off-frequency listening), and which limits the range of filter center frequencies over which the signal-to-masker ratio is sufficiently high enough to be useful. The masker should also prevent beats from being used as a cue. This can be achieved using a noise masker with a spectral notch around the signal frequency. For such a masker, the highest signal-to-masker ratio occurs for a filter which is centered reasonably close to the signal frequency, and performance is not improved (or is improved very little) by combining information over filters covering a range of center frequencies (Patterson, 1976; Patterson and Moore, 1986). Also, audible beats are not produced by adding a sinusoidal signal to such a noise. The filter shape can be estimated by measuring signal threshold as a function of the width of the notch.

For moderate noise levels, the auditory filter is almost symmetrical on a linear frequency scale (Moore and Glasberg, 1987b; Patterson, 1974, 1976; Patterson and Nimmo-Smith, 1980). Hence, the auditory filter shape can be estimated using a notched-noise masker with the notch placed symmetrically about the signal frequency. The method is illustrated in Fig. 2. For a masker with a notch width of $2\Delta f$, and a center frequency f_c, Equation (3) becomes:

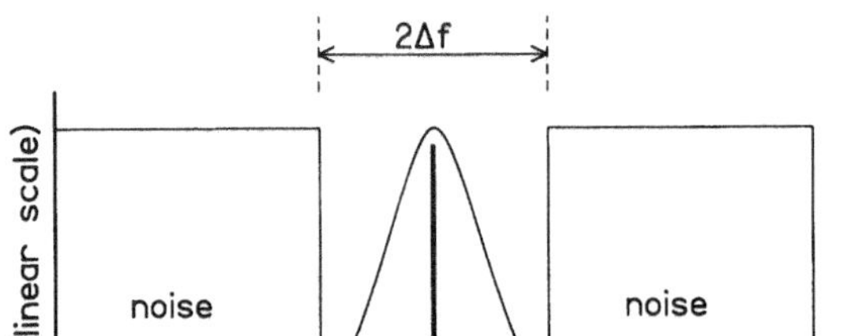

FIG. 2. Schematic illustration of the technique used by Patterson (1976) to determine the shape of the auditory filter. The threshold of the sinusoidal signal is measured as a function of the width of a spectral notch in the noise masker. The amount of noise passing through the auditory filter centered at the signal frequency is proportional to the ruled areas.

$$P_s = K \int_{f_c-\Delta f-B}^{f_c-\Delta f} W(f)N_0 df + K \int_{f_c+\Delta f}^{f_c+\Delta f+B} W(f)N_0 df \qquad (4)$$

where N_0 is the power spectral density of the noise in its passbands and B is the bandwidth of each noise band. Often, the lower noise band extends down to zero frequency (as illustrated), in which case, $f_c - \Delta f - B = 0$. The two integrals on the right-hand side of Equation (4) represent the respective areas in Fig. 2 where the lower and upper noise bands overlap the filter. Since both the filter and the masker are symmetrical about the signal frequency, these two areas are equal. Thus, the function relating P_s to the width of the notch provides a measure of the integral of the auditory filter. Hence, the value of $W(f)$ at a given deviation Δf from the center frequency is given by the slope of the threshold function at a notch width of $2\Delta f$.

When the auditory filter is asymmetric, as it is at high levels (see later text), then the filter shape can still be measured using a notched-noise masker if some reasonable assumptions are made, and if the range of measurements is extended to include conditions where the notch is placed asymmetrically about the signal frequency. It is necessary first to assume that the auditory filter shape can be approximated by a simple mathematical expression with a small number of free parameters. Patterson *et al.* (1982) suggested a family of such expressions, all having the form of an exponential with a rounded top, called rounded-exponential or "roex" for brevity. The simplest of these expressions was called the roex(p) filter shape. It is convenient to measure frequency in terms of the absolute value of the deviation from the center frequency of the filter, f_c, and to normalize this frequency variable by dividing by the center frequency of the filter. The new frequency variable, g, is:

$$g = |f - f_c|/f_c. \qquad (5)$$

The roex(p) filter shape is then given by:

$$W(g) = (1 + pg)\exp(-pg), \qquad (6)$$

where p is a parameter which determines both the bandwidth and the slope of the skirts of the auditory filter. The higher the value of p, the more sharply tuned is the filter. The equivalent rectangular bandwidth (ERB) is equal to $4f_c/p$. When the filter is assumed to be asymmetric, then p is allowed to have different values on the two sides of the filter: p_l for the lower branch and p_u for the upper branch. The ERB in this case is $2f_c/p_l + 2f_c/p_u$.

Having assumed this general form for the auditory filter shape, the values of p_l and p_u for a particular experiment can be determined by re-writing Equation (4) in terms of the variable g, and substituting the above expression for W; the value of p_l is used for the first integral, and the value of p_u for the second. The equation can then be solved analytically; for full details see Patterson *et al.* (1982) and Glasberg *et al.* (1984b). Starting values of p_l and p_u are assumed, and the equation is used to predict the threshold for each condition (for notches placed both symmetrically and asymmetrically about the signal frequency). The center frequency of the filter is allowed to shift for each condition so as to find the center frequency giving the highest signal-to-masker ratio; this center frequency is assumed in making the prediction for that condition. Standard least-squares minimization procedures are then used to find the values of p_l and p_u which minimize the mean-squared deviation between the obtained and predicted values. The minimization is done with the thresholds expressed in decibels.

The roex(p) filter shape is usually quite successful in predicting the data from notched-noise experiments, except when the thresholds cover a wide range of levels or when the masked thresholds approach absolute threshold. In such cases there is a decrease in the slope of the function relating threshold to notch width, a decrease which is not predicted by the roex(p) filter shape. This can be accommodated in two ways. The first involves limiting the dynamic range of the filter, using a second parameter, r. This gives the roex(p,r) filter shape of Patterson *et al.* (1982):

$$W(g) = (1 - r)(1 + pg)\exp(-pg) + r. \tag{7}$$

As before, p can have different values for the upper and lower branches of the filter. However, the data can generally be well predicted using the same value of r for the two sides of the filter (Glasberg and Moore, 1986; Tyler *et al.*, 1984). The method of deriving filter shapes using this expression is exactly analogous to that previously described.

When the noise level used is relatively high, and when a large range of notch widths is used, there may be systematic deviations of the data from values predicted by the roex(p,r) model. In such cases, a better fit to the data can be obtained using a model in which the slope of the filter is assumed to decrease once its attenuation exceeds a certain value. This is achieved using the roex(p,w,t) model suggested by Patterson *et al.* (1982). The filter is assumed to be the sum of two exponentials, both of which are rounded:

$$W(g) = (1 - w)(1 + pg)\exp(-pg) + w(1 + tg)\exp(-tg) \tag{8}$$

The parameter t determines the slope of the filter at large deviations from the center frequency, and the parameter w determines the point at which the shallower "tail" takes over from the steeper central passband. The shallower tail may correspond to the "passive" filter determined by the hydromechanics of the cochlea, while the sharper tip filter may correspond to the operation of the "active" mechanism in the cochlea (see Chapter 5).

One limitation of the notched-noise method occurs when the auditory filter is markedly asymmetric, as it is, for example, at high sound levels. In such cases, the method does not define the sharper side of the filter very well. As a rule of thumb, when the value of p for one side of the filter is more than twice that for the other, the slope of the steeper side is very poorly determined.

A second potential problem with the method is that components within the upper band of noise may interact to produce combination products whose frequencies lie within the notch in the noise. Such combination products are produced by non-linear processes within the cochlea, and they occur even when the input is at low to moderate sound levels (Goldstein, 1967; Greenwood, 1971; Smoorenburg, 1972a,b). The effect of this is that the upper band of noise may produce more masking than would be the case if no combination products were present. This can result in a derived filter shape with a shallower upper skirt. However, the effect on the derived filter shape is usually small (Moore *et al.*, 1995).

Computer programs for deriving auditory filter shapes from notched-noise data using the above models can be downloaded from http://hearing.psychol.cam.ac.uk/Demos/demos.html.

C. The Rippled-Noise Method

Several researchers have estimated auditory filter shapes using rippled (or comb-filtered) noise as a masker. This is produced by adding white noise to a copy of itself which has been delayed by T seconds. The resulting spectrum has peaks spaced at $1/T$ Hz, with minima in between. When the delayed version of the noise is added to the original in phase, the first peak in the spectrum of the noise occurs at 0 Hz; this noise is referred to as cosine+. When the polarity of the delayed noise is reversed, the first peak is at $0.5/T$ Hz; this is referred to as cosine−. The sinusoidal signal is usually fixed in frequency, and the values of T are chosen so that the signal falls at either a maximum or minimum in the masker spectrum; the signal threshold is measured for both cosine+ and cosine− noise for various ripple densities (different values of T).

The auditory filter shape can be derived from the data either by approximating the auditory filter as a Fourier series (Houtgast, 1977; Pick, 1980; Pick *et al.*, 1977) or by a method similar to that described for the notched-noise method (Glasberg *et al.*, 1984a; Patterson and Moore, 1986). The filter shapes obtained in this way are generally similar to those obtained using the notched-noise method, although they tend to have a slightly broader and flatter top (Glasberg *et al.*, 1984a). The method seems to be quite good for defining the shape of the tip of the auditory filter, but it does not allow the auditory filter shape to be measured over a wide dynamic range.

D. Allowing for the Transfer Function of the Outer and Middle Ear

The transfer function of the outer and middle ear varies markedly with frequency, particularly at very low and high frequencies (Aibara *et al.*, 2001; Puria *et al.*, 1997; Shaw, 1974). Clearly this can have a significant influence on measures of frequency selectivity. For example, if one of the bands of noise in a notched-noise experiment is very low or high in center frequency, it will be strongly attenuated by the middle ear, and so will not do much masking. The auditory filter is usually thought of as resulting from processes occurring after the outer/middle ear. The effect of the outer and middle ear can be treated as a fixed frequency-dependent attenuation applied to all stimuli before auditory filtering occurs (Glasberg and Moore, 1990) (see also, Chapter 5).

The outer/middle ear transfer function should be taken into account in the fitting procedure for deriving filter shapes. Essentially, the spectra of the stimuli at the input to the cochlea have to be calculated by assuming a certain form for the frequency-dependent transfer. The fitting procedure then has to work on the basis of these "corrected" spectra. In practice, this means that the integral in Eq. 4 cannot be solved analytically, but has to be evaluated numerically. Several different forms for the outer/middle ear transfer function have been proposed, depending of the method of delivery of the stimuli (e.g., free field or headphone) and the assumptions made about the middle ear transfer function (Glasberg and Moore, 1990, 2000; Moore *et al.*, 1997).

IV. Summary of the Characteristics of the Auditory Filter Derived Using Simultaneous Masking

A. Variation with Center Frequency

Moore and Glasberg (1983a) presented a summary of experiments measuring auditory filter shapes using symmetric notched-noise maskers. All of the data were obtained at moderate noise levels and were analyzed using the roex(p,r) filter shape. The ERBs of the filters derived from the data available in 1983 are shown as asterisks in Fig. 3. Other symbols show ERBs estimated in more recent experiments, including data from experiments using asymmetric notches. The solid line in Fig. 3 provides a good fit to the ERB values over the whole frequency range tested. It is described by the following equation:

$$\mathrm{ERB_N} = 24.7(0.00437f + 1), \tag{9}$$

where f is center frequency in Hz. The suffix "N" is used to emphasize that this is the mean value of the auditory filter bandwidth determined using normal-hearing listeners at moderate sound levels. Equation 9 is a modification of one originally suggested by Greenwood (1961a) to describe the variation of the CB with center frequency. He based it on the assumption that each CB corresponds to a constant distance along the BM. Although the constants in Equation 9 differ

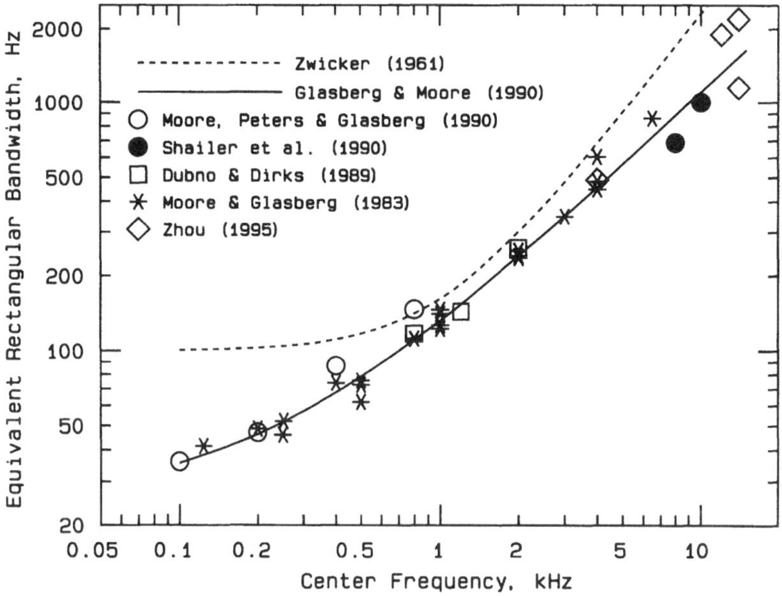

FIG. 3. Estimates of the auditory filter bandwidth from a variety of experiments, plotted as a function of center frequency. The dashed line shows the CB function proposed by Zwicker (1961). The solid line shows the function proposed by Glasberg and Moore (1990), and specified by Equation 9.

from those given by Greenwood, the form of the equation is the same. Each ERB_N corresponds to a distance of about 0.89 mm on the BM.

It should be noted that the function specified by Equation 9 differs somewhat from the "traditional" CB function (Zwicker, 1961) which flattens off below 500 Hz at a value of about 100 Hz. This function is shown as the dashed line in Fig. 3. The traditional function was obtained by combining data from a variety of experiments. However, the data were sparse at low frequencies, and the form of the function was strongly influenced by measures of the critical ratio. As described earlier, the critical ratio does not provide a good estimate of the CB, particularly at low frequencies. It seems clear that the CB does continue to decrease below 500 Hz (Moore and Sek, 1995).

It is sometimes useful to plot experimental data and theoretical functions on a frequency-related scale based on units of the ERB_N of the auditory filter. Essentially, the ERB_N is used as the unit of frequency. For example, the value of the ERB_N for a center frequency of 1 kHz is about 130 Hz, so an increase in frequency from 935 to 1065 Hz represents a step of one ERB_N. A scale based on the ERB_N of the auditory filter, and derived from Equation 9, is:

$$ERB_N \text{ number} = 21.4\log_{10}(0.00437f + 1) \quad (10)$$

Auditory filter bandwidths for young normal-hearing subjects do not vary greatly across subjects; the standard deviation of the ERB_N is typically about 10% of its mean value (Moore, 1987). However, the variability tends to increase at very low frequencies (Moore *et al.*, 1990) and at very high frequencies (Patterson *et al.*, 1982; Shailer *et al.*, 1990; Zhou, 1995).

B. Variation with Level

If the auditory filter were linear, then its shape would not vary with the level of the noise used to measure it. However, this is not the case. One way of conceptualizing the changes with level is based on the idea of a shallow "tail" filter, determined by the "passive" properties of the BM and surrounding structures, and a sharper "tip" filter dependent on the "active" mechanism in the cochlea (see Chapter 5). At low sound levels, the active component has a strong influence, and sharp tuning is observed, at least around the "tip" of the tuning curve. As the level is increased, the gain of the active tip filter decreases and the passive filter plays a greater role. Thus, the tuning becomes broader with increasing level, especially for frequencies below the CF of the place on the BM whose response is being measured.

There has been some controversy about what aspect of the stimulus level determines the filter shape. Often, auditory filter shapes have been measured using a fixed noise level. When several fixed-noise levels are used, the data for each noise level being analyzed separately, the derived filter tends to become broader on the low-frequency side as the level increases (Glasberg and Moore, 1990; Lutfi and Patterson, 1984; Moore and Glasberg, 1987b; Patterson and Moore, 1986; Rosen *et al.*, 1992, 1998); changes on the high-frequency side are smaller, and are not consistent across studies.

Rosen and colleagues (Baker *et al.*, 1998; Rosen *et al.*, 1992, 1998) have proposed a method for deriving auditory filter shapes from notched-noise data in which several different levels are used within the same experiment. The analysis of the results was based on the assumption that each side of the filter could be described as the sum of two rounded exponential functions, but the parameters describing the filter shape were made functions of level. They found that the data were fitted better when the parameters were made to depend on the signal level than when they were made to depend on the masker spectrum level. The broadening of the low-frequency side of the filter with increasing level could be well described by a change in the gain of the tip filter relative to that of the tail filter. Glasberg and Moore (2000) conducted a similar analysis, where the filter parameters were made to depend either on the signal level or on the masker level per ERB_N. They used a special type of notched noise, called "uniformly exciting noise"; within its two passbands the level of the noise reaching the cochlea was intended to be constant within each ERB_N. The gain of the tip filter was assumed to vary with level in a similar way to the BM input-output function for a frequency close to CF (Ruggero *et al.*, 1997). The data were fitted best when the gain of the tip filter was assumed to be a function of the signal level. The filter shapes showed a level dependence that qualitatively resembled the level dependence of filtering on the BM. The maximum gain of the tip filter tended to increase with increasing center frequency up to 1 kHz, but to remain roughly constant for higher frequencies.

In their analyses, Rosen and colleagues (Baker *et al.*, 1998; Rosen *et al.*, 1992, 1998) assumed that the signal level at masked threshold was closely related to the magnitude of the output of the auditory filter; the output was assumed to be equal to the signal level at threshold plus a constant (in dB). Thus, they characterized their results as indicating that the filter shape depends on the level at its output. However, if gain of the tip filter depends on the signal level (Baker *et al.*, 1998; Glasberg and Moore, 2000; Rosen *et al.*, 1998), then the output power of the filter is not directly proportional to the signal power. Thus, the results should not be taken as indicating that the filter shape depends simply on the power at its output. Rather, the filter shape depends on the power at the output divided by the gain (or, in dB terms, the level at the output minus the gain). This dependence may arise because filtering on the BM involves a feedback mechanism, which means that it is unrealistic to characterize the shape as dependent only on the input level or only on the output level (see also Patterson *et al.*, 2003). Another possibility is that the tail and tip filters effectively act in series, with the output of the tail filter controlling the gain of the tip filter (Zhang *et al.*, 2001).

Figure 4 is based on the analyses of Glasberg and Moore (2000). It illustrates how the shape of the auditory filter centered at 2 kHz varies with the frequency of a sinusoidal input, for several different input levels. The upper panel shows the responses of the filters plotted as normalized gain, i.e., the gain is assumed to be 0 dB at the tip. The lower panel shows filter responses in terms of gain, assuming that the gain of the tip filter approaches 0 dB for a signal at its center frequency for high input levels. Note that the calculation of these filter shapes required an iterative procedure, as the filter shape was assumed to be determined by the normalized output level (the output level when the filter is scaled to have 0 dB gain at its center frequency), and this output level depends on the assumed filter shape (for details see Glasberg and Moore, 2000).

V. Masking Patterns and Excitation Patterns

In the experiments described thus far, the frequency of the signal was held constant, while the masker was varied. These experiments are most appropriate for estimating the shape of the auditory filter at a given center frequency.

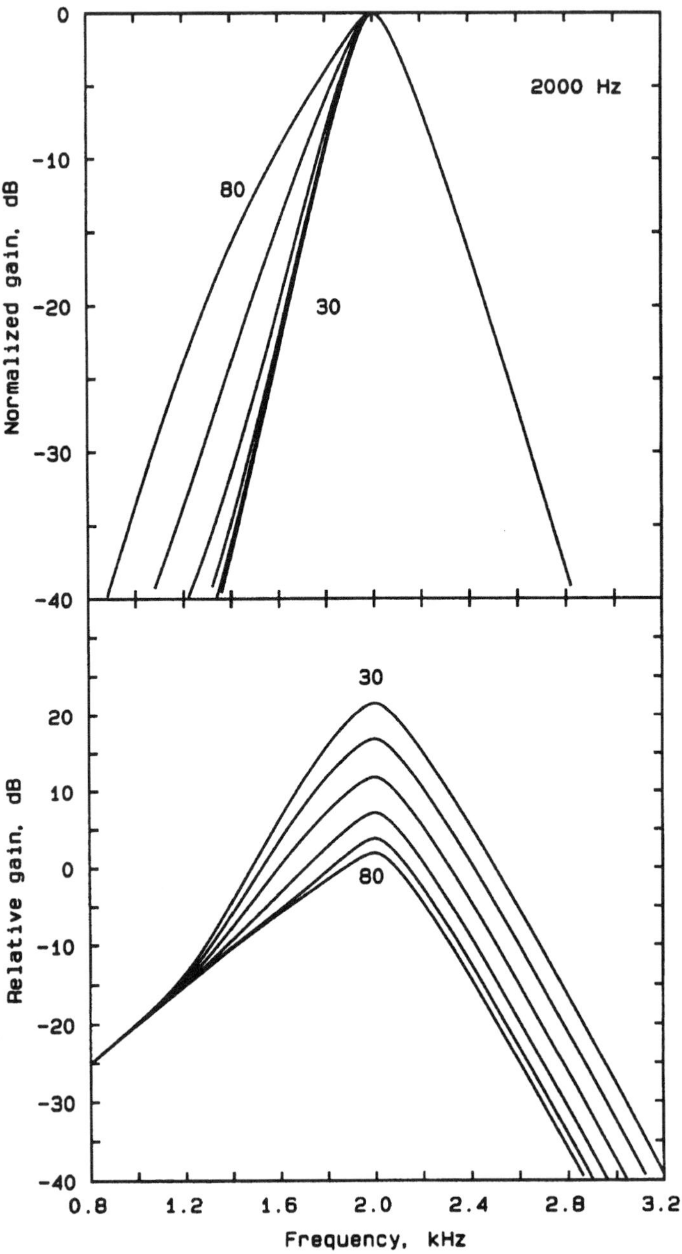

FIG. 4. The shape of the auditory filter centered at 2 kHz, for levels of a sinusoidal input ranging from 30 to 80 dB. The upper panel shows the output of the filter when the output is normalized to have a gain at the tip of 0 dB for every input level. The lower panel shows the filter shapes as gains without this normalization.

However, many of the early experiments on masking did the opposite; the signal frequency was varied while the masker was held constant. Wegel and Lane (1924) published the first systematic investigation of the masking of one pure tone by another. They determined the threshold of a signal with adjustable frequency in the presence of a masker with fixed frequency and intensity. The function relating masked threshold to the signal frequency is known as a masking pattern, or sometimes as a masked audiogram. The results of Wegel and Lane were complicated by the occurrence of beats when the signal and masker were close together in frequency. To avoid this problem later experimenters (Egan and Hake, 1950; Moore *et al.*, 1998; Zwicker and Fastl, 1999) have used a narrow band of noise as either the signal or the masker. The masking patterns obtained in these experiments show steep slopes on the low-frequency side. The slopes on the high-frequency side are less steep and decrease with increasing masker level. Thus, if the level of a low-frequency masker is increased by, say, 10 dB, the masked threshold of a high-frequency signal is elevated by more than 10 dB; the amount of masking grows non-linearly on the high-frequency side. This has been called the "upward spread of masking."

The masking patterns do not reflect the use of a single auditory filter. Rather, for each signal frequency the listener uses a filter centered close to the signal frequency. Thus, the auditory filter is shifted as the signal frequency is altered. One way of interpreting the masking pattern is as a crude indicator of the excitation pattern of the masker. The excitation pattern of a sound is a representation of the activity or excitation evoked by that sound as a function of CF (Zwicker, 1970). In the case of a masking pattern, one might assume that the signal is detected when the excitation it produces is some constant proportion of the excitation produced by the masker in the frequency region of the signal. If this were the case, the threshold of the signal as a function of frequency would be proportional to the masker's excitation level. The masking pattern would thus be parallel to the excitation pattern of the masker, but shifted vertically by a small amount. In practice, the situation is not so straightforward, since the shape of the masking pattern is influenced by factors such as off-frequency listening (Johnson-Davies and Patterson, 1979), the detection of combination tones produced by the interaction of the signal and the masker (Greenwood, 1971; Moore *et al.*, 1998), and cues produced by the temporal interaction of the masker and signal, which can occur even when a noise masker is used (Alcántara *et al.*, 2000; Moore *et al.*, 1998).

A. Relationship of the Auditory Filter to the Excitation Pattern

Moore and Glasberg (1983a) have described a way of deriving the shapes of excitation patterns using the concept of the auditory filter. They suggested that the excitation pattern of a given sound can be thought of as the output of the auditory filters as a function of their center frequency. This idea is illustrated in Fig. 5. The upper portion of the figure shows auditory filter shapes for five center frequencies. Note that the filter shapes are plotted in terms of normalized gain, i.e., with the gain at the tip of each filter scaled to be 0 dB. Each filter was assumed to be symmetrical on the linear frequency scale used, but the bandwidths of the filters rise with increasing center frequency, as illustrated in Fig. 3. The dashed line represents a 1-kHz sinusoidal signal whose excitation pattern is to be derived. The lower panel shows the output from each filter in response to the 1-kHz signal, plotted as a function of the center frequency of each filter; this is the desired excitation pattern.

To see how this pattern is derived, consider the output from the filter with the lowest center frequency. This has a relative output in response to the 1-kHz tone of about −40 dB, as indicated by point "a" in the upper panel. In the lower panel, this gives rise to the point "a" on the excitation pattern; the point has an ordinate value of −40 dB and is positioned on the abscissa at a frequency corresponding to the center frequency of the lowest filter illustrated. The relative outputs of the other filters are indicated, in order of increasing center frequency, by points "b" to "e," and each leads to a corresponding point on the excitation pattern. The complete excitation pattern was actually derived by calculating the filter outputs for filters spaced at 10-Hz intervals. In deriving the excitation pattern, excitation levels were expressed relative to the level at the tip of the pattern, which was arbitrarily labeled as 0 dB.

Note that, although the auditory filters were assumed to be symmetric on a linear frequency scale, the derived excitation pattern is asymmetric. This happens because the bandwidth of the auditory filter increases with increasing center frequency. As pointed out by Patterson (1974), the increase in auditory filter bandwidth with frequency can also explain why masking patterns are asymmetric when the auditory filter itself is roughly symmetric.

B. Changes in Excitation Patterns with Level

One problem in calculating excitation patterns from filter shapes is how to deal with the level dependence of the auditory filter. Some methods for calculating excitation patterns are based on the assumption that it is the input level to the filter (or the level per ERB_N for a broadband signal) that determines its shape (Glasberg and Moore, 1990; Moore and Glasberg, 1987b; Zwicker, 1970). However, more recent data, as previously described, suggest that this is not accurate. The filter shape may be determined by the normalized level at the output of the filter (Glasberg and Moore, 2000; Rosen *et al.*, 1998), or by the output level of the tail filter (Glasberg and Moore, 2000; Zhang *et al.*, 2001).

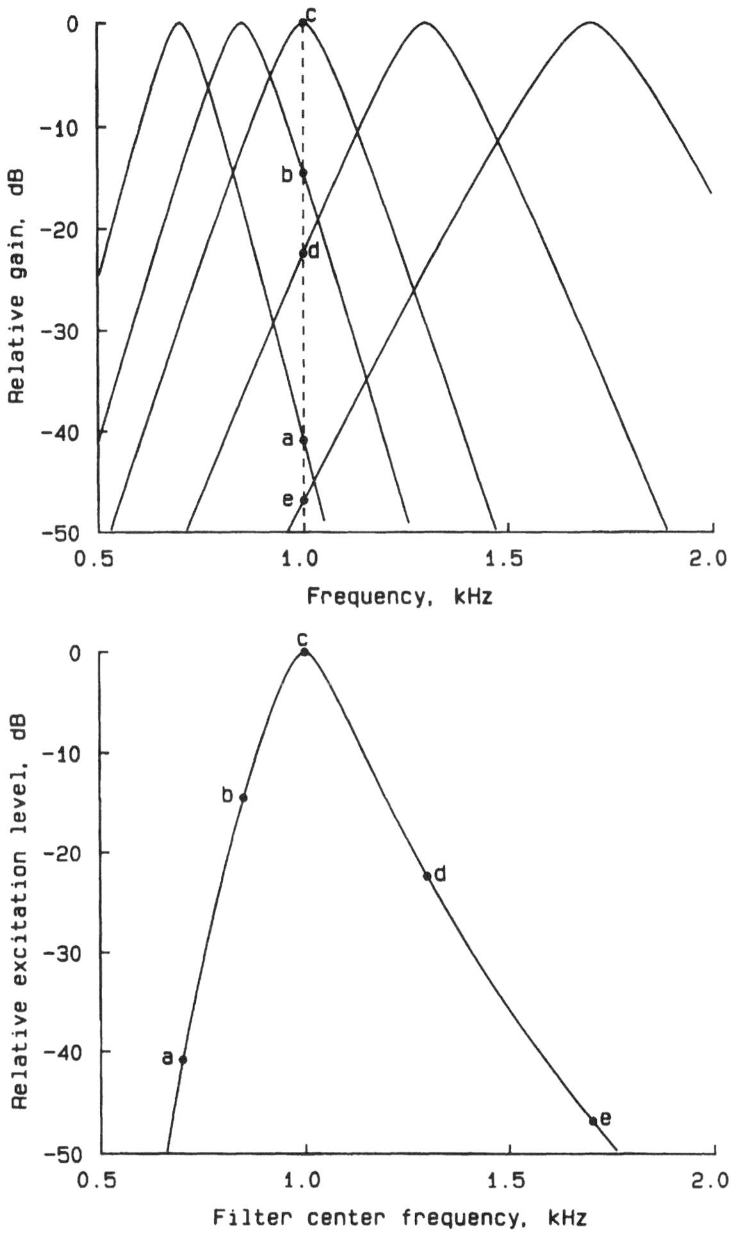

FIG. 5. An illustration of how the excitation pattern of a 1-kHz sinusoid can be derived by calculating the outputs of the auditory filters as a function of their center frequency. The top half shows five auditory filters, centered at different frequencies, and the bottom half shows the calculated excitation pattern. Adapted from Moore and Glasberg (1983a).

Calculation of excitation patterns using the assumption that the filter shape is controlled by its normalized output level requires an iterative procedure, and this can be computationally intensive when the input signal is complex (Baker *et al.*, 1998; Glasberg and Moore, 2000). However, the resulting excitation patterns, at least for some stimuli, are remarkably similar to those obtained using the simpler method of Glasberg and Moore (1990), which is based on the assumption that the filter shape is determined by the input level per ERB_N.

Other approaches to modeling level-dependent changes in auditory filters and excitation patterns have recently been developed, and show good potential in accounting for a wide variety of data (Irino and Patterson, 2001; Lopez-Poveda and Meddis, 2001; Meddis and O'Mard, 2001; Patterson *et al.*, 2003; Zhang *et al.*, 2001).

VI. Non-Simultaneous Masking

A. Basic Properties of Non-Simultaneous Masking

Simultaneous masking describes situations where the masker is present for the whole time that the signal occurs. Masking can also occur when a brief signal is presented just before or after the masker; this is called non-simultaneous masking. Two basic types of non-simultaneous masking can be distinguished: (1) backward masking, in which the signal precedes the masker (also known as pre-stimulatory masking); and (2) forward masking, in which the signal follows the masker (also known as post-stimulatory masking).

Although many studies of backward masking have been published, the phenomenon is poorly understood. The amount of backward masking depends strongly on how much practice the subjects have received, and practiced subjects often show little or no backward masking (Miyazaki and Sasaki, 1984; Oxenham and Moore, 1994). The larger masking effects found for unpracticed subjects may reflect some sort of "confusion" of the signal with the masker. In contrast, forward masking can be substantial even in highly practiced subjects. The main properties of forward masking are as follows:

1. Forward masking is greater the nearer in time to the masker that the signal occurs. When the delay D of the signal after the end of the masker is plotted on a logarithmic scale, the data fall roughly on a straight line. In other words, the amount of forward masking, in dB, is a linear function of $\log(D)$.

2. The rate of recovery from forward masking is greater for higher masker levels. Thus, regardless of the initial amount of forward masking, the masking decays to 0 after 100 to 200 ms.

3. Increments in masker level do not produce equal increments in amount of forward masking. For example, if the masker level is increased by 10 dB, the masked threshold may only increase by 3 dB. This contrasts with simultaneous masking, where, at least for wideband maskers, threshold usually corresponds to a constant signal-to-masker ratio. This effect can be quantified by plotting the signal threshold as a function of masker level. The resulting function is called a growth-of-masking function. In simultaneous masking such functions would have slopes close to 1 (on a dB versus dB scale). In forward masking the slopes are usually <1, and the slopes decrease as the value of D increases.

4. The amount of forward masking increases with increasing masker duration up to at least 20 ms. The results for greater masker durations vary somewhat across studies. Some studies show an effect of masker duration for durations up to 200 ms (Kidd and Feth, 1982), while others show little effect for durations beyond 50 ms (Fastl, 1976).

The mechanisms underlying forward masking are not clear. It could be explained in terms of a reduction in sensitivity of recently stimulated neurons, or in terms of persistence in the pattern of neural activity evoked by the masker. Both points of view can be found in the literature (Oxenham, 2001). In addition, the response of the BM to the masker takes a certain time to decay, and for small intervals between the signal and the masker this may result in forward masking (Duifhuis, 1973). This effect is probably most important for low center frequencies.

The main properties of forward masking, and particularly the non-linear growth of forward masking with masker level, can be explained in terms of the compressive input-output function of the BM; for details see Moore (1998, 2003), Moore and Oxenham (1998), and Plack and Oxenham (1998).

B. Evidence for Suppression Effects from Non-Simultaneous Masking

Measurements of BM motion (Ruggero *et al.*, 1992), or in single neurons (Arthur *et al.*, 1971; Hind *et al.*, 1967; Sachs and Kiang, 1968), show that the response to a tone of a given frequency can sometimes be suppressed by a tone with a different frequency, a phenomenon known as two-tone suppression. For other complex signals, similar phenomena occur and are given the general name of lateral suppression or suppression. This can be characterized in the following way. Strong activity at a given CF can suppress weaker activity at adjacent CFs. The question now arises as to why effects of suppression are not usually seen in experiments on simultaneous masking.

Houtgast (1972) has argued that simultaneous masking is not an appropriate tool for detecting the effects of suppression. His argument is as follows. In simultaneous masking, the masking stimulus and the signal are processed simultaneously in the same channel (the same auditory filter). Thus, any suppression in that channel will affect the neural activity caused by both the signal and the masker. In other words, the signal-to-masker ratio in a given frequency region will be unaffected by suppression, and thus the threshold of the signal will remain unaltered. This argument may not be correct in all cases. Suppression might change the signal-to-masker ratio at the output of a given auditory filter if the signal and masker are widely separated in frequency. However, Houtgast's argument certainly seems reasonable when applied to cases where the signal and masker have coincident or near coincident frequency components. Say, for example, that the masker consists of two frequency components, at 500 and 2000 Hz, while the signal is a sinusoid at 2000 Hz. If the signal and masker are presented simultaneously, then the 500-Hz component in the masker may produce suppression at the place on the BM tuned to 2000 Hz, but that suppression will be applied equally to the 2000-Hz component of the masker and to the 2000-Hz signal, without changing their ratio. If the masking of the signal is determined mainly by the 2000-Hz component of the masker, then the suppression produced by the 500-Hz component will have no measurable effects.

Houtgast suggested that this difficulty could be overcome by presenting the masker and the signal successively, for example, by using forward masking. If suppression does occur, then its effects will be seen in forward masking provided that: (1) in the chain of levels of neural processing, the level at which the suppression occurs is not later than the level at which most of the forward masking effect arises; and (2) the suppression built up by the masker has decayed by the time that the signal is presented (otherwise the problems described for simultaneous masking will be encountered).

Following the pioneering work of Houtgast (1972, 1973, 1974), many workers have reported that there are systematic differences between the results obtained using simultaneous and non-simultaneous masking techniques. An extensive review is provided by Moore and O'Loughlin (1986). One major difference is that non-simultaneous masking reveals effects that can be directly attributed to suppression. A good demonstration of this involves a psychophysical analog of neural two-tone suppression. Houtgast (1973, 1974) measured the threshold for detecting a 1-kHz signal following a 1-kHz masker (forward masking). He then added a second tone to the masker (synchronous with the masker) and measured the threshold again. He found that sometimes the addition of this second tone produced a reduction in the signal threshold, and attributed this to a suppression of the 1-kHz component in the masker by the second component. If the 1-kHz component is suppressed, then there will be less activity in the frequency region around 1 kHz, producing a drop in the threshold for detecting the signal. The second tone was most effective as a "suppressor" when it was somewhat more intense than the 1-kHz component, and above it in frequency. Similar results have been obtained by Shannon (1976).

Under some circumstances, the reduction in threshold (unmasking) produced by adding one or more extra components to a forward masker can be partly explained in terms of additional cues provided by the added components, rather than in terms of suppression. Specifically, in forward masking the added components may reduce "confusion" of the signal with the masker by indicating exactly when the masker ends and the signal begins (Moore, 1980, 1981; Moore and Glasberg, 1982, 1985; Moore and O'Loughlin, 1986; Neff, 1985). This may have led some researchers to overestimate the magnitude of suppression as indicated in non-simultaneous masking experiments. However, not all unmasking can be explained in this way.

C. The Enhancement of Frequency Selectivity Revealed in Non-Simultaneous Masking

A second major difference between simultaneous and non-simultaneous masking is that the frequency selectivity revealed in non-simultaneous masking is greater than that revealed in simultaneous masking. A well-studied example of this is the PTC. PTCs determined in forward masking are typically sharper than those obtained in simultaneous masking (Moore, 1978). Several other methods of estimating frequency selectivity have indicated sharper tuning in non-simultaneous masking than in simultaneous masking. For example, auditory filter shapes estimated in forward masking using a notched-noise masker, have smaller bandwidths and greater slopes than those estimated in simultaneous masking (Glasberg and Moore, 1982; Moore and Glasberg, 1981; Moore *et al.*, 1987; Oxenham and Shera, 2003).

The cause of the difference in frequency selectivity between simultaneous and forward masking is not clear. According to Houtgast (1974) the difference arises because the internal representation of the masker (its excitation pattern) is sharpened by a suppression process, with the greatest sharpening occurring on the low-frequency side. In simultaneous masking, the effects of suppression are not seen, since any reduction of the masker activity in the frequency region of the signal is accompanied by a similar reduction in signal-evoked activity. In other words, the signal-to-masker ratio in the frequency region of the signal is unaffected by the suppression. In forward masking, on the other hand, the suppression does not affect the signal. For maskers with frequencies above that of the signal, the effect of suppression is to sharpen the excitation pattern of the masker, resulting in an increase of the masker level required to mask the signal.

Thus, the suppression is revealed as an increase in the slopes of the PTC or derived filter.

An alternative explanation is that, in simultaneous masking, the low-level signal may be suppressed by the masker (Delgutte, 1988, 1990; Pickles, 1984; Weber, 1983; Wightman *et al.*, 1977). The neural data indicate that tones falling outside of the region bounded by the neural tuning curve can produce suppression. For masker frequencies remote from the signal frequency, simultaneous masking may be caused mainly by suppression, rather than by the spread of masker excitation to the signal frequency (Delgutte, 1988, 1990, 1996). However, Moore and Vickers (1997) presented evidence that the masking of a sinusoid by a masker of much lower frequency was not caused solely by suppression.

Whichever explanation is correct, one thing is clear: auditory filters estimated from simultaneous masking data are likely to be less sharp than the underlying neural filters or filters on the BM. Similarly, excitation patterns estimated from simultaneous masking patterns (Moore and Glasberg, 1983b; Zwicker, 1970) or calculated using auditory filter shapes derived from simultaneous masking (Glasberg and Moore, 1990; Moore and Glasberg, 1983a), may be less sharp than the underlying neural excitation patterns.

Excitation patterns form the basis of several perceptual models, including models of loudness perception (Moore and Glasberg, 1996; Moore *et al.*, 1997; Zwicker and Scharf, 1965) and models for the detection of frequency and amplitude modulation (Zwicker, 1952; Zwicker and Fastl, 1999). If the excitation patterns on which those models are based are too broad, then the models themselves must be inaccurate. In this context, it is of interest that Moore and Sek (1994) showed that an excitation-pattern model could account for thresholds for the detection of amplitude modulation, frequency modulation, and mixtures of the two (for a modulation rate of 10 Hz), but only if the assumed excitation patterns were sharper than specified by Zwicker and coworkers or by Glasberg and Moore (1990). Similarly, Moore and Tan (2004) were able to predict perceptual judgments of the naturalness of speech or music subjected to spectral distortion (tilts and/or ripples in frequency response) using a model that assumed sharper auditory filters than determined using simultaneous masking.

Finally, it should be noted that the effects of suppression seem to be stronger at high frequencies than at low (Hicks and Bacon, 1999), probably because suppression depends on the cochlear active mechanism (Ruggero *et al.*, 1992) and this mechanism plays a greater role at high frequencies than at low (Hicks and Bacon, 1999; Moore *et al.*, 1999; Yates, 1995). Therefore, the discrepancy between frequency resolution measured in simultaneous masking and the underlying cochlear filtering may be greater at high frequencies than at low frequencies (Oxenham and Shera, 2003).

VII. The Audibility of Partials in Complex Tones

According to Ohm's (1843) acoustical law, the ear is able to hear pitches corresponding to the individual sinusoidal components in a complex sound. In other words, we can "hear out" the individual partials. For periodic complex sounds, the most prominent pitch is usually the residue pitch or virtual pitch associated with the sound as a whole (Plomp, 1976), and we are not normally aware of hearing pitches corresponding to individual partials (Moore, 2003). Nevertheless, such pitches can be heard if attention is directed appropriately (Hartmann and Doty, 1996; Helmholtz, 1863).

Plomp (1964) and Plomp and Mimpen (1968) used complex tones with 12 sinusoidal components to investigate the limits of this ability. The listener was presented with two comparison tones, one of which was of the same frequency as a partial in the complex; the other lay halfway between that frequency and the frequency of the adjacent higher or lower partial. The listener was allowed to switch freely between the complex tone and the comparison tones, and was required to decide which of the two comparison tones coincided with the partial in the complex tone. The score (varying between 50% and 100%) was used as an index of how well the partial could be heard out from the complex tone. For harmonic complex tones, only about the first five to seven harmonics could be heard out. However, using a technique in which the harmonic of interest was pulsed on and off, Bernstein and Oxenham (2003) found that harmonics up to about the 10th could be heard out to some extent.

Plomp and Mimpen (1968) suggested that a component can be heard out only when its frequency separation from adjacent components exceeds the CB. The spacing of the components in a harmonic complex is uniform on a linear frequency scale, but, relative to the CB, the upper harmonics are more closely spaced than the lower harmonics. Harmonics above about the 8th are separated by less than a CB and cannot be heard out. Results for two subjects using complex tones where the frequency ratios between components were "compressed" relative to a harmonic complex gave basically the same result: only the lower components could be heard out.

The concept that the CB was the main factor limiting the audibility of partials in complex tones was questioned by Soderquist (1970). He used a task similar to that of Plomp (1964), but compared the results for musicians and non-musicians. He found that musicians performed markedly better than non-musicians. A possible explanation for this is that musicians have narrower CBs than non-musicians. However, Fine and Moore (1993) estimated auditory filter bandwidths in musicians and non-musicians, using the notched-noise method, and found that ERBs did not differ for the two groups. An alternative possibility is that performance depends on some factor or factors other than the CB.

Moore and Ohgushi (1993) examined the ability of musically trained subjects to hear individual partials in complex tones with partials uniformly spaced on an ERB_N scale. ERB_N spacings of 0.75, 1.0, 1.25, 1.5, and 2 were used, and the central component always had a frequency of 1000 Hz. On each trial, subjects heard a pure tone (the "probe") followed by a complex tone. The probe was close in frequency to one of the partials in the complex, but was mistuned downwards by 4.5% on half the trials (at random) and mistuned upwards by 4.5% on the other half. The task of the subject was to indicate whether the probe was higher or lower in frequency than the nearest partial in the complex. The partial that was "probed" varied randomly from trial to trial. If auditory filtering were the only factor affecting performance on this task, then scores for a given ERB_N spacing should be similar for each component in the complex sound.

Scores for the highest and lowest components in the complexes were generally high for all component spacings, although they worsened somewhat for ERB_N spacings of 0.75 and 1.0. Scores for the inner components were close to chance level at 0.75-ERB_N spacing, and improved progressively as the ERB_N spacing was increased from 1 to 2 ERB_Ns. For ERB_N spacings of 1.25 or less, the scores did not change smoothly with component frequency; marked irregularities were observed, as well as systematic errors. Moore and Ohgushi suggested that these resulted from irregularities in the transmission of sound through the middle ear; such irregularities could change the relative levels of the components, making some components more prominent than others, and therefore easier to hear out.

Performance for the inner components tended to be worse for component frequencies above 1000 Hz than below 1000 Hz. This is consistent with the pattern of results found by Plomp and Mimpen (1968) and Hartmann *et al.* (1990), and indicates that some factor other than auditory filtering influences the audibility of partials in complex tones. Moore and Ohgushi (1993) suggested that the pitches of individual components may be partly coded in the time patterns of neural activity (phase locking) in the auditory nerve (AN), as has also been suggested by previous researchers (Hartmann *et al.*, 1990; Moore *et al.*, 1984b; Moore and Glasberg, 1990; Ohgushi, 1978; Srulovicz and Goldstein, 1983). Phase locking is more precise below 1000 Hz than above 1000 Hz. This can explain why, for partials uniformly spaced on an ERB_N scale, the identification of partials was better for components with frequencies below 1000 Hz than above 1000 Hz.

The influence of phase locking can also explain the superior identification of the lowest and highest components in the complex tones. Generally, neurons with CFs close to the frequency of a given partial will phase lock to that partial, provided the frequency separation of partials is sufficient, and provided the frequency of the partial is not too high. For an inner partial in a complex tone, the pattern of phase locking in neurons with CFs close to the frequency of the partial will be disturbed by the partials on either side, making it more difficult to extract the pitch of that partial, especially when the components are closely spaced. In contrast, neurons tuned just below the lower edge frequency, or just above the higher edge frequency, will show a pattern of phase locking that is less disturbed by the other components. A similar explanation can be offered for Plomp's (1964) finding that the partials in a two-tone complex could be heard out for smaller frequency separations than were found for multi-tone complexes.

In summary, it seems clear that auditory filtering plays a strong role in limiting the ability to hear partials in complex tones. However, it is probably not the only factor involved. Specifically, the pitches of individual partials may partly be coded in the patterns of phase locking of neurons in the AN. This coding is more accurate at low frequencies than at high.

VIII. Frequency Resolution in Impaired Ears

This section considers the consequences of cochlear hearing loss for frequency resolution. Such loss is often associated with damage to the OHCs and inner hair cells (IHCs) in the cochlea, the former being more common for cases of mild to moderate loss (Borg *et al.*, 1995; Wright *et al.*, 1987). Damage to the OHCs results in reduced functioning of the active mechanism in the cochlea (Moore, 1998; Ruggero, 1992). Damage to the IHCs results in reduced transduction efficiency or no transduction at all of BM vibration. When the IHCs and/or neurons are completely non-functioning over a certain region of the BM, this is referred to as a dead region (Moore, 2001, 2004; Moore and Glasberg, 1997; Moore *et al.*, 2000).

A. Complicating Factors

Comparisons of frequency selectivity in normally hearing and hearing-impaired people are complicated by several factors. One factor is the sound level of the stimuli used. As previously mentioned, the auditory filters of normally hearing subjects broaden on the low-frequency side with increasing level, an effect which probably depends on the active mechanism in the cochlea. The active mechanism is usually damaged or completely non-functioning in ears with cochlear damage. Hence, changes in frequency selectivity with level are absent or much less pronounced (Moore *et al.*, 1985; Nelson and Schroder, 1997; Stelmachowicz *et al.*, 1987). As a result, the differences between normally hearing and hearing-impaired subjects tend to decrease at high sound levels.

A second complicating factor is off-frequency listening. Some measures of frequency selectivity, especially PTCs, can be strongly influenced by off-frequency listening. Although these problems can be reduced by gating a notched noise with the variable masker, as described earlier, this has not usually been done in studies with hearing-impaired subjects.

B. Psychophysical Tuning Curves

There have been many studies comparing PTCs in normally hearing subjects and subjects with cochlear hearing loss (Bonding, 1979; Carney and Nelson, 1983; Festen and Plomp, 1983; Florentine *et al.*, 1980; Hoekstra and Ritsma, 1977; Kluk and Moore, 2005; Leshowitz *et al.*, 1975; Nelson, 1991; Stelmachowicz *et al.*, 1985; Tyler *et al.*, 1982; Zwicker and Schorn, 1978). Although the studies differ in detail, their results are in general agreement that PTCs are broader than normal in hearing-impaired subjects. However, it is difficult to quantify the differences from normal, owing to the problems discussed above. Most studies have found that the sharpness of tuning of PTCs decreases with increasing absolute threshold, although the correlation between threshold and sharpness of tuning varies markedly across studies. No systematic differences in PTCs have been reported between cochlear losses of different origin, such as noise-induced, Ménière's, aging, and hereditary losses. In some cases, PTCs have been found that are W-shaped rather than V-shaped (Hoekstra and Ritsma, 1977; Kluk and Moore, 2005).

Sometimes, PTCs have been found whose tips are shifted well away from the signal frequency (Florentine and Houtsma, 1983; Kluk and Moore, 2005; Moore, 2001; Moore and Alcántara, 2001; Moore *et al.*, 2000; Thornton and Abbas, 1980; Turner *et al.*, 1983). This can happen when the signal frequency falls in a dead region. The most-studied situation of this type is when there is a low-frequency dead region, i.e., when there is severe damage to the IHCs at the apical end of the cochlea. The detection of low-frequency tones is then mediated by neurons with high CFs. If the signal to be detected has a frequency corresponding to the dead region, the tip of the tuning curve lies well above the signal frequency. In other words, a masker centered well above the signal in frequency is more effective than a masker centered close to the signal frequency. This happens because the higher-frequency masker lies closer to the CFs of the neurons mediating detection of the signal. PTCs can be used as a tool for diagnosing dead regions (Kluk and Moore, 2005; Moore and Alcántara, 2001; Sek *et al.*, 2005), although the test described by Moore *et al.* (2000, 2004) is easier to use in clinical practice.

Kluk and Moore (2005) showed that W-shaped PTCs often occurred when the signal frequency fell inside or close to a high-frequency dead region, and that the W-shape resulted from the detection of beats and/or simple difference tones produced by the interaction of the signal and the masker. When extra components were added to the masker to prevent beat and difference tone detection, the PTCs showed a single broad tip, and the frequency at the tip was often below the signal frequency.

C. Auditory Filter Shapes Measured with Notched Noise

Auditory filter shapes of subjects with cochlear impairments have been estimated in several studies using notched-noise maskers (Dubno and Dirks, 1989; Glasberg and Moore, 1986; Laroche *et al.*, 1992; Leek and Summers, 1993; Leeuw and Dreschler, 1994; Peters and Moore, 1992; Sommers and Humes, 1993; Stone *et al.*, 1992; Tyler *et al.*, 1984). The results generally agree in showing that auditory filters are broader than normal in hearing-impaired subjects and that, on average, the degree of broadening increases with increasing hearing loss.

Figure 6 summarizes values of the ERB of the auditory filter obtained from subjects with cochlear hearing loss in my laboratory and in the laboratory of Robert Peters. The summary was prepared by Michael Stone. The upper panel shows ERB values plotted relative to the value for young normally hearing subjects at moderate sound levels, as given by Equation 9. The lower panel shows ERB values plotted relative to the value for young normal subjects tested at the same sound pressure level as the impaired subjects, assuming that the ERB for normal subjects varies with level as described by Glasberg and Moore (1990). In both cases, the ERB values are plotted as a function of the absolute threshold (dB HL) at the test frequency. There is a clear trend for the ERB to increase with increasing absolute threshold. The increase is less in the lower panel, since the auditory filters for normal subjects broaden with increasing level of the test stimuli. However, the trend is still quite clear. There is also considerable scatter in the data, indicating that the ERB of the auditory filter cannot be predicted reliably from the absolute threshold.

In summary, all methods of estimating frequency selectivity suggest that it decreases with increasing hearing loss. However, for a given elevation in absolute threshold, there can be considerable scatter both in the degree of broadening and in the asymmetry of the auditory filter.

The scatter in the data shown in Figure 6 may reflect different patterns of hair cell loss within the cochlea (Liberman and Dodds, 1984; Liberman *et al.*, 1986; Moore *et al.*, 1999).

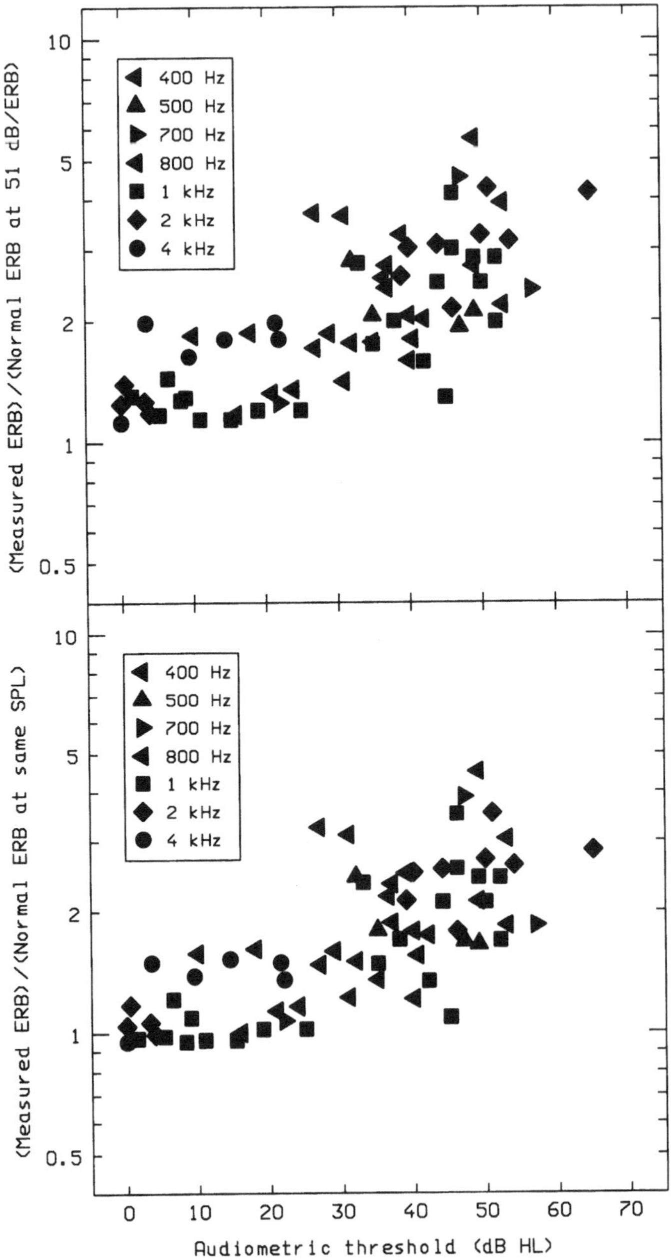

FIG. 6. Values of the ERB of the auditory filter for subjects with cochlear hearing loss, plotted as a function of absolute threshold (dB HL) at the test frequency. The ERB values are plotted either relative to the values given by Equation 9, or relative to the ERBs for young normally hearing subjects tested at the same sound level.

D. Perceptual Consequences of Reduced Frequency Resolution

Reduced frequency resolution affects many aspects of auditory perception; for a review, see Moore (1998). Perhaps one of the most important consequences is a reduced ability to understand speech in background noise (Baer and Moore, 1993, 1994; ter Keurs *et al.*, 1992, 1993).

Acknowledgments

The author thanks Brian Glasberg for his contributions to this work, and for assistance with figures. He also thanks Brian Glasberg and Thomas Stainsby for helpful comments on an earlier version of this chapter.

References

Aibara, R., Welsh, J. T., Puria, S., and Goode, R. L. (2001). Human middle-ear sound transfer function and cochlear input impedance. *Hear. Res.* **152,** 100–109.

Alcántara, J. I., Moore, B. C. J., and Vickers, D. A. (2000). The relative role of beats and combination tones in determining the shapes of masking patterns at 2 kHz: I. Normal-hearing listeners. *Hear. Res.* **148,** 63–73.

Alcántara, J. I., Moore, B. C. J., Glasberg, B. R., Wilkinson, A. J. K., and Jorasz, U. (2003). Phase effects in masking: Within- versus across-channel processes. *J. Acoust. Soc. Am.* **114,** 2158–2166.

Arthur, R. M., Pfeiffer, R. R., and Suga, N. (1971). Properties of 'two-tone inhibition' in primary auditory neurones. *J. Physiol.* **212,** 593–609.

Baer, T., and Moore, B. C. J. (1993). Effects of spectral smearing on the intelligibility of sentences in the presence of noise. *J. Acoust. Soc. Am.* **94,** 1229–1241.

Baer, T., and Moore, B. C. J. (1994). Effects of spectral smearing on the intelligibility of sentences in the presence of interfering speech. *J. Acoust. Soc. Am.* **95,** 2277–2280.

Baker, R. J., Rosen, S., and Darling, A. M. (1998). An efficient characterisation of human auditory filtering across level and frequency that is also physiologically reasonable. *In* "Psychophysical and Physiological Advances in Hearing" (A. R. Palmer, A. Rees, A. Q. Summerfield, and R. Meddis, Eds.), pp. 81–87. Whurr, London.

Bernstein, J. G., and Oxenham, A. J. (2003). Pitch discrimination of diotic and dichotic tone complexes: Harmonic resolvability or harmonic number? *J. Acoust. Soc. Am.* **113,** 3323–3334.

Bernstein, R. S., and Raab, D. H. (1990). The effects of bandwidth on the detectability of narrow- and wideband signals. *J. Acoust. Soc. Am.* **88,** 2115–2125.

Bonding, P. (1979). Frequency selectivity and speech discrimination in sensorineural hearing loss. *Scand. Audiol.* **8,** 205–216.

Borg, E., Canlon, B., and Engström, B. (1995). Noise-induced hearing loss—literature review and experiments in rabbits. Morphological and electrophysiological features, exposure parameters and temporal factors, variability and interactions. *Scand. Audiol.* **24**(Suppl. 40), 1–147.

Buus, S. (1985). Release from masking caused by envelope fluctuations. *J. Acoust. Soc. Am.* **78,** 1958–1965.

Buus, S., Schorer, E., Florentine, M., and Zwicker, E. (1986). Decision rules in detection of simple and complex tones. *J. Acoust. Soc. Am.* **80,** 1646–1657.

Carney, A. E., and Nelson, D. A. (1983). An analysis of psychophysical tuning curves in normal and pathological ears. *J. Acoust. Soc. Am.* **73,** 268–278.

Chistovich, L. A. (1957). Frequency characteristics of masking effect. *Biofizika* **2,** 743–755.

Delgutte, B. (1988). Physiological mechanisms of masking. *In* "Basic Issues in Hearing" (H. Duifhuis, J. W. Horst, and H. P. Wit, Eds.), pp. 204–214. Academic Press, London.

Delgutte, B. (1990). Physiological mechanisms of psychophysical masking: Observations from auditory-nerve fibers. *J. Acoust. Soc. Am.* **87,** 791–809.

Delgutte, B. (1996). Physiological models for basic auditory percepts. *In* "Auditory Computation" (H. L. Hawkins, T. A. McMullen, A. N. Popper, and R. R. Fay, Eds.), pp. 157–220. Springer, New York.

Dubno, J. R., and Dirks, D. D. (1989). Auditory filter characteristics and consonant recognition for hearing-impaired listeners. *J. Acoust. Soc. Am.* **85,** 1666–1675.

Duifhuis, H. (1973). Consequences of peripheral frequency selectivity for nonsimultaneous masking. *J. Acoust. Soc. Am.* **54,** 1471–1488.

Egan, J. P., and Hake, H. W. (1950). On the masking pattern of a simple auditory stimulus. *J. Acoust. Soc. Am.* **22,** 622–630.

Evans, E. F., Pratt, S. R., and Cooper, N. P. (1989). Correspondence between behavioural and physiological frequency selectivity in the guinea pig. *Br. J. Audiol.* **23,** 151–152.

Fastl, H. (1976). Temporal masking effects: I. Broad band noise masker. *Acustica* **35,** 287–302.

Festen, J. M., and Plomp, R. (1983). Relations between auditory functions in impaired hearing. *J. Acoust. Soc. Am.* **73,** 652–662.

Fine, P. A., and Moore, B. C. J. (1993). Frequency analysis and musical ability. *Music Percept.* **11,** 39–53.

Fletcher, H. (1940). Auditory patterns. *Rev. Mod. Phys.* **12,** 47–65.

Florentine, M., Buus, S., Scharf, B., and Zwicker, E. (1980). Frequency selectivity in normally-hearing and hearing-impaired observers. *J. Speech Hear. Res.* **23,** 643–669.

Florentine, M., and Houtsma, A. J. M. (1983). Tuning curves and pitch matches in a listener with a unilateral, low-frequency hearing loss. *J. Acoust. Soc. Am.* **73,** 961–965.

Glasberg, B. R., and Moore, B. C. J. (1982). Auditory filter shapes in forward masking as a function of level. *J. Acoust. Soc. Am.* **71,** 946–949.

Glasberg, B. R., Moore, B. C. J., and Nimmo-Smith, I. (1984a). Comparison of auditory filter shapes derived with three different maskers. *J. Acoust. Soc. Am.* **75,** 536–544.

Glasberg, B. R., Moore, B. C. J., Patterson, R. D., and Nimmo-Smith, I. (1984b). Dynamic range and asymmetry of the auditory filter. *J. Acoust. Soc. Am.* **76,** 419–427.

Glasberg, B. R., and Moore, B. C. J. (1986). Auditory filter shapes in subjects with unilateral and bilateral cochlear impairments. *J. Acoust. Soc. Am.* **79,** 1020–1033.

Glasberg, B. R., and Moore, B. C. J. (1990). Derivation of auditory filter shapes from notched-noise data. *Hear. Res.* **47,** 103–138.

Glasberg, B. R., and Moore, B. C. J. (2000). Frequency selectivity as a function of level and frequency measured with uniformly exciting notched noise. *J. Acoust. Soc. Am.* **108,** 2318–2328.

Gockel, H., Moore, B. C. J., and Patterson, R. D. (2002). Asymmetry of masking between complex tones and noise: The role of temporal structure and peripheral compression. *J. Acoust. Soc. Am.* **111,** 2759–2770.

Goldstein, J. L. (1967). Auditory non-linearity. *J. Acoust. Soc. Am.* **41,** 676–689.

Greenwood, D. D. (1961a). Critical bandwidth and the frequency coordinates of the basilar membrane. *J. Acoust. Soc. Am.* **33,** 1344–1356.

Greenwood, D. D. (1961b). Auditory masking and the critical band. *J. Acoust. Soc. Am.* **33,** 484–501.

Greenwood, D. D. (1971). Aural combination tones and auditory masking. *J. Acoust. Soc. Am.* **50,** 502–543.

Hall, J. W., Haggard, M. P., and Fernandes, M. A. (1984). Detection in noise by spectro-temporal pattern analysis. *J. Acoust. Soc. Am.* **76,** 50–56.

Hall, J. W., Grose, J. H., and Mendoza, L. (1995). Across-channel processes in masking. *In* "Hearing" (B. C. J. Moore, Ed.), pp. 243–266. Academic Press, San Diego.

Hamilton, P. M. (1957). Noise masked thresholds as a function of tonal duration and masking noise bandwidth. *J. Acoust. Soc. Am.* **29,** 506–511.

Hartmann, W. M., McAdams, S., and Smith, B. K. (1990). Hearing a mistuned harmonic in an otherwise periodic complex tone. *J. Acoust. Soc. Am.* **88,** 1712–1724.

Hartmann, W. M., and Doty, S. L. (1996). On the pitches of the components of a complex tone. *J. Acoust. Soc. Am.* **99,** 567–578.

Helmholtz, H. L. F. (1863). "Die Lehre von den Tonempfindungen als physiologische Grundlage für die Theorie der Musik." F. Vieweg, Braunschweig.

Hicks, M. L., and Bacon, S. P. (1999). Psychophysical measures of auditory nonlinearities as a function of frequency in individuals with normal hearing. *J. Acoust. Soc. Am.* **105,** 326–338.

Hind, J. E., Rose, J. E., Brugge, J. F., and Anderson, D. J. (1967). Coding of information pertaining to paired low-frequency tones in single auditory nerve fibers of the squirrel monkey. *J. Neurophysiol.* **30,** 794–816.

Hoekstra, A., and Ritsma, R. J. (1977). Perceptive hearing loss and frequency selectivity. *In* "Psychophysics and Physiology of Hearing" (E. F. Evans and J. P. Wilson, Eds.), pp. 263–271. Academic, London, England.

Houtgast, T. (1972). Psychophysical evidence for lateral inhibition in hearing. *J. Acoust. Soc. Am.* **51,** 1885–1894.

Houtgast, T. (1973). Psychophysical experiments on 'tuning curves' and 'two-tone inhibition'. *Acustica* **29,** 168–179.

Houtgast, T. (1974). Lateral suppression in hearing. Ph.D. thesis, Free University of Amsterdam.

Houtgast, T. (1977). Auditory-filter characteristics derived from direct-masking data and pulsation-threshold data with a rippled-noise masker. *J. Acoust. Soc. Am.* **62,** 409–415.

Irino, T., and Patterson, R. D. (2001). A compressive gammachirp auditory filter for both physiological and psychophysical data. *J. Acoust. Soc. Am.* **109,** 2008–2022.

Johnson-Davies, D., and Patterson, R. D. (1979). Psychophysical tuning curves: Restricting the listening band to the signal region. *J. Acoust. Soc. Am.* **65,** 765–770.

Kidd, G., and Feth, L. L. (1982). Effects of masker duration in pure-tone forward masking. *J. Acoust. Soc. Am.* **72,** 1384–1386.

Kluk, K., and Moore, B. C. J. (2004). Factors affecting psychophysical tuning curves for normally hearing subjects. *Hear. Res.* **194,** 118–134.

Kluk, K., and Moore, B. C. J. (2005). Factors affecting psychophysical tuning curves for hearing-impaired subjects. *Hear. Res.* **200,** 115–131.

Kohlrausch, A., and Sander, A. (1995). Phase effects in masking related to dispersion in the inner ear. II. Masking period patterns of short targets. *J. Acoust. Soc. Am.* **97,** 1817–1829.

Kohlrausch, A., Fassel, R., van der Heijden, M., Kortekaas, R., Van de Par, S., Oxenham, A. J., and Püschel, D. (1997). Detection of tones in low-noise noise: Further evidence for the role of envelope fluctuations. *Acustica - Acta Acustica* **83,** 659–669.

Laroche, C., Hétu, R., Quoc, H. T., Josserand, B., and Glasberg, B. (1992). Frequency selectivity in workers with noise-induced hearing loss. *Hear. Res.* **64,** 61–72.

Leek, M. R., and Summers, V. (1993). Auditory filter shapes of normal-hearing and hearing-impaired listeners in continuous broadband noise. *J. Acoust. Soc. Am.* **94,** 3127–3137.

Leeuw, A. R., and Dreschler, W. A. (1994). Frequency-resolution measurements with notched noise for clinical purposes. *Ear Hear.* **15,** 240–255.

Leshowitz, B., Linstrom, R., and Zurek, P. (1975). Psychophysical tuning curves in normal and impaired ears. *J. Acoust. Soc. Am.* **58,** s71.

Liberman, M. C., and Dodds, L. W. (1984). Single neuron labeling and chronic cochlea pathology. III. Stereocilia damage and alterations in threshold tuning curves. *Hear. Res.* **16,** 54–74.

Liberman, M. C., Dodds, L. W., and Learson, D. A. (1986). Structure-function correlation in noise-damaged ears: A light and electron-microscopic study. *In* "Basic and Applied Aspects of Noise Induced Hearing Loss" (R. J. Salvi, D. Henderson, R. P. Hamernik, and V. Colletti, Eds.), pp. 163–176. Plenum, New York.

Lopez-Poveda, E. A., and Meddis, R. (2001). A human non-linear cochlear filterbank. *J. Acoust. Soc. Am.* **110,** 3107–3118.

Lutfi, R. A., and Patterson, R. D. (1984). On the growth of masking asymmetry with stimulus intensity. *J. Acoust. Soc. Am.* **76,** 739–745.

Meddis, R., and O'Mard, L. (2001). A computational algorithm for computing non-linear auditory frequency selectivity. *J. Acoust. Soc. Am.* **109,** 2852–2861.

Miyazaki, K., and Sasaki, T. (1984). Pure-tone masking patterns in non-simultaneous masking conditions. *Jap. Psychol. Res.* **26,** 110–119.

Moore, B. C. J. (1978). Psychophysical tuning curves measured in simultaneous and forward masking. *J. Acoust. Soc. Am.* **63,** 524–532.

Moore, B. C. J. (1980). Detection cues in forward masking. *In* "Psychophysical, Physiological and Behavioural Studies in Hearing" (G. van den Brink and F. A. Bilson, Eds.), pp. 222–229. Delft University Press, Delft.

Moore, B. C. J. (1981). Interactions of masker bandwidth with signal duration and delay in forward masking. *J. Acoust. Soc. Am.* **70,** 62–68.

Moore, B. C. J., and Glasberg, B. R. (1981). Auditory filter shapes derived in simultaneous and forward masking. *J. Acoust. Soc. Am.* **70,** 1003–1014.

Moore, B. C. J., and Glasberg, B. R. (1982). Contralateral and ipsilateral cueing in forward masking. *J. Acoust. Soc. Am.* **71,** 942–945.

Moore, B. C. J., and Glasberg, B. R. (1983a). Suggested formulae for calculating auditory-filter bandwidths and excitation patterns. *J. Acoust. Soc. Am.* **74,** 750–753.

Moore, B. C. J., and Glasberg, B. R. (1983b). Masking patterns of synthetic vowels in simultaneous and forward masking. *J. Acoust. Soc. Am.* **73,** 906–917.

Moore, B. C. J., Glasberg, B. R., and Roberts, B. (1984a). Refining the measurement of psychophysical tuning curves. *J. Acoust. Soc. Am.* **76,** 1057–1066.

Moore, B. C. J., Glasberg, B. R., and Shailer, M. J. (1984b). Frequency and intensity difference limens for harmonics within complex tones. *J. Acoust. Soc. Am.* **75,** 550–561.

Moore, B. C. J., and Glasberg, B. R. (1985). The danger of using narrowband noise maskers to measure suppression. *J. Acoust. Soc. Am.* **77,** 2137–2141.

Moore, B. C. J., Laurence, R. F., and Wright, D. (1985). Improvements in speech intelligibility in quiet and in noise produced by two-channel compression hearing aids. *Br. J. Audiol.* **19,** 175–187.

Moore, B. C. J. (1986). Parallels between frequency selectivity measured psychophysically and in cochlear mechanics. *Scand. Audiol.* Suppl. **25,**139–152.

Moore, B. C. J., and O'Loughlin, B. J. (1986). The use of nonsimultaneous masking to measure frequency selectivity and suppression. *In* "Frequency Selectivity in Hearing" (B. C. J. Moore, Ed.), pp. 179–250. Academic, London.

Moore, B. C. J. (1987). Distribution of auditory-filter bandwidths at 2 kHz in young normal listeners. *J. Acoust. Soc. Am.* **81,** 1633–1635.

Moore, B. C. J., and Glasberg, B. R. (1987a). Factors affecting thresholds for sinusoidal signals in narrow-band maskers with fluctuating envelopes. *J. Acoust. Soc. Am.* **82,** 69–79.

Moore, B. C. J., and Glasberg, B. R. (1987b). Formulae describing frequency selectivity as a function of frequency and level and their use in calculating excitation patterns. *Hear. Res.* **28,** 209–225.

Moore, B. C. J., Poon, P. W. F., Bacon, S. P., and Glasberg, B. R. (1987). The temporal course of masking and the auditory filter shape. *J. Acoust. Soc. Am.* **81,** 1873–1880.

Moore, B. C. J. (1988). Dynamic aspects of auditory masking. *In* "Auditory Function: Neurobiological Bases of Hearing" (G. Edelman, W. Gall, and W. Cowan, Eds.), pp. 585–607. Wiley, New York.

Moore, B. C. J., and Glasberg, B. R. (1990). Frequency discrimination of complex tones with overlapping and non-overlapping harmonics. *J. Acoust. Soc. Am.* **87,** 2163–2177.

Moore, B. C. J., Peters, R. W., and Glasberg, B. R. (1990). Auditory filter shapes at low center frequencies. *J. Acoust. Soc. Am.* **88,** 132–140.

Moore, B. C. J., and Ohgushi, K. (1993). Audibility of partials in inharmonic complex tones. *J. Acoust. Soc. Am.* **93,** 452–461.

Moore, B. C. J., and Sek, A. (1994). Effects of carrier frequency and background noise on the detection of mixed modulation. *J. Acoust. Soc. Am.* **96,** 741–751.

Moore, B. C. J., Glasberg, B. R., van der Heijden, M., Houtsma, A. J. M., and Kohlrausch, A. (1995). Comparison of auditory filter shapes obtained with notched-noise and noise-tone maskers. *J. Acoust. Soc. Am.* **97,** 1175–1182.

Moore, B. C. J., and Sek, A. (1995). Auditory filtering and the critical bandwidth at low frequencies. *In* "Advances in Hearing Research" (G. A. Manley, G. M. Klump, C. Köppl, H. Fastl, and H. Oeckinghaus, Eds.), pp. 425–436. World Scientific, Singapore.

Moore, B. C. J., and Glasberg, B. R. (1996). A revision of Zwicker's loudness model. *Acustica - Acta Acustica* **82,** 335–345.

Moore, B. C. J., and Glasberg, B. R. (1997). A model of loudness perception applied to cochlear hearing loss. *Auditory Neurosci.* **3,** 289–311.

Moore, B. C. J., Glasberg, B. R., and Baer, T. (1997). A model for the prediction of thresholds, loudness and partial loudness. *J. Audio Eng. Soc.* **45,** 224–240.

Moore, B. C. J., and Vickers, D. A. (1997). The role of spread of excitation and suppression in simultaneous masking. *J. Acoust. Soc. Am.* **102,** 2284–2290.

Moore, B. C. J. (1998). "Cochlear Hearing Loss." Whurr, London.

Moore, B. C. J., Alcántara, J. I., and Dau, T. (1998). Masking patterns for sinusoidal and narrowband noise maskers. *J. Acoust. Soc. Am.* **104,** 1023–1038.

Moore, B. C. J., and Oxenham, A. J. (1998). Psychoacoustic consequences of compression in the peripheral auditory system. *Psych. Rev.* **105,** 108–124.

Moore, B. C. J., Vickers, D. A., Plack, C. J., and Oxenham, A. J. (1999). Inter-relationship between different psychoacoustic measures assumed to be related to the cochlear active mechanism. *J. Acoust. Soc. Am.* **106,** 2761–2778.

Moore, B. C. J., Huss, M., Vickers, D. A., Glasberg, B. R., and Alcántara, J. I. (2000). A test for the diagnosis of dead regions in the cochlea. *Br. J. Audiol.* **34,** 205–224.

Moore, B. C. J. (2001). Dead regions in the cochlea: Diagnosis, perceptual consequences, and implications for the fitting of hearing aids. *Trends Amplif.* **5,** 1–34.

Moore, B. C. J., and Alcántara, J. I. (2001). The use of psychophysical tuning curves to explore dead regions in the cochlea. *Ear Hear.* **22,** 268–278.

Moore, B. C. J. (2003). "An Introduction to the Psychology of Hearing, 5th Ed." Academic Press, San Diego.

Moore, B. C. J. (2004). Dead regions in the cochlea: Conceptual foundations, diagnosis and clinical applications. *Ear Hear.* **25,** 98–116.

Moore, B. C. J., Glasberg, B. R., and Stone, M. A. (2004). New version of the TEN test with calibrations in dB HL. *Ear Hear.* **25,** 478–487.

Moore, B. C. J., and Tan, C. T. (2004). Development and validation of a method for predicting the perceived naturalness of sounds subjected to spectral distortion. *J. Audio Eng. Soc.* **52,** 900–914.

Neff, D. L. (1985). Stimulus parameters governing confusion effects in forward masking. *J. Acoust. Soc. Am.* **78,** 1966–1976.

Nelson, D. A. (1991). High-level psychophysical tuning curves: Forward masking in normal-hearing and hearing-impaired listeners. *J. Speech Hear. Res.* **34,** 1233–1249.

Nelson, D. A., and Schroder, A. C. (1996). Release from upward spread of masking in regions of high-frequency hearing loss. *J. Acoust. Soc. Am.* **100,** 2266–2277.

Nelson, D. A., and Schroder, A. C. (1997). Linearized response growth inferred from growth-of-masking slopes in ears with cochlear hearing loss. *J. Acoust. Soc. Am.* **101,** 2186–2201.

Ohgushi, K. (1978). On the role of spatial and temporal cues in the perception of the pitch of complex tones. *J. Acoust. Soc. Am.* **64,** 764–771.

Ohm, G. S. (1843). Über die Definition des Tones, nebst daran geknüpfter Theorie der Sirene und ähnlicher tonbildender Vorrichtungen. *Annalen der Physik und Chemie* **59,** 513–565.

O'Loughlin, B. J., and Moore, B. C. J. (1981a). Off-frequency listening: Effects on psychoacoustical tuning curves obtained in simultaneous and forward masking. *J. Acoust. Soc. Am.* **69,** 1119–1125.

O'Loughlin, B. J., and Moore, B. C. J. (1981b). Improving psychoacoustical tuning curves. *Hear. Res.* **5,** 343–346.

Oxenham, A. J., and Moore, B. C. J. (1994). Modeling the additivity of nonsimultaneous masking. *Hear. Res.* **80,** 105–118.

Oxenham, A. J. (2001). Forward masking: Adaptation or integration? *J. Acoust. Soc. Am.* **109,** 732–741.

Oxenham, A. J., and Shera, C. A. (2003). Estimates of human cochlear tuning at low levels using forward and simultaneous masking. *J. Assoc. Res. Otolaryngol.* **4,** 541–554.

Patterson, R. D. (1974). Auditory filter shape. *J. Acoust. Soc. Am.* **55,** 802–809.

Patterson, R. D. (1976). Auditory filter shapes derived with noise stimuli. *J. Acoust. Soc. Am.* **59,** 640–654.

Patterson, R. D., and Henning, G. B. (1977). Stimulus variability and auditory filter shape. *J. Acoust. Soc. Am.* **62,** 649–664.

Patterson, R. D., and Nimmo-Smith, I. (1980). Off-frequency listening and auditory filter asymmetry. *J. Acoust. Soc. Am.* **67,** 229–245.

Patterson, R. D., Nimmo-Smith, I., Weber, D. L., and Milroy, R. (1982). The deterioration of hearing with age: Frequency selectivity, the critical ratio, the audiogram, and speech threshold. *J. Acoust. Soc. Am.* **72,** 1788–1803.

Patterson, R. D., and Moore, B. C. J. (1986). Auditory filters and excitation patterns as representations of frequency resolution. *In* "Frequency Selectivity in Hearing" (B. C. J. Moore, Ed.), pp. 123–177. Academic, London.

Patterson, R. D., Unoki, M., and Irino, T. (2003). Extending the domain of center frequencies for the compressive gammachirp auditory filter. *J. Acoust. Soc. Am.* **114,** 1529–1542.

Peters, R. W., and Moore, B. C. J. (1992). Auditory filter shapes at low center frequencies in young and elderly hearing-impaired subjects. *J. Acoust. Soc. Am.* **91,** 256–266.

Pick, G., Evans, E. F., and Wilson, J. P. (1977). Frequency resolution in patients with hearing loss of cochlear origin. *In* "Psychophysics and Physiology of Hearing" (E. F. Evans and J. P. Wilson, Eds.), pp. 273–281. Academic Press, London.

Pick, G. (1980). Level dependence of psychophysical frequency resolution and auditory filter shape. *J. Acoust. Soc. Am.* **68,** 1085–1095.

Pickles, J. O. (1984). Frequency threshold curves and simultaneous masking functions in single fibers of the guinea pig auditory nerve. *Hear. Res.* **14,** 245–256.

Plack, C. J., and Oxenham, A. J. (1998). Basilar-membrane nonlinearity and the growth of forward masking. *J. Acoust. Soc. Am.* **103,** 1598–1608.

Plomp, R. (1964). The ear as a frequency analyzer. *J. Acoust. Soc. Am.* **36,** 1628–1636.
Plomp, R., and Mimpen, A. M. (1968). The ear as a frequency analyzer II. *J. Acoust. Soc. Am.* **43,** 764–767.
Plomp, R. (1976). "Aspects of Tone Sensation." Academic Press, London.
Puria, S., Rosowski, J. J., and Peake, W. T. (1997). Sound-pressure measurements in the cochlear vestibule of human-cadaver ears. *J. Acoust. Soc. Am.* **101,** 2754–2770.
Robles, L., and Ruggero, M. A. (2001). Mechanics of the mammalian cochlea. *Physiol. Rev.* **81,** 1305–1352.
Rosen, S., Baker, R. J., and Kramer, S. (1992). Characterizing changes in auditory filter bandwidth as a function of level. *In* "Auditory Physiology and Perception" (Y. Cazals, K. Horner, and L. Demany, Eds.), pp. 171–177. Pergamon Press, Oxford.
Rosen, S., Baker, R. J., and Darling, A. (1998). Auditory filter nonlinearity at 2 kHz in normal hearing listeners. *J. Acoust. Soc. Am.* **103,** 2539–2550.
Ruggero, M. A. (1992). Responses to sound of the basilar membrane of the mammalian cochlea. *Curr. Opin. Neurobiol.* **2,** 449–456.
Ruggero, M. A., Robles, L., and Rich, N. C. (1992). Two-tone suppression in the basilar membrane of the cochlea: Mechanical basis of auditory-nerve rate suppression. *J. Neurophysiol.* **68,** 1087–1099.
Ruggero, M. A., Rich, N. C., Recio, A., Narayan, S. S., and Robles, L. (1997). Basilar-membrane responses to tones at the base of the chinchilla cochlea. *J. Acoust. Soc. Am.* **101,** 2151–2163.
Sachs, M. B., and Kiang, N. Y. S. (1968). Two-tone inhibition in auditory nerve fibers. *J. Acoust. Soc. Am.* **43,** 1120–1128.
Scharf, B. (1970). Critical bands. *In* "Foundations of Modern Auditory Theory" (J. V. Tobias, Ed.), pp. 157–202. Academic Press, New York.
Sek, A., Alcántara, J. I., Moore, B. C. J., Kluk, K., and Wicher, A. (2005). Development of a fast method for determining psychophysical tuning curves. *Int. J. Audiol.* **44**, 408–420.
Shailer, M. J., Moore, B. C. J., Glasberg, B. R., Watson, N., and Harris, S. (1990). Auditory filter shapes at 8 and 10 kHz. *J. Acoust. Soc. Am.* **88,** 141–148.
Shannon, R. V. (1976). Two-tone unmasking and suppression in a forward masking situation. *J. Acoust. Soc. Am.* **59,** 1460–1470.
Shaw, E. A. G. (1974). Transformation of sound pressure level from the free field to the eardrum in the horizontal plane. *J. Acoust. Soc. Am.* **56,** 1848–1861.
Small, A. M. (1959). Pure-tone masking. *J. Acoust. Soc. Am.* **31,** 1619–1625.
Smoorenburg, G. F. (1972a). Audibility region of combination tones. *J. Acoust. Soc. Am.* **52,** 603–614.
Smoorenburg, G. F. (1972b). Combination tones and their origin. *J. Acoust. Soc. Am.* **52,** 615–632.
Soderquist, D. R. (1970). Frequency analysis and the critical band. *Psychon. Sci.* **21,** 117–119.
Sommers, M. S., and Humes, L. E. (1993). Auditory filter shapes in normal-hearing, noise-masked normal, and elderly listeners. *J. Acoust. Soc. Am.* **93,** 2903–2914.
Spiegel, M. F. (1981). Thresholds for tones in maskers of various bandwidths and for signals of various bandwidths as a function of signal frequency. *J. Acoust. Soc. Am.* **69,** 791–795.
Srulovicz, P., and Goldstein, J. L. (1983). A central spectrum model: A synthesis of auditory-nerve timing and place cues in monaural communication of frequency spectrum. *J. Acoust. Soc. Am.* **73,** 1266–1276.
Stelmachowicz, P. G., Jesteadt, W., Gorga, M. P., and Mott, J. (1985). Speech perception ability and psychophysical tuning curves in hearing-impaired listeners. *J. Acoust. Soc. Am.* **77,** 621–627.
Stelmachowicz, P. G., Lewis, D. E., Larson, L. L., and Jesteadt, W. (1987). Growth of masking as a measure of response growth in hearing-impaired listeners. *J. Acoust. Soc. Am.* **81,** 1881–1887.
Stone, M. A., Glasberg, B. R., and Moore, B. C. J. (1992). Simplified measurement of impaired auditory filter shapes using the notched-noise method. *Br. J. Audiol.* **26,** 329–334.

Summers, V., and Leek, M. R. (1998). Masking of tones and speech by Schroeder-phase harmonic complexes in normally hearing and hearing-impaired listeners. *Hear. Res.* **118,** 139–150.

ter Keurs, M., Festen, J. M., and Plomp, R. (1992). Effect of spectral envelope smearing on speech reception. I. *J. Acoust. Soc. Am.* **91,** 2872–2880.

ter Keurs, M., Festen, J. M., and Plomp, R. (1993). Effect of spectral envelope smearing on speech reception. II. *J. Acoust. Soc. Am.* **93,** 1547–1552.

Thornton, A. R., and Abbas, P. J. (1980). Low-frequency hearing loss: Perception of filtered speech, psychophysical tuning curves, and masking. *J. Acoust. Soc. Am.* **67,** 638–643.

Turner, C. W., Burns, E. M., and Nelson, D. A. (1983). Pure tone pitch perception and low-frequency hearing loss. *J. Acoust. Soc. Am.* **73,** 966–975.

Tyler, R. S., Wood, E. J., and Fernandes, M. A. (1982). Frequency resolution and hearing loss. *Br. J. Audiol.* **16,** 45–63.

Tyler, R. S., Hall, J. W., Glasberg, B. R., Moore, B. C. J., and Patterson, R. D. (1984). Auditory filter asymmetry in the hearing impaired. *J. Acoust. Soc. Am.* **76,** 1363–1368.

Vogten, L. L. M. (1974). Pure-tone masking: A new result from a new method. *In* "Facts and Models in Hearing" (E. Zwicker and E. Terhardt, Eds.), pp. 142–155. Springer-Verlag, Berlin.

Weber, D. L. (1977). Growth of masking and the auditory filter. *J. Acoust. Soc. Am.* **62,** 424–429.

Weber, D. L. (1983). Do off-frequency simultaneous maskers suppress the signal? *J. Acoust. Soc. Am.* **73,** 887–893.

Wegel, R. L., and Lane, C. E. (1924). The auditory masking of one sound by another and its probable relation to the dynamics of the inner ear. *Phys. Rev.* **23,** 266–285.

Wightman, F. L., McGee, T., and Kramer, M. (1977). Factors influencing frequency selectivity in normal and hearing-impaired listeners. *In* "Psychophysics and Physiology of Hearing" (E. F. Evans and J. P. Wilson, Eds.), pp. 295–306. Academic Press, London.

Wright, A., Davis, A., Bredberg, G., Ulehlova, L., and Spencer, H. (1987). Hair cell distributions in the normal human cochlea. *Acta Otolaryngol.* **444**(Suppl.), 1–48.

Yates, G. K. (1995). Cochlear structure and function. *In* "Hearing" (B. C. J. Moore, Ed.), pp. 41–73. Academic Press, San Diego.

Zhang, X., Heinz, M. G., Bruce, I. C., and Carney, L. H. (2001). A phenomenological model for the responses of auditory-nerve fibers: I. Nonlinear tuning with compression and suppression. *J. Acoust. Soc. Am.* **109,** 648–670.

Zhou, B. (1995). Auditory filter shapes at high frequencies. *J. Acoust. Soc. Am.* **98,** 1935–1942.

Zwicker, E. (1952). Die Grenzen der Hörbarkeit der Amplitudenmodulation und der Frequenzmodulation eines Tones (The limits of audibility of amplitude modulation and frequency modulation of a pure tone). *Acustica* **2,** 125–133.

Zwicker, E. (1961). Subdivision of the audible frequency range into critical bands (Frequenzgruppen). *J. Acoust. Soc. Am.* **33,** 248.

Zwicker, E., and Scharf, B. (1965). A model of loudness summation. *Psych. Rev.* **72,** 3–26.

Zwicker, E. (1970). Masking and psychological excitation as consequences of the ear's frequency analysis. *In* "Frequency Analysis and Periodicity Detection in Hearing" (R. Plomp and G. F. Smoorenburg, Eds.), pp. 376–394. Sijthoff, Leiden.

Zwicker, E., and Schorn, K. (1978). Psychoacoustical tuning curves in audiology. *Audiology* **17,** 120–140.

Zwicker, E., and Fastl, H. (1999). "Psychoacoustics—Facts and Models, Second Edition." Springer-Verlag, Berlin.

ACROSS-CHANNEL SPECTRAL PROCESSING

John H. Grose, Joseph W. Hall III, and Emily Buss

Department of Otolaryngology/Head & Neck Surgery
University of North Carolina at Chapel Hill
Chapel Hill, North Carolina 27599, USA

Although one of the first and primary functions of the ear is to filter sounds into separate frequency bands, it is equally clear that subsequent stages of the auditory system are very adept at combining information from across the audio-frequency range. The purpose of this chapter is to provide an overview of the various ways in which the processing of acoustic information across frequencies can be understood, and to describe some of the auditory phenomena that reflect across-frequency analysis. The chapter begins with aspects of wideband analysis that result in unitary percepts, namely spectral integration of complex signals and virtual pitch. Next, other concepts of across-channel processing are introduced using the phenomenon of monaural envelope correlation perception, and the discussion then examines those processes that are in some sense disadvantageous for signal detection. These include overshoot, modulation detection interference, comodulation detection differences, across-channel masking, and informational masking. Advantageous across-channel processes are examined, including profile

INTERNATIONAL REVIEW OF
NEUROBIOLOGY, VOL. 70
DOI: 10.1016/S0074-7742(05)70003-9

0074-7742/05 $35.00

analysis and comodulation masking release. The chapter then considers modeling issues related to across-channel processing. It concludes with a brief discussion of how across-channel processing relates to perceptual organization in general.

I. Introduction

Most natural sounds—particularly those that are important to communication—are spectrally complex. Yet even before these sounds are represented as neural code, the ear separates them into narrow frequency bands. As described in Chapter 2, one of the most striking characteristics of the auditory system is the exquisite frequency selectivity associated with this filtering in the healthy ear. It is not surprising, therefore, that a great deal of auditory research has been devoted to understanding spectral resolution and the internal representation of sounds contained within single frequency channels. Even though one of the main initial functions of the ear is to filter the incoming sounds into separate frequency bands, it is equally clear that the auditory system is very adept at combining information from across the audio-frequency range. The propensity of the auditory system to process spectrally broad stimuli has been termed 'wide-band analysis' or 'spectral integration,' but there has been a growing realization that the manner in which information is combined across frequencies is not necessarily driven by simple, obligatory rules. Specifically, the combination across frequencies of auditory information contained within discrete frequency channels can depend on complicated analyses that take into consideration, for example, the comparative behavior of responses within those channels over time. The purpose of this chapter is to provide an overview of the various ways in which the processing of acoustic information across frequencies can be understood, and to provide a description of some of the auditory phenomena that reflect across-frequency analysis. The chapter will begin by examining spectral integration for both non-speech and speech sounds, and will then examine the various phenomena that exemplify both advantageous and disadvantageous aspects of across-channel processing. The implications of these phenomena for models of spectral processing will be discussed. The chapter will conclude with a brief coverage of the role of across-channel processing in perceptual organization.

II. Spectral Integration

Stimulus energy falling within a single auditory filter is summed such that the power at the output of the filter is directly proportional to stimulus bandwidth. This means that a doubling of stimulus bandwidth within the filter should translate to about a 3-dB increase in power. Assuming that the detection of a signal passed by an auditory filter depends largely upon the signal-to-noise ratio within the filter, a doubling of signal bandwidth should lead to a 3-dB reduction in detection threshold in terms of spectrum level. Early work by Gässler (1954), for example, supported this in that masked thresholds for narrow-band complex tones decreased by 3-dB per doubling of component number when expressed as level per component. Later work showed this to hold for both diotic and dichotic signal configurations (Langhans and Kohlrausch, 1992). The question of interest here, however, is what happens to the detection of a complex signal when its energy is spread out over multiple (independent) auditory filters. In this case, what appears to be combined, or integrated, is not some quantity directly related to the signal energy but rather the information concerning the detectability of the signal at the output of each frequency channel. This detectability can be expressed in units of d′ (d prime). For complex signals comprised of equally detectable components passed by independent frequency channels, the detectability of the complex signal in terms of d′ should increase as the square root of the number of components (Green and Swets, 1966). Because in hearing the detection of a signal is tied to its intensity, this relationship can be expressed in dB as $10 \log \sqrt{n}$, where n equals the number of components (Green, 1958).

Although this relation appears to hold for signals longer in duration than approximately 100 ms (Buus *et al.*, 1986), it does not accurately describe spectral integration for short-duration signals (van den Brink and Houtgast, 1990a,b). For brief signals of the order of approximately 10 ms, spectral integration is better described by a $16 \log \sqrt{n}$ rule. This has been interpreted as demonstrating a more efficient spectral integration for brief signals. This more efficient spectral integration does not appear to be due to more shallow psychometric functions for brief stimuli, nor to a listening strategy that calls for a monitoring of multiple auditory filters for brief stimuli (Wright and Dai, 1994). However, because this more efficient spectral integration is sensitive to the phase relations among the spectrally dispersed signal components, it is possible that it may reflect a dependence on the degree of neural synchrony across the neural elements responding to the complex signal (Hicks and Buus, 2000).

The propensity of the auditory system to perform spectral integration raises the question of the bandwidth over which this integration is feasible. Although it has been speculated that the range of spectral integration may be limited only by the audibility curve of the ear (Spiegel, 1979), there has been little work aimed directly at probing the limits of spectral integration. However, there are several findings from related lines of investigation that are informative on this issue. In one study that examined the detectability of a complex signal composed of three equally-detectable tones masked by an unmodulated masker, the expected change in threshold using the $10 \log \sqrt{n}$ rule held even when the tones were distributed over a four-octave range (Bacon *et al.*, 2002). Uniform spectral integration over a yet greater range has been seen in the study of gap detection: When detecting a silent interval, or gap, in a narrow band of noise, it has been shown that detection improves if a second independent band of noise is present that also contains the same (synchronous) gap (Hall *et al.*, 1996). This improvement, which follows the expected combined d′ rule, holds for pairs of noise bands that are separated by up to five octaves. However, even these frequency ranges cannot be taken as a limit of spectral integration, at least for these stimuli, because wider frequency separations were not examined in the respective studies.

For other types of stimuli there does appear to be a frequency limit to functional spectral integration. One such stimulus type is a two-tone complex with a slight intensity difference between the two tones. The waveform of this complex exhibits a slow fluctuation (the envelope) and a fine structure that is characterized by frequency modulation. The instantaneous direction of the frequency modulation depends on which of the two components is the more intense, whereas the envelope is impervious to this manipulation. Listeners are sensitive to the pitch shifts associated with the fine structure variations, but only if the two tones are separated by $<\sim5$ equivalent rectangular bandwidths (ERBs) (Feth and O'Malley, 1977).[1] The spectral limit of about 5 ERBs (or 3.5 barks) for static tone pairs appears to be a limit for more speech-like stimuli as well. When listening to synthetic vowels, two formants that are relatively close in frequency may be perceived as a single peak whose frequency forms the center of gravity of the two actual formant frequencies (Schwartz and Escudier, 1989). This convergence of spectral information, however, holds only for frequency separations between the two actual formant peaks of $<\sim3.5$ barks (Xu *et al.*, 2004).

The issue of spectral integration for speech sounds is a broad topic, and an exhaustive treatment is beyond the scope of this chapter. Instead, the issue of spectral integration in speech will be restricted to the synergistic effects seen across bands of speech (e.g., Grant and Braida, 1991; Pollack, 1948; Warren *et al.*, 1995). For example, Warren *et al.* (1995) showed that if speech material is filtered into a single narrow band, e.g., a third-octave band centered at either 370 or 6000 Hz, the speech will be hardly intelligible. However, if both third-octave bands are now played simultaneously, intelligibility will markedly increase. In their study, intelligibility increased from about 22% correct for a single band to about 80% correct for both bands together. This synergistic effect is illustrated schematically in Fig. 1. In the same vein, if speech sentences are filtered into four separate third-octave bands, with each band separated by over an octave, the intelligibility of each band alone will be very poor whereas the intelligibility of all

[1]Earlier work in this area related to speech used the bark scale rather than the ERB scale. For the geometric center frequency of 1000 Hz used here, a 5-ERB separation corresponds to about 3.5 barks, or a frequency separation of about 700 Hz. For further information on the ERB scale, see Glasberg and Moore (1990); for further information on the bark scale, see Zwicker (1980).

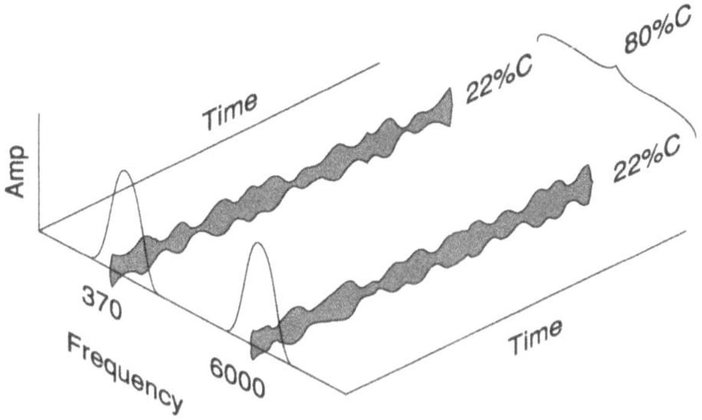

FIG. 1. Schematic showing synergistic effect for filtered speech passed by non-overlapping auditory filters. When either third-octave-wide speech band is presented alone (centered at 370 Hz or 6000 Hz) intelligibility is about 22% correct. When both bands are presented together, intelligibility rises to about 80% correct.

four bands together will be close to 90% correct (Greenberg *et al.*, 1998). This synergistic effect across wide frequency separations indicates that spectral integration of (non-independent[2]) bands is an essential characteristic of the speech perception mechanism. A complementary line of investigation has endeavored to establish frequency importance functions for different components of speech by introducing sharp notches in the speech spectrum that vary in number and notchwidth (Kasturi *et al.*, 2002; Lippmann, 1996). Such work has shown that the frequency-importance function for consonants is fairly flat across frequency, whereas vowels exhibit unequal frequency weighting. However, the process of spectral integration must preserve the temporal alignment of the individual speech bands. In the four-band experiment described earlier, an asynchrony between bands of about 50 ms results in a significant reduction in intelligibility. For two-band synergistic speech perception, performance begins to decline if more than a 12.5-ms asynchrony is introduced across the two narrow bands of speech (Healy and Bacon, 2002). This detrimental effect of asynchrony across speech bands must be distinguished from the case where subsets of the speech bands are alternately present or absent. When speech material is filtered into four (or more) contiguous bands, and alternate bands are gated on and off at a low rate (10 or 20 Hz), speech intelligibility is higher when the interleaved, alternating

[2]The degree to which speech bands filtered from the same token can be considered dependent or independent—at least in terms of their temporal envelopes—appears to be a function of the frequency separation between the bands (Crouzet and Ainsworth, 2001; Apoux and Bacon, 2004).

bands are all present than when just the subset of one or other commonly gated bands is present (Buss *et al.*, 2004a; Howard-Jones and Rosen, 1993).

In summary, spectral integration refers to the propensity of the auditory system to undertake a wide-band analysis of sound even though the initial stage of auditory processing involves a filtering of that sound into discrete frequency bands. Spectral integration, then, is one aspect of across-(frequency)-channel processing. The sorts of stimuli that have been discussed thus far, such as complex tones and speech, have the common characteristic that the percept resulting from the spectral integration is singular. That is, the complex tone is perceived as a unitary signal and the spectrally rich speech token is perceived as a single semantic unit. In this respect, spectral integration is obligatory in that it combines individual components to form a whole, and it is the 'whole' that the listener is interested in. There is another aspect in which a spectrally rich stimulus can be perceived in a unitary fashion, and this is in terms of complex pitch. The next section briefly covers this topic.

III. Virtual Pitch Perception

The perception of virtual pitch (or pitch of the missing fundamental) is *de facto* evidence that the auditory system is adept at linking together information about spectrally distributed components to synthesize a more global construct. Although it is beyond the scope of this chapter to discuss virtual pitch perception in detail, the focus here will be on some aspects of the phenomenon that relate to spectral integration. When a number of harmonics of a fundamental frequency are presented together, listeners generally hear a single, coherent pitch (corresponding to the fundamental frequency) rather than a multitude of pitches associated with the individual harmonics. This coherent percept is particularly robust when many harmonics are present and the harmonics are peripherally resolved (Moore *et al.*, 1985b; Plomp, 1967; Ritsma, 1967). It has been pointed out that virtual pitch perception represents an important example of perceptual organization that permits the across-frequency energy arising from a particular source to be tied together on the basis of harmonicity (Darwin and Gardner, 1986; Duifhuis *et al.*, 1982; Hartmann, 1988; Moore *et al.*, 1985a; Yost, 1991). This type of spectral integration not only contributes to the formation of auditory objects, but also enhances our ability to process those objects in the presence of competing sounds (Assmann and Summerfield, 1990; Brokx and Nooteboom, 1982; Scheffers, 1983). Several types of models have been proposed to account for the synthesis of across-channel information in virtual pitch perception. Some models center around the ability of the central auditory system to estimate the fundamental frequency from the pattern of activity across a range of peripheral channels (Goldstein, 1973; Terhardt, 1974; Wightman, 1973). These are often termed spectral pattern or template-matching models. Models of virtual pitch that are based heavily or exclusively upon specifically temporal factors have long been of interest. A model based upon the combination of temporal autocorrelation functions across auditory channels (summary autocorrelation function or SACF) has attracted particular attention in recent years (Meddis and O'Mard, 1997). Other models require precise coding, both spectrally and temporally, across a range of auditory spectral channels (Moore, 1982; Shamma and Klein, 2000; Srulovicz and Goldstein, 1983).

Virtual pitch formation, like the portrayal of spectral integration earlier, represents a type of across-channel processing that is obligatory in that it combines individual components to form a unitary percept. However, there are other aspects of across-channel processing where obligatory wideband analysis is not particularly beneficial to the listener, or where the combination of components from across the spectrum is dependent on the existence of other concurrent factors. The next sections examine such aspects of across-channel processing.

IV. Monaural Envelope Correlation Perception

Energy remote from a target frequency can have material effects on the ability to process stimulus features at a target frequency. Such effects can be advantageous or disadvantageous, depending upon the particular stimulus characteristics, as will be expanded on in the following text. A more fundamental issue pertains to the sensitivity of the auditory system to across-frequency differences between/among spectral components. Richards (1987) examined this question using frequency-separated bands of noise, testing the sensitivity of listeners to across-frequency differences in waveform envelope. She found that listeners were sensitive to such differences across a wide range of noise bandwidths and frequency separations, and termed this phenomenon "monaural envelope correlation perception." In general, performance on this type of task improves with *increases* in the bandwidth, duration, number, or spectral region of the noise band stimuli, and with *decreases* in the spectral separation between noise bands (Hall and Grose, 1993; Moore and Emmerich, 1990; Richards, 1987). Richards (1990) noted that within-channel cues might be used in the task when noise bands are separated by less than one-third of an octave. Moore and Emmerich (1990) speculated that the better performance for wider noise bandwidths might result from "multiple looks," as the number of fluctuations per time period increases with increasing noise bandwidth. Sensitivity to across-frequency envelope timing (for stimuli separated by an octave or more) has also been demonstrated for a *delay* between the envelopes of sinusoidally amplitude modulated tones (Strickland *et al.*, 1989; Wakefield and Edwards, 1987; Yost and Sheft, 1989). The sensitivity of the auditory system to across-frequency coherence of temporal envelope is generally consistent with the notion that one of the cues that the auditory system uses to accomplish auditory perceptual organization is across-frequency coherence of temporal envelope (Bregman *et al.*, 1985; Rappold *et al.*, 1993; Yost and Sheft, 1989). Therefore, it is possible that performance on monaural envelope correlation perception may provide insights into more complex auditory processes that are involved in the separation of auditory sources in complex sound environments. As previously noted, many phenomena associated with across-channel processing can be categorized (somewhat heuristically) as representing either advantageous or disadvantageous effects. The discussion begins by considering disadvantageous effects.

V. Disadvantageous Effects Related to Energy Remote from the Target Frequency

Results from several different paradigms demonstrate that sensitivity to features of an auditory signal can sometimes be blunted by the presence of energy remote from the target frequency. Although some such effects can be accounted for in terms of simple, peripheral masking (e.g., through spread of excitation), other effects appear to involve a comparison or combination of information that is processed in quasi-independent peripheral auditory channels.

A. Overshoot

Overshoot refers to the finding that, under some conditions, threshold for a brief signal presented just after the onset of a gated masker is elevated relative to threshold for the same signal presented after a longer delay from masker onset (Elliott, 1965; Zwicker, 1965). This effect has been shown to depend upon masker energy remote from the signal frequency (McFadden, 1989), particularly at mid- and low-signal frequencies (Carlyon and White, 1992). Various mechanisms have been proposed to account for overshoot, including neural adaptation (Smith and Zwislocki, 1975), dynamic changes in the effective auditory bandwidth (Wright, 1995), and modulation of the cochlear active mechanism (Rhode, 1971; Ruggero and Rich, 1991; Sellick *et al.*, 1982) by the olivo-cochlear efferent system (Strickland, 2001; Schmidt and Zwicker, 1991; Turner and Doherty, 1997; von Klitzing and Kohlrausch, 1994; Zeng *et al.*, 2000). It is likely that overshoot arises due to a combination of peripheral and central factors.

B. Modulation Detection/Discrimination Interference

The ear is remarkably sensitive to the presence of fluctuations in the amplitude of a sound (Riesz, 1928). However, the sensitivity to amplitude modulation (AM) at a target frequency can be diminished severely by remote energy possessing significant amounts of AM (Yost and Sheft, 1988, 1989). This is illustrated schematically in Fig. 2. This phenomenon is referred to as modulation detection interference in cases where the listener detects the presence of modulation, and modulation discrimination interference in cases where the listener detects a change in modulation depth (in either case, the acronym is MDI). Basic features of MDI include: (1) robustness with regard to the spectral separation between target and interferer; (2) relatively poor tuning with respect to modulation phase and frequency; (3) effectiveness of modulation carriers either below or above the frequency of the target carrier; and (4) effectiveness of modulation carriers presented to either the ear ipsilateral to or contralateral to the target ear (Bacon and Opie, 1994; Hall and Grose, 1991b; Mendoza *et al.*, 1995; Moore, 1991; Moore and Jorasz, 1992; Yost and Sheft, 1988, 1990; Yost *et al.*, 1989). MDI has been modeled in terms of auditory grouping processes associated with modulation and in terms of modulation filter banks (Dau, 1996; Dau *et al.*, 1997) that operate across a large range of carrier frequencies. At this time, models capable of quantitatively predicting MDI have not been developed.

A finding reminiscent of MDI has been reported in a temporal gap detection paradigm. Here, it is found that it is difficult to detect the presence of a temporal gap in a band of noise when a second band of noise (at a remote frequency) is present but does *not* contain a gap (Grose and Hall, 1993a; Moore *et al.*, 1993). Again, there is an inability to independently process the temporal information in a particular auditory channel when fluctuating energy is present in remote channels. As noted earlier in this chapter, the auditory system appears to integrate temporal envelope information over a very large bandwidth in gap-detection paradigms. It is possible that effects related to MDI are driven by the same underlying processes, where across-frequency energy is "lumped together" in temporal envelope analysis.

C. Co-Modulation Detection Differences

McFadden (1986) and Cohen and Schubert (1987a) reported an interesting phenomenon involving the detectability of a target narrow band of noise presented in a background consisting of one or more narrow bands of noise. When the background consisted of multiple bands, those bands had comodulated envelopes. The key result was that the detection threshold was poorer if the target band had an envelope that was the same as that of the background set than

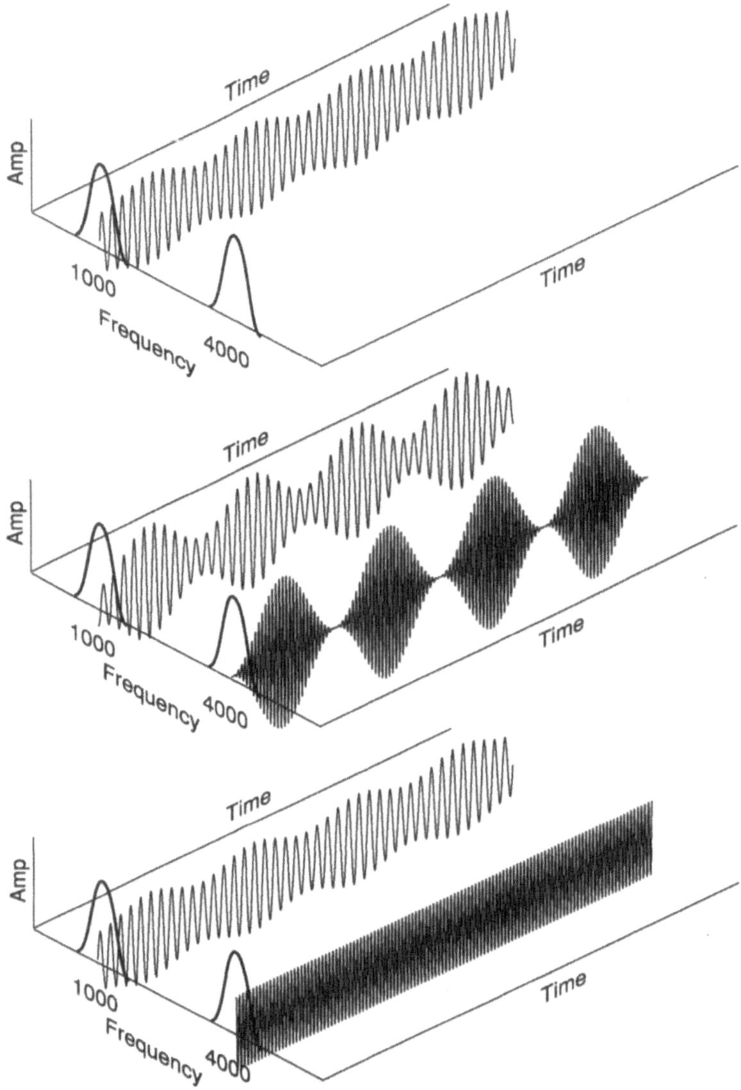

FIG. 2. Typical stimulus configuration for modulation detection interference (MDI). The upper panel depicts modulation depth at detection threshold for a 1000-Hz carrier presented alone. The middle panel illustrates the decline in modulation detection for the 1000-Hz carrier when a modulated 4000-Hz carrier is also present. The non-overlapping filters indicate independence of the two frequency regions. The lower panel shows the recovery of modulation detection threshold when the 4000-Hz carrier is unmodulated.

if the target band had an independent envelope. The magnitude of this effect has been found to increase with the increasing number of background noise bands (McFadden, 1987), but has been shown to diminish as the number of target bands increases (fixed number of background masker bands) (Wright, 1990). An early interpretation of comodulation detection differences (CDD) was in terms of auditory grouping (McFadden, 1987). When the target band has a comodulated envelope, it is difficult to hear because it "blends in" with the more intense comodulated set; when the target band has an independent envelope, it is easier to hear, because it emerges as a separate auditory object. This interpretation is of interest from the standpoint of auditory perceptual organization because it suggests that across-frequency comodulation of envelope may play an important role in the formation of auditory objects. Wright (1990), using a CDD paradigm that included multiple *target* bands, reported results that are pertinent to this issue. Here, some thresholds for multiple, comodulated signal bands were substantially better than predicted by statistical summation or independent threshold models. Wright speculated that signal detection in these conditions might be aided by auditory grouping processes related to comodulation. Recently, Borrill and Moore (2002) and Moore and Borrill (2002) challenged centrally based interpretations of CDD, noting that many of the findings from CDD experiments may be wholly compatible with peripheral accounts based upon masking and suppression (Arthur *et al.*, 1971; Sachs and Kiang, 1968). Such peripheral accounts are attractive from the standpoint of parsimony and future research is needed to establish whether more complex, central explanations are warranted.

D. Across-Channel Masking (ACM)

For pure-tone signals masked by a narrowband of noise centered on the signal, an increase in masking is often obtained when additional, random bands of narrowband noise are present (Cohen and Schubert, 1987b; Eddins and Wright, 1994; Hall *et al.*, 1984; McFadden, 1986; Schooneveldt and Moore, 1987). In modeling such data, Eddins and Wright (1994) noted that the additional masking could not always be accounted for in terms of direct masking, but instead appeared to be attributable to some kind of interference. Moore *et al.* (1990) has termed such effects "across-channel masking" (ACM). Moore *et al.* noted that this type of effect can occur even when the additional noise bands are comodulated with the on-signal band. Whereas comodulated flanking energy can result in an improvement in signal threshold under many conditions (see the following text), such energy results in a worsening of threshold when the signal is brief and occurs entirely in the envelope *peak* of the fluctuating noise masker. Delahaye *et al.* (2001) suggested that the auditory processes affecting the ability to detect a signal presented in a masker peak might be related to the same processes that contribute to MDI (see the previous text).

E. Informational Masking and Across-Channel Interference (ACI)

The detection of a fixed-frequency pure-tone signal can sometimes be extremely poor in the context of a moderate number of masking tones that vary in frequency composition on a trial-by-trial basis (Watson, 1987). In such cases, the poor performance does not appear to result from peripheral processes (most paradigms employ a "protected" region around the signal where masking components are excluded), but rather is thought to reflect an "informational masking" wherein the auditory system fails to separate the excitation arising from the signal from the excitation arising from the masker. Although individual differences in the amount of informational masking can be remarkable, it is not uncommon for listeners to show reliable informational masking effects as great as 30 dB or more (Kidd *et al.*, 1994; Lutfi, 1990; Leek *et al.*, 1991; Neff and Callaghan, 1987, 1988; Neff and Green, 1987; Richards *et al.*, 2002). The amount of informational masking is a non-monotonic function of the number of components comprising the masker, first increasing as the number of masker components increases, and then decreasing as the spectral density of the masker approaches that of noise.

The component-relative entropy (CoRE) model developed by Lutfi (1993) can account for many of the basic features of informational masking using two free parameters (the size of the presumed attentional band contributing to detection and the number of auditory filters over which the listener integrates sound energy). The model is generally successful in accounting for conditions where the signal and masker do not differ from each other except in terms of frequency and amplitude. The model is not as successful when the signal and masker take on different properties (for example, if they differ in terms of harmonicity [Oh and Lutfi, 2000]). Other researchers have shown that acoustical manipulations that result in perceptual segregation of the target and masker can greatly reduce informational masking (Kidd *et al.*, 1994; Leek *et al.*, 1991; Neff, 1995). Durlach *et al.* (2003) have suggested that an effective approach to the study of informational masking should involve stimulus uncertainty, target-masker similarity, and the interaction between these two factors. The failure of the CoRE approach to account for some aspects of informational masking that are related to auditory perceptual organization should not necessarily be taken as serious shortcomings of the model. For example, the magnitude of the masking level difference (MLD) can be affected appreciably by the presence of remote energy that is gated synchronously with a narrow band of noise centered on the signal (as in binaural interference [Bernstein, 1991]), but such effects do not necessarily undermine models of the MLD for more basic stimuli.

An effect similar to that reported in many informational masking paradigms occurs when intensity discrimination for a pure tone or narrow band of noise is measured in the presence of flanking tones or bands whose levels are roved from interval to interval (Buss *et al.*, 2004b; Fantini and Moore, 1994a). From the standpoint of conventional auditory processing based upon auditory filter analysis, the presence of such flanking energy would not be expected to have an appreciable effect: the listener could simply base intensity discrimination upon the output of the auditory filter centered on the signal. In contrast to this conventional expectation, the effect of the flanking energy is to greatly increase the intensity difference limen, a phenomenon that has been termed across-channel interference (ACI) (Buss *et al.*, 2004b; Fantini and Moore, 1994a). As has been found in many informational masking paradigms, providing cues that aid perceptual segregation of the target and flanking energy can significantly diminish the disadvantageous effects of the flanking energy (Buss *et al.*, 2004a).

In general, many of the disadvantageous effects occurring with spectrally distributed flanking energy can be interpreted in terms of a failure of the auditory system to determine what is signal and what is noise. In acoustically rich, real environments, where multiple cues often exist to aid the separation of signal from noise, such problems may be relatively minimal. However, in laboratory situations where stimulus information within many dimensions may be relatively impoverished, there may be a strong tendency for the auditory system to "lump" the signal in with the noise, particularly when both signal and noise share a common temporal gating pattern.

F. Informational Masking and Speech Perception

Most studies of informational masking concern a reduced ability to perceive a tone due to masking effects that cannot be accounted for by peripheral processes (e.g., basilar membrane [BM] suppression or neural swamping/masking). Instead, informational masking is attributed to more complex processes that may depend critically upon the way in which the central auditory system organizes the input from the periphery. Similar masking effects have been noted for speech perception. For example, Carhart *et al.* (1969) found that a two-talker masker was more effective in masking a speech signal than a noise masker, and termed the added masking effect "perceptual masking." Several other investigators have also reported various conditions in which the processing of a speech target appears to be interfered with by the presence of a speech background (Brungart, 2001; Freyman *et al.*, 2001, 2004; Hall *et al.*, 2002). In paradigms where both the speech signal and masker are broadband, it is a challenge to distinguish between energetic masking and informational masking. One approach has been to filter the target and masking speech into non-overlapping frequency bands, and to further control energetic masking by including similarly filtered non-speech maskers (Arbogast *et al.*, 2002). Despite manifest similarities between informational masking for tones and for speech, there are probably essential and important differences, particularly in terms of the level of auditory processing at which the failure in selective analysis occurs. For example, in a recent positron emission tomography (PET) study by Scott *et al.* (2004) it was found that auditory areas associated with linguistic meaning appear to be involved in the disadvantageous effects related to the informational masking of speech.

VI. Advantageous Effects Related to Energy Remote from the Target Frequency

In addition to across-channel effects reducing sensitivity, there is a second class of across-channel effects for which thresholds improve with inclusion of stimulus components falling outside the auditory filter centered on the signal. The most widely studied phenomena in this class are profile analysis and comodulation masking release.

A. Profile Analysis

In the typical profile analysis paradigm, the observer's task is to discriminate between a standard stimulus, comprised of spectral components distributed across frequency, and a target stimulus, where one or more of those spectral components is incremented in amplitude relative to the other components (Green, 1988). The overall level of the stimulus is roved on an interval-by-interval basis, interfering with the reliability of a within-channel cue. This is illustrated in Fig. 3. The expected threshold for performance based on a within-channel across-interval level cue can be computed, and thresholds below this value can be interpreted as reflecting across-channel processing. The detection advantage in profile analysis is greatest for large numbers of components falling into independent auditory channels (Green, 1992), and increases as a function of stimulus duration, particularly for stimuli with close frequency spacing (Dai and Green, 1993). The presence of AM has been shown to affect profile analysis. Whereas coherent AM of stimulus components does not interfere with the effect, discrepancies in AM pattern across frequency reduce it (Bacon and Smith, 1996;

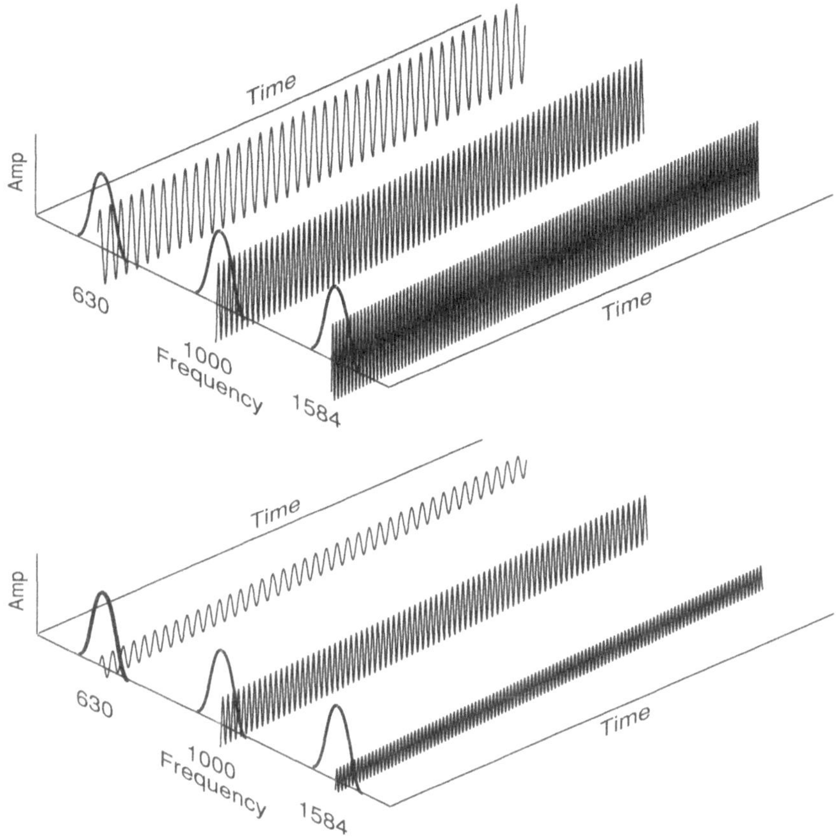

FIG. 3. Typical stimulus configuration for profile analysis. The upper panel depicts a non-signal stimulus consisting of three equal-amplitude pure tones passed by independent auditory filters. The lower panel depicts a signal stimulus consisting of the same three frequencies at a lower overall amplitude, but with the 1000-Hz tone having a greater relative amplitude than the two flanking tones. Detection of the increment in the 1000-Hz tone is facilitated by its prominence in the spectral profile rather than by its absolute level.

Green and Nguyen, 1988).[3] One way to interpret this result is in terms of the stability of the spectral cue as a function of time: incoherent AM introduces dynamic changes in the spectral profile, whereas coherent AM preserves the spectral profile from one epoch to the next. Results of profile analysis studies are often discussed in terms of the level-invariant spectral features characteristic of naturally occurring sounds, such as speech.

[3]Stimuli used in these studies bear a strong resemblance to those of CMR (Section VI.B). Similar to CMR, AM is applied to all components in the standard 'no-signal' interval. In contrast to CMR, however, AM is also applied to the incremented signal component in the profile analysis paradigm.

B. Co-Modulation Masking Release

Whereas the cue in profile analysis can be characterized in terms of long-term spectral shape, the detection cue in comodulation masking release (CMR) is inherently dynamic. Typically, CMR can be defined as the advantage of a coherent masker AM across frequency for the detection of a tonal signal. One example of this advantage can be demonstrated in a band-widening paradigm. In this paradigm, detection thresholds are measured for a pure-tone signal in the presence of a bandpass masker of variable bandwidth, spectrally centered on the signal. For a band-limited Gaussian noise masker, thresholds rise with increases in the masker bandwidth up to a critical bandwidth, and then thresholds asymptote. In contrast, when the masker is amplitude-modulated coherently across frequency, detection thresholds rise with increases in masker bandwidth up to a critical bandwidth, and then begin to decrease with further masker band-widening (Hall *et al.*, 1984). This effect is more pronounced with perfectly coherent patterns of amplitude modulation, although CMR occurs for stimuli with imperfectly coherent patterns of amplitude modulation as well (e.g., Moore and Schooneveldt, 1990).

The advantage of coherent masker AM across frequency for the detection of a tonal signal can also be demonstrated with multiple narrowband maskers (e.g., Hall *et al.*, 1984; Schooneveldt and Moore, 1987) and with multiple amplitude-modulated tonal maskers (e.g., Grose and Hall, 1989). In general, thresholds with multiple maskers are lowest when the masker centered on the signal is comodulated with the flanking noisebands, and highest when the on-signal masker and flanking noisebands have random (uncorrelated) patterns of AM; the case of on-signal masker alone lies somewhere intermediate to those two cases (Moore and Schooneveldt, 1990). Stimuli typical of the paradigm incorporating multiple narrow bands of noise are illustrated in Fig. 4.

The CMR measured using both the band-widening and the multiple noise-band paradigms is often argued to reflect the same underlying phenomenon. Sensitivity to the presence of a pure tone signal is improved if the masker energy in the region of the signal and the masker energy distant from the signal are comodulated, regardless of whether the on-signal and flanking maskers are separated by spectral gaps or whether they are contiguous in frequency. The magnitude of CMR can be quantified in two ways: (1) the decrease in threshold for detection of a tonal signal comparing the case of the on-signal masker alone versus a comodulated masker, or (2) the decrease in threshold comparing the case of a randomly modulated masker versus a comodulated masker.

One feature of CMR that is considered central to the effect is the importance of signal energy coincident with masker modulation minima. Using sinusoid amplitude modulated (SAM) tones for the central and flanking maskers, studies have found a large CMR when the signal pips coincided with modulation

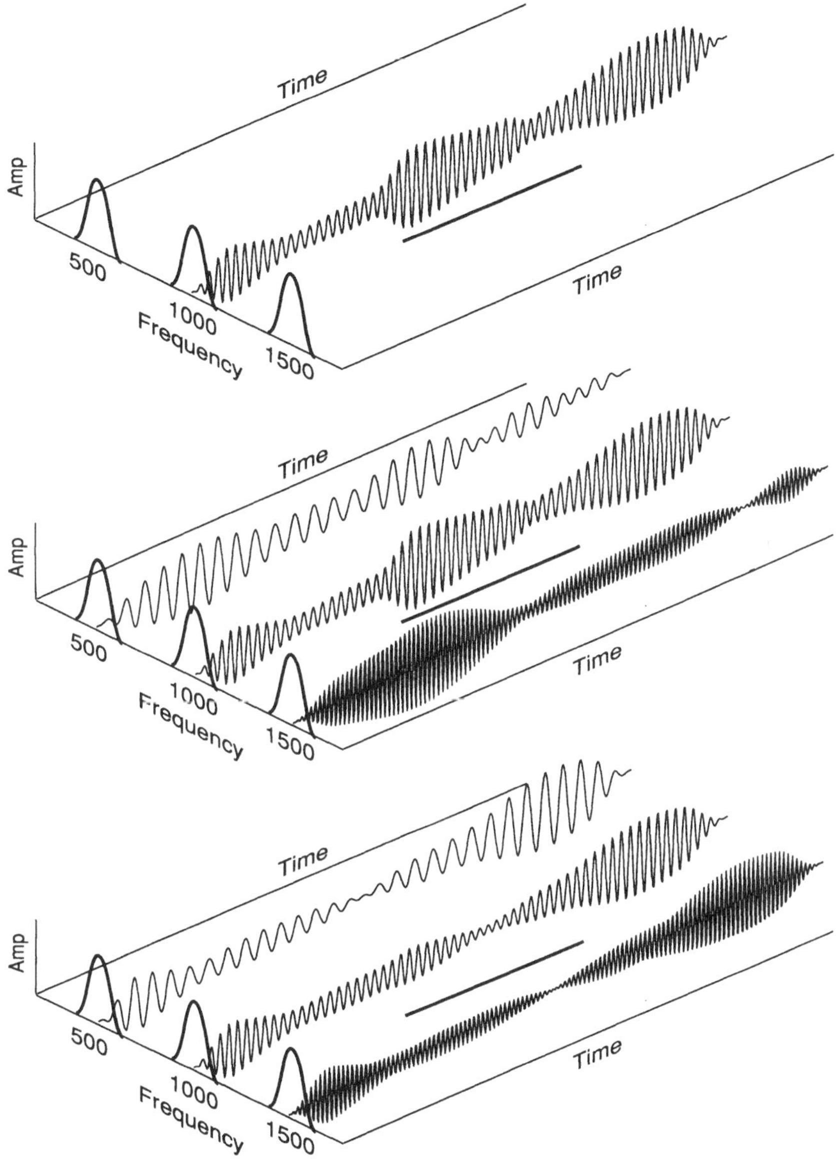

FIG. 4. Typical stimulus configuration for comodulation masking release (CMR) obtained with multiple narrow bands of noise. The upper panel shows a narrow band noise passed by an auditory filter centered at 1000 Hz. A pure-tone signal has been added to the on-signal masker during the period marked by the line segment. The middle panel shows the same on-signal masker accompanied by two independent narrow bands of noise passed by non-overlapping auditory filters. The detection level of the pure-tone signal has not benefited by the presence of these uncorrelated bands. The lower panel shows the same on-signal masker accompanied by two comodulated narrow bands of noise. The detection level of the added signal has now been reduced, demonstrating a masking release.

minima, and no detection advantage when the signal pips coincided with modulation maxima (Grose and Hall, 1989; Moore *et al.*, 1990). Buus *et al.* (1996) quantified this effect for narrowband noise stimuli in terms of temporal weights applied as a function of envelope modulation, and found higher weights were applied during AM minima than AM maxima. These findings are of particular relevance to modeling CMR, a topic considered in the following text.

As in profile analysis, CMR is more pronounced for greater numbers of maskers falling into independent auditory filters, either in the same or opposite ear (Carlyon *et al.*, 1989; Haggard *et al.*, 1990; Hall *et al.*, 1990; Schooneveldt and Moore, 1989b). In addition, longer duration stimuli are associated with larger effects, particularly for stimuli with close frequency spacing (Hatch *et al.*, 1995; Schooneveldt and Moore, 1989a). Unlike profile analysis, the detection advantage in CMR is not diminished with reductions in masker spacing, as neighboring components begin to activate common auditory channels. On the contrary, within-channel effects resembling CMR are quite robust, and are thought to rely on the introduction of temporal cues to individual auditory channels. A great deal of attention has been given to discriminating within-versus across-channel effects (e.g., Schooneveldt and Moore, 1987), with the designation of 'true CMR' reserved for the latter.

VII. Model(s) of Across-Frequency-Channel Processing

In some ways, phenomena demonstrating a processing disadvantage due to stimulus components falling into independent channels do not require a radical departure from the traditional, power-spectrum model of masking. According to that model, the listener bases psychophysical judgments on the output of the auditory channel conveying the most favorable detection or discrimination cue. For example, for a pure-tone signal in noise, detection can be modeled as a constant signal-to-noise ratio at the output of a filter centered at or near the frequency of that signal (Fletcher, 1940), irrespective of the outputs of other auditory channels. The finding of reduced sensitivity with the addition of stimulus components falling into independent auditory channels (e.g., ACM) could be explained in terms of central inefficiency; the listener may fail to make optimal use of information that is available peripherally due to distraction or reduced attention consequent to more stringent task demands. Assuming that efficiency could be restored, the limit on performance would again be based on the quality of the information carried in a single channel. Modifying the power spectrum model via inclusion of a central processor of limited capacity lowers the quantitative predictive power of the model, but such a modification seems justifiable, and in this way the power spectrum model is not fundamentally inconsistent with findings of phenomena of disadvantageous across-channel effects.

In the case of *advantageous* across-channel effects, however, the power spectrum model is more clearly insufficient to account for the data. Cases where psychophysical performance improves with inclusion of non-signal components implies a more elaborate process whereby information across auditory channels is combined. While such an elaborate process may seem obvious given our subjective experience of naturally occurring, spectrally complex stimuli as unitary and cohesive, simple psychophysical paradigms such as profile analysis and CMR provide a unique opportunity to formulate and test models of across-channel processing. Because of the more far-reaching implications for advantageous (as opposed to disadvantageous) across-channel effects, the discussion of across-channel models will center on profile analysis and CMR.

Historically, there have been two approaches to modeling performance in profile analysis. One approach is based on the insight that spectral shape can be characterized in terms of the relative intensities at the output of a bank of auditory filters. This approach, often referred to as the 'channel model,' has been highly successful at capturing some of the basic findings of profile analysis. A second approach is based on the pitch cue associated with complex stimuli, a cue that changes with modifications in spectral profile. Early work in this vein attempted to apply models of pitch developed for narrowband stimuli to the spectrally broad stimuli of profile analysis. While initial attempts to model profile analysis in terms of a change in perceived pitch were largely unsuccessful, more recent work considering the role of pitch strength (Gockel *et al.*, 2004) has revived interest in this approach.

The most well-known formulation of the channel model was introduced by Durlach and colleagues at MIT (Durlach *et al.*, 1986). In this model, illustrated in Fig. 5, a bank of bandpass filters resolves the stimulus into frequency channels, and information in these channels is combined optimally, with the exact formula characterizing this stage depending on the task demands the listener is faced with. The combination of information across auditory filter outputs can be represented as a weighted sum of intensity across channels, with weights computed based on the signal to be detected and the variance across channels relative to the variance common to all channels. In a typical, single-target profile analysis stimulus, the detection cue is based on the average difference in intensities between target and non-target components. This model predicts no effect of adding distractor stimuli that are not predictive of the presence of a signal, because the associated channels would be assigned a weight of zero. As such, the original form of the model is capable of predicting only advantageous across-channel effects, not disadvantageous effects.

Despite the fact that the channel model has been relatively successful in predicting data in the profile analysis paradigm, there are some data that have proven more challenging. First is the finding of poorer performance for

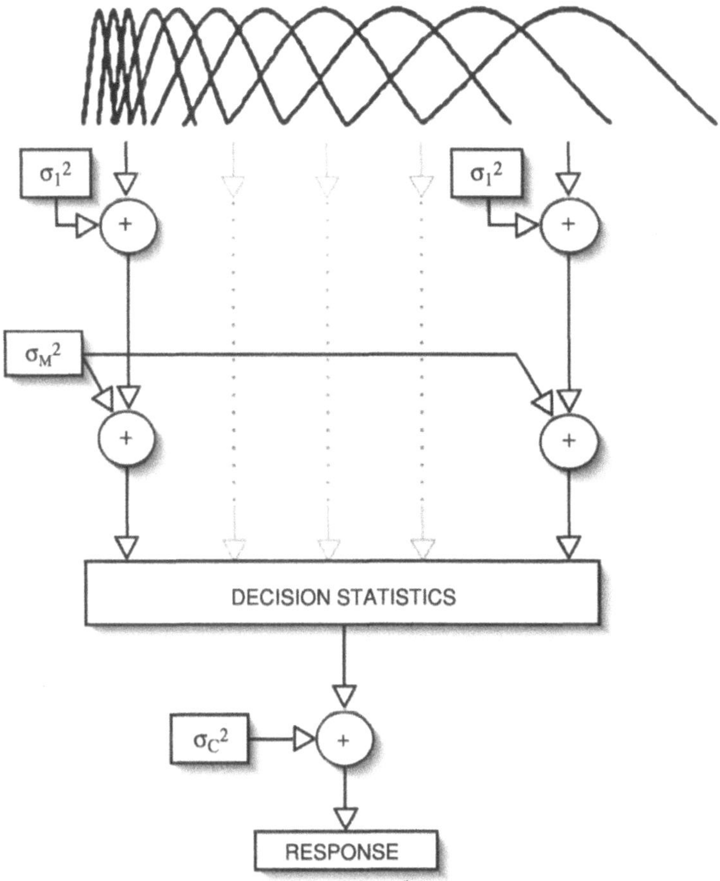

FIG. 5. Schematic of Durlach's channel model. A bank of bandpass filters resolves the stimulus into frequency channels, with channel-specific information combined optimally in a task-dependent manner.

decrement as opposed to increment detection. This is an interesting result because the channel model predicts equal sensitivity for increments and decrements. It has been argued that the observed difference between increments and decrements is due to the greater degree of spread of excitation into channels corresponding to spectral dips than peaks (Ellermeier, 1996), an effect that could be captured in the filtering stage at the input of the model. A second challenge for the channel model is the reduction in the profile effect with asynchronous component gating, a result that is sometimes discussed in terms of the segregation effects of asynchronous gating. It is not obvious how the channel model could be adapted to incorporate the effects of grouping.

The channel model was adapted by Richards and Lentz (1998) to characterize performance with dynamic stimuli. Instead of the more typical profile task of detecting changes in amplitude as a function of frequency, observers in this experiment were asked to detect discrepancies in the AM phase, or dynamic changes in the pattern of amplitudes across frequency. The channel model was modified to compute the short-term spectral profile as a function of time, with the stimuli segmented into 'boxcar' sub-samples. This approach was somewhat successful at capturing performance in that study. Further, it is consistent with the results of Green and Nguyen (1988), which showed that level fluctuations which affect short-term profile interfere with detection, and on these grounds argued that the detection cue in profile analysis could be based on multiple short-term spectral cues. This adaptation of the channel model is intriguing because it provides a framework for defining the relationship between the static cues of the traditional profile analysis and the dynamic cues of other phenomena, such as CMR.

Perhaps the most compelling reason to consider a pitch cue in modeling profile analysis is the subjective report from listeners that their judgments are based on changes in pitch. One formulation of the complex pitch cue is the envelope-weighted instantaneous frequency (EWAIF) model proposed by Feth (1974). Whereas this model was originally designed to compute estimates of pitch for spectrally narrow dynamic stimuli, the fact that it does not incorporate a frequency-selective filter at the input allows its application to spectrally wide complex stimuli, such as those used in profile analysis. Richards *et al.* (1989) investigated the predictive power of this model for profile analysis by assessing the effects of randomizing the rate at which a digital-to-analog converter (DAC) presented the stimulus, a manipulation that introduces frequency uncertainty across intervals. Observers' performance was not as impaired by this randomization as expected based on the model, nor was the EWAIF cue correlated with performance for different signal configurations. More recently, pitch strength has been proposed as a cue in profile analysis (Gockel *et al.*, 2004). Change in pitch strength has yet to be tested quantitatively as a cue in profile analysis.

There has been no shortage of theories to explain CMR, although few have enjoyed the quantitative success of the channel model for profile analysis. An overview of several of these approaches most often referred to in recent work on CMR is provided in the following text.

The first approach posits that CMR is not actually the result of an across-channel process, but reflects a broad filter in which energy from both the on-signal band and the flanker bands sums (Berg, 1996; Verhey *et al.*, 1999). Models of this type have been successful at modeling performance in a band-widening CMR paradigm, but tend to fail for multiple discrete masker bands distributed across frequency or masker bands presented to the ear contralateral to the signal.

Another approach posits that across-channel envelope comparisons are the basis of CMR. The addition of a pure tone signal to a narrowband masker changes the envelope of that stimulus, typically increasing the mean and smoothing out the pattern of AM. In a spectrally broad, coherently modulated masker, the presence of a signal is reflected as a reduction in depth of AM at the output of the auditory filter centered on the signal (Hall, 1986), a reduction in cross-covariance or cross-correlation across auditory channels (Richards, 1987; van de Par and Kohlrausch, 1998), and a remainder at the output of a modified version of Durlach's 1963 equalization-cancellation model (Buus, 1985; Green and Dai, 1992).

Another so-called '*listening in the dips*' hypothesis posits that the flanking comodulated noisebands cue the auditory system to give more perceptual weight to the output of the auditory channel centered on the signal during epochs of low masker energy, as indicated by the envelope pattern of masker AM (Buus, 1985; Hall and Grose, 1991a). Buus *et al.* (1996) suggested that this emphasis on the dips could be a by-product of BM compression. While this approach is consistent with many basic findings of CMR, to date there is relatively little known about how such cuing mechanisms might be implemented.

Some approaches to CMR have been firmly rooted in physiological mechanisms (e.g., Hofer and Klump, 2003). There have been several recent attempts to account for some basic aspects of CMR in cochlear nucleus (Pressnitzer *et al.*, 2001) or cortical (Nelken *et al.*, 1999) physiology. For example, the cochlear nucleus model of Pressnitzer *et al.* (2001) constructs a neural circuit containing a fast-acting inhibitory cell having a wide-bandwidth receptive field and an excitatory cell with a narrow-bandwidth receptive field. The wideband cell responds to the entire array of comodulated flanking bands whereas the narrowband cell responds only to the localized frequency region occupied by the signal and on-signal masker. Interactions between these cell types can simulate some of the beneficial effects of comodulated flanking bands. Whereas these approaches have the advantage of being based upon specific neural mechanisms, their predictive power is still relatively coarse.

A final approach to modeling across-channel processing of AM stimuli that will be mentioned here is the suggestion that the detection cue is based on a dynamic profile. This has received relatively little attention in the CMR literature despite the fact that the basic idea was suggested in the first report of CMR (Hall *et al.*, 1984). Studies by Eddins (2001) and by Fantini and Moore (1994a,b) have produced some intriguing comparisons between these phenomena, but neither modeled these results formally. There is little substantive difference between across-frequency temporal cues and dynamic short-term spectral cues (as proposed here), and recent advances in the literature on profile analysis provide important insights into how to model CMR. Most notable is the Richards and Lentz (1998) adaptation of the channel model to characterize the effects of short-term changes in spectral profile.

VIII. Perceptual Organization and Across-Channel Spectral Processing

All of the across-channel effects described in the previous text, both advantageous and disadvantageous, imply an inclusion of information across auditory channels in the decision process. This grouping of channel outputs into a single processing stream is often described in terms of a *synthetic* listening mode. In contrast, cases where components of a complex sound are segregated and processed separately are said to reflect an *analytic* listening mode. The conditions under which listeners employ the *synthetic* and the *analytic* listening modes have been a topic of great interest, in part because the combination of information across frequencies is central to the perceptual organization of an auditory scene (see Bregman, 1990). Many of the stimulus features associated with synthetic listening can be characterized as reflecting a single auditory source or event, such as common temporal properties across frequency or common spectral patterns, such as harmonic structures. This topic is dealt with more completely in Chapter 11. However, several key features of across-channel processing as related to auditory scene analysis will be mentioned here.

Temporal cues to stream segregation are often discussed in terms of the temporal coherence of auditory stimulus components produced by discrete events in the environment. Sounds that ramp on together and fluctuate coherently are likely to be produced by a unitary source and are most effectively processed synthetically. On the other hand, components that follow distinct time courses are likely to be produced by different events, calling for a more analytic listening mode. Cues of gating synchrony and AM coherence are thought to be based on this association in the natural world.

Synchronous gating of stimulus components across frequency has been argued to be one of the most powerful grouping cues, having a larger effect than either frequency proximity, coherence of AM or ear of presentation (Turgeon *et al.*, 2002). This has led some to interpret gating effects as a hallmark of across-channel processing, reflecting central rather than peripheral combination of information across frequencies (e.g., Dau *et al.*, 2004). In general, introduction of onset asynchrony across stimulus components falling into independent auditory filters is thought to reduce synthetic listening. Under conditions where synthetic listening is disadvantageous, sensitivity can be improved with introduction of onset asynchrony: this has been demonstrated for MDI[4] (Hall and Grose, 1991b), ACM (Hicks and Bacon, 1995), and informational masking (Neff, 1995).

[4]Asynchronous onset of the stimulus components, as discussed here, should be distinguished from asynchronous onset of modulation on an already-present carrier. In the later case, there is evidence that MDI is actually larger for an asynchronously gated masker AM than for masker AM that is gated on synchronously with the target AM (Gockel *et al.*, 2002).

Under conditions where synthetic listening is advantageous, sensitivity can be degraded by introduction of onset asynchrony: this has been demonstrated for profile analysis (Green and Dai, 1992; Hill and Bailey, 1997), CMR (Grose and Hall, 1993b), and for virtual pitch and vowel identification (Hukin and Darwin, 1995).

Coherence of AM rate and across-frequency phase has been shown to affect grouping (e.g., Bregman *et al.*, 1985, 1990; Rappold *et al.*, 1993), with some degree of segregation resulting from mismatches in AM rate, AM phase, or both. In some cases, coherent AM closely resembles sequential onsets and offsets, suggesting that the mechanism by which coherent AM promotes synthetic listening is similar to that associated with synchronous onsets. Coherence of AM tends to be strongly associated with some across-channel effects. Even small mismatches in AM rate across frequency dramatically reduce or eliminate CMR (Haggard *et al.*, 1985). Similarly, manipulation of AM coherence has been shown to play a role in profile analysis; coherent AM of component tones does not interfere with the effect, but discrepancies of AM pattern across frequencies reduce it (Bacon and Smith, 1996; Green and Nguyen, 1988). Coherence of AM rate does not have a uniform effect in the across-channel effects considered here, however. Studies of MDI suggest that this effect is relatively broadly tuned for AM rate and *in*sensitive to the phase of AM across frequencies (Yost *et al.*, 1989). This has led some to question the role of AM coherence as a causal factor in MDI (e.g., Moore and Jorasz, 1996).

One finding of particular relevance to the grouping effects of AM coherence on CMR is that of Hall and Grose (1990). In that study the CMR achieved with a set of co-modulated narrowband maskers was reduced by the inclusion of an additional masker band with a different modulation pattern (a deviant masker). The disruptive influence of the deviant was reduced by the inclusion of additional deviant bands with the same modulation pattern (co-deviants). Hall and Grose argued that the auditory system groups the co-deviant maskers together on the basis of common AM, segregating them from the other components and decreasing their interference in signal detection; with a single deviant, however, a distinct deviant "group" is not formed. This explanation was bolstered by the finding that asynchronous onset of deviant and comodulated maskers further decreased the disruptive effect of those deviant maskers.

In addition to temporal cues, lateralization cues also feature prominently in across-channel processes that underlie perceptual organization. Contralateral presentation of stimulus components results in a reduced, although significant, effect relative to monotic presentation for both profile analysis (Green and Kidd, 1983) and CMR (Schooneveldt and Moore, 1987). Contralateral presentation of the masker in MDI results in a less robust, but still significant interference (Bacon and Opie, 1994; Yost and Sheft, 1990). Presenting the masker diotically and the signal monotically (NoSm) introduces a reliable binaural cue; Kidd *et al.* (1994) found that observers are able to use this cue to reduce informational masking, but that there were large individual differences even among well-practiced adult observers.

Harmonicity has been argued to promote a synthetic listening mode. Moore *et al.* (1986) showed that mistuning one of a series of harmonic stimuli could cause that tone to be heard as a separate sound, segregated from the remaining harmonically related tones. The threshold level of mistuning necessary to obtain this effect was negatively correlated with the duration of the stimuli: 50-ms stimuli required more mistuning than 1.6-s stimuli (with thresholds of 5% to 10% and 0.5% to 3%, respectively). This was interpreted as showing that the common temporal features of the background and target components in the short-duration stimuli dominated the harmonicity cue, such that the components were grouped. For the long-duration stimuli, the steady portion of the stimulus allowed segregation in response to harmonicity to occur. Harmonicity of components does not seem to impact the magnitude of the profile analysis effect (Zera *et al.*, 1993), nor does the anharmonicity of maskers seem to have much of an effect in CMR (Grose and Hall, 1993b). Hall and Grose (1991b) showed that the MDI effect can be substantially reduced by the inclusion of 'captor' tones, synchronously gated with and harmonically related to the masker. This result was interpreted in terms of the grouping effects acting between the captors and the masker, effectively segregating them from the target tone; either harmonicity and/or spectral proximity of the masker and captor tones could be responsible for this effect.

The case for spectral proximity as a grouping cue relates particularly to sequential sounds (e.g., the 'galloping' ABA, ABA paradigm of van Noorden [1975]). However, one problem in assessing the effect of spectral proximity on across-channel effects for simultaneous components is the possibility of introducing within-channel cues. One heuristic for ensuring that within-channel effects do not influence results is to test for proximity effects using binaural stimuli. Studies of binaural fusion have shown proximity to play a role in grouping stimulus components presented to the two ears (Perrott and Nelson, 1969), a finding that justifies looking for spectral proximity effects across ears. Mendoza *et al.* (1995) presented maskers contralateral to the target, thus eliminating concern about within-channel effects, and demonstrated a robust proximity effect for synchronously gated MDI stimuli: maskers closer to the signal frequency caused more MDI than those more distant in frequency. In contrast, Cohen and Schubert (1987a) failed to show an orderly effect of stimulus component proximity on CDD.

In summary, it is clear that there is a natural extension between studies of across-frequency processing and issues related to auditory scene analysis. Understanding the physiological implementations of these mechanisms, and the effects of impaired auditory systems on their function, is a current challenge.

IX. Conclusions and Future Research Directions

The ear is remarkably sensitive to across-frequency properties of sound, with a number of phenomena indicating that the processing of a stimulus at a target frequency can be facilitated by the presence of energy that is well removed from the target frequency. Off-frequency energy can also be remarkably detrimental to the processing of stimulus features at a target frequency (e.g., MDI and ACI). A clear challenge for future research will be to determine whether these advantageous and disadvantageous effects, although divergent, might nevertheless be accounted for in terms of a set of common underlying processes. A striking characteristic of the across-frequency auditory processes considered here is that they can be influenced markedly by stimulus presentation parameters that influence auditory perceptual organization. For example, in cases where energy remote from the target frequency is *detrimental* to performance, the disadvantage can be minimized by providing auditory cues that promote perceptual segregation of the target and remote energy; in cases where energy remote from the target frequency is *beneficial* to performance, the advantage can be reduced by providing auditory cues that promote perceptual segregation of the target and remote energy. Informational masking effects are generally large when cues for perceptual segregation are impoverished but can be substantially reduced when such cues are ample. An important challenge for future research will be the development of across-frequency auditory processing models that incorporate features that can account for effects related to perceptual organization.

Acknowledgments

Preparation of this chapter was supported by grants from the NIDCD (DC001507 and DC00418).

References

Apoux, F., and Bacon, S. P. (2004). Relative importance of temporal information in various frequency regions for consonant identification in quiet and in noise. *J. Acoust. Soc. Am.* **116,** 1671–1680.

Arbogast, T. L., Mason, C. R., and Kidd, G., Jr. (2002). The effect of spatial separation on informational and energetic masking of speech. *J. Acoust. Soc. Am.* **112,** 2086–2098.

Arthur, R. M., Pfeiffer, R. R., and Suga, N. (1971). Properties of 'two-tone inhibition' in primary auditory neurones. *J. Physiol.* **212,** 593–609.

Assmann, P., and Summerfield, Q. (1990). Modeling the perception of concurrent vowels: Vowels with different fundamental frequency. *J. Acoust. Soc. Am.* **88,** 680–697.

Bacon, S. P., and Opie, J. M. (1994). Monotic and dichotic modulation detection interference in practiced and unpracticed subjects. *J. Acoust. Soc. Am.* **95,** 2637–2641.

Bacon, S. P., and Smith, M. A. (1996). Profile analysis with amplitude-modulated nontarget components. *J. Acoust. Soc. Am.* **99,** 1653–1659.

Bacon, S. R., Grimault, N., and Lee, J. (2002). Spectral integration in bands of modulated or unmodulated noise. *J. Acoust. Soc. Am.* **112,** 219–226.

Berg, B. G. (1996). On the relation between comodulation masking release and temporal modulation transfer functions. *J. Acoust. Soc. Am.* **100,** 1013–1023.

Bernstein, L. R. (1991). Spectral interference in a binaural detection task. *J. Acoust. Soc. Am.* **89,** 1306–1313.

Borrill, S. J., and Moore, B. C. (2002). Evidence that comodulation detection differences depend on within-channel mechanisms. *J. Acoust. Soc. Am.* **111,** 309–319.

Bregman, A. S. (1990). "Auditory Scene Analysis: The Perceptual Organisation of Sound." Bradford Books, Cambridge, Mass.

Bregman, A. S., Abramson, J., Doehring, P., and Darwin, C. J. (1985). Spectral integration based on common amplitude modulation. *Percept. & Psychophys.* **37,** 483–493.

Bregman, A. S., Levitan, R., and Liao, C. (1990). Fusion of auditory components: Effects of the frequency of amplitude modulation. *Percept. & Psychophys.* **47,** 68–73.

Brokx, J. P. L., and Nooteboom, S. G. (1982). Intonation and the perceptual separation of simultaneous voices. *J. Phonet.* **10,** 23–36.

Brungart, D. S. (2001). Informational and energetic masking effects in the perception of two simultaneous talkers. *J. Acoust. Soc. Am.* **109,** 1101–1109.

Buss, E., Hall, J. W., 3rd, and Grose, J. H. (2004a). Spectral integration of synchronous and asynchronous cues to consonant identification. *J. Acoust. Soc. Am.* **115,** 2278–2285.

Buss, E., Hall, J. W., and Grose, J. H. (2004b). Across-channel interference and grouping. *J. Acoust. Soc. Am.* **115,** 2459.

Buus, S. (1985). Release from masking caused by envelope fluctuations. *J. Acoust. Soc. Am.* **78,** 1958–1965.

Buus, S., Schorer, E., Florentine, M., and Zwicker, E. (1986). Decision rules in detection of simple and complex tones. *J. Acoust. Soc. Am.* **80,** 1646–1657.

Buus, S., Zhang, L., and Florentine, M. (1996). Stimulus-driven, time-varying weights for comodulation masking release. *J. Acoust. Soc. Am.* **99,** 2288–2297.

Carhart, R., Tillman, T., and Greetis, R. (1969). Perceptual masking in multiple sound backgrounds. *J. Acoust. Soc. Am.* **45,** 694–703.

Carlyon, R., and White, L. (1992). Effect of signal frequency and masker level on the frequency regions responsible for the overshoot effect. *J. Acoust. Soc. Am.* **91,** 1034–1344.

Carlyon, R. P., Buus, S., and Florentine, M. (1989). Comodulation masking release for three types of modulators as a function of modulation rate. *Hear. Res.* **42,** 37–46.

Cohen, M. F., and Schubert, E. D. (1987a). The effect of cross-spectrum correlation on the detectability of a noise band. *J. Acoust. Soc. Am.* **81,** 721–723.

Cohen, M. F., and Schubert, E. D. (1987b). Influence of place synchrony on detection of a sinusoid. *J. Acoust. Soc. Am.* **81,** 452–458.

Crouzet, O., and Ainsworth, W. A. (2001). "On the Various Influences of Envelope Information on the Perception of Speech in Adverse Conditions: An Analysis of Between-Channel Envelope Correlation." Aalborg, Denmark.

Dai, H., and Green, D. M. (1993). Discrimination of spectral shape as a function of stimulus duration. *J. Acoust. Soc. Am.* **93,** 957–965.

Darwin, C., and Gardner, R. (1986). Mistuning a harmonic of a vowel: Grouping and phase effects in vowel quality. *J. Acoust. Soc. Am.* **79,** 838–845.

Dau, T. (1996). Modeling auditory processing of amplitude modulation. Ph.D. thesis, Oldenburg University.

Dau, T., Ewert, S. D., and Oxenham, A. J. (2004). Effects of concurrent and sequential streaming in comodulation masking release. *In* "Auditory Signal Processing: Physiology, Psychoacoustics, and Models" (D. Pressnitzer, A. de Cheveigne, S. McAdams, and L. Collet, Eds.). Springer Verlag.

Dau, T., Kollmeier, B., and Kohlrausch, A. (1997). Modeling auditory processing of amplitude modulation. I. Detection and masking with narrow-band carriers. *J. Acoust. Soc. Am.* **102,** 2892–2905.

Delahaye, R., Fantini, D. A., and Meddis, R. (2001). Effects of flanking component spectral position and modulation pattern on thresholds for signals presented in the peaks of a modulated tonal masker. *J. Acoust. Soc. Am.* **109,** 426–429.

Duifhuis, H., Willems, L. F., and Sluyter, R. J. (1982). Measurement of pitch in speech: An implementation of Goldstein's theory of pitch perception. *J. Acoust. Soc. Am.* **71,** 1568–1580.

Durlach, N. I., Braida, L. D., and Ito, Y. (1986). Towards a model for discrimination of broadband signals. *J. Acoust. Soc. Am.* **80,** 63–72.

Durlach, N. I., Mason, C. R., Kidd, G., Jr., Arbogast, T. L., Colburn, H. S., and Shinn-Cunningham, B. G. (2003). Note on informational masking. *J. Acoust. Soc. Am.* **113,** 2984–2987.

Eddins, D. A. (2001). Monaural masking release in random-phase and low-noise noise. *J. Acoust. Soc. Am.* **109,** 1538–1549.

Eddins, D. A., and Wright, B. A. (1994). Comodulation masking release for single and multiple rates of envelope fluctuation. *J. Acoust. Soc. Am.* **96,** 3432–3442.

Ellermeier, W. (1996). Detectability of increments and decrements in spectral profiles. *J. Acoust. Soc. Am.* **99,** 3119–3125.

Elliott, L. L. (1965). Changes in the simultaneous masked threshold of brief tones. *J. Acoust. Soc. Am.* **38,** 738–746.

Fantini, D. A., and Moore, B. C. (1994a). A comparison of the effectiveness of across-channel cues available in comodulation masking release and profile analysis tasks. *J. Acoust. Soc. Am.* **96,** 3451–3462.

Fantini, D. A., and Moore, B. C. J. (1994b). Profile analysis and comodulation detection differences using narrow bands of noise and their relation to comodulation masking release. *J. Acoust. Soc. Am.* **95,** 2180–2191.

Feth, L. L. (1974). Frequency discrimination of complex periodic tones. *Percept. & Psychophys.* **15,** 375–379.

Feth, L. L., and O'Malley, H. (1977). Frequency discrimination of complex periodic tones. *J. Acoust. Soc. Am.* **62,** 940–947.

Fletcher, H. (1940). Auditory patterns. *Rev. Mod. Phys.* **12,** 47–65.

Freyman, R. L., Balakrishnan, U., and Helfer, K. S. (2001). Spatial release from informational masking in speech recognition. *J. Acoust. Soc. Am.* **109,** 2112–2122.

Freyman, R. L., Balakrishnan, U., and Helfer, K. S. (2004). Effect of number of masking talkers and auditory priming on informational masking in speech recognition. *J. Acoust. Soc. Am.* **115,** 2246–2256.

Gässler, G. (1954). Über die Hörschwelle für Schallereignisse mit verschieden breitem Frequencz-spektrum. *Acustica* **4,** 408–414.

Glasberg, B. R., and Moore, B. C. J. (1990). Derivation of auditory filter shapes from notched-noise data. *Hear. Res.* **47,** 103–138.

Gockel, H., Carlyon, R. P., and Deeks, J. M. (2002). Effect of modulator asynchrony of sinusoidal and noise modulators on frequency and amplitude modulation detection interference. *J. Acoust. Soc. Am.* **112,** 2975–2984.

Gockel, H., Carlyon, R. P., and Plack, C. J. (2004). Across-frequency interference effects in fundamental frequency discrimination: Questioning evidence for two pitch mechanisms. *J. Acoust. Soc. Am.* **116,** 1092–1104.

Goldstein, J. L. (1973). An optimum processor theory for the central formation of the pitch of complex tones. *J. Acoust. Soc. Am.* **54,** 1496–1516.

Grant, K. W., and Braida, L. D. (1991). Evaluating the articulation index for auditory-visual input. *J. Acoust. Soc. Am.* **89,** 2952–2960.

Green, D. M. (1958). Detection of multiple component signals in noise. *J. Acoust. Soc. Am.* **30,** 904–911.

Green, D. M. (1988). "Profile Analysis." Oxford University Press, Oxford.

Green, D. M. (1992). The number of components in profile analysis tasks. *J. Acoust. Soc. Am.* **91,** 1616–1623.

Green, D. M., and Dai, H. (1992). Temporal relations in profile comparisons. *In* "Auditory Physiology and Perception" (Y. Cazals, L. Demany, and K. Horner, Eds.). Pergamon Press, New York.

Green, D. M., and Kidd, G., Jr. (1983). Further studies of auditory profile analysis. *J. Acoust. Soc. Am.* **73,** 1260–1265.

Green, D. M., and Nguyen, Q. (1988). Profile analysis: Detecting dynamic spectral changes. *Hear. Res.* **32,** 147–163.

Green, D. M., and Swets, J. A. (1966). "Signal Detection Theory and Psychophysics." Wiley, New York.

Greenberg, S., Arai, T., and Silipo, R. (1998). "Speech Intelligibility Derived from Exceedingly Sparse Spectral Information," Proceedings of the Int. Conf. on Spoken Lang. Proc., Sydney, Australia.

Grose, J. H., and Hall, J. W. (1989). Comodulation masking release using SAM tonal complex maskers: Effects of modulation depth and signal position. *J. Acoust. Soc. Am.* **85,** 1276–1284.

Grose, J. H., and Hall, J. W. (1993a). Gap detection in a narrow band of noise in the presence of a flanking band of noise. *J. Acoust. Soc. Am.* **93,** 1645–1648.

Grose, J. H., and Hall, J. W., III. (1993b). Comodulation masking release: Is comodulation sufficient? *J. Acoust. Soc. Am.* **93,** 2896–2902.

Haggard, M. P., Hall, J. W., and Grose, J. H. (1990). Comodulation masking release as a function of bandwidth and test frequency. *J. Acoust. Soc. Am.* **88,** 113–118.

Haggard, M. P., Harvey, A. D. G., and Carlyon, R. P. (1985). Peripheral and central components of comodulation masking release. *J. Acoust. Soc. Am.* **78,** S63.

Hall, J. W. (1986). The effect of across-frequency differences in masking level on spectro-temporal pattern analysis. *J. Acoust. Soc. Am.* **79,** 781–787.

Hall, J. W., 3rd, Grose, J. H., and Joy, S. (1996). Gap detection for pairs of noise bands: Effects of stimulus level and frequency separation. *J. Acoust. Soc. Am.* **99,** 1091–1095.

Hall, J. W., and Grose, J. H. (1990). Comodulation masking release and auditory grouping. *J. Acoust. Soc. Am.* **88,** 119–125.

Hall, J. W., and Grose, J. H. (1991a). Relative contributions of envelope maxima and minima to comodulation masking release. *Quart. J. Exp. Psych.* **43A,** 349–372.

Hall, J. W., and Grose, J. H. (1991b). Some effects of auditory grouping factors on modulation detection interference (MDI). *J. Acoust. Soc. Am.* **90,** 3028–3036.

Hall, J. W., and Grose, J. H. (1993). Monaural envelope correlation perception in listeners with normal hearing and cochlear impairment. *J. Speech Hear. Res.* **36,** 1306–1314.

Hall, J. W., Grose, J. H., Buss, E., and Dev, M. B. (2002). Spondee recognition in a two-talker masker and a speech-shaped noise masker in adults and children. *Ear. Hear.* **23,** 159–165.

Hall, J. W., Grose, J. H., and Haggard, M. P. (1990). Effects of flanking band proximity, number, and modulation pattern on comodulation masking release. *J. Acoust. Soc. Am.* **87,** 269–283.

Hall, J. W., Haggard, M. P., and Fernandes, M. A. (1984). Detection in noise by spectro-temporal pattern analysis. *J. Acoust. Soc. Am.* **76,** 50–56.

Hartmann, W. M. (1988). Pitch perception and the segregation and integration of auditory entities. *In* "Auditory Function - Neurobiological Bases of Hearing" (G. M. Edelman, W. E. Gall, and W. M. Cowen, Eds.). Wiley, New York.

Hatch, D. R., Arne, B. C., and Hall, J. W. (1995). Comodulation masking release (CMR): Effects of gating as a function of number of flanking bands and masker bandwidth. *J. Acoust. Soc. Am.* **97,** 3768–3774.

Healy, E. W., and Bacon, S. P. (2002). Across-frequency comparison of temporal speech information by listeners with normal and impaired hearing. *J. Speech Lang. Hear. Res.* **45,** 1262–1275.

Hicks, M. L., and Bacon, S. P. (1995). Some factors influencing comodulation masking release and across-channel masking. *J. Acoust. Soc. Am.* **98,** 2504–2514.

Hicks, M. L., and Buus, S. (2000). Efficient across-frequency integration: Evidence from psychometric functions. *J. Acoust. Soc. Am.* **107,** 3333–3342.

Hill, N. I., and Bailey, P. J. (1997). Profile analysis with an asynchronous target: Evidence for auditory grouping. *J. Acoust. Soc. Am.* **102,** 477–481.

Hofer, S. B., and Klump, G. M. (2003). Within- and across-channel processing in auditory masking: A physiological study in the songbird forebrain. *J. Neurosci.* **23,** 5732–5739.

Howard-Jones, P. A., and Rosen, S. (1993). Uncomodulated glimpsing in "checkerboard" noise. *J. Acoust. Soc. Am.* **93,** 2915–2922.

Kasturi, K., Loizou, P. C., Dorman, M., and Spahr, T. (2002). The intelligibility of speech with "holes" in the spectrum. *J. Acoust. Soc. Am.* **112,** 1102–1111.

Kidd, G., Jr., Mason, C. R., Deliwala, P. S., Woods, W. S., and Colburn, H. S. (1994). Reducing informational masking by sound segregation. *J. Acoust. Soc. Am.* **95,** 3475–3480.

Langhans, A., and Kohlrausch, A. (1992). Spectral integration of broadband signals in diotic and dichotic masking experiments. *J. Acoust. Soc. Am.* **91,** 317–326.

Leek, M. R., Brown, M. E., and Dorman, M. F. (1991). Informational masking and auditory attention. *Percept & Psychophys.* **50,** 205–214.

Lippmann, R. P. (1996). Accurate consonant perception without mid-frequency speech energy. *IEEE Trans. Speech Audio Proc.* **4,** 66–69.

Lutfi, R. A. (1990). How much masking is informational masking. *J. Acoust. Soc. Am.* **88,** 2607–2610.

Lutfi, R. A. (1993). A model of auditory pattern analysis based on component-relative-entropy. *J. Acoust. Soc. Am.* **94,** 748–758.

McFadden, D. (1986). Comodulation masking release: Effects of varying the level, duration, and time delay of the cue band. *J. Acoust. Soc. Am.* **80,** 1658–1667.

McFadden, D. (1987). Comodulation detection differences using noise-band signals. *J. Acoust. Soc. Am.* **81,** 1519–1527.

McFadden, D. (1989). Spectral differences in the ability of temporal gaps to reset the mechanisms underlying overshoot. *J. Acoust. Soc. Am.* **85,** 254–261.

Meddis, R., and O'Mard, L. (1997). A unitary model of pitch perception. *J. Acoust. Soc. Am.* **102,** 1811–1820.

Mendoza, L., Hall, J. W., and Grose, J. H. (1995). Within- and across-channel processes in modulation detection interference. *J. Acoust. Soc. Am.* **97,** 3072–3079.

Moore, B., Peters, R., and Glasberg, B. (1985a). Thresholds for the detection of inharmonicity in complex tones. *J. Acoust. Soc. Am.* **77,** 1861–1867.

Moore, B. C., and Borrill, S. J. (2002). Tests of a within-channel account of comodulation detection differences. *J. Acoust. Soc. Am.* **112,** 2099–2109.

Moore, B. C. J. (1982). "An Introduction to the Psychology of Hearing." Academic, New York.

Moore, B. C. J. (1991). Modulation discrimination interference and auditory grouping. *Royal Society Discussion Meeting-Auditory Processing of Complex Sounds.*

Moore, B. C. J., and Emmerich, D. S. (1990). Monaural envelope correlation perception, revisited: Effects of bandwidth, frequency separation, duration, and relative level of the noise bands. *J. Acoust. Soc. Am.* **87,** 2628–2633.

Moore, B. C. J., Glasberg, B. R., and Peters, R. W. (1985b). Relative dominance of individual partials in determining the pitch of complex tones. *J. Acoust. Soc. Am.* **77,** 1853–1860.

Moore, B. C. J., Glasberg, B. R., and Peters, R. W. (1986). Thresholds for hearing mistuned partials as separate tones in harmonic complexes. *J. Acoust. Soc. Am.* **80,** 479–483.

Moore, B. C. J., Glasberg, B. R., and Schooneveldt, G. P. (1990). Across-channel masking and comodulation masking release. *J. Acoust. Soc. Am.* **87,** 1683–1694.

Moore, B. C. J., and Jorasz, U. (1992). Detection of changes in modulation depth of a target sound in the presence of other modulated sounds. *J. Acoust. Soc. Am.* **91,** 1051–1061.

Moore, B. C. J., and Jorasz, U. (1996). Modulation discrimination and comodulation masking release as a function of the number and spectral placement of narrow-band noise modulators. *J. Acoust. Soc. Am.* **100,** 2373–2381.

Moore, B. C. J., and Schooneveldt, G. P. (1990). Comodulation masking release as a function of bandwidth and time delay between on-frequency and flanking maskers. *J. Acoust. Soc. Am.* **88,** 725–731.

Moore, B. C. J., Shailer, M. J., and Black, M. J. (1993). Dichotic interference effects in gap detection. *J. Acoust. Soc. Am.* **93,** 2130–2133.

Neff, D. L. (1995). Signal properties that reduce masking by simultaneous, random-frequency maskers. *J. Acoust. Soc. Am.* **98,** 1909–1920.

Neff, D. L., and Callaghan, B. P. (1987). Simultaneous masking by small numbers of sinusoids under conditions of uncertainty. *In* "Auditory Processing of Complex Sounds" (W. A. Yost and C. S. Watson, Eds.). Lawrence Erlbaum Associates, New Jersey.

Neff, D. L., and Callaghan, B. P. (1988). Effective properties of multicomponent simultaneous maskers under conditions of uncertainty. *J. Acoust. Soc. Am.* **83,** 1833–1838.

Neff, D. L., and Green, D. M. (1987). Masking produced by spectral uncertainty with multicomponent maskers. *Percept & Psychophys.* **41,** 409–415.

Nelken, I., Rotman, Y., and Bar Yosef, O. (1999). Responses of auditory-cortex neurons to structural features of natural sounds. *Nature* **397,** 154–157.

Oh, E. L., and Lutfi, R. A. (2000). Effect of masker harmonicity on informational masking. *J. Acoust. Soc. Am.* **108,** 706–709.

Perrott, D. R., and Nelson, M. A. (1969). Limits for the detection of binaural beats. *J. Acoust. Soc. Am.* **46,** 1477–1481.

Plomp, R. (1967). Pitch of complex tones. *J. Acoust. Soc. Am.* **41,** 1526–1533.

Pollack, I. (1948). Effects of high pass and low pass filtering on the intelligibility of speech in noise. *J. Acoust. Soc. Am.* **20,** 259–266.

Pressnitzer, D., Meddis, R., Delahaye, R., and Winter, I. M. (2001). Physiological correlates of comodulation masking release in the mammalian ventral cochlear nucleus. *J. Neurosci.* **21,** 6377–6386.

Rappold, P. W., Mendoza, L., and Collins, M. J. (1993). Measuring the strength of auditory fusion for synchronously and nonsynchronously fluctuating narrow-band noise pairs. *J. Acoust. Soc. Am.* **93,** 1196–1199.

Rhode, W. S. (1971). Observations of the vibration of the basilar membrane in squirrel monkeys using the Mössbauer technique. *J. Acoust. Soc. Am.* **49,** 1218–1231.

Richards, V. M. (1987). Monaural envelope correlation perception. *J. Acoust. Soc. Am.* **82,** 1621–1630.

Richards, V. M. (1990). The role of single channel cues in synchrony perception: The summed waveform. *J. Acoust. Soc. Am.* **88,** 786–795.

Richards, V. M., and Lentz, J. J. (1998). Sensitivity to changes in level and envelope patterns across frequency. *J. Acoust. Soc. Am.* **104,** 3019–3029.

Richards, V. M., Onsan, Z., and Green, D. M. (1989). Auditory profile analysis: Potential pitch cues. *Hear. Res.* **39,** 27–36.

Richards, V. M., Tang, Z., and Kidd, G. D., Jr. (2002). Informational masking with small set sizes. *J. Acoust. Soc. Am.* **111,** 1359–1366.

Riesz, R. R. (1928). Differential intensity sensitivity of the ear for pure tones. *Physics Rev.* **31,** 867–875.

Ritsma, R. (1967). Frequencies dominant in the perception of the pitch of complex sounds. *J. Acoust. Soc. Am.* **42,** 191–198.

Ruggero, M. A., and Rich, N. C. (1991). Furosemide alters organ of corti mechanics: Evidence for feedback of outer hair cells upon the basilar membrane. *J. Neurosci.* **11,** 1057–1067.

Sachs, M. B., and Kiang, N. Y. S. (1968). Two-tone inhibition in auditory nerve fibers. *J. Acoust. Soc. Am.* **43,** 1120–1128.

Scheffers, M. (1983). Sifting vowels: Auditory pitch analysis and sound segregation. Ph.D. thesis, University of Groningen, the Netherlands.

Schmidt, S., and Zwicker, E. (1991). The effect of masker spectral asymmetry on overshoot in simultaneous masking. *J. Acoust. Soc. Am.* **89,** 1991.

Schooneveldt, G. P., and Moore, B. C. J. (1987). Comodulation masking release (CMR): Effects of signal frequency, flanking band frequency, masker bandwidth, flanking band level, and monotic vs dichotic presentation flanking bands. *J. Acoust. Soc. Am.* **82,** 1944–1956.

Schooneveldt, G. P., and Moore, B. C. J. (1989a). Comodulation masking release (CMR) as a function of masker bandwidth, modulator bandwidth and signal duration. *J. Acoust. Soc. Am.* **85,** 273–281.

Schooneveldt, G. P., and Moore, B. C. J. (1989b). Comodulation masking release (CMR) for various monaural and binaural combinations of the signal, on-frequency, and flanking bands. *J. Acoust. Soc. Am.* **85,** 262–272.

Schwartz, J.-L., and Escudier, P. (1989). A strong evidence for the existence of a large-scale integrated spectral representation in vowel perception. *Speech Commun.* **8,** 235–259.

Scott, S. K., Rosen, S., Wickham, L., and Wise, R. J. (2004). A positron emission tomography study of the neural basis of informational and energetic masking effects in speech perception. *J. Acoust. Soc. Am.* **115,** 813–821.

Sellick, P. M., Patuzzi, R., and Johnstone, B. M. (1982). Measurement of basilar membrane movement in the guinea pig using the Mossbauer technique. *J. Acoust. Soc. Am.* **72,** 131–141.

Shamma, S., and Klein, D. (2000). The case of the missing pitch templates: How harmonic templates emerge in the early auditory system. *J. Acoust. Soc. Am.* **107,** 2631–2644.

Smith, R. L., and Zwislocki, J. J. (1975). Short-term adaptation and incremental responses of single auditory-nerve fibers. *Biol. Cyber.* **17,** 169–182.

Spiegel, M. F. (1979). The range of spectral integration. *J. Acoust. Soc. Am.* **66,** 1356–1363.

Srulovicz, P., and Goldstein, J. (1983). A central spectrum model: A synthesis of auditory-nerve timing and place cues in monaural communication of frequency spectrum. *J. Acoust. Soc. Am.* **73,** 1266–1276.

Strickland, E. A. (2001). The relationship between frequency selectivity and overshoot. *J. Acoust. Soc. Am.* **109,** 2062–2073.

Strickland, E. A., Viemeister, N. F., Fantini, D. A., and Garrison, M. A. (1989). Within- versus cross-channel mechanisms in detection of envelope phase disparity. *J. Acoust. Soc. Am.* **86,** 2160–2166.

Terhardt, E. (1974). Pitch, consonance, and harmony. *J. Acoust. Soc. Am.* **55,** 1061–1069.

Turgeon, M., Bregman, A. S., and Ahad, P. A. (2002). Rhythmic masking release: Contribution of cues for perceptual organization to the cross-spectral fusion of concurrent narrow-band noises. *J. Acoust. Soc. Am.* **111,** 1819–1831.

Turner, C. W., and Doherty, K. A. (1997). Temporal masking and the 'active' process in normal and hearing-impaired listeners. *In* "Modeling Sensorineural Hearing Loss" (W. Jesteadt, Ed.). Erlbaum, Hillsdale.

van de Par, S., and Kohlrausch, A. (1998). Comparison of monaural (CMR) and binaural (BMLD) masking release. *J. Acoust. Soc. Am.* **103,** 1573–1579.

van den Brink, W. A., and Houtgast, T. (1990a). Efficient across-frequency integration in short-signal detection. *J. Acoust. Soc. Am.* **87,** 284–291.

van den Brink, W. A., and Houtgast, T. (1990b). Spectro-temporal integration in signal detection. *J. Acoust. Soc. Am.* **88,** 1703–1711.

van Noorden, L. P. A. S. (1975). "Temporal Coherence in the Perception of Tone Sequences," Ph.D. thesis, Eindhoven University of Technology, the Netherlands.

Verhey, J. L., Dau, T., and Kollmeier, B. (1999). Within-channel cues in comodulation masking release (CMR): Experiments and model predictions using a modulation-filterbank model. *J. Acoust. Soc. Am.* **106,** 2733–2745.

von Klitzing, R., and Kohlrausch, A. (1994). Effect of masker level on overshoot in running- and frozen-noise maskers. *J. Acoust. Soc. Am.* **95,** 2192–2201.

Wakefield, G. H., and Edwards, B. (1987). Discrimination of envelope phase disparity. *J. Acoust. Soc. Am.* **82,** S41.

Warren, R. M., Riener, K. R., Bashford, J. A., Jr., and Brubaker, B. S. (1995). Spectral redundancy: Intelligibility of sentences heard through narrow spectral slits. *Percept & Psychophys.* **57,** 175–182.

Watson, C. S. (1987). Uncertainty, informational masking, and the capacity of immediate auditory memory. *In* "Auditory Processing of Complex Sounds" (W. A. Yost and C. S. Watson, Eds.). Lawrence Erlbaum Associates, New Jersey.

Wightman, F. L. (1973). The pattern-transformation model of pitch. *J. Acoust. Soc. Am.* **54,** 407–416.

Wright, B. A. (1990). Comodulation detection differences with multiple signal bands. *J. Acoust. Soc. Am.* **87,** 293–303.

Wright, B. A. (1995). Detectability of simultaneously masked signals as a function of signal bandwidth for different signal delays. *J. Acoust. Soc. Am.* **98,** 2493–2503.

Wright, B. A., and Dai, H. (1994). Detection of unexpected tones with short and long durations. *J. Acoust. Soc. Am.* **95,** 931–938.

Xu, Q., Jacewicz, E., and Feth, L. L. (2004). Bandwidth of spectral resolution for two-formant synthetic vowels and two-tone complex signals. *J. Acoust. Soc. Am.* **115,** 1653–1664.

Yost, W. A. (1991). Auditory image perception and analysis: The basis for hearing. *Hear. Res.* **56,** 8–18.

Yost, W. A., and Sheft, S. (1988). Detecting amplitude modulation of sinusoidal carriers. *J. Acoust. Soc. Am.* **83,** S35.

Yost, W. A., and Sheft, S. (1989). Across-critical-band processing of amplitude-modulated tones. *J. Acoust. Soc. Am.* **85,** 848–857.

Yost, W. A., and Sheft, S. (1990). A comparison among three measures of cross-spectral processing of amplitude modulation with tonal signals. *J. Acoust. Soc. Am.* **87,** 897–900.

Yost, W. A., Sheft, S., and Opie, J. (1989). Modulation interference in detection and discrimination of amplitude modulation. *J. Acoust. Soc. Am.* **86,** 2138–2147.

Zeng, F. G., Martino, K. M., Linthicum, F. H., and Soli, S. D. (2000). Auditory perception in vestibular neurectomy subjects. *Hear. Res.* **142,** 102–112.

Zera, J., Onsan, Z. A., Nguyen, Q. T., and Green, D. M. (1993). Auditory profile analysis of harmonic signals. *J. Acoust. Soc. Am.* **93,** 3431–3441.

Zwicker, E. (1965). Temporal effects in simultaneous masking by white-noise bursts. *J. Acoust. Soc. Am.* **37,** 653–663.

Zwicker, E., and Terhardt, E. (1980). Analytical expressions for critical-band rate and critical bandwidth as a function of frequency. *J. Acoust. Soc. Am.* **68,** 1523–1525.

SPEECH AND MUSIC HAVE DIFFERENT REQUIREMENTS FOR SPECTRAL RESOLUTION

Robert V. Shannon

House Ear Institute, Los Angeles, California 90057, USA

The prevailing view of auditory processing for the last 50 years is of a hierarchical system, with "primitive" processing occurring in the periphery and brainstem and more sophisticated processing taking place in the cortex. Many arguments have been made about the similarities in the complex processing needs of speech and music, suggesting common mechanisms for extracting pitch and temporal structure, for example. However, recent research on cochlear implants has shown that speech and music are quite different in their demands on peripheral information. Speech can be understood in the presence of severe degradation of spectral and temporal cues, while music recognition and appreciation are damaged by even mild degradation. This chapter reviews the evidence that speech and music are quite different in terms of their requirements for spectral resolution, suggesting that it is important to understand the demands of the listening task in order to understand the importance of spectral resolution.

I. Introduction

Auditory research for the last 50 years has been dominated by the study of the cochlea. The general feeling has been that the cochlea is the "bottleneck" through which all auditory information must pass, so we could not understand further processing until we fully understood the end organ. And what a beautiful end organ it is—complex hydrodynamics and nonlinearities; the cochlea is a marvel of evolution. But in this chapter it will be argued that this cochlear

INTERNATIONAL REVIEW OF
NEUROBIOLOGY, VOL. 70
DOI: 10.1016/S0074-7742(05)70004-0

fixation has cost us dearly in terms of understanding hearing in all its complexity: some hearing tasks are dominated and limited by the cochlea while others are not. The obsessive focus on the cochlea has resulted in a relatively poor understanding of central and cognitive aspects of hearing. The difference in processing required for speech and music presents an excellent contrast between hearing limited by central and by cochlear processing.

Of course, one of the most elegant features of the cochlea is its ability to process sound energy of different frequencies in different tonotopic locations; that feature is the basic theme of this entire book. Sound energy is separated by the hydromechanical resonance system so that low-frequency energy is primarily represented in the apex and high-frequency energy is primarily represented in the base of the cochlea. The tonotopic organization of the cochlea is reproduced at all levels up the auditory system from the cochlear nucleus complex to the auditory cortex. Considerable detail is known about the mechanics and physiology of this tonotopic representation. Models and theories of hearing have based higher-order hearing on fine features of the tonotopic and temporal representation of sound in the cochlea and auditory nerve. The design of electrodes and signal processing for cochlear implants (CIs) is based on reproducing the tonotopic pattern of information in the cochlea. However, research on CIs has demonstrated that auditory perception is complex and multidimensional. Auditory perception is not completely determined by the properties of the cochlea and auditory nerve, it only starts there.

The basic premise of this chapter is that the study of auditory perception needs to consider the demands of the perceptual task and the requirements of the brain for information from the auditory periphery. Some complex perceptual tasks can be accomplished with coarse and distorted information from the periphery while other tasks require precise detailed information. This distinction is most clearly demonstrated in the difference between speech and music.

II. What Is a Spectral Channel?

The idealized concept of spectral processing is that there is a narrowly tuned representation in the nervous system that responds only to acoustic stimulation in a narrow frequency range. In an extreme form, each neuron would respond to a single acoustic frequency. This extreme case cannot be realized in a biological system, but it is common to think of sound being analyzed into narrow independent bands in the cochlea. However, most representations of spectral selectivity in the cochlea are based on neural or basilar membrane (BM) tuning curves, which are quite sharp at the tip. This representation is misleading, because that sharp tonotopic selectivity exists only near the absolute threshold. Kim and Molnar

(1979) measured responses in large populations of auditory nerve (AN) fibers and showed that at even moderate sound levels a large portion of the cochlea is activated by a single sinusoid. Neurons that have low thresholds and a high spontaneous rate (HSR) of activity are generally saturated at moderate sound levels and may not contribute to rate-based activity patterns at medium to high sound levels. However, high-threshold neurons with low spontaneous rate (LSR) can maintain a sharp spectral tuning and good dynamic range even at moderate to loud sound levels. The zeitgeist of auditory research emphasizes temporal coding to utilize the contributions of rate-saturated neurons; their phase-locked temporal responses are still tonotopically selective even at stimulation levels where their rate responses are saturated.

Recent psychophysical studies (e.g., co-modulation masking release and profile analysis) have shown a surprising ability of listeners to integrate information across a wide spectral region when necessary for a detection task. So while the peripheral auditory system is sharply tuned in the spectral domain near threshold, central perceptual processing appears to be capable of integrating information across a wide frequency region. Clearly, there is more to spectral resolution than the tuning of neurons near the threshold.

In this chapter we will only consider spectral resolution in the simple case—excitation produced by activation in a single region in the cochlea, whether from acoustic stimulation with a sinusoid or electric stimulation of a single electrode.

III. Comparison of Spectral Resolution in Normal Hearing and Cochlear Implants

In a CI the residual AN in deaf listeners is activated by presenting electrical pulses on a series of electrodes positioned inside the scala tympani. The selectivity of activation can be controlled to some extent by the configuration of the active and return electrodes, but it appears that the main factor determining selectivity is the distance between the electrode and the stimulable neurons. When the distance is small, local, and selective activation can be achieved, while a large distance means that the current fields will be broad and diffuse at the location of the stimulatable neurons. The early CIs contained only a single or a small number of electrodes positioned along the scala tympani. Despite the limited control they could achieve in the spectral dimension, patients found the devices useful and a few could understand a limited amount of speech with them.

How does tonotopic selectivity in a CI compare with that of normal acoustic hearing? Chatterjee and Shannon (1998) measured excitation patterns in CI listeners with forward masking. A masker was placed on one electrode and the threshold was measured on all other electrodes for a brief signal presented after the offset of the masker (forward masking). If masker and signal activate independent neural populations we would not expect the masker to cause an increase in the threshold of the signal. But if masker and signal were activating the same neural population then the adaptation produced in that population by the masker would cause an increase in the threshold for the following signal. Figure 1 presents a comparison of the forward-masked excitation patterns from three CI listeners with a similar curve obtained from three normal-hearing (NH) listeners using the same technique (from Shannon, 1976). The curves have been equated in terms of relative masking and millimeters along the cochlea. One CI listener's function was only slightly more broadly tuned than the NH listeners' functions, while one CI listener's function was considerably broader than the NH functions. In this small sample there appeared to be a link between tonotopic selectivity and performance because the CI listener with the sharpest selectivity had the best speech recognition of the three and the one with the widest selectivity had the poorest speech recognition.

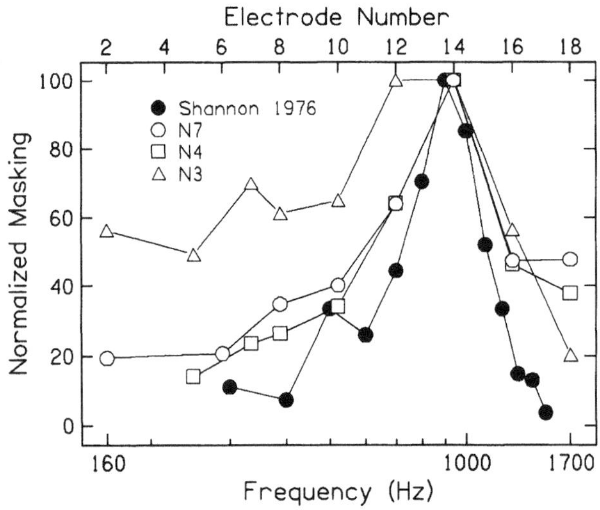

FIG. 1. Forward masking curves from normal hearing (from Shannon, 1976) and cochlear implant listeners. Masking is plotted relative to the peak masking value. The normalization in amplitude was done in dB for the acoustic data and in microamps for the implant data. The abscissa has been equated for the two datasets by converting electrode spacing and frequency to cochlear distance in mm (using the function of Greenwood, 1990) and then aligning the peaks. Reproduced with permission from Chatterjee and Shannon, 1998.

IV. Effects of Spectral Resolution and Distortion on Speech

When CIs were gaining popularity in the 1970s it was widely thought that they would never be able to restore good speech understanding because they could not reproduce the fine structure of the normal cochlea either in the temporal or in the spectral domain. With such a crude signal produced by the implant most people thought it could not produce useful perceptual results. When multichannel implants were introduced some patients were able to recognize speech well enough to converse on the telephone, which challenged the zeitgeist of the day. Shannon *et al.* (1995) demonstrated that neither fine detail in the spectral nor temporal domain were necessary for understanding speech. They processed speech with a noise-band vocoder that was similar to the processing used in CI speech processors. They found nearly 100% correct identification of words in sentences with only four modulated noise bands representing speech. Four, one-octave bands of noise cannot represent formant structures in vowels, but was sufficient for nearly 100% correct recognition of vowels in a 12-vowel set. Despite the arguments about the necessity for good spectral resolution for speech understanding, this study showed remarkable speech understanding with quite coarse spectral resolution. Figure 2 shows a meta-analysis of sentence and music

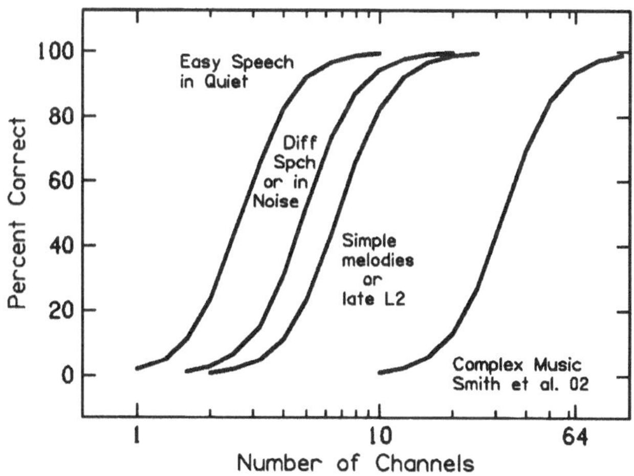

FIG. 2. Speech and music recognition as a function of the number of spectral channels in a noise band vocoder. This meta-analysis averaged data from multiple published studies as described in the text. Note that the curves maintain a similar shape and simply shift to the right (requiring more spectral channels) with increasing task difficulty, increasing noise interference, and decreasing experience with English. Modified from Shannon *et al.*, 2004.

recognition results as a function of the number of spectral channels in a noise-band vocoder. The results are averaged from several studies in the literature, as cited in the following text. Sigmoidal functions were fit to the combined datasets of several studies and are presented in Fig. 2. When listeners were required to identify simple sentences in quiet listening conditions (Dorman *et al.*, 1997b; Fishman *et al.*, 1997; Friesen *et al.*, 2001; Loizou *et al.*, 1999; Padilla, 2003; Shannon *et al.*, 1995; Smith *et al.*, 2002) recognition improved as the number of spectral channels increased, reaching 75% correct identification at about three channels. The curve maintains the same shape and moves to the right if the task is to recognize more difficult sentences (Smith *et al.*, 2002), or recognize simple sentences in noise (Fu *et al.*, 1998; Friesen *et al.*, 2001), or sentence recognition by five-year-old children (Eisenberg *et al.*, 2000), or sentence recognition by people in a second language (Padilla, 2003). In these cases 75% correct was only achieved with six channels of spectral information. When the task is even more difficult the curve shifts further to the right, requiring eight to nine channels for 75% recognition of simple melodies (Burns *et al.*, 2001) or English sentence recognition by non-native listeners who learned English as adults (Padilla, 2003). Finally 40 or more channels of spectral information are required for 75% recognition of simple melodies when presented in the presence of competing melodies (Smith *et al.*, 2002). The overall picture is that auditory pattern recognition requires more spectral resolution depending on the difficulty of the materials, the difficulty of the listening conditions, and the experience of the listener. The similarity of the curves across these different conditions suggests that a single metric, the number of spectral channels required for a criterion level of correct recognition (like the Speech Recognition Threshold), could be used to equate the difficulty of materials, the difficulty of listening in noise, and the inexperience of the listener (Shannon *et al.*, 2004).

How can we recognize speech with such minimal spectral resolution? The answer appears to be that speech recognition is primarily a top-down process of pattern recognition that is highly overlearned from a lifetime of experience. Children are exposed to millions of words in directed interactive conversation in the first few years of life and to tens of millions of words in incidental exposure (Hart and Risley, 1995). Included in this massive experience set are many different talkers and listening situations. The listener is required to extract the linguistic message regardless of whether the talker is male or female, the utterance is spoken fast or slow, heard in quiet or in noise, mumbled, distorted, and so on. There is no acoustic "template" for a given word or phoneme. Despite a decades-long search for acoustic invariance in speech none has been found (Stevens and Blumstein, 1981). There is no set of specific acoustic properties that uniquely specifies a given phoneme. The acoustic representation of a phoneme can vary depending on the talker and listening conditions. So the brain must rely on context-dependent pattern recognition to identify words. This learning process requires many years and many thousands, probably millions, of training "trials."

Consider how much training is required to become robust at recognizing distorted speech. An adult who has a lifetime of training can recognize speech when represented by only three to four bands of modulated noise. A five-year-old child with NH has a modest vocabulary and appears to have no trouble understanding speech in his or her native language. Would a five-year-old child be able to understand speech when degraded to four bands of modulated noise? Eisenberg *et al.* measured speech recognition with noise-band vocoders in five- to seven-year-old children and in 11- to 12-year-olds as well as in adults. The results show that 11- to 12-year-olds are similar to adults, but the five- to seven-year-olds required better spectral resolution for equivalent recognition, i.e., five- to seven-year-olds required six to seven spectral channels to recognize sentences as well as adults using three to four channels (see Fig. 2, second curve). This suggests that five- to seven-years-olds are not fully mature in their pattern recognition ability. Even five to seven years of training with NH is not enough to train the brain to the adult level of robust pattern recognition. Speech recognition is not unique in this regard. Many complex perceptual and motor tasks are not fully mature until ages 12 to 15. An example in visual perception is size constancy, where the child is able to correctly judge the size of a distant object. Younger children tend to judge a distant object as smaller. Recent anatomical evidence (Moore, 2002; Moore and Guan, 2001) shows that the deep layers in human auditory cortex are not fully myelinated until ages 12 to 15, suggesting that experience continues to influence brain development into the mid-teens. It is likely that the long maturation time reflects the long training time necessary for complex perceptual and motor tasks.

Once speech pattern recognition is fully established in the normal-hearing brain, speech recognition can be accomplished with highly distorted and degraded speech signals. Several studies have demonstrated that the spectral information can be degraded by spectral smearing to a surprising degree before speech recognition is adversely effected (Boothroyd *et al.*, 1996; Shannon *et al.*, 1995, 1998; ter Keurs *et al.*, 1992, 1993).

In contrast with spectral smearing, speech recognition is relatively sensitive to spectral shifting and warping. Baskent and Shannon (2003, 2005) have shown that linear distortions of the frequency-place mapping in the cochlear, by compression and expansion, are highly detrimental to speech recognition both in CIs and in simulations of CIs. In their expansion condition the cochlear representation of the most important frequency region for speech was expanded to cover a large portion of the cochlea. Although this manipulation provided better resolution of the relative frequency components within speech, speech recognition was poor with even modest degrees of expansion. This result highlights the fact that speech patterns in normal ears are learned from their original representations in the cochlea; manipulations that should theoretically provide improved resolution are not recognizable because, although the representation is "enhanced," it does not match the pattern recognized by more central processing.

Early attempts to shift the frequency region from an area of impaired hearing into a frequency region of better hearing (frequency transposition) were quite unsuccessful (Braida *et al.*, 1979). Recent studies have shown that the tonotopic representation can be shifted up to one-half octave, but further shifts are detrimental (Dorman *et al.*, 1997a; Fu and Shannon, 1999; Hogan and Turner, 1998; Shannon *et al.*, 1998). One explanation for such sensitivity to a tonotopic shift is implied by data from Fu and Shannon. In their study, recognition of children talkers by adults survived downward shifts of up to one octave, but could not tolerate an upward shift. Recognition of male talkers was just the opposite; it could tolerate a one octave upward shift, but almost no downward shift. This pattern suggests that the existing sensitivity to spectral shift is an indication of the region of tolerance of the brain's pattern recognition system for speech. In an NH brain there is no need to develop a tolerance for extreme spectral shifts because the cochlea is "hard-wired" to the lower levels of the auditory system. However, the brain does have to accommodate to speech spoken by a wide range of talkers, accommodating a range of fundamental frequencies from <100 Hz to >300 Hz and accommodating to formants generated by the resonances of oral cavities that differ in volume over an order of magnitude (children to adult males). To accomplish this feat the pattern recognition mechanisms in the brain, learned during the millions of words heard as children, must be flexible enough to accommodate the natural variation experienced in the world. However, there is no need for the pattern recognition mechanisms to be flexible in terms of extreme tonotopic shift, because this does not happen in the normal acoustic world. The brain can count on information about the energy at 1000 Hz coming from the same group of neurons from the cochlea. Speech pattern recognition mechanisms need to be able to adapt to slightly different distributions of information from different talkers, but it always gets information about 1000 Hz energy from the same peripheral neurons. With a CI we can present frequency information to any electrode in any tonotopic location. These results demonstrate that to get the best speech recognition it is important to understand the limits of the pattern recognition system and program the device to be within those limits.

How is speech information coordinated across frequencies? Are speech patterns dependent on events that are coordinated or synchronized across frequencies? In a normal ear, sharp releases in speech, like a plosive consonant (p, t, k), create a short burst of energy across a wide frequency range. Such a synchronous event might serve as a time synchronization marker so the central processor could coordinate temporal patterns of speech information across different frequency bands. A series of experiments has shown that speech pattern recognition is remarkably insensitive to cross-spectral temporal asynchrony. Arai and Greenberg (1998) and Fu and Galvin (2001) manipulated the temporal delay in each frequency band. To investigate this issue, these studies analyzed speech into a number of spectral bands (ranging from 1 to 16). Within each band a group time delay was added to disrupt the cross-spectral synchrony. Arai and Greenberg added a random group delay and Fu *et al.* added a delay that maximized the temporal disparity between adjacent bands. Both studies found that the overall delay across frequency bands had little effect on speech recognition until the delay reached more than 200 ms, the length of a speech syllable or phoneme. In the Fu *et al.* study they additionally removed all spectral fine structure within each frequency band by using a noise-band vocoder to simulate a CI. Even in this case, temporal asynchrony had little effect on speech recognition up to time delays of 80 ms for the 16-band case.

Saberi and Perrott (1999) presented an even more dramatic demonstration that reversed speech segments in time. When speech is cut into time segments of equal duration and each segment is reversed in time, the result is highly intelligible until the reversed segments are longer than 100 ms. This result suggests that rapid transitions between and within phonemes are relatively unimportant for understanding running speech. Vowels are of relatively long duration and reversal in time does not alter the harmonic or formant structure of these long-duration segments. Similarly sibilant consonants are relatively long and are relatively unaltered by time reversal. This result is consistent with an effective temporal window for speech analysis of 50 to 100 ms.

These results suggest that speech pattern recognition is highly insensitive to the relative phase or group delay across frequencies. This fact probably results from the physics of acoustic signals in complex listening environments. In enclosed spaces (rooms, caves) the phase and time delay of the acoustic signal reaching each of the listener's ears will depend on the relative placement of the talker and listener, as well as the reflective and absorptive properties of the space. If speech recognition were dependent on the relative phase across frequencies it would cause intelligibility to change as we change position in a room. Such cross-spectral temporal information is not important for speech so the pattern recognition system does not appear to use it.

V. Effects of Spectral Resolution on Music

Speech and music are the two most important cultural acoustic signals. Are they similar in their demands on auditory processing? We have just reviewed studies that show that speech can be recognized when represented by only four bands of modulated noise. As the speech and listening conditions become more difficult more spectral resolution is required. How does music depend on spectral resolution? Is it similar to speech? The answer is clearly no. Speech and music have completely different demands on the auditory system. Studies have shown that speech remains intelligible even when the spectral fine structure and temporal fine structure are removed. For speech pattern recognition the goal is to identify the linguistic content, and the acoustic variation and detail are not critical. However, music identification and appreciation depend critically on the spectral and temporal fine structure. To highlight the difference between speech and music, let us examine factors that are known to degrade music that are not important for speech.

The CI research has shown that there are two distinct and independent types of pitch: periodicity pitch and place pitch. In addition, there is probably a third category of harmonic pitch that combines periodicity and place pitch information. Pitch has traditionally been defined as that aspect of sound that defines a melody. It was thought that tonotopic location contributed to the timbre of a sound but that pitch was primarily temporal. However, it is now clear that place information can carry melodic information and so should be considered pitch in the classic sense. Both place pitch and periodicity pitch are available to people with CIs (Carlyon *et al.*, 2002; McDermott and McKay, 1994; McKay and McDermott, 1996; Shannon, 1983; Tong *et al.*, 1982; Townshend *et al.*, 1987). In general, CI listeners can detect changes in pitch related to periodicity at least up to 300 Hz, a result that is consistent with that observed in NH listeners when spectral cues are eliminated (Burns and Viemeister, 1976, 1981). Place pitch is evident in those with CIs as the location of the stimulated electrode is changed. Most CI listeners can detect a difference in pitch between adjacent electrodes, typically spaced <1 mm apart.

However, CIs do not generally support music appreciation. Most CI listeners report that music sounds muddled and distorted. They can detect the rhythmic elements in music and may be able to recognize simple familiar melodies (McDermott and McKay, 1997; Pijl and Schwarz, 1995), but they usually report that even the recognized melodies have poor quality. When multiple instruments are present, or complex unfamiliar melodies, implant listeners generally show an inability to separate instruments or to follow the melody.

Why are CIs so poor at representing music when implant listeners appear to be able to utilize periodicity pitch and place pitch? The difference appears to be in the representation of complex harmonic pitch. Harmonic pitch is determined from spectral regions where harmonics are fully or partially resolved. Harmonic pitch is weak in spectral regions where several successive harmonics fall within an auditory filter. CI listeners have reduced spectral selectivity and so have little if any perception of harmonic pitch (Carlyon *et al.*, 2002).

One experiment with CI listeners and with simulations of implant processing (Burns *et al.*, 2001) shows that listeners can identify simple familiar isochronous melodies with only 16 channels of spectral resolution. This level of spectral resolution is not sufficient to allow harmonic pitch, but does give coarse place pitch. Recognition of simple melodies is possible using only coarse place pitch mechanisms and non-spectral temporal pitch mechanisms. Thus, simple familiar melodies could be thought of as an intermediate between speech and complex music. The simple patterns of familiar notes can be recognized with only limited spectral resolution, using primarily cochlear place information usually called timbre. But even simple familiar melodies require more spectral resolution than speech.

A recent experiment demonstrates that recognition of complex music requires many more spectral channels of information (Smith *et al.*, 2002). Smith *et al.* constructed "chimeras" in which music or speech were deconstructed into a multi-band noise band vocoder. Instead of a noise band as a carrier signal for a given spectral band the fine spectral information from a different signal was used, e.g., the fine structure from speech was modulated by the envelope signal from music for that band. These chimeric signals consisted of N frequency-specific bands of envelope information from one signal (speech or music) modulating the same bands of fine structure information from another signal (noise, speech, or music). Speech-noise chimeras, in which the speech envelope modulated noise bands were understandable with only three to four spectral bands, replicates Shannon *et al.* (1995) (left curve in Fig. 2). Speech-speech chimeras required more spectral bands to overcome the conflicting message between the envelope and fine structure. Music-music chimeras required about 50 bands for recognition of simple melodies. Informal listening tests indicate that complex music recognition and quality are poor even with noise-band vocoders of up to 100 bands. Thus, even high spectral resolution is not adequate to provide good complex pitch and quality for music recognition. This observation suggests that envelope information from 100 bands is not adequate for conveying complex harmonic pitch. Music recognition and quality require not only many spectral channels but also fine temporal information within and across spectral channels to allow harmonic pitch extraction.

A recent study (Oxenham *et al.*, 2004) demonstrated that the temporal information for complex pitch also has to be in the correct tonotopic location in the cochlea. If correct temporal information is shifted to a slightly different frequency region complex pitch is lost. This study suggests that complex pitch models based completely on temporal properties cannot work unless there is a place-specific mechanism as well. Since place cues, even up to 100 channels, are not sufficient, and temporal cues alone are not sufficient, it appears that complex pitch is coded by place-specific temporal information.

VI. Conclusions

The number of spectral channels needed depends on the difficulty of the listening task and situation.

Speech recognition, because it is a highly trained pattern recognition process, requires only four spectral channels of envelope information in the appropriate tonotopic place. Six to eight spectral channels are required for speech recognition in noisy listening conditions or for difficult speech materials or for people who are not native listeners in the language. Even more spectral channels may be required for fine voice quality distinctions. In contrast, music requires at least 16 spectral channels even for identification of simple familiar melodies played with a single stream of notes. Recognition and enjoyment of more complex music and music with multiple instruments require at least 64 channels of spectral resolution and possibly many more. This large difference between music and speech highlights the difference in how the brain utilizes information from the auditory periphery. For tasks in which fine spectral detail is necessary for complex pitch and quality, reduced spectral information, as occurs in hearing loss or in a CI will result in a loss of that function. However, in tasks in which the overall spectral pattern recognition is more important, like speech recognition in quiet or recognition of simple and familiar melodies, a loss of spectral resolution will only produce a mild reduction in the ability. To advance our understanding of processing in the auditory system it is important to understand the relative roles of fine detail from the periphery and top-down pattern processing by the brain for different tasks.

CIs and simulations of CIs present powerful tools for evaluating possible mechanisms of complex pitch coding. It appears that purely place pitch and purely temporal pitch are only a portion of the total picture. Complex harmonic pitch requires a combination of temporal and place cues that must occur in the appropriate tonotopic location. It remains to be seen if prosthetic devices are capable of producing the combination of temporal and place cues necessary.

References

Arai, T., and Greenberg, S. (1998). Speech intelligibility in the presence of cross-channel spectral asynchrony. *Proc. IEEE/ICASSP* 933–936, Seattle, WA.

Baskent, D., and Shannon, R. V. (2003). Speech recognition under conditions of frequency-place compression and expansion. *J. Acoust. Soc. Amer.* **113,** 2064–2076.

Baskent, D. E., and Shannon, R. V. (2005). Interactions between cochlear implant electrode insertion depth and frequency-place mapping. *J. Acoust. Soc. Amer.* **117,** 1405–1416.

Boothroyd, A., Mulhearn, B., Gong, J., and Ostroff, J. (1996). Effects of spectral smearing on phoneme and word recognition. *J. Acoust. Soc. Am.* **100,** 1807–1818.

Braida, L. D., Durlach, N. I., Lippmann, R. P., Hicks, B. L., Rabinowitz, W. M., and Reed, C. M. (1979). Hearing aids—a review of past research on linear amplification, amplitude compression, and frequency lowering. *ASHA Monogr.* **19,** 1–114.

Burns, E. M., and Viemeister, N. F. (1976). Nonspectral pitch. *J. Acoust. Soc. Amer.* **60,** 863–869.

Burns, E. M., and Viemeister, N. F. (1981). Played-again SAM: Further observations on the pitch of amplitude-modulated noise. *J. Acoust. Soc. Amer.* **70,** 1655–1660.

Burns, E. M., Sanborn, E. S., Shannon, R. V., and Fu, Q.-J. (2001). Perception of familiar melodies by cochlear implant users. *2001 Conference on Implantable Auditory Prostheses*, Asilomar, CA, August 19–24 (poster).

Carlyon, R., van Wieringen, A., Long, C. J., Deeks, J. M., and Wouters, J. (2002). Temporal pitch mechanisms in acoustic and electric hearing. *J. Acoust. Soc. Amer.* **112,** 621–633.

Chatterjee, M., and Shannon, R. V. (1998). Forward masking excitation patterns in multi-electrode cochlear implants. *J. Acoust. Soc. Amer.* **103**(5), 2565–2572.

Dorman, M. F., Loizou, P. C., and Rainey, D. (1997a). Simulating the effect of cochlear-implant electrode insertion depth on speech understanding. *J. Acoust. Soc. Amer.* **102,** 2993–2996.

Dorman, M. F., Loizou, P. C., and Rainey, D. (1997b). Speech intelligibility as a function of the number of channels of stimulation for signal processors using sine-wave and noise-band outputs, *J. Acoust. Soc. Amer.* **102,** 2403–2411.

Eisenberg, L., Shannon, R. V., Martinez, A. S., Wygonski, J., and Boothroyd, A. (2000). Speech recognition with reduced spectral cues as a function of age. *J. Acoust. Soc. Amer.* **107**(5), 2704–2710.

Fishman, K., Shannon, R. V., and Slattery, W. H. (1997). Speech recognition as a function of the number of electrodes used in the SPEAK cochlear implant speech processor. *J. Speech Hearing Res.* **40,** 1201–1215.

Friesen, L., Shannon, R. V., Baskent, D., and Wang, X. (2001). Speech recognition in noise as a function of the number of spectral channels: Comparison of acoustic hearing and cochlear implants. *J. Acoust. Soc. Amer.* **110,** 1150–1163.

Fu, Q.-J., and Galvin, J. (2001). Recognition of spectrally asynchronous speech by normal-hearing listeners and Nucleus-22 cochlear implant users. *J. Acoust. Soc. Amer.* **109**(3), 1166–1172.

Fu, Q.-J., and Shannon, R. V. (1999). Recognition of spectrally degraded and frequency-shifted vowels in acoustic and electric hearing. *J. Acoust. Soc. Amer.* **105,** 1889–1900.

Fu, Q-J., Shannon, R. V., and Wang, X. (1998). Effects of noise and number of channels on vowel and consonant recognition: Acoustic and electric hearing. *J. Acoust. Soc. Amer.* **104,** 3586–3596.

Greenwood, D. D. (1990). A cochlear frequency-position function for several species - 29 years later. *J. Acoust. Soc. Amer.* **87,** 2592–2605.

Hart, B., and Risley, T. R. (1995). Meaningful Differences in the Everyday Experience of Young American Children. Brookes Publishing, Baltimore, MD.

Hogan, C. A., and Turner, C. W. (1998). High-frequency audibility: Benefits for hearing-impaired listeners. *J. Acoust. Soc. Amer.* **104,** 432–441.

Kim, D. O., and Molnar, C. E. (1979). A population study of cochlear nerve fibers: Comparison of spatial distributions of average-rate and phase-locking measures of responses to single tones. *J. Neurophysiol.* **42,** 16–30.

Loizou, P. C., Dorman, M., Poroy, O., and Spahr, T. (1999). Speech recognition by normal-hearing and cochlear implant listeners as a function of intensity resolution. *J. Acoust. Soc. Amer.* **108,** 2377–2387.

McDermott, H. J., and McKay, C. M. (1994). Pitch ranking with nonsimultaneous dual-electrode electrical stimulation of the cochlea. *J. Acoust. Soc. Amer.* **96**(1), 155–162.

McDermott, H. J., and McKay, C. M. (1997). Musical pitch perception with electrical stimulation of the cochlea. *J. Acoust. Soc. Amer.* **101**(3), 1622–1631.

McKay, C. M., and McDermott, H. J. (1996). The perception of temporal patterns for electrical stimulation presented at one or two intracochlear sites. *J. Acoust. Soc. Amer.* **100**(2 Pt 1), 1081–1092.

Moore, J. K. (2002). Maturation of human auditory cortex: Implications for speech perception. *Ann Otol. Rhinol. Laryngol.* **189**(Suppl.), 7–10.

Moore, J. K., and Guan, Y. L. (2001). Cytoarchitectural and axonal maturation in human auditory cortex. *J. Assoc. Res. Otolaryngol.* **2**(4), 297–311.

Oxenham, A. J., Bernstein, J. G. W., and Panagos, H. (2004). Correct tonotopic representation is necessary for complex pitch perception. *PNAS* **101,** 1421–1425.

Padilla, M. (2003). English phoneme and word recognition by nonnative English speakers as a function of spectral resolution and English experience, Department of Biomedical Engineering, University of Southern California, Unpublished Doctoral Dissertation.

Pijl, S., and Schwarz, D. W. F. (1995). Melody recognition and musical interval perception by deaf subjects stimulated with electrical pulse trains through single cochlear implant electrodes. *J. Acoust. Soc. Am.* **98,** 886–895.

Saberi, K., and Perrott, D. R. (1999). Cognitive restoration of reversed speech. *Nature* **398,** 760.

Shannon, R. V. (1976). Two-tone unmasking and suppression in a forward masking situation. *J. Acoust. Soc. Amer.* **59,** 1460–1470.

Shannon, R. V. (1983). Multichannel electrical stimulation of the auditory nerve in man: I. Basic psychophysics. *Hear. Res.* **11,** 157–189.

Shannon, R. V., Zeng, F.-G., Kamath, V., Wygonski, J., and Ekelid, M. (1995). Speech recognition with primarily temporal cues. *Science* **270,** 303–304.

Shannon, R. V., Zeng, F.-G., and Wygonski, J. (1998). Speech recognition with altered spectral distribution of envelope cues. *J. Acoust. Soc. Amer.* **104,** 2467–2476.

Shannon, R. V., Fu, Q.-J., and Galvin, J. (2004). The number of spectral channels required for speech recognition depends on the difficulty of the listening situation. *Acta Oto-Laryngologica Suppl.* **552,** 50–54.

Smith, Z. M., Delgutte, B., and Oxenham, A. J. (2002). Chimaeric sounds reveal dichotomies in auditory perception. *Nature* **416,** 87–90.

Stevens, K. N., and Blumstein, S. E. (1981). The search for invariant acoustic correlates of phonetic features. *In* "Perspectives on the Study of Speech" (P. D. Eimas and J. L. Miller, Eds.). Lawrence Erlbaum Assoc., NJ.

ter Keurs, M., Festen, J. M., and Plomp, R. (1992). Effect of spectral envelope smearing on speech reception. I. *J. Acoust. Soc. Amer.* **91,** 2872–2880.

ter Keurs, M., Festen, J. M., and Plomp, R. (1993). Effect of spectral envelope smearing on speech reception. II. *J. Acoust. Soc. Amer.* **93,** 1547–1552.

Tong, Y. C., Clark, G. M., Blamey, P. J., Busby, P. A., and Dowell, R. C. (1982). Psychophysical studies for two multiple-channel cochlear implant patients. *J. Acoust. Soc. Amer.* **71,** 153–160.

Townshend, B., Cotter, N., Van Compernolle, D., and White, R. L. (1987). Pitch perception by cochlear implant subjects. *J. Acoust. Soc. Am.* **82,** 106–115.

NON-LINEARITIES AND THE REPRESENTATION OF AUDITORY SPECTRA

Eric D. Young,* Jane J. Yu,[†] and Lina A. J. Reiss[‡]

*Department of Biomedical Engineering, Johns Hopkins University
Baltimore, Maryland 21205, USA
[†]Department of Biomedical Engineering, Center for Hearing Sciences
Johns Hopkins University, Baltimore, Maryland 21205, USA
[‡]Department of Speech Pathology and Audiology, University of Iowa
Iowa City, Iowa 52242, USA

The analysis of sounds by their frequency content is a fundamental property of auditory neurons. In auditory nerve (AN) fibers, this process can be approximated with a bandpass filter accompanied by some non-linearities. In central auditory neurons, non-linearities make the description of frequency analysis more difficult. This chapter describes the effects of non-linearities on frequency analysis, using neurons in the dorsal cochlear nucleus (DCN) as an example. The physiological characteristics that make the principal cells of the DCN non-linear are described, along with a weight-function model that allows description of non-linearities in the DCN to second-order. The model is based on responses to noise stimuli with a wide variety of pseudorandom spectral shapes and explicitly separates frequency analysis from temporal response characteristics which are

INTERNATIONAL REVIEW OF
NEUROBIOLOGY, VOL. 70
DOI: 10.1016/S0074-7742(05)70005-2

0074-7742/05 $35.00

not considered. The quality of the model is evaluated by its ability to predict the responses of DCN neurons to spectral shapes other than those used to derive the model. A second-order model performs well at this task in ventral cochlear nucleus (VCN) and less well in DCN. Similar to AN fibers, VCN neurons are dominated by a bandpass filtering operation, with the difference of showing more inhibitory response. These properties suggest that DCN neurons behave like feature detectors as opposed to AN fibers and VCN neurons, which are generalized spectrum analyzers. Despite their non-linearity, DCN neurons encode approximately the same amount of information about the stimulus spectrum as do VCN neurons.

I. Introduction: Tuning Curves, Response Maps, and Non-Linearity

The concept of frequency tuning is fundamental to understanding the auditory system at both perceptual and neurophysiological levels. In perceptual studies, the idea of a critical band or an auditory filter underlies most models of perceptual processing (Fletcher, 1938; Moore, 2004). In the auditory filter model, the peripheral auditory system is represented as a parallel bank of linear bandpass filters (Fig. 1A) and most processing is done on the separate outputs of the filters. Although information may be combined across filters, the frequency separation in peripheral filtering is fundamental; often the quantitative properties of perceptual phenomena depend on the properties of the filters. As examples, there are differences in the pitch of a harmonic complex between cases in which the harmonics are or are not separated by the filters (Goldstein, 1973; Schouten *et al.*, 1962) and there are differences in the properties of loudness summation for energy within one filter versus energy spread across filters (Zwicker *et al.*, 1957).

For auditory neurons, frequency selectivity is a fundamental property; measures of frequency tuning are usually the first aspect studied in auditory neurons. Frequency selectivity can be summarized in auditory neurons by tuning curves (Fig. 1B); the tuning curve plots the sound levels needed to produce a criterion increase in the neuron's discharge rate at various frequencies. For AN fibers, the tuning curve has a V-shape centered on the best frequency (BF). Such tuning could be produced by a bandpass filter followed by a suitable neuron model. Although this concept is oversimplified even for AN fibers, the idea that fibers consist of a filter augmented with some non-linearities has still been useful in representing peripheral neurons (Meddis *et al.*, 2001; Zhang *et al.*, 2001). For AN fibers, this model provides approximately correct answers for the responses to speech-like stimuli (Bruce *et al.*, 2003; Holmes *et al.*, 2004) and it has the advantage of corresponding to the psychophysical concept of an auditory filter.

Neurons in the central nervous system (CNS) provide increasingly more complex analysis of frequency, involving both excitatory and inhibitory responses to single tones (Sutter *et al.*, 1999; Young and Brownell, 1976). Figure 1C shows the discharge rate of a so-called type IV neuron, a principal cell in the dorsal cochlear nucleus (DCN), responding to tones of different frequencies, across a range of sound levels. The responses are both excitatory (shown as black) and inhibitory (shown as shaded). Such neurons receive excitatory and inhibitory synaptic inputs from a variety of neurons with different effective peripheral filters (Young and Davis, 2001). However, the relationship of the inhibitory and excitatory regions to peripheral filters is at best indirect. For example, the excitatory area marked by the arrow in Fig. 1C does not correspond to any actual filter in the system. Instead, this excitatory region results from the partial overlap of excitatory and inhibitory inputs to the neuron. Because of considerations like this, it is inappropriate to call response characteristics, like those in Fig. 1C, tuning curves; they do not correspond to a filtering process like the bandpass filter implicit in the AN tuning curves in Fig. 1B. Instead here they are called response maps.

In this chapter, spectral processing of peripheral auditory neurons is discussed in the AN and cochlear nucleus (CN) of the unanesthetized cat. An important aspect of this topic is the non-linearity of auditory neurons. Strictly, linearity and non-linearity are defined with reference to the superposition test for linearity: a system is linear if the response to the sum of two input signals is the sum of the responses to the two signals presented separately. However, that description does not lead to an easy understanding of the problems posed by non-linearities in the auditory system. We will show how non-linearities complicate the process of describing frequency selectivity and describe an approach to developing better models of non-linear auditory frequency selectivity.

II. Examples of the Effects of Non-Linearities on Measures of Spectral Processing

A. Auditory Nerve Tuning Curves

An example of the problems posed by non-linearity for AN fibers is shown in Fig. 2. Suppose that the fiber responds to the velocity of motion of the basilar membrane (BM) such that there is a monotonic and fixed relationship between BM velocity and discharge rate in the fiber. Although this assumption is not strictly correct (Ruggero and Rich, 1987), it simplifies the following argument and allows a point to be made. With this assumption, the tuning curve corresponds to constant BM velocity, because the threshold criterion for tuning curves is a fixed rate change from the neuron. Figure 2A shows schematic plots of BM velocity versus sound level, based on the data of Ruggero *et al.* (1997). The BM

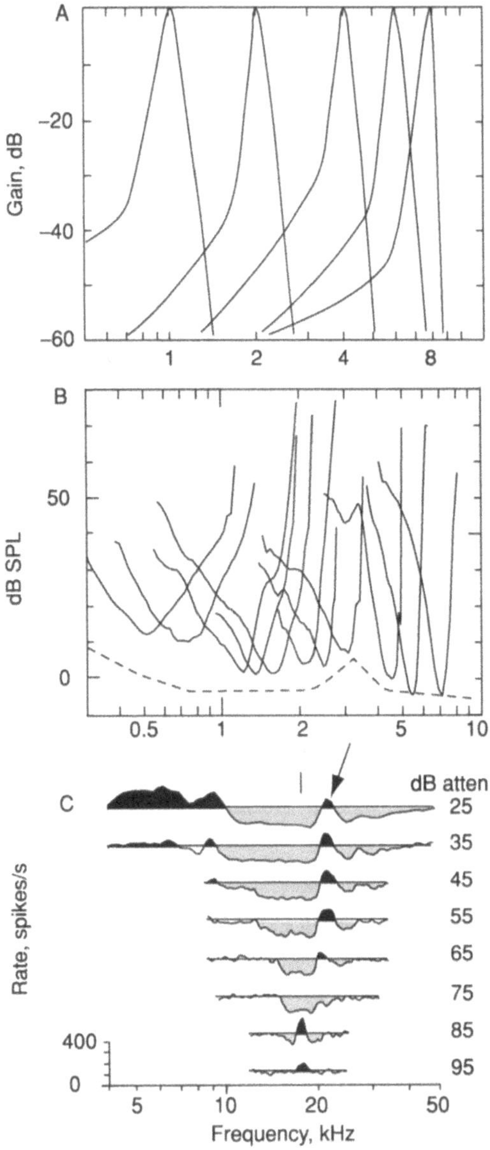

FIG. 1. Examples of perceptual and physiological tuning. (A) Auditory filters derived from masking experiments in human observers. These were derived using a forward masking paradigm for probe tones at 1, 2, 4, 6, and 8 kHz. After measuring the levels of notch-noise maskers required to mask each of the probe tones, filters that would account for the masking data were computed using the roex (*p, w, t, p*) model (Glasberg *et al.*, 1984). (Redrawn from Oxenham and Shera, 2003). (B) Tuning curves from cat auditory nerve fibers. Each plot shows the threshold of one fiber as a function of frequency. The frequency at which threshold is lowest is called the best frequency (BF).

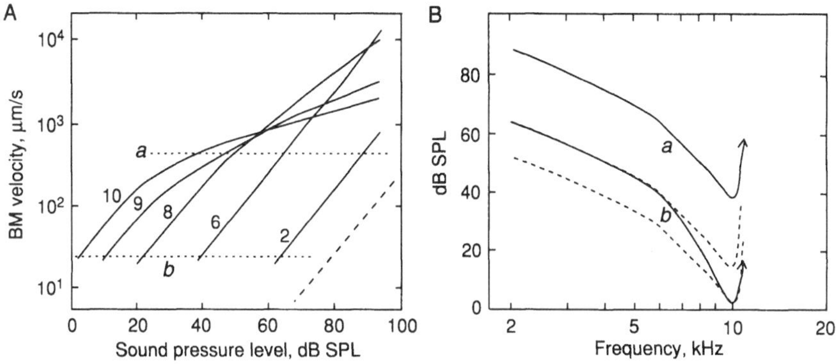

FIG. 2. An illustration of the effects of non-linearity on measures of tuning. (A) Hypothetical BM response-growth functions based on data of Ruggero *et al.* (1997). The plots show BM velocity (ordinate) versus sound pressure level for tones at various frequencies, given by the numbers on the curves. The dashed line shows linear BM response growth, a slope of 1 in these log-log coordinates. The dotted horizontal lines show two constant-velocity criteria (*a* and *b*) that correspond to the threshold responses for two tuning curves, shown in B. (B) The solid lines show tuning curves derived with the two isovelocity criteria in A. The low-frequency side of the tuning curves are derived from the data in A and the high frequency sides are assumed to be steep and simply shown with arrows. Because of the compressive growth of BM velocity, curve *a* is broader than *b*, as shown by the two dashed curves which are repeats of *a* shifted vertically to align the lowest frequency point and the BF point between *a* and *b*.

velocity is linearly related to sound level at frequencies below the BF, shown by the lines with slope 1 for frequencies of 8 kHz and below. Near BF, the slopes are linear at low sound levels, but have a compressive relationship (slope <1, non-linear on these log-log coordinates) at higher levels. Above BF, the slopes are also compressive, but are not shown in Fig. 2A.

Tuning curves can be constructed from the two iso-velocity conditions shown by the dashed lines labeled *a* and *b*. These could correspond to a low-threshold fiber *b* and a high-threshold fiber *a*, for example, or to tuning curves from the same fiber with different response criteria. The resulting tuning curves, shown in Fig. 2B, are constructed by plotting the sound levels at which the BM response functions intersect the iso-velocity contours at various frequencies. Because of the compressive growth of BM response, the isovelocity tuning curve *b* is sharper than *a*, as shown by comparison with the dashed repetitions of *a*. Thus, the

The lowest thresholds in the most sensitive cats are given by the dashed line. (C) Response map for a type IV neuron in the DCN. Each plot shows discharge rate versus the frequency of a test tone; successive plots show rates at various sound levels, given as attenuations at right (0 dB attenuation is approximately 100 dB SPL, but varies with frequency). The horizontal lines are spontaneous rate; excitatory regions are marked in black and inhibitory regions are shaded.

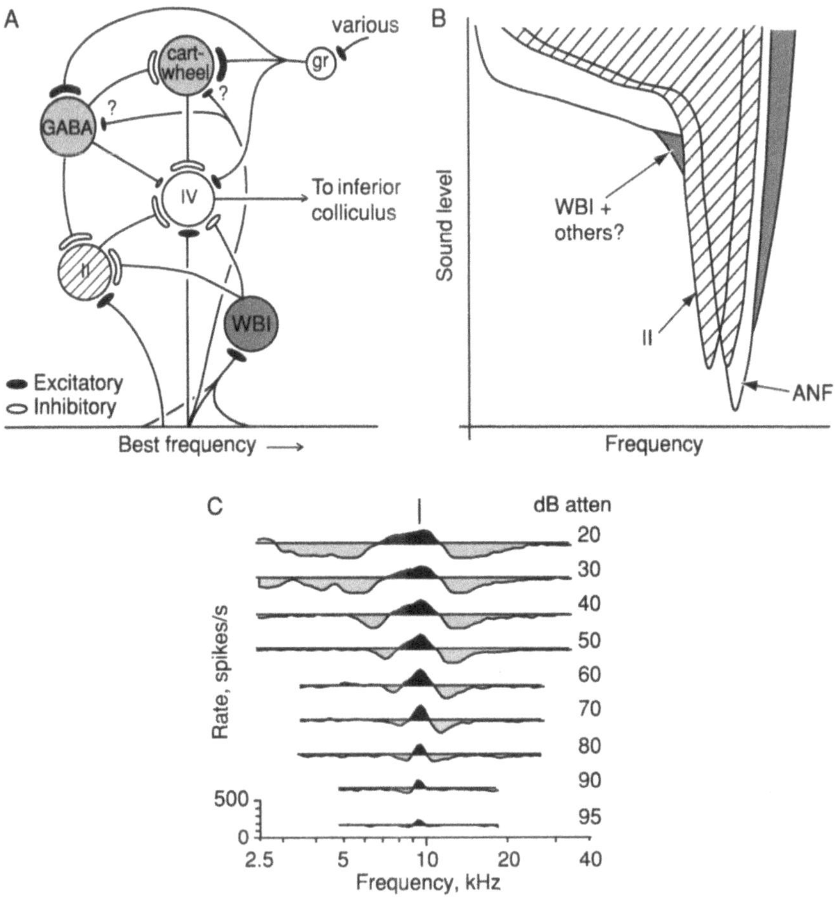

FIG. 3. Construction of the response map of DCN type IV neurons. (A) A schematic of the wiring diagram of the DCN. The principal cells are type IV neurons (IV). Excitatory and inhibitory connections are indicated by the filled and unfilled terminals, respectively, and the size of the terminal is an indication of the strength of the connection. The horizontal line at bottom shows the tonotopic array of AN fibers and vertical lines radiating upward from the horizontal line show AN connections of different BFs to the cells of the DCN. The shaded and hatched neurons are inhibitory. Type II neurons (II) are anatomical vertical cells and the wideband inhibitor (WBI) is anatomically the D-stellate or radiate neuron in the ventral cochlear nucleus. These two interneurons dominate responses to sound, although there are other sources whose anatomical identity is not known (Reiss and Young, 2005). The three cell types at the top of the diagram convey non-auditory information to the principal cells (various), such as somatosensory information about the state of the ears (Kanold and Young, 2001) and also give weak auditory responses (Parham and Kim, 1995). Granule cells (gr) are excitatory input neurons, cartwheel cells are inhibitory neurons, and the GABAergic neuron is an unknown inhibitory source for which there is only pharmacological evidence (Davis and Young, 2000). (B) Schematic of a type IV response map for tones. The plot has the coordinates of a tuning curve and shows the overlapped tuning curves of presumed excitatory and inhibitory inputs to a DCN

frequency selectivity of this system is not well defined by a single tuning curve and depends on sound level. This example is a specific case of the general rule that characterizing a non-linear system provides different results, depending on how it is done. For example, measuring the response rate when holding the input amplitude (sound level) constant (e.g., an iso-intensity rate plot, Geisler *et al.*, 1974) gives different tuning than varying in the input amplitude to hold the output response constant (i.e., a standard tuning curve, Kiang and Moxon, 1974).

The tuning of AN fibers is affected by other non-linearities like cochlear suppression (Delgutte, 1990; Robles and Ruggero, 2001; Sachs and Kiang, 1968). Most simply, a tone outside the tuning curve, which does not produce any change in rate when presented alone, can suppress the responses to tones within the tuning curve. Suppressive interactions also occur among tones within the tuning curve. Presumably suppression occurs on the BM, although this is not fully understood (Robles and Ruggero, 2001). Suppression is a non-linear effect that cannot be captured by a linear filter model. It probably does not have an effect on the tuning curve, which is measured with a single tone, but it certainly does affect responses to broadband stimuli and causes those responses to differ from the predictions of a linear filtering model. For example, the responses of fibers tuned to the second formant in a vowel sound are suppressed by the first formant, leading ultimately, at high sound levels, to a loss of response to the second formant component of the stimulus (Sachs and Young, 1979; Wong *et al.*, 1998), an effect completely at odds with a linear model.

B. DCN Principal Neurons

The response maps of DCN type IV neurons (Fig. 1C) provide an example that illustrates sources of non-linearity in neural circuits. The responses of these cells can be partially understood based on the known wiring diagram of the nucleus (reviewed in Young and Davis, 2001). Figure 3A shows the model as it applies to the unanesthetized cat DCN. The neurons receive excitatory input from AN fibers and inhibitory inputs from several sources (shaded and hatched areas). For tone responses, the most important source is the so-called type II neuron, which corresponds to anatomical vertical cells, a glycinergic inhibitory

type IV neuron. The excitatory input is from an auditory nerve fiber (ANF). There are strong inhibitory inputs from type II neurons (hatched) and weaker inhibitory inputs, shown mostly hidden, from other inhibitory sources which include the WBI. (C) Two-tone response map of a type II neuron, in the same format as Fig. 1C. These neurons have no spontaneous activity; to show inhibitory areas, a weak tone a few dB above threshold at BF was added to produce background activity. The horizontal lines are the response rate to this tone (redrawn from Young and Davis, 2001).

neuron in DCN (Rhode, 1999; Saint Marie *et al.*, 1991; Voigt and Young, 1990). The type II neurons have a strong inhibitory effect on the type IV neurons, so that tones within the type II response area overwhelm the excitatory inputs and inhibit the type IV cell. This provides the basic response map shape illustrated schematically in Fig. 3B, which should be compared with the actual response map in Fig. 1C. Two properties of the type II-type IV connection are relevant. 1) the type II neurons have high thresholds, about 10 to 20 dB above type IV neurons (Young and Brownell, 1976). The threshold difference allows a small excitatory area to appear in type IV neurons near BF threshold. 2) The BFs of the type II inhibitory input are shifted downward in frequency from the BFs of the excitatory input, exposing the excitatory inputs along a vertical strip at the high-frequency side of their tuning curves (Voigt and Young, 1990). This interpretation of the excitatory strip justifies the statement made previously that this narrow excitatory region (arrow in Fig. 1C) should not be regarded as a narrowly tuned excitatory area. Instead, it is merely the difference between two neural responses that have tuning widths typical of auditory neurons.

There are additional inhibitory inputs, including the so-called wideband inhibitor (WBI) which responds strongly to broadband noise stimuli (Nelken and Young, 1994) as well as others that are from uncertain sources (Davis and Young, 2000; Reiss and Young, 2005). These are shown as the shaded inhibitory tuning curve hidden behind the AN fiber and type II curves in Fig. 3B. The qualitative model in Fig. 3 has been made quantitative (Blum and Reed, 1998; Hancock and Voigt, 1999; Reed and Blum, 1995) and has been shown to account for many of the properties of DCN type IV neurons.

The response maps of the type II inhibitory inputs are shown in Fig. 3B as simple excitatory areas. However, these neurons also receive substantial inhibitory inputs, shown in the circuit diagram in Fig. 3A and in the response map in Fig. 3C. Their excitatory input is probably from AN fibers; the inhibitory input seems to be from D-multipolar neurons of the VCN, called the WBI in Fig. 3A (Smith and Rhode, 1989; Winter and Palmer, 1995). The WBI has been shown to respond strongly to broadband noise but weakly to tones (Jiang *et al.*, 1996; Palmer *et al.*, 1995). Anatomically these neurons have dendrites that distribute across the isofrequency sheets in the CN (Doucet and Ryugo, 1997; Oertel *et al.*, 1990), thus bringing them into contact with AN fibers over a wide range of BFs. Presumably, they require activity across the BF range in order to respond. The inhibitory input from the WBI to type II neurons causes the latter to give weak or no responses to broadband stimuli, despite their strong responses to tones. The inhibitory inputs show up in type II response maps (Fig. 3C) as strong inhibitory areas, especially at frequencies above BF.

The non-linearity of DCN neurons can be qualitatively understood as resulting from the lack of spontaneous activity in type II neurons and most WBI neurons. Thus, when tones are presented, the responses of a type IV neuron are predominantly inhibitory (Fig. 1C) because of the strong inhibition from the type II inputs and the WBI contributes little. For broadband stimuli, however, type II neurons are themselves inhibited and effectively disappear from the circuit, because of their lack of spontaneous activity, and the major inhibition is from the WBI. Type IV neurons usually provide net excitatory responses to broadband noise; apparently the direct inhibitory inputs to type IV neurons from the WBI and other noise-responsive neurons are weak (Nelken and Young, 1994; Spirou and Young, 1991). Because the inhibitory circuits are different for narrow and broadband stimuli, the properties of spectral integration are also different: predominantly inhibitory for tones, weakly excitatory for noise. This arrangement of inputs leads to inconsistencies between responses to tones and noise, so that energy that is inhibitory when presented as a tone is excitatory when presented as part of wideband noise.

III. Why Is Non-Linearity a Problem?

The non-linearities discussed in the previous text cause difficulties in defining the tuning or frequency selectivity of an auditory neuron. In the simplest case of AN fibers, there is an ambiguity about the sharpness of tuning caused by the level-dependence of tuning. In the more complex case of the DCN, the whole concept of frequency selectivity is difficult, in that the neuron's response is a non-linear combination of individual filters, the structure of which depends on the stimulus, with different properties for narrow and broadband stimuli.

Most importantly, the non-linearities make it difficult to develop a general model that predicts responses to arbitrary acoustic stimuli (Eggermont *et al.*, 1983; Machens *et al.*, 2004; Sen *et al.*, 2001; Versnel and Shamma, 1998; Yamada and Lewis, 1999; Yeshurun *et al.*, 1989; Yu and Young, 2000). In the case of the DCN for example, responses to tones cannot be used to predict responses to broadband stimuli and vice-versa (Nelken *et al.*, 1997; Spirou and Young, 1991). Because of the lack of predictive power, models of neurons are incomplete and inadequate and may be misleading. At minimum, it is not clear what is the benefit of detailed study of models that do not predict a significant fraction of the response properties of the neuron to relevant stimuli. In the cochlear nucleus, it has been possible to develop circuit models that overcome this problem. However, in higher nuclei, where the details of circuits are not available, it is necessary to use signal processing models, in which the responses to some set of stimuli are used to construct a model that one hopes will adequately characterize the neuron (Eggermont, 1993; Escabi and Read, 2003; Klein *et al.*, 2000). At minimum, such a model should predict responses to a significant class of stimuli other than those used to derive the model.

IV. Defining Tuning Through Non-Linear System Theory

A. Volterra-Wiener Theory

A general approach to non-linear systems is provided by the Volterra and Wiener theories (Johnson, 1980; Marmarelis and Marmarelis, 1978; Rugh, 1981). It is not possible to develop those theories in this chapter; however Fig. 4

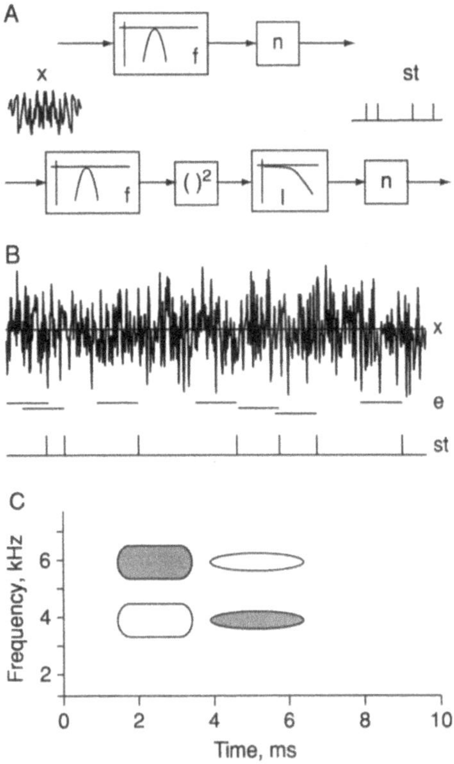

FIG. 4. (A) Example of a first- (top) and a second-order (bottom) system. The input acoustic signal (x) is at the left. Both systems contain a bandpass filter (*f*) for their tuning. The second-order system contains a squaring module $(( ))^2$ and a low-pass filter (*l*). The final element is a neuron model (*n*) that converts the analog signal into a spike train (*st*). (B) Reverse correlation means averaging some function of the stimulus preceding spikes. The stimulus is shown at top (*x*) and the spike train (*st*) at bottom. The epochs (*e*) preceding spikes during which the stimulus is analyzed are shown by the horizontal lines in the middle. (C) A schematic example of a STRF containing two excitatory areas (unfilled) and two inhibitory areas (shaded). The abscissa is time preceding spikes and corresponds to the horizontal pre-spike epochs in part B, except reversed in time. The ordinate is stimulus frequency.

illustrates some approaches to auditory neurons based on the Volterra-Wiener theory. In the simplest case, a neuron can be approximated by the first-order system as shown at the top of Fig. 4A. The system consists of a linear bandpass filter *f* followed by a neuron model *n*. The neuron model can be a threshold device in which an action potential is produced whenever the filtered signal crosses a threshold (de Boer and Kuyper, 1968) or it can be a time-varying Poisson process that produces a spike with a probability that depends on the filtered signal (Colburn, 1973). It can also have more complex properties including refractoriness. The filter in this model provides an unequivocal definition of the tuning of the neuron, to the extent that the model is an accurate (meaning predictive) one that can be estimated from data.

The bottom part of Fig. 4A shows a second-order model in which the filter is followed by a squaring device so that the input to the neuron model is the square of the filtered signal. For AN fibers, a low-pass filter *l* is included to prevent phase-locking (van Dijk and Wit, 1994), but such a filter is not a necessary part of the model. The second-order model can have a tuning curve similar to the first-order model, given the same input bandpass filter, but its other properties are different. The first-order model will phase-lock to the stimulus, for example, but the second-order model will not. Again, the filter provides an unequivocal definition of the frequency selectivity of a neuron to which the second-order model can be fit.

The second-order model in Fig. 4B is the simplest of a range of possible second-order models. In general, a second-order model will have several such filter/squarer combinations in parallel (Lewis *et al.*, 2002, 2003; Qiu *et al.*, 2003; Rugh, 1981). In addition, it is usually necessary to add multiple orders of models (i.e., first-order, second-order, third-order, and so on). Models of this kind have mainly been estimated up to second-order for AN fibers (Recio-Spinoso *et al.*, 2005; van Dijk and Wit, 1994; Yamada and Lewis, 1999); the first-order model works for low BF fibers and the second-order model works for higher BF fibers that do not phase-lock.

B. Reverse Correlation

The filters in Volterra-Wiener models can be recovered from data by reverse correlation, as illustrated in Fig. 4B. The method is based on the theory of Wiener kernels (Johnson, 1980; Marmarelis and Marmarelis, 1978; Rugh, 1981), in which the system is approximated as a sum of non-linear systems of increasing order, zero plus first plus second, and so on. The nth order component of the system is constructed from an nth-order cross-correlation of the input and output of the system, for a white noise input. For the auditory system, the input stimulus is broadband Gaussian noise and the output is a spike train. Given a spike-train output, the cross-correlations become averages of some property of the stimulus preceding the spikes. For the first-order model, the noise waveform preceding spikes is averaged, i.e., the noise waveforms during the epochs (*e*) marked by horizontal lines in Fig. 4B. The resulting average is the impulse response of the filter in the model. de Boer and Kuyper (1968) were the first to apply this technique and showed that the filter is recovered more or less exactly depending on the nature of the neuron model. de Boer and de Jongh (1978) showed that this model predicts many aspects of responses to noise of low-BF AN fibers; in particular, the filter is consistent with the fiber's tuning curve.

For the second-order model some version of the autocorrelation of the noise preceding spikes is averaged. A variety of methods have been used. If the autocorrelation is averaged, then the result is the so-called second Wiener kernel, from which the filter in the second-order model in Fig. 4A can be computed (again assuming that the model is accurate).

C. The Spectrotemporal Receptive Field (STRF)

For neurons in the CNS, a different technique is usually used (Aertsen and Johannesma, 1981; Eggermont, 1993; Klein *et al.*, 2000; Theunissen *et al.*, 2000; van Gisbergen *et al.*, 1975). In this case, a frequency-time distribution computed from the stimulus epochs is averaged. Intuitively, these methods compute the average power spectrum of the stimulus as a function of time-preceding spikes. The result is called a spectrotemporal receptive field or STRF, as illustrated schematically in Fig. 4C. The STRF has a time axis, which corresponds to time-preceding spikes, and a frequency axis. Positive areas in the STRF (shown as unshaded) are excitatory regions and negative areas (shown as shaded) are inhibitory regions. The idea is that the STRF is the spectral and temporal pattern of stimulus energy that is a particularly strong stimulus for the neuron, i.e., the neuron is more likely to produce a spike when the particular combination of features in the STRF occurs. For the example in Fig. 4C, there are four regions. The two regions at the shortest latency along the time axis (2 to 3 ms) show that the neuron is excited by energy at 4 kHz and inhibited by 6 kHz. The two regions with an opposite sign at longer latencies are usually interpreted as effects of refractoriness, in the sense that the neuron responds most strongly if its best stimulus (energy at 4 kHz, none at 6 kHz) is preceded by its inverse (energy at 6 kHz, none at 4 kHz), which guarantees that no spike occurred in the recent past to make the neuron refractory.

The STRF can be derived from the second Wiener kernel (Aertsen and Johannesma, 1981; Klein *et al.*, 2000; Lewis and van Dijk, 2004), but the opposite calculation requires additional assumptions, because the STRF does not contain phase information. Nevertheless, the STRF is a second-order characterization of neurons.

As discussed in the previous text, the measure of the quality of a neural model is its ability to predict responses to other stimuli. STRFs have generally not done well in this regard, even in tests in which the stimuli being predicted are similar to those used to derive the STRF (Eggermont *et al.*, 1983; Machens *et al.*, 2004; Sen *et al.*, 2001; Versnel and Shamma, 1998). The poor performance shows up in a second way, in that the STRFs derived from different stimulus sets (e.g., broadband noise and natural sounds, Escabi and Schreiner, 2002; Nelken *et al.*, 1997; Theunissen *et al.*, 2000) are often quite different. This is analogous to the difference in the responses of DCN neurons to tones and broadband stimuli as previously discussed. Several reasons have been put forth as a cause for the poor performance of STRFs in response prediction. Most obvious is that the neurons may have significant components of their response which cannot be modeled by a second-order non-linearity. Evidence for this idea is given in later sections of this chapter. Other possibilities that have been studied include static non-linearities (such as saturation) in the neuron model, jitter in spike production, and time-varying properties of the spike production process (such as adaptation).

D. The Optimal Stimulus

An important concept that has emerged from the study of STRFs is the idea that the STRF represents the optimal stimulus for the neuron (deCharms *et al.*, 1998). For example, the schematic STRF in Fig. 4C suggests that this neuron might respond strongly to a 1-ms tone pip at 4 kHz preceded by a 1 ms tone at 6 kHz. This stimulus is optimal if it produces the highest probability of spiking for all possible stimuli of the same total energy. Although some evidence to support this idea has been offered, a systematic study of the problem has not been done. One difficulty is determining the exact definition of the optimal stimulus. The optimum can be the stimulus producing the maximum discharge rate, but it could also be the stimulus to which the neuron is most sensitive, in the sense that rate changes are largest for stimulus changes away from the optimal, analogous to the difference between the maximum of a function and its maximum slope. The optimum could be the stimulus about which the neuron proves the most information in some sense. Regardless of the definition, an obvious question is whether the optimal stimulus can be derived from a response measure that does not do well in predicting responses to other stimuli.

A property of STRFs that is important in the next section is separability, that the STRF can be expressed as the product of a function of time and a function of frequency $STRF(t, f) = g_{time}(t) \cdot g_{freq}(f)$. Generally neurons in the auditory system at the levels of the inferior colliculus and below have separable STRFs; this question has been studied quantitatively (Lewis and van Dijk, 2004;

Qiu *et al.*, 2003), but separability is also evident from the inspection of STRFs published for various brainstem auditory nuclei (Kim and Young, 1994; Nelken *et al.*, 1997). The schematic STRF in Fig. 4C is separable and can be created by multiplying time and frequency functions that each consist of a positive peak followed by a negative peak.

V. The Weight Function Model

The previous section suggests that it should be profitable to explore the properties of auditory neurons at orders of non-linearity above 2, i.e., more non-linear than STRFs. Because most auditory neurons in the CNS do not phase-lock to the stimulus waveform, except at very low frequencies, this means estimating only even-order non-linearities (four, six, and so on). However, a large number of parameters must be estimated to construct Wiener kernels of those orders, a process that does not seem feasible. An alternative approach is to take advantage of the separability of STRFs by assuming that such separability will apply to system functions of all orders. That is, the time and frequency sensitivity of neurons can be studied separately, which reduces the parameter estimation problem. The remainder of this chapter describes the properties of the spectral part of the spectro-temporal response characteristics of auditory neurons, using DCN as an example.

The spectral response of neurons can be considered as a relationship between the power spectrum of the stimulus $S(f)$, a function of frequency f, and the average discharge rate of the neuron r. The stimuli are assumed to be steady (unmodulated) noise with various spectral shapes, as described in the following text. They are presented for periods of 100 to 400 ms and the average discharge rate of the neuron is computed over this time period. Such a model is only reasonable for neurons that provide steady spike trains in this stimulus situation, e.g., excluding onset units. For neurons in the brainstem auditory system, this is a reasonable assumption for the majority of neurons at levels up to the inferior colliculus, especially in unanesthetized animals.

With these assumptions, the discharge rate of the neuron is modeled as:

$$r = R_0 + \sum_{i=n_1}^{n_2} w_i S(f_i) + \sum_{i=n_1}^{n_2} \sum_{j=i}^{n_2} w_{ij} S(f_i) S(f_j), \tag{1}$$

where r is the neuron's response rate, $S(f_i)$ is the stimulus level at frequency f_i in dB relative to a reference sound level, R_0 is the rate in response to an all-0-dB stimulus (i.e., with $S(f)$ equal to the reference level for all frequencies), and w_i and w_{ij} are sets of weights that are adjusted to fit the model to data from a neuron. The w_i are first-order weights; they describe the response of the neuron to energy at a particular frequency, and are positive for excitatory responses and negative for inhibitory responses. The w_{ij} are second-order weights; they describe the interaction of energies at two different frequencies or, for terms like $w_{jj}S(f_j)^2$, the quadratic dependence of rate on energy at the frequency f_j. The weights are significant over a range of frequencies with indexes from n_1 to n_2.

Note that the terminology has changed here, in that the first-order weights correspond to the $g_{freq}(f)$ part of the STRF (see the following text) and, therefore, are second-order in the Volterra-Wiener sense. Similarly, the second-order weights correspond to aspects of the fourth-order Wiener representation.

The choice of a logarithmic measure of stimulus power, $S(f)$ in dB, was made because data on the responses of AN fibers to complex stimuli of the kind assumed here show an approximately linear relationship between the average discharge rate of the fibers and the dB level of the stimulus at BF (Conley and Keilson, 1995; Escabi *et al.*, 2003; May and Huang, 1997; May *et al.*, 1996). The relationship holds for positive and negative values of S, i.e., for S above and below the average dB value of the stimulus across frequency, within the dynamic range of the fiber.

The first-order weights w_j are related to the frequency part of the separable STRF of the neuron, i.e., $g_{freq}(f)$ above. To see this, write the STRF model in the form of a convolution (Theunissen *et al.*, 2001) as:

$$r(t) = R_0 + \sum_{k=n_1}^{n_2} \sum_{j=0}^{t_{max}} STRF(j,k) \cdot x(t-j,k). \tag{2}$$

The j summation is over the time axis, which takes the form of a convolution, and the k variable is over the frequency axis. Implicit in Eq. 2 is a binning of the time and frequency axes into suitably small bins. The frequency bins are the same ones used in Eq. 1. $x(t, f)$ is a suitable spectro-temporal characterization of the stimulus, i.e., x is the energy in each time-frequency bin; various choices of x have been used, but the frequency-specific envelope of the stimulus is most common. $r(t)$ is the time-varying discharge rate in the time bins and R_0 is the mean rate. The STRF is assumed to be non-zero over a range of frequencies indexed by n_1 to n_2 and over a range of time delays from 0 to t_{max}.

Now compute the average discharge rate over a stimulus duration T, assumed to be large compared to t_{max}. In doing this, STRF in Eq. 2 is written in its separable form.

$$r_{avg} = \frac{1}{T}\sum_{t=0}^{T} r(t) = R_0 + \frac{1}{T}\sum_{t=0}^{T}\sum_{k=n_1}^{n_2} g_{freq}(k)\sum_{j=0}^{t_{max}} g_{time}(j) \cdot x(t-j,k) \tag{3}$$

$$= R_0 + \sum_{k=n_1}^{n_2} g_{freq}(k) \cdot \sum_{j=0}^{t_{max}} g_{time}(j)\frac{1}{T}\sum_{t=1}^{T} x(t-j,k)$$

$$= R_0 + \sum_{k=n_1}^{n_2} g_{freq}(k) \cdot X(k)\sum_{j=0}^{t_{max}} g_{time}(j) \tag{4}$$

In Eq. 4, the stimulus appears as $X(k)$, which is the average amplitude of $x(t, k)$ integrated over 0 to T, and the temporal part of the STRF appears as a summation of g_{time} over its duration, which is a constant. Thus, the average rate from the STRF takes the same form as the first-order weight summation from the weight model. There is a possible difference in that $X(k)$ and $S(f_k)$ are not defined the same way. S is a logarithmic measure of stimulus energy and X is usually defined in linear units. However, Escabi *et al.* (2003) have argued that STRFs should be defined with reference to a logarithmic measure of stimulus power, so this difference is not fatal. With these assumptions, the first-order weights equal the frequency-dependent component of the STRF, $g_{freq}(f)$ multiplied by the integrated value of the time-dependent component.

VI. Estimating Weights from Random Spectral Shapes (RSS)

The stimuli used to estimate the weights in Eq. 1 are shown in Fig. 5. These RSS stimuli are designed to correspond to the definition of stimulus spectrum $S(f_i)$ given in the previous text. The stimuli consist of tones logarithmically spaced at 1/64 octave over a frequency range varying from 2 to 8 octaves, depending on the application. Because the tones are logarithmically spaced, the stimuli are not periodic in time and they have minimal envelope. Logarithmic spacing also corresponds roughly to the layout of frequencies along the BM, so the RSS stimuli provide a constant energy density along the tonotopic axis at the inputs to neural circuits.

Tones are gathered into bins of (usually) 8 tones, corresponding to 1/8 octave per bin. The tones in each bin have the same amplitude, and the energy in each bin is the sum of the energies in the tones making up the bin (about 9 dB above the level of individual tones for 1/8 octave bins). As described in connection with Eq. 1, the level in a bin $S(f)$ is measured logarithmically, as dB relative to a reference level. The reference level is changed to set the overall sound level of the whole stimulus. The dB value in each bin is drawn from a truncated approximate Gaussian distribution with a standard deviation of 12 dB. The line plot at right in Fig. 5 shows the tones making up three bins.

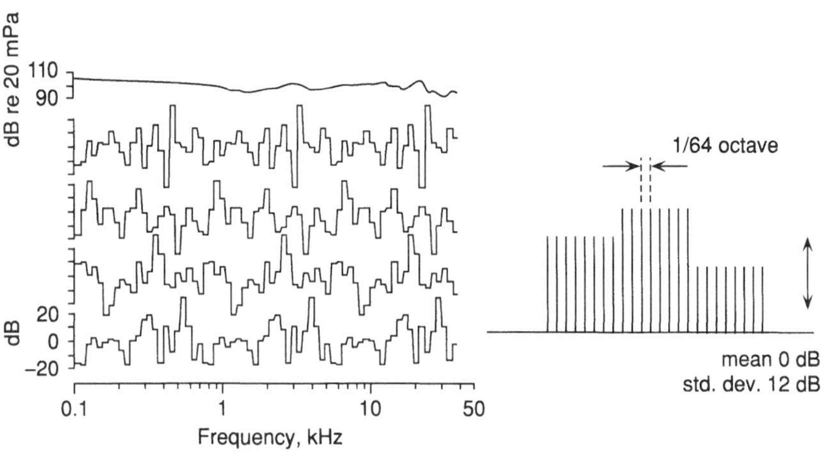

FIG. 5. The spectra of the RSS stimuli used to estimate weights. At the left are the spectra of four examples of RSS stimuli plotted as the dB value in each bin relative to the reference level. The stimuli are periodic along the frequency axis; this was done in some cases for technical reasons having to do with the stimulus synthesis, but one period was always wider than the frequency-integrating range of the neurons (n_1 to n_2 in Eq. 1). A typical acoustic calibration for the stimulus source is shown above the stimuli. The stimuli are not corrected for the calibration. Instead the calibration is added to the spectral shapes when weights are computed. The inset plot at right shows the tones making up three bins in an RSS stimulus; the abscissa of this plot is log frequency and the ordinate is dB.

A large number of stimuli (100 to 400 depending on the dataset) are generated with different spectral shapes, i.e., different amplitudes in their bins. Examples of the spectra of four such stimuli are shown at left in Fig. 5. The RSS stimuli are synthesized to have duration 0.4 s; they are presented to the neuron under study and the neuron's average discharge rates are measured. This gives 100 to 400 equations of the form of Eq. 1 with R_0 and the weights w_i and w_{ij} as the unknowns. The equations are linear in R_0 and the weights and robust solutions are obtained for up to approximately one-half as many weights as there are equations. Bootstrapping is done to estimate the standard deviations of the weights.

VII. Weight Functions for a DCN Neuron: Interpretation of Second-Order Weights

The first- and second-order weights for an example neuron, a type IV neuron from the DCN, are shown in Fig. 6. The first-order weights in Fig. 6A are negative at and below BF and positive above BF. This behavior is typical of one group of type IV neurons in the DCN (Yu, 2003). Note the qualitative similarities between these weights and the type IV response map shown in Fig. 1C, especially the large inhibitory area at and below BF. However, other features of type IV response maps, like the inhibitory area above BF, do not appear in the weight function.

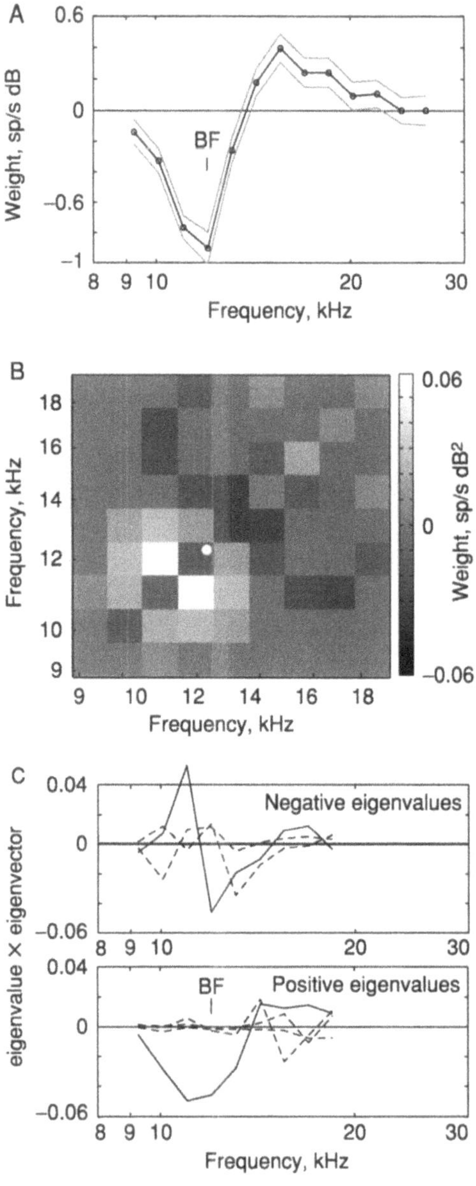

FIG. 6. Weights for a type IV neuron in the DCN. The weights were computed as the best fit of Eq. 1 to the rates produced by 400 RSS stimuli; the reference sound level was 20 dB SPL per bin. (A) First-order weights plotted versus the frequency of the center of the bin, i.e., w_j versus f_j. Shaded lines show 1 SD confidence limits based on repeated (bootstrap) calculation of the weights. The BF of the neuron, 12.3 kHz, is marked by the short vertical line. (B) Second-order weights from the same fit, plotted versus two frequencies, i.e., w_{jk} versus f_j and f_k. The weight values are shown using a gray scale,

The second-order weights of the same neuron are shown in Fig. 6B. The weight values are shown in grayscale as a function of two frequencies. The BF of the neuron is given by the white dot. Weights along the diagonal are for quadratic terms, where the two frequencies are equal, i.e., $w_{jj}S(f_j)^2$. These weights are consistently negative in this group of type IV neurons, for frequencies near BF. The negative diagonal weights provide a significant inhibitory effect like the one in the first-order weights. Most of the off-diagonal weights are positive in this example, which is also typical. Note that the off-diagonal weights are symmetric around the diagonal, because the interactions between frequencies are symmetric; that is, $w_{ij}S(f_i)S(f_j)$ is the same as $w_{ji}S(f_j)S(f_i)$ and these terms are entered only once in Eq. 1.

The second-order weight matrix can be interpreted by changing the way in which the weights are specified. Eq. 1 can be written in matrix notation as:

$$r = R_0 + \vec{S}^T \vec{w} + \vec{S}^T \mathbf{M} \vec{S}, \tag{5}$$

where r and R_0 are as before, $\vec{S}$ is a column vector containing the stimulus values $S(f_i)$, $\vec{w}$ is a column vector of first-order weights, and $\mathbf{M}$ is a matrix of second-order weights. Because $\mathbf{M}$ is an $N \times N$ symmetric matrix ($N = n_2 - n_1 + 1$), it has real eigenvalues and orthogonal eigenvectors; thus, $\mathbf{M}$ can be written in an eigenvalue expansion as:

$$\mathbf{M} = \sum_{j=1}^{N} \lambda_j \vec{e_j} \vec{e_j}^T, \tag{6}$$

where λ_j are the eigenvalues and $\vec{e_j}$ are the eigenvectors of $\mathbf{M}$. The second-order term in Eqs. 1 and 5 then becomes

$$\vec{S}^T \mathbf{M} \vec{S} = \sum_{j=1}^{N} \lambda_j (\vec{S}^T \vec{e_j})^2. \tag{7}$$

Equation 7 shows that the second-order weights behave as a sum of quadratic terms. The vector $\vec{S}$ describing the stimulus is weighted by the eigenvectors of $\mathbf{M}$, in an inner product similar to the first-order weighting in Eq. 5 and the

---

as in the calibration bar at right. The white dot marks the BF. Weights <1 SD from 0, based on bootstrap, are set to 0. The weights are symmetric around the diagonal and are plotted both above and below the diagonal, even though only weights above the diagonal are included in Eq. 1. (C) Eigenvalue decomposition of the second-order weight matrix. The plots show the eigenvectors of the weight matrix plotted versus frequency, on the same scale as the first-order weights in A. Eigenvectors are multiplied by the corresponding eigenvalue to show their relative importance in Eq. 7, and eigenvectors corresponding to positive and negative eigenvalues are plotted separately, as labeled. The vectors corresponding to the largest positive and negative eigenvalues are plotted with solid lines, vectors with smaller eigenvalues are plotted with dashed lines. Three eigenvectors corresponding to very small eigenvalues are not plotted.

result is squared and multiplied by the corresponding eigenvalue. Because the square of $\vec{S}^T \vec{e_j}$ is always positive, excitatory effects occur for positive eigenvalues (i.e., $\lambda_j > 0$) and inhibitory effects for negative eigenvalues. The eigenvectors define spectral shapes of the stimuli that have either positive or negative second-order effects. Suppose, for example, that a stimulus happens to be parallel to a particular eigenvector, i.e. that $\vec{S} = (\text{const})\,\vec{e_k}$. Because the eigenvectors are orthogonal, the only non-zero term in Eq. 7 will be the kth one and the effect will be excitatory or inhibitory depending on the sign of λ_k. In general, the stimuli are not aligned with a particular eigenvector, so a mixture of excitatory and inhibitory effects occur.

The largest eigenvectors of the second-order matrix of Fig. 6B are plotted in Fig. 6C as eigenvalue times eigenvector, with separate plots for negative eigenvalues (shown at the top) and positive eigenvalues (shown at the bottom). The eigenvector corresponding to the largest eigenvalue in each set is plotted with a solid line and the next few are plotted as dashed lines. The largest positive eigenvalue corresponds to an eigenvector (Fig. 6C, bottom plot) that is similar in shape to the first-order weight vector in Fig. 6A. Thus, stimuli that align with the first-order weights (have a spectral shape similar to the first-order weights in Fig. 6A) will have an additional excitatory response from the second-order weights. This aspect of the second-order weights is just a quadratic term in the response with the same frequency selectivity as the first-order weights. The largest negative eigenvalue corresponds to an eigenvector that weights the stimulus spectrum somewhat like a derivative of the frequency selectivity of the first-order weights (Fig. 6C, top plot).

VIII. Prediction of Responses to Arbitrary Stimuli as a Test of the Weight Model

The way the first- and second-order weights interact in the rate responses of DCN neurons can be seen by considering some examples of prediction of responses using the weight model. Figure 7 shows an example, for responses of the DCN neuron in Fig. 6 to spectral edges. This neuron is typical of one group of DCN type IV neurons that show a strong rate response to a rising spectral edge centered at or above BF (Reiss and Young, 2005). The gray lines in Fig. 7 show the discharge rate of this DCN neuron to a 1/2 octave-wide noise notch (Fig. 7A) or a 1/2 octave-wide noise band (Fig. 7B) centered at various frequencies. There is a prominent peak of discharge rate with the notch centered at about 10.5 kHz (Fig. 7A) and with the band centered at about 16 kHz (Fig. 7B). The spectra of the stimuli at the these rate peaks are plotted schematically at the top of the figure. In both cases, the peak in rate occurs with the rising spectral edge in the stimulus just above BF (marked by the double arrow).

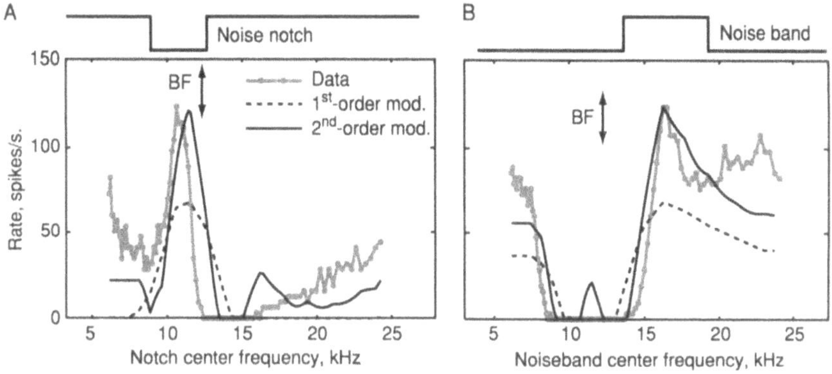

FIG. 7. Discharge rate of a DCN type IV neuron in response to a 1/2 octave notch in wideband noise (A) or a 1/2 octave noise band (B). The notch and band were centered at various frequencies, as indicated on the x-axis. The heavy gray line with points shows the data. The dashed line shows the predictions of the weight-function model using only R_0 and the first-order weights. The weights are the ones shown in Fig. 6. The solid line shows the prediction using the full model. The schematics at the top of the figure show the spectra of the noise notch and the noise band that give the maximum discharge rate. The passband level of the noise is 12 dB re 20 μPa/Hz$^{1/2}$.

The shape of the weight functions for this neuron (Fig. 6) are consistent with a rate maximum for a spectral edge just above BF. In the first-order weights, such an alignment of the stimulus spectrum and the weights puts the stimulus stop-band in the inhibitory response area (i.e., where the first-order weights are negative) and the passband in the excitatory area at higher frequencies. The positive-eigenvalue second-order weights show the same effect. By the opposite argument, a falling spectral edge should be inhibitory for the first-order weights, which is observed in the data. Figure 7 shows comparisons of the experimental rates and the predictions of the weight model using first-order weights only (shown as a dashed line) and using both first- and second-order weights (shown as a solid line). Although the first-order weights make qualitatively correct rate predictions, adding the second-order weights improves the prediction, especially near the rate peak, and produces a better match to the overall discharge rates. Note, however, that the second-order weights predict a small erroneous rate peak at the falling spectral edge (at a notch center frequency near 16 kHz and a band center frequency near 11 kHz); this effect occurs because the second-order terms do not reverse sign (Eq. 7) when the sign of the spectrum reverses; presumably this error would be corrected by higher order non-linear terms, if present.

The quality of the predictions made by the weight model varies among units. Generally, neurons in the VCN show good predictions and neurons in DCN, especially type IV neurons, show poor predictions (Yu and Young, 2000). The example in Fig. 7 is a relatively good prediction for a type IV neuron.

Prediction has been studied extensively for noise stimuli consisting of broadband noise filtered by head-related transfer functions (HRTFs) from cats (see Chapter 12 by Carlile *et al.*). Examples of cat HRTFs are shown in Fig. 8A. HRTFs are acoustic transfer functions between free field and a location near the eardrum (Musicant *et al.*, 1990; Rice *et al.*, 1992); they describe the changes in the spectrum of a sound made by the acoustic effects of the head and pinna. As is typical for the cat, HRTFs have large notches in the frequency range from 8 to 20 kHz. The notch and its associated spectral features vary in location along the frequency axis as the azimuth and elevation of the sound source vary, as illustrated for elevation in Fig. 8A. The notch frequency is a cue for sound location in cats (Huang and May, 1996; Tollin and Yin, 2003). Actually, the important feature may be the upper-frequency edge of the notch, rather than its center frequency (Middlebrooks, 1992; Reiss and Young, 2005). Regardless, what is important for the present is that the HRTFs provide spectral cues that both cats and humans (Middlebrooks and Green, 1991) use for sound localization.

The response maps of a chopper neuron from the VCN and a type IV neuron from the DCN are shown in Fig. 8B and C. The VCN neuron has a typical V-shaped excitatory area centered on BF with inhibitory areas at frequencies away from BF. VCN neurons of all types, except for onset neurons, have response maps like this. The DCN neuron has one large inhibitory area which is probably the union of the usual inhibitory areas at and above BF. Although the response map in Fig. 1C is more typical, response maps like the one in Fig. 8C are common in the DCN.

The responses of these two neurons to a set of 99 HRTF-filtered noises are shown in Fig. 8D and E. The stimuli were generated by filtering a broadband noise with 99 different HRTFs, measured from locations throughout the frontal field of the cat. Because of the interest in the notches, the discharge rates are plotted against the center frequency of the notch in the HRTFs. To make best use of the notches, the spectra were translated along the frequency axis to equate the BF of the neuron to the center of the range of notch frequencies in the stimulus set; this was done by adjusting the sampling rate of the stimuli. Both neurons encode the notches as a minimum in discharge rate when the notch is at the neuron's BF. In the case of the VCN neuron, it is easy to see why such a minimum should occur, because when the notch is at BF, there is less stimulus energy in the excitatory area of the neuron (Poon and Brugge, 1993). However, this explanation does not work for the type IV neuron. The neuron's tone response map (Fig. 8C) suggests that removing energy from the stimulus at frequencies near BF should be excitatory, not inhibitory (Spirou and Young, 1991). The existence of a rate minimum in type IV responses to notches has been explained qualitatively on the basis of the DCN circuit in Fig. 3 (Blum and Reed, 1998; Hancock and Voigt, 1999); however, that model has recently been shown to be inadequate to explain all of the features of the notch responses in Fig. 7 (Reiss and Young, 2005).

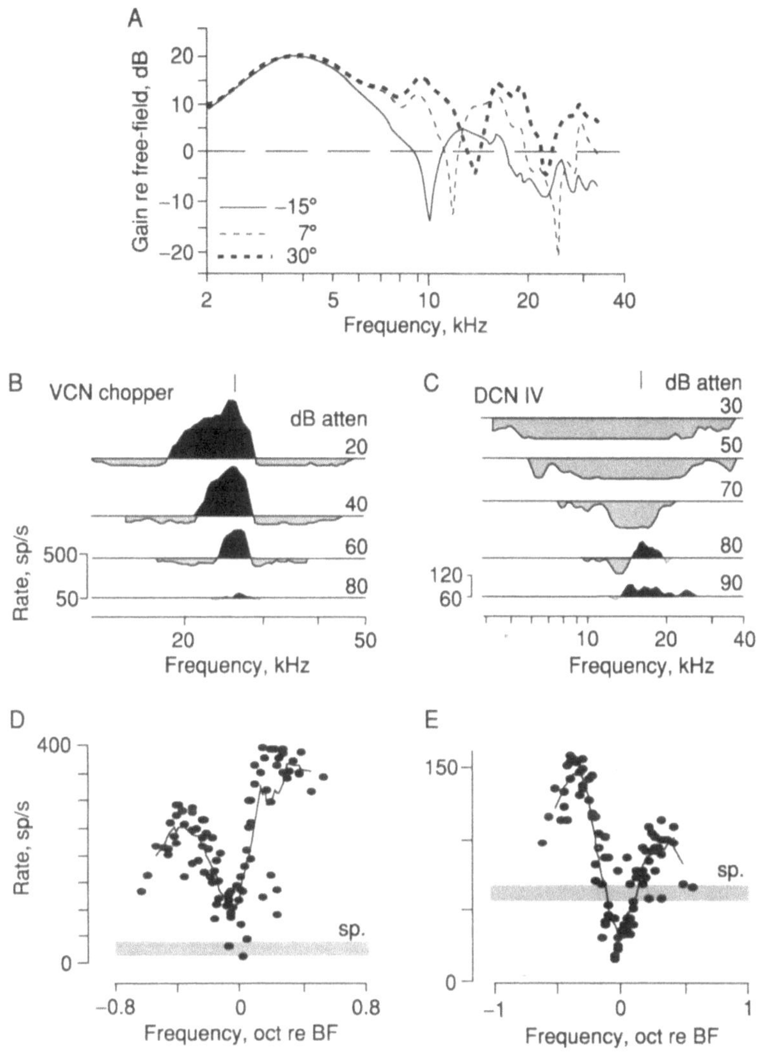

FIG. 8. (A) HRTFs for a cat from three different elevations (legend) at the same azimuth (0°, straight ahead). The HRTFs are plotted as gain versus frequency, where gain is the amplitude of the signal in the ear canal near the eardrum divided by the amplitude in free-field, i.e., with the cat removed. Note the prominent spectral notch at 10 to 15 kHz which moves higher in frequency as the elevation increases. (B, C) Response maps for a VCN chopper neuron and a DCN type IV neuron. These are plotted as in Fig. 1C. (D, E) Discharge rate versus the frequency of the spectral notch for the same two neurons responding to 99 HRTF-filtered noises. The notch frequencies are plotted as octaves relative to the neuron's BF, i.e., as $20 \cdot \log_{10}$(notch frequency/BF). The stimuli are broadband (0 to 50 kHz) noise filtered with HRTFs. Stimuli were shifted by adjusting their playback sampling rate to center the BF in the frequency region occupied by notches. The horizontal shaded bars show spontaneous discharge rate. Stimuli were 400 ms in duration and were ≈30 dB (D) or ≈40 dB (E) above threshold (Rice *et al.*, 1992; redrawn from Yu and Young, 2000).

Figure 9 shows how the weight model does at predicting responses of a VCN neuron (left column) and a DCN type IV neuron (right column) to HRTF stimuli. These are different neurons than in Fig. 8, but their properties are similar. The first- and second-order weights are shown in the top two rows. The VCN neuron has a simple first-order weight function (Fig. 9A) consisting of an excitatory area (positive weights) centered on BF with inhibitory weights both below and above BF. The second-order weights (Fig. 9C), when decomposed as in Fig. 6C, yield an excitatory eigenvalue whose eigenvector is similar to the first-order weights, plus smaller eigenvalues. Thus, the second-order weights contribute mainly a quadratic non-linearity. The DCN neuron's weights (Fig. 9B and D) have a similar shape, but with larger inhibitory weights. DCN neurons like this one, with trimodal first-order weights and excitation at BF, are often quite non-linear, meaning that prediction of rate responses works poorly (Yu, 2003). That is true in this case.

Figure 9E and F show predictions of the discharge rates of these neurons in response to 99 different HRTF spectral shapes. The Xs and the shaded line show the discharge rates produced by these two neurons, plotted against the center frequency of the notch. Again, both neurons show a substantial rate minimum with the notch centered on BF. In the type IV neuron (Fig. 9F), the response is again inhibitory (below spontaneous rate) with the notch at BF. The unfilled circles and the dashed line show the predictions of the first-order model and the filled circles and the solid line show the predictions of the second-order model. The Q values given in the figure legends are a measure of the quality of the fit related to fraction of variance (see Appendix 1). Q varies from 0 (poor fit) to 1 (perfect fit); Q_1 is the quality of the fit with first-order terms only and Q_2 is the quality of the fit for the full model. In the VCN neuron, both models do well, with only slight improvement from adding the second-order terms. The general shape of the notch response is captured by the model and the main error is a constant offset. In the DCN neuron, both models do poorly with only a small notch response. Although the second-order model does better than the first-order model in that it has a larger Q, both models fail in that the inhibition produced by the notch is not predicted by either.

IX. Non-Linearity Does Not Mean Less Information About the Stimulus

The poor performance of prediction in the DCN neuron in Fig. 9 is a common observation. Generally the weight model predicts responses of VCN neurons well but predicts DCN neurons poorly (Yu, 2003; Yu and Young, 2000). One possible explanation for this result is that DCN neurons are more noisy in their responses, so that a larger fraction of their response is random and unpredictable. However, direct comparison of the variance of spike rate between VCN

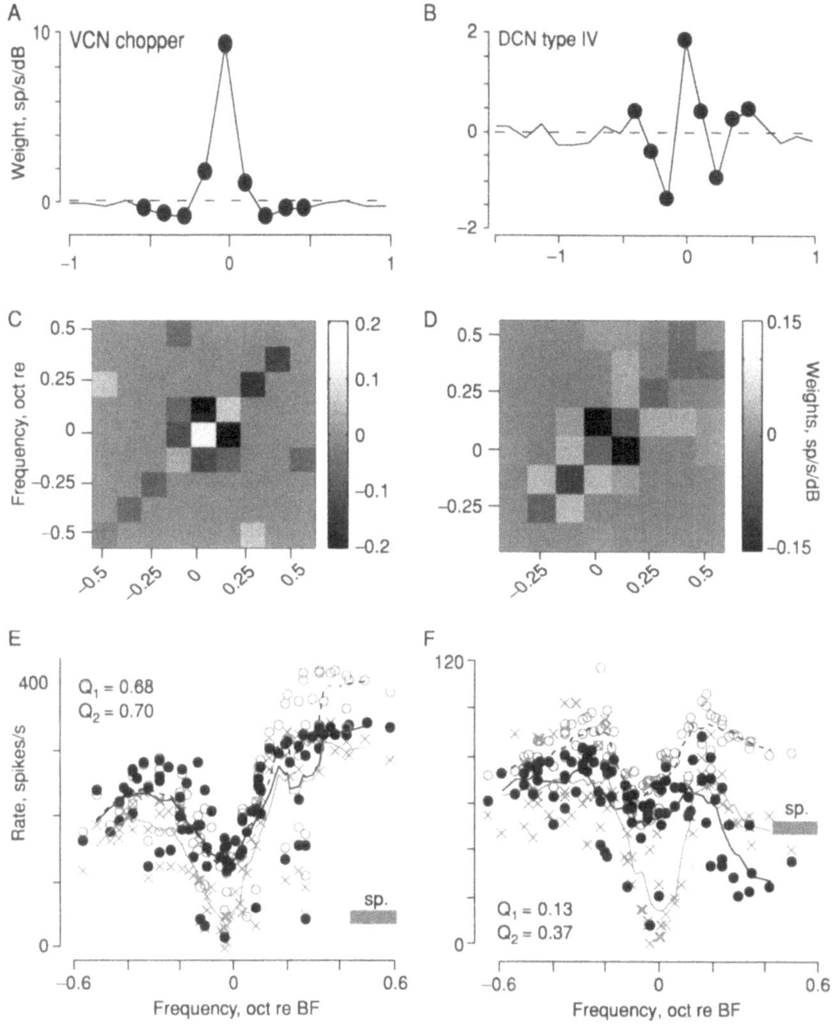

FIG. 9. Prediction of responses to HRTF-filtered stimuli for a VCN chopper neuron (left column) and a DCN type IV neuron (right column). (A, B) First-order weights for the two neurons, plotted against bin center frequency (f_i in Eq. 1) in octaves relative to BF. Filled circles are weights that are more than 1 SD from 0, based on bootstrap estimates of SD. (C, D) Second-order weights of the same neurons, plotted against two frequencies in octaves re BF. Weights <1 SD from 0 are set to 0. (E, F) Discharge rates of the neurons to 99 HRTF-filtered noise stimuli, plotted against the frequency of the prominent mid-frequency notch, again in terms of octaves relative to BF. Data were obtained from 400 ms stimuli at about 40 dB above threshold. Data are shown as Xs and the shaded line shows a moving average of the data. Predictions based on the first-order weight model (R_0 plus the first-order weight summation only) are shown as unfilled circles and a dashed line. Predictions based on the full model are shown by filled circles and a heavy solid line. Spontaneous rates are shown by the shaded bars at right, marked "sp." The values of Q are a measure of the quality of the fit with the first-order model only (Q_1) and with the full model (Q_2) (redrawn from Yu and Young, 2000).

and DCN neurons show that they are similar. Also, repeat determinations of the model from different datasets yield weight functions that are very similar; again there is no difference between VCN and DCN. Another possible explanation is that DCN neurons simply do not encode spectral shape, so that both RSS responses and responses to HRTF stimuli are not controlled by the stimulus parameters in the model. Two arguments against this idea can be given. First, the discharge rates of DCN neurons vary over as large a range, in response to RSS or HRTF stimuli, as do the discharge rates of VCN neurons (e.g., Figs. 8D, E, 9E, and F). Both sets of neurons show large and consistent rate modulations as the stimulus spectrum varies. Second, a basic information analysis, discussed in the following text, shows that DCN neurons convey, by their discharge rates, as much information about the stimuli as VCN neurons.

For the information analysis, the question asked is whether the discharge rate of the neuron allows identification of the different stimuli. Figure 10A shows histograms of the number of spikes produced in response to the 100 stimuli of an RSS set by a DCN neuron. These data are typical of responses to both the RSS and HRTF stimulus sets for both DCN and VCN neurons. Each row in the matrix shows, in grayscale, the distribution of spike counts in response to nine repetitions of one of the stimuli in the set. The stimuli have been placed along the ordinate in the order of the mean number of spikes they produce. Although there is a clear relationship between stimulus number and spike count, there is a fair amount of scatter; the overlap between the responses to nearby stimuli prevents unique identification of the stimulus producing a particular spike count. Nevertheless, there is information about which stimulus was presented. The quality of the representation can be quantified by the mutual information $I(R;S)$ between spike count and stimulus number (Borst and Theunissen, 1999; Cover and Thomas, 1991). In the simplest case, $I(R;S)$ is $\log_2(n)$ where n is the effective number of classes into which the stimuli can be divided based on the spike counts. For the case in Fig. 10A, $I(R;S) = 1.96$, suggesting that spike count can specify about four different stimulus classes.

Because these data were not taken with an information analysis in mind, they are not optimized for calculating mutual information and there are many fewer repetitions of the stimuli than are actually needed to allow accurate estimation (Treves and Panzeri, 1995). However, the degree of bias should be similar for different neurons and $I(R;S)$ should provide a reliable comparative measure of the quality of the representation across populations of neurons.

The important point to be derived from these data is that information about stimulus identity is not related to the linearity of the representation or to the location of the neurons in DCN or VCN. Figure 10B shows $I(R;S)$ plotted against Q, the measure of the quality of the weight model as previously discussed, for a population of 12 VCN neurons (mostly choppers) and 12 DCN neurons (mostly type IV, but all DCN principal cell types). Results are shown for responses to both

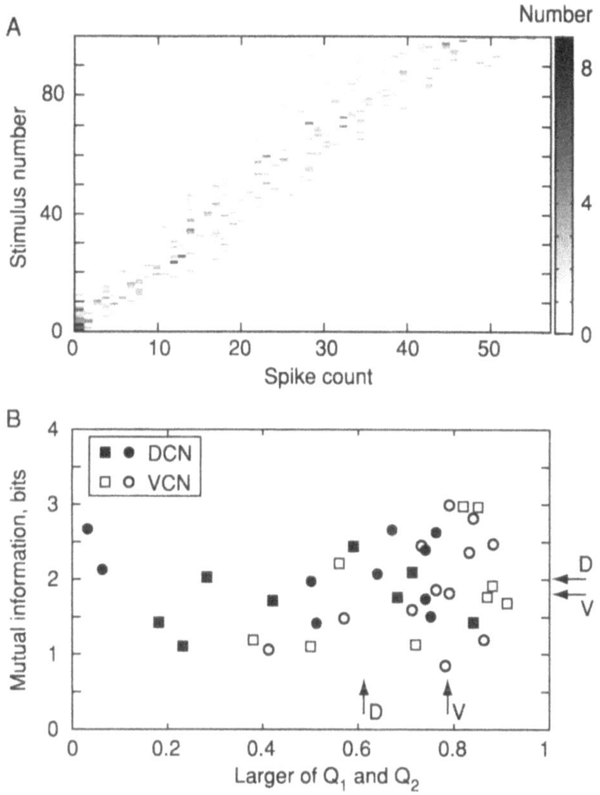

FIG. 10. (A) Matrix of histograms of the responses of a DCN type IV neuron to 100 RSS stimuli. Each row in this matrix is one histogram, showing the distribution of the number of spikes produced in response to one of the stimuli in 9 repetitions of 400 ms stimuli. The spike count is on the abscissa and the number of occurrences of each spike count is given by the grayscale. The stimuli are placed on the ordinate in order of their mean spike counts. (B) Mutual information calculated from data like that in A for 12 DCN neurons and 12 VCN neurons, identified in the legend. Circles are responses to RSS stimuli and squares to HRTF stimuli. $I(R;S)$ was calculated as in Appendix 2. The abscissa is the larger of Q_1 and Q_2 measures of the fit of the weight model when used to predict rate responses to RSS or HRTF stimuli. Usually the second-order model provides the better fit, larger Q. For the RSS stimuli, 3/4 of the data were used to compute the weights and the remaining 1/4 was used for prediction, providing a cross-validation test. For the HRTF stimuli, the model was computed from RSS stimuli as in previous figures. The arrows show medians of the DCN (D) and VCN (V) data along the two axes. There is no significant difference between DCN and VCN neurons in $I(R;S)$ nor is there a correlation between Q and $I(R;S)$.

RSS stimuli (circles) and HRTF stimuli (squares). Non-linear neurons, in the sense that the weight model does not predict their responses, have low values of Q and linear neurons have high values of Q. The arrows just above the abscissa and to the right of the plot show the medians for DCN and VCN neurons. The median Q values are significantly different ($p \approx 0.002$ rank-sum) but the median values of $I(R;S)$ are not different ($p \approx 0.7$). Moreover, there is no significant relationship between $I(R;S)$ and Q in either or both populations ($r = 0.19$ for the whole population, $p =$ NS). Thus, the information provided about the stimulus is not linked to the linearity of the representation, as measured by the weight model. The non-linear representation provided by DCN neurons is as good, in the sense of mutual information, as the linear representation provided by VCN neurons.

X. Conclusions: Implications for Models of Spectral Representation in the Auditory System

These results provide useful insights into the nature of the representation of stimulus spectra in auditory neurons. By separating the calculation of spectral and temporal properties of neurons, it is possible to estimate a non-linear response characteristic, in this case the second-order weights. In many cases, the resulting weight model provides accurate prediction of the responses to stimuli not used in the construction of the model, i.e., the spectral edges in Fig. 7, the HRTF stimuli in Fig. 9, and the cross-validation RSS stimuli in Fig. 10. When the weight model accurately predicts the responses of the neuron, then the model provides useful information about the neuron's properties. However, when the model does not accurately predict responses, considerable caution should be exercised in using the model to study the neuron's properties since the neuron must have significant response characteristics that are not observable in the model.

The possibility of separating neurons into linear and non-linear groups, where "linear" means that responses are predictable from a first- or second-order model, provides some insights into neural coding. In Fig. 8, for example, the notches in the stimuli are represented by a rate minimum in both cases. However, the nature of the two responses is rather different. The VCN neuron is operating in a fashion that is well-described by its response map (Fig. 8B) or by its weight functions (not shown). It is excited by energy near BF and inhibited by energy over a broader frequency range. Thus, a relatively simple concept of summation of excitatory and inhibitory effects is sufficient to account for this neuron's properties and this concept can be made quantitative and explicit by the weight model. For the DCN neuron, however, a successful model of this type is not available. As a consequence, it is misleading to conclude from the response map in Fig. 8C that the neuron is inhibited by energy near BF for sound levels 20 dB or so above threshold. The responses to the notch noise show that the opposite is true when that energy is present in the context of a broad band of noise. In the case of the DCN, it is possible to provide a partial explanation for this non-linearity in terms of the known circuitry of the nucleus (Reiss and Young, 2005).

Despite the existence of non-linearities that cannot be captured using the quadratic model considered here, the DCN provides good information about the stimulus spectrum, in fact as much information as is provided by VCN neurons (Fig. 10B). This dichotomy is an example of the range of spectral representations in the peripheral auditory system, ranging from generalized spectral encoders, like VCN neurons, that represent the energy in the stimulus in a straightforward frequency analysis versus nascent feature detectors, like DCN neurons, that represent various aspects of the spectrum in ways that are not directly related to energy integration.

An important question is the extent to which better models of spectral integration can be developed for non-linear auditory neurons. In fact, if the RSS analysis is repeated with lower-contrast stimuli (i.e., RSS stimuli with 3 dB instead of 12 dB standard deviations of their bin levels), the quadratic model provides good prediction of similar low-contrast stimuli (L. A. Reiss and S. Bandyopadhyay, unpublished). Preliminary work suggests that it will be possible to "stitch" such local models together into non-linear models that will provide a good representation of the properties of auditory neurons over a wide range of sound levels (S. Bandyopadhyay, unpublished).

Acknowledgments

Preparation of this chapter was supported by NIH (NIDCD) grant DC000115. The content of the chapter benefited from discussions with Kechen Zhang (especially Section VII), Sharba Bandyopadhyay, and Steven Chase

Appendix 1. The Q Measure of the Quality of Fit

The quality of the fit was measured by the variable Q, given by

$$Q = 1 - \frac{\sum_{j=1}^{N} (r_{Aj} - r_{Bj})^2}{N(\bar{r}_A - \bar{r}_B)^2 + \sum_{j=1}^{N} (r_{Aj} - \bar{r}_A)^2}. \tag{8}$$

Ordinarily, the r_{Aj} are the experimental rates, the r_{Bj} are the rates predicted by the model, and $\widehat{r}_A$ and $\widehat{r}_B$ are the mean values of the two sets of rates. However, the A and B values are interchanged if var(r_{Bj}) > var(r_{Aj}); that is, A and B are chosen to make var(r_{Aj}) > var(r_{Bj}). In the case where $\widehat{r}_A = \widehat{r}_B$, Q is exactly equal to the commonly used fraction of variance measure. Q has the advantage over fraction of variance in that the latter is only guaranteed to be <1 and can be negative if there are substantial differences between $\widehat{r}_A$ and $\widehat{r}_B$, whereas Q is constrained to vary between 0 and 1 regardless. If there are large differences in the mean rates, $Q \approx 1/(2 - fv)$.

Appendix 2. Calculation of Mutual Information

The mutual information was calculated from histograms like the one shown in Fig. 10A. The following formula was used (Cover and Thomas, 1991):

$$I(R;S) = H(R) - H(R|S) = -\sum_r \widehat{p}(r)\log_2[\widehat{p}(r)] + \sum_s p(s)\sum_r \widehat{p}(r|s)\log_2[\widehat{p}(r|s)]. \quad (9)$$

$H(R)$ and $H(R|S)$ are, respectively, estimates of the entropy of the spike counts r and the conditional entropy of the spike counts given the stimulus s. The estimates of the probabilities $\widehat{p}\,(r)$ and $\widehat{p}\,(r|s)$ were computed from the histograms as the spike counts in the appropriate bins divided by the total spike count in the histogram. For $\widehat{p}\,(r)$, the matrix in Fig. 10A was collapsed vertically to one histogram of spike counts. For $\widehat{p}\,(r|s)$ each row of the matrix contributed a probability function, conditioned on the corresponding stimulus. These estimates were corrected for bins containing zero spikes as follows: for each histogram, the minimum and maximum counts were found. All bins between the minimum and maximum containing zero counts were set to 0.5 (K-T correction) and the normalization of the histogram to a probability estimate $\widehat{p}$ was corrected by adding the appropriate number of 0.5s. The probability of the stimuli $p(s)$ was 0.01, because each stimulus was presented exactly once.

The estimate in previous text is biased and, because the number of repeat observations of spike-counts was small (7 to 11), it was not possible to eliminate the bias. Empirically, the method of computation used contains less bias than other methods. Bias was further reduced by bootstrapping the computation (Efron and Tibshirani, 1993). Because of the bias, the absolute values of the mutual information in Fig. 10B should be interpreted with caution. However, the bias should be roughly the same across neurons, so the comparisons between populations in Fig. 10B are probably safe.

References

Aertsen, A. M. H. J., and Johannesma, P. I. M. (1981). The spectro-temporal receptive field. A functional characteristic of auditory neurons. *Biol. Cybern.* **42,** 133–143.

Blum, J. J., and Reed, M. C. (1998). Effects of wide band inhibitors in the dorsal cochlear nucleus. II. Model calculations of the responses to complex tones. *J. Acoust. Soc. Am.* **103,** 2000–2009.

Borst, A., and Theunissen, F. E. (1999). Information theory and neural coding. *Nat. Neurosci.* **2,** 947–957.

Bruce, I. C., Sachs, M. B., and Young, E. D. (2003). An auditory-periphery model of the effects of acoustic trauma on auditory nerve responses. *J. Acoust. Soc. Am.* **113,** 369–388.

Colburn, H. S. (1973). Theory of binaural interaction based on auditory-nerve data. I. General strategy and preliminary results on interaural discrimination. *J. Acoust. Soc. Am.* **54,** 1458–1470.

Conley, R. A., and Keilson, S. E. (1995). Rate representation and discriminability of second formant frequencies for /ε/-like steady-state vowels in cat auditory nerve. *J. Acoust. Soc. Am.* **98,** 3223–3234.

Cover, T. M., and Thomas, J. A. (1991). "Elements of Information Theory." John Wiley & Sons, Inc., New York.

Davis, K. A., and Young, E. D. (2000). Pharmacological evidence of inhibitory and disinhibitory neural circuits in dorsal cochlear nucleus. *J. Neurophysiol.* **83,** 926–940.

de Boer, E., and Kuyper, P. (1968). Triggered correlation. *IEEE Trans. BME* **15,** 169–179.

de Boer, E., and de Jongh, H. R. (1978). On cochlear encoding: Potentialities and limitations of the reverse-correlation technique. *J. Acoust. Soc. Am.* **63,** 115–135.

deCharms, R. C., Blake, D. T., and Merzenich, M. M. (1998). Optimizing sound features for cortical neurons. *Science* **280,** 1439–1443.

Delgutte, B. (1990). Two-tone rate suppression in auditory-nerve fibers: Dependence on suppressor frequency and level. *Hearing Res.* **49,** 225–246.

Doucet, J. R., and Ryugo, D. K. (1997). Projections from the ventral cochlear nucleus to the dorsal cochlear nucleus in rats. *J. Comp. Neurol.* **385,** 245–264.

Efron, B., and Tibshirani, R. J. (1993). "An Introduction to the Bootstrap." Chapman and Hall, New York.

Eggermont, J. J. (1993). Wiener and Volterra analyses applied to the auditory system. *Hearing Res.* **66,** 177–201.

Eggermont, J. J., Aertsen, A. M. H. J., and Johannesma, P. I. M. (1983). Prediction of the responses of auditory neurons in the midbrain of the grass frog based on the spectro-temporal receptive field. *Hearing Res.* **10,** 191–202.

Escabi, M. A., Miller, L. M., Read, H. L., and Schreiner, C. E. (2003). Naturalistic auditory contrast improves spectrotemporal coding in the cat inferior colliculus. *J. Neurosci.* **23,** 11489–11504.

Escabi, M. A., and Read, H. L. (2003). Representation of spectrotemporal sound information in the ascending auditory pathway. *Biol. Cybern.* **89,** 350–362.

Escabi, M. A., and Schreiner, C. E. (2002). Nonlinear spectrotemporal sound analysis by neurons in the auditory midbrain. *J. Neurosci.* **22,** 4114–4131.

Fletcher, H. (1938). The mechanism of hearing as revealed through experiment on the masking effect of thermal noise. *PNAS* **24,** 265–274.

Geisler, C. D., Rhode, W. S., and Kennedy, D. T. (1974). Responses to tonal stimuli of single auditory nerve fibers and their relationship to basilar membrane motion in the squirrel monkey. *J. Neurophysiol.* **37,** 1156–1172.

Glasberg, B. R., Moore, B. C., Patterson, R. D., and Nimmo-Smith, I. (1984). Dynamic range and asymmetry of the auditory filter. *J. Acoust. Soc. Am.* **76,** 419–427.

Goldstein, J. L. (1973). An optimum processor theory for the central formation of the pitch of complex tones. *J. Acoust. Soc. Am.* **111,** 492–498.

Hancock, K. E., and Voigt, H. F. (1999). Wideband inhibition of dorsal cochlear nucleus type IV units in cat: A computational model. *Ann. Biomed. Eng.* **27,** 73–87.

Holmes, S. D., Sumner, C. J., O'Mard, L. P., and Meddis, R. (2004). The temporal representation of speech in a nonlinear model of the guinea pig cochlea. *J. Acoust. Soc. Am.* **116,** 3534–3545.

Huang, A. Y., and May, B. J. (1996). Sound orientation behavior in cats. II. Mid-frequency spectral cues for sound localization. *J. Acoust. Soc. Am.* **100,** 1070–1080.

Jiang, D., Palmer, A. R., and Winter, I. M. (1996). Frequency extent of two-tone facilitation in onset units in the ventral cochlear nucleus. *J. Neurophysiol.* **75,** 380–395.

Johnson, D. H. (1980). Applicability of white-noise nonlinear system analysis to the peripheral auditory system. *J. Acoust. Soc. Am.* **68,** 876–884.

Kanold, P. O., and Young, E. D. (2001). Proprioceptive information from the pinna provides somatosensory input to cat dorsal cochlear nucleus. *J. Neurosci.* **21,** 7848–7858.

Kiang, N. Y.-S., and Moxon, E. C. (1974). Tails of tuning curves of auditory-nerve fibers. *J. Acoust. Soc. Am.* **55,** 620–630.

Kim, P. J., and Young, E. D. (1994). Comparative analysis of spectro-temporal receptive fields, reverse correlation functions, and frequency tuning curves of auditory-nerve fibers. *J. Acoust. Soc. Am.* **95,** 410–422.

Klein, D. J., Depireux, D. A., Simon, J. Z., and Shamma, S. A. (2000). Robust spectrotemporal reverse correlation for the auditory system: Optimizing stimulus design. *J. Comput. Neurosci.* **9,** 85–111.

Lewis, E. R., Henry, K. R., and Yamada, W. M. (2002). Tuning and timing of excitation and inhibition in primary auditory nerve fibers. *Hearing Res.* **171,** 13–31.

Lewis, E. R., and van Dijk, P. (2004). New variations on the derivation of spectro-temporal receptive fields for primary auditory afferent axons. *Hearing Res.* **189,** 120–136.

Linden, J. F., Liu, R. C., Sahani, M., Schreiner, C. E., and Merzenich, M. M. (2003). Spectrotemporal structure of receptive fields in areas AI and AAF of mouse auditory cortex. *J. Neurophysiol.* **90,** 2660–2675.

Machens, C. K., Wehr, M. S., and Zador, A. M. (2004). Linearity of cortical receptive fields measured with natural sounds. *J. Neurosci.* **24,** 1089–1100.

Marmarelis, P. Z., and Marmarelis, V. Z. (1978). "Analysis of Physiological Systems: The White-Noise Approach." Plenum, New York.

May, B. J., Huang, A., LePrell, G., and Hienz, R. D. (1996). Vowel formant frequency discrimination in cats: Comparison of auditory nerve representations and psychophysical thresholds. *Aud. Neurosci.* **3,** 135–162.

May, B. J., and Huang, A. Y. (1997). Spectral cues for sound localization in cats: A model for discharge rate representations in the auditory nerve. *J. Acoust. Soc. Am.* **101,** 2705–2719.

Meddis, R., O'Mard, L. P., and Lopez-Poveda, E. A. (2001). A computational algorithm for computing nonlinear auditory frequency selectivity. *J. Acoust. Soc. Am.* **109,** 2852–2861.

Middlebrooks, J. C. (1992). Narrow-band sound localization related to external ear acoustics. *J. Acoust. Soc. Am.* **92,** 2607–2624.

Middlebrooks, J. C., and Green, D. M. (1991). Sound localization by human listeners. *Annu. Rev. Psychol.* **42,** 135–159.

Moore, B. C. J. (2004). "An Introduction to the Psychology of Hearing." Elsevier Ltd., Amsterdam.

Musicant, A. D., Chan, J. C. K., and Hind, J. E. (1990). Direction-dependent spectral properties of cat external ear: New data and cross-species comparisons. *J. Acoust. Soc. Am.* **87,** 757–781.

Nelken, I., Kim, P. J., and Young, E. D. (1997). Linear and non-linear spectral integration in type IV neurons of the dorsal cochlear nucleus: II. Predicting responses using non-linear methods. *J. Neurophysiol.* **78,** 800–811.

Nelken, I., and Young, E. D. (1994). Two separate inhibitory mechanisms shape the responses of dorsal cochlear nucleus type IV units to narrowband and wideband stimuli. *J. Neurophysiol.* **71,** 2446–2462.

Oertel, D., Wu, S. H., Garb, M. W., and Dizack, C. (1990). Morphology and physiology of cells in slice preparations of the posteroventral cochlear nucleus of mice. *J. Comp. Neurol.* **295,** 136–154.

Oxenham, A. J., and Shera, C. A. (2003). Estimates of human cochlear tuning at low levels using forward and simultaneous masking. *J. Assoc. Res. Otolaryngol.* **4,** 541–554.

Palmer, A. R., Winter, I. M., Jiang, D., and James, N. (1995). Across-frequency integration by neurones in the ventral cochlear nucleus. *In* "Advances in Hearing Research" (G. A. Manley, G. M. Klump, C. Köppl, H. Fastl, and H. Oeckinghaus, Eds.), pp. 250–261. World Scientific, Singapore.

Parham, K., and Kim, D. O. (1995). Spontaneous and sound-evoked discharge characteristics of complex-spiking neurons in the dorsal cochlear nucleus of the unanesthetized decerebrate cat. *J. Neurophysiol.* **73,** 550–561.

Poon, P. W. F., and Brugge, J. F. (1993). Sensitivity of auditory nerve fibers to spectral notches. *J. Neurophysiol.* **70,** 655–666.

Qiu, A., Schreiner, C. E., and Escabi, M. A. (2003). Gabor analysis of auditory midbrain receptive fields: Spectro-temporal and binaural composition. *J. Neurophysiol.* **90,** 456–476.

Recio-Spinoso, A., Temchin, A. N., van Dijk, P., Fan, Y. H., and Ruggero, M. A. (2005). Wiener-kernel analysis of responses to noise of chinchilla auditory-nerve fibers. *J. Neurophysiol.* **93,** 3615–3634.

Reed, M. C., and Blum, J. J. (1995). A computational model for signal processing by the dorsal cochlear nucleus, I: Responses to pure tones. *J. Acoust. Soc. Am.* **97,** 425–438.

Reiss, L. A., and Young, E. D. (2005). Spectral edge sensitivity in neural circuits of the dorsal cochlear nucleus. *J. Neurosci.* **25,** 3680–3691.

Rhode, W. S. (1999). Vertical cell responses to sound in cat dorsal cochlear nucleus. *J. Neurophysiol.* **82,** 1019–1032.

Rice, J. J., May, B. J., Spirou, G. A., and Young, E. D. (1992). Pinna-based spectral cues for sound localization in cat. *Hearing Res.* **58,** 132–152.

Robles, L., and Ruggero, M. A. (2001). Mechanics of the mammalian cochlea. *Physiol. Rev.* **81,** 1305–1352.

Ruggero, M. A., and Rich, N. C. (1987). Timing of spike initiation in cochlear afferents: Dependence on site of innervation. *J. Neurophysiol.* **58,** 379–403.

Ruggero, M. A., Rich, N. C., Recio, A., Narayan, S. S., and Robles, L. (1997). Basilar-membrane responses to tones at the base of the chinchilla cochlea. *J. Acoust. Soc. Am.* **101,** 2151–2163.

Rugh, W. J. (1981). "Nonlinear System Theory: The Volterra/Wiener Approach." Johns Hopkins Univ. Press (Web version at http://www.ece.jhu.edu/~rugh/214/me/rugh.html), Baltimore.

Sachs, M. B., and Kiang, N. Y. (1968). Two-tone inhibition in auditory-nerve fibers. *J. Acoust. Soc. Am.* **43,** 1120–1128.

Sachs, M. B., and Young, E. D. (1979). Encoding of steady-state vowels in the auditory nerve: Representation in terms of discharge rate. *J. Acoust. Soc. Am.* **66,** 470–479.

Saint Marie, R. L., Benson, C. G., Ostapoff, E. M., and Morest, D. K. (1991). Glycine immunoreactive projections from the dorsal to the anteroventral cochlear nucleus. *Hearing Res.* **51,** 11–28.

Schouten, J. F., Ritsma, R. J., and Cardozo, B. L. (1962). Pitch of the residue. *J. Acoust. Soc. Am.* **34,** 1418–1424.

Sen, K., Theunissen, F. E., and Doupe, A. J. (2001). Feature analysis of natural sounds in the songbird auditory forebrain. *J. Neurophysiol.* **86,** 1445–1458.

Smith, P. H., and Rhode, W. S. (1989). Structural and functional properties distinguish two types of multipolar cells in the ventral cochlear nucleus. *J. Comp. Neurol.* **282,** 595–616.

Spirou, G. A., and Young, E. D. (1991). Organization of dorsal cochlear nucleus type IV unit response maps and their relationship to activation by bandlimited noise. *J. Neurophysiol.* **65,** 1750–1768.

Sutter, M. L., Schreiner, C. E., McLean, M., O'Connor, K. N., and Loftus, W. C. (1999). Organization of inhibitory frequency receptive fields in cat primary auditory cortex. *J. Neurophysiol.* **82,** 2358–2371.

Theunissen, F. E., David, S. V., Singh, N. C., Hsu, A., Vinje, W. E., and Gallant, J. L. (2001). Estimating spatio-temporal receptive fields of auditory and visual neurons from their responses to natural stimuli. *Network: Comput. Neural Syst.* **12,** 289–316.

Theunissen, F. E., Sen, K., and Doupe, A. J. (2000). Spectral-temporal receptive fields of nonlinear auditory neurons obtained using natural sounds. *J. Neurosci.* **20,** 2315–2331.

Tollin, D. J., and Yin, T. C. T. (2003). Spectral cues explain illusory elevation effects with stereo sounds in cats. *J. Neurophysiol.* **90,** 525–530.

Treves, A., and Panzeri, S. (1995). The upward bias in measures of information derived from limited data samples. *Neural Comp.* **7,** 399–407.

van Dijk, P., and Wit, H. P. (1994). Wiener kernel analysis of inner ear function in the American bullfrog. *J. Acoust. Soc. Am.* **95,** 904–919.

van Gisbergen, J. A. M., Grashuis, J. L., Johannesma, P. I. M., and Vendrik, A. J. H. (1975). Neurons in the cochlear nucleus investigated with tone and noise stimuli. *Exp. Brain Res.* **23,** 387–406.

Versnel, H., and Shamma, S. A. (1998). Spectral-ripple representation of steady-state vowels in primary auditory cortex. *J. Acoust. Soc. Am.* **103,** 2502–2514.

Voigt, H. F., and Young, E. D. (1990). Cross-correlation analysis of inhibitory interactions in dorsal cochlear nucleus. *J. Neurophysiol.* **64,** 1590–1610.

Winter, I. M., and Palmer, A. R. (1995). Level dependence of cochlear nucleus onset unit responses and facilitation by second tones or broadband noise. *J. Neurophysiol.* **73,** 141–159.

Wong, J. C., Miller, R. L., Calhoun, B. M., Sachs, M. B., and Young, E. D. (1998). Effects of high sound levels on responses to the vowel /e/ in cat auditory nerve. *Hearing Res.* **123,** 61–77.

Yamada, W. M., and Lewis, E. R. (1999). Predicting the temporal responses of non-phase-locking bullfrog auditory units to complex acoustic waveforms. *Hearing Res.* **130,** 155–170.

Yeshurun, Y., Wollberg, Z., and Dyn, N. (1989). Prediction of linear and non-linear responses of MGB neurons by system identification methods. *Bull. Math. Biol.* **51,** 337–346.

Young, E. D., and Brownell, W. E. (1976). Responses to tones and noise of single cells in dorsal cochlear nucleus of unanesthetized cats. *J. Neurophysiol.* **39,** 282–300.

Young, E. D., and Davis, K. A. (2001). Circuitry and function of the dorsal cochlear nucleus. *In* "Integrative Functions in the Mammalian Auditory Pathway" (D. Oertel, A. N. Popper, and R. R. Fay, Eds.), pp. 160–206. Springer Verlag, New York.

Yu, J. J. (2003). Spectral information encoding in the cochlear nucleus and inferior colliculus: A study based on the random spectral shape method. Ph.D. thesis. Johns Hopkins Univ., Baltimore.

Yu, J. J., and Young, E. D. (2000). Linear and nonlinear pathways of spectral information transmission in the cochlear nucleus. *PNAS* **97,** 11780–11786.

Zhang, X., Heinz, M. G., Bruce, I. C., and Carney, L. H. (2001). A phenomenological model for the responses of auditory-nerve fibers: I. Nonlinear tuning with compression and suppression. *J. Acoust. Soc. Am.* **109,** 648–670.

Zwicker, E., Flottorp, G., and Stevens, S. S. (1957). Critical bandwidth in loudness summation. *J. Acoust. Soc. Am.* **29,** 548–557.

SPECTRAL PROCESSING IN THE INFERIOR COLLICULUS

Kevin A. Davis

Departments of Biomedical Engineering & Neurobiology and
Anatomy & Center for Navigation and Communication Sciences
University of Rochester, Rochester, New York 14642, USA

The central nucleus of the inferior colliculus (ICC) receives direct inputs from all of the major auditory nuclei in the brainstem and, in turn, provides nearly all of the input to the auditory forebrain. The manner in which the brainstem projections to the ICC combine creates functionally distinct synaptic domains within the ICC and presumably basic differences in the spectral sensitivity of ICC neurons. In this chapter, the anatomy of the mammalian ICC is reviewed. Then, the frequency response areas of ICC units are described, followed by a review of what is known about the roles of the various excitatory and inhibitory inputs to the ICC in shaping these response areas. Finally, the functional roles of the ICC in the processing of spectral information are discussed.

I. Introduction

The processing of sounds depends on the analysis of their frequency content by the cochlea and by the central auditory system. The central nucleus of the inferior colliculus (ICC) occupies a pivotal position in the central auditory system;

INTERNATIONAL REVIEW OF
NEUROBIOLOGY, VOL. 70
DOI: 10.1016/S0074-7742(05)70006-4

0074-7742/05 $35.00

it receives converging projections from most, if not all, of the auditory nuclei in the brainstem and, in turn, provides nearly all of the input to the auditory forebrain (reviewed by Irvine, 1986; Oliver and Huerta, 1992; Oliver and Shneiderman, 1991). Some of the ascending projections to the ICC are excitatory (e.g., cochlear nucleus: Oliver, 1987; Semple and Aitkin, 1980; superior olive: Glendenning *et al.*, 1992), whereas others are inhibitory. Moreover, the inhibitory projections are both GABAergic (e.g., dorsal nucleus of the lateral lemniscus: Adams and Mugnaini, 1984; Roberts and Ribak, 1987; Shneiderman and Oliver, 1989; Shneiderman *et al.*, 1988, 1993) and glycinergic (e.g., ipsilateral lateral superior olive: Saint-Marie *et al.*, 1989; ipsilateral ventral nucleus of the lateral lemniscus: Saint-Marie *et al.*, 1997). The manner in which these inputs combine creates functionally distinct synaptic domains within the ICC and presumably basic differences in ICC frequency response areas (Aitkin and Schuck, 1985; Brunso-Bechtold *et al.*, 1981; Loftus *et al.*, 2004; Maffi and Aitkin, 1987; Oliver and Huerta, 1992; Oliver *et al.*, 1997; Roth *et al.*, 1978; Ryugo *et al.*, 1981; Shneiderman and Henkel, 1987; Zook and Casseday, 1987).

In this chapter, the anatomy of the mammalian ICC is reviewed briefly, paying particular attention to features that are common across species. Then, the spectral response properties of ICC neurons are described, followed by a review of what is known about the roles of the different excitatory and inhibitory projections to the ICC in shaping the frequency response areas of these units. Finally, the functional roles of the ICC in the processing of spectral information are discussed.

II. Neuronal Architecture of the IC

The inferior colliculus (IC), located on the roof of the mid brain, is divided into three main subdivisions based on cytoarchitectural grounds (Fig. 1) (Morest and Oliver, 1984; Oliver and Morest, 1984). The central nucleus (ICC) is the largest and most prominent of these subdivisions. It occupies the ventrolateral region of the IC (C, M, V, and L) and is defined by the presence of laminae. Above it in the second subdivision lies the dorsal cortex. This subdivision is distinguished by its layered structure (layers I to IV) which resembles that observed in cortex. The third subdivision is composed of the remaining paracentral nuclei, which are located dorsomedial (DM) and ventrolateral (VL, LN, B) of the central nucleus. Consistent with this tripartite parcellation, each subdivision in the IC has a unique pattern of afferent and efferent connections (reviewed by Huffman and Henson, 1990); it is the ICC which is the primary target of the ascending auditory system.

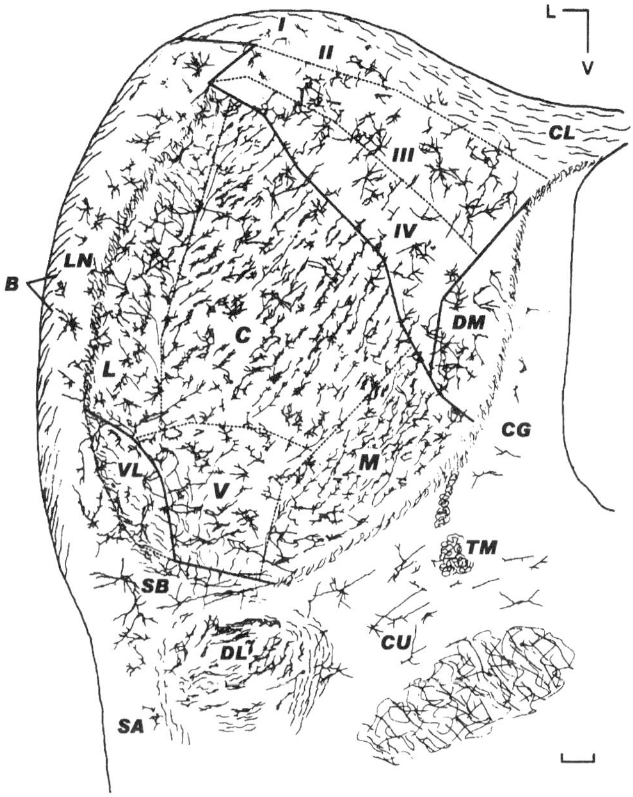

FIG. 1. Neuronal architecture of the inferior colliculus (IC). The middle of the IC in transverse plane. All three major subdivisions of the IC are seen, including the central nucleus (ICC) (C = central; L = lateral; V = ventral; M = medial), the dorsal cortex (I-IV) and the paracentral nuclei (DM = dorsomedial; VL = ventrolateral; LN = lateral; B = brachium of the IC). SA = sagulum; SB = subcollicular area; DL = dorsal nucleus of the lateral lemniscus; CU = cuneiform nucleus; TM = trigeminal nerve; CG = central gray; CL = lateral commissural nucleus. Golgi-Cox method; 2-month old cat; scale = 0.5 mm. Orientation: L = lateral; V = ventral. (From Morest and Oliver (1984), reprinted by permission from Wiley-Liss, Inc., a subsidiary of John Wiley and Sons, Inc.)

The axons of the afferents to the ICC (called the lateral lemniscus) enter the ventrolateral aspect of the central nucleus and course in a dorsomedial direction. The marked laminar organization of the ICC is formed by the dendrites of one of the two principal cell types in the ICC and axonal plexus (Fig. 2). In particular, the laminae are composed of neurons with disc-shaped dendritic fields (***c***) that are arranged in parallel with each other and to the incoming lemniscal afferents (***a,b***). These fibrodendritic laminae are thought to provide the structural basis for

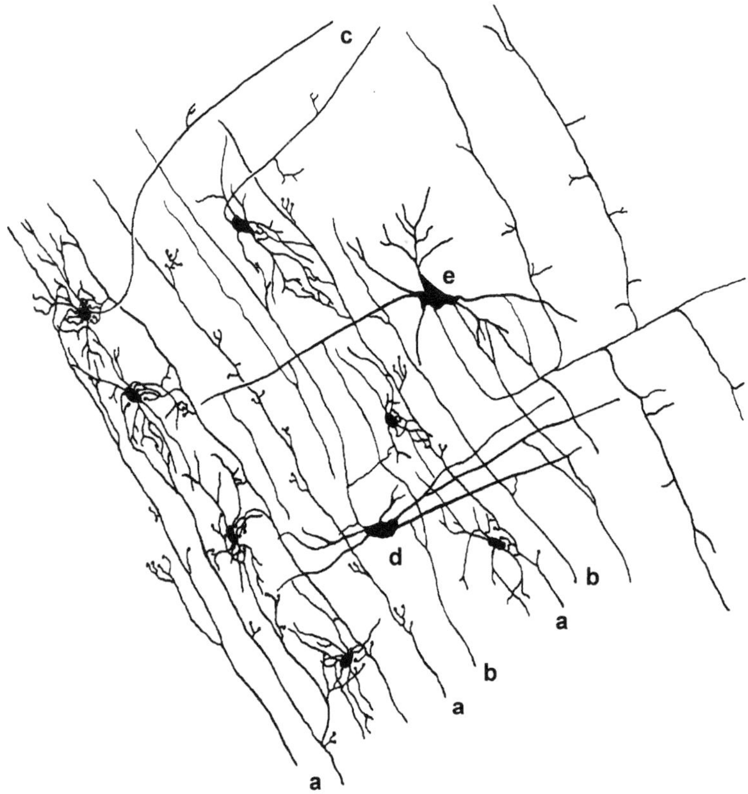

FIG. 2. Scheme of the intrinsic organization of the ICC. Disc-shaped cells (c) have dendrites that are oriented in parallel to the incoming lemniscal afferents (a, b) to form fibrodendritic laminae. Stellate cells (d, e) have dendrites that traverse several laminae. (From Oliver and Morest (1984), reprinted with permission from Wiley-Liss, Inc., a subsidiary of John Wiley and Sons, Inc.)

the tonotopic organization observed in the ICC (e.g., Merzenich and Reid, 1974; Semple and Aitkin, 1979); thus, disc-shaped cells are presumed to receive inputs from afferent neurons with similar best frequencies (BF; the most sensitive frequency). Stellate cells (***d,e***) are the second principal cell type in the ICC. These cells have broad dendritic fields that traverse several adjacent laminae; thus, stellate cells are presumed to integrate inputs across a wide range of frequencies. Despite this limited number of cell types, however, a wide diversity of response properties are recorded in the ICC (reviewed by Aitkin, 1986; Irvine, 1986, 1992). These results suggest that functional differences between cells reflect heterogeneity in their inputs.

III. Ascending Pathways to the ICC

The ICC receives highly convergent input from most, if not all, of the brainstem auditory nuclei, the contralateral IC, and perhaps direct or indirect descending input from the thalamus and cortex (reviewed by Huffman and Henson, 1990; Irvine, 1986; Oliver and Huerta, 1992; Oliver and Shneiderman, 1991). All of the ascending pathways to the ICC originate from the cochlear nucleus (CN) (Fig. 3). The CN is not a homogeneous structure; rather, it can be divided into three major divisions: the anteroventral (AVCN), posteroventral (PVCN), and dorsal cochlear nucleus (DCN) based on anatomical grounds (reviewed by Cant, 1992). Each of these subdivisions receives input from the

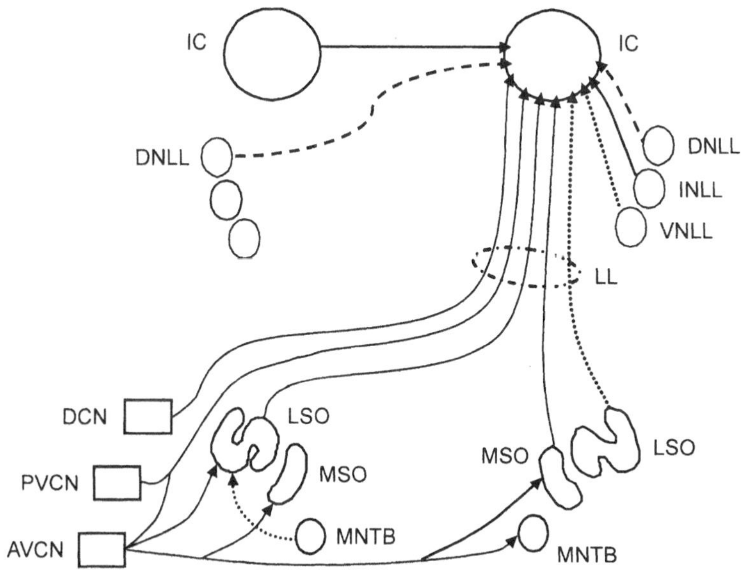

FIG. 3. The principal ascending pathways from the cochlear nucleus to the ICC. Two direct pathways arise from the contralateral ventral (AVCN = anteroventral; PVCN = posteroventral) and dorsal cochlear nuclei (DCN). The medial superior olive (MSO) receives bilateral input from the AVCN and projects ipsilaterally to the ICC. The lateral superior olive (LSO) receives bilateral input from the AVCN, where the contralateral input is via a synapse in the medial nucleus of the trapezoid body (MNTB), and projects bilaterally to the ICC. All of the ascending fibers run in the tract known as the lateral lemniscus (LL), where collaterals from various of these sources contact the ventral (VNLL), intermediate (INLL), and dorsal nuclei of the lateral lemniscus (DNLL). The VNLL and INLL project to the ipsilateral ICC, whereas the DNLL projects bilaterally. Putative excitatory pathways are shown with solid lines; glycinergic inhibitory pathways are indicated by dotted lines; and GABAergic inhibitory pathways are represented by dashed lines.

auditory nerve (AN) (Osen, 1970) and contains a variety of morphologically distinct cell types. Many of these cell types give rise to efferent pathways that ultimately terminate in the ICC.

The main ascending pathways from the CN to the ICC are shown in Fig. 3, where putative excitatory pathways are shown with solid lines and inhibitory pathways are shown with dashed (GABAergic) or dotted (glycinergic) lines. The pathways are direct, indirect, and multisynaptic. Two direct, largely monaural, pathways to the ICC emerge from the stellate cells in the ventral cochlear nucleus (VCN) and the fusiform and giant cells in the DCN. Each of these projections terminates in the contralateral ICC in a banded, tonotopic manner (Oliver, 1984, 1987) and is likely to be excitatory (Oliver, 1984, 1985, 1987; Semple and Aitkin, 1980). (For additional details on the anatomy and physiology of these cell types in the CN, please refer to Chapter 5 in this volume.)

Two indirect binaural pathways to the ICC are via the superior olive. These pathways originate from spherical and globular bushy cells in the AVCN (e.g., Cant and Casseday, 1986; Smith *et al.*, 1991, 1993; Warr, 1966, 1972). The axons of spherical bushy cells provide bilateral input to the medial superior olive (MSO) and ipsilateral input to the lateral superior olive (LSO). Globular bushy cells project to the contralateral LSO via a synapse in the medial nucleus of the trapezoid body (MNTB). In turn, the projection from the MSO to the ICC is primarily ipsilateral, while the projections from the LSO are bilateral (Brunso-Bechtold *et al.*, 1981; Roth *et al.*, 1978; Shneiderman *et al.*, 1988; Zook and Casseday, 1982). Both the MSO and LSO projections to the ICC terminate in a banded, tonotopic manner (Casseday and Covey, 1987; Henkel and Spangler, 1983; Loftus *et al.*, 2004; Shneiderman and Henkel, 1987; Zook and Casseday, 1987). The MSO and crossed LSO projections are excitatory, whereas the ipsilateral LSO projection is mostly inhibitory and glycinergic (Glendenning *et al.*, 1992; Oliver *et al.*, 1995; Saint-Marie and Baker, 1990; Saint-Marie *et al.*, 1989).

Multisynaptic pathways from the CN to the ICC include a synapse in the ventral (VNLL), intermediate (INLL), or dorsal nuclei of the lateral lemniscus (DNLL). Connectional studies indicate that these nuclei receive different sets of inputs and have different output targets (reviewed by Helfert and Aschoff, 1997; Irvine, 1986; Oliver and Shneiderman, 1991; Schwartz, 1992). The VNLL receives most of its input from the contralateral VCN and, in turn, provides a diffuse input to the ipsilateral ICC. The INLL receives input from the contralateral CN, the ipsilateral MNTB and VNLL, and projects to the ipsilateral ICC. In the case of the DNLL, it receives bilateral input from both the CN and superior olivary nuclei, as well as input from the contralateral DNLL. In turn, the DNLL provides bilateral input to the ICC which is banded (Shneiderman *et al.*, 1988; Zook and Casseday, 1987). Evidence suggests that the projections from the VNLL to the ICC are largely inhibitory and glycinergic (Saint-Marie *et al.*,

1997), whereas they are predominantly excitatory from the INLL (Saint-Marie *et al.*, 1997; Winer *et al.*, 1995), and mostly inhibitory and GABAergic from the DNLL (Adams and Mugnaini, 1984; Moore and Moore, 1987; Roberts and Ribak, 1987; Saint-Marie *et al.*, 1997; Thompson *et al.*, 1985).

IV. Synaptic Domains in the ICC

The evidence for banded afferents in the ICC suggests that axons from different sources may overlap at some locations, but remain segregated in other regions, forming unique synaptic domains (Oliver and Huerta, 1992). The synaptic domain hypothesis implies that neurons within a single domain will respond to sound in a similar manner, but that neurons in different domains will exhibit different physiological response properties. Such discrete functional zones could explain observations that neurons with similar monaural or binaural response properties are often clustered together within isofrequency contours (Roth *et al.*, 1978; Semple and Aitkin, 1979).

The most direct and comprehensive dataset in support of the synaptic domain hypothesis has been obtained from studies in cats using anterograde tract-tracing techniques (Loftus *et al.*, 2004; Malmierca *et al.*, 1999; Oliver *et al.*, 1997; Shneiderman and Henkel, 1987). In these experiments, two different tracers were injected in tonotopically matched areas in two brainstem nuclei and the resultant distributions of labeled axons in the ICC were compared. The results of these experiments in chronological order are as follows: ipsilateral MSO and LSO inputs converge in low-frequency ICC; VCN and DCN inputs overlap extensively, but not completely; contralateral DCN and LSO projections overlap in the ventral part of the ICC, but not in the dorsal part where axons from the DCN alone are found; and contralateral and ipsilateral LSO inputs to the ICC do not overlap, but instead terminate in side-by-side layers.

The results of these tracing experiments are summarized in Fig. 4 (Oliver, 2004), which shows a view of three isofrequency lamina (labeled low, mid, and high) from a low-frequency perspective. Each lamina is divided into two side-by-side layers. Inputs to the ICC enter at the base of each layer and course upwards until delimited; excitatory inputs are shown as blocks and inhibitory inputs are indicated by circles. The textured blocks represent zones that receive ipsilateral MSO input, while the solid gray blocks indicate zones that receive contralateral LSO input. The rostral part of the MSO block at low frequencies is a paler shade because the density of MSO input decreases in rostral ICC. All of the white blocks receive input from the contralateral CN; the zone where the DCN in particular terminates is hatched. Finally, the white dots represent input from the ipsilateral LSO, whereas the black dots represent input from the DNLL. It should

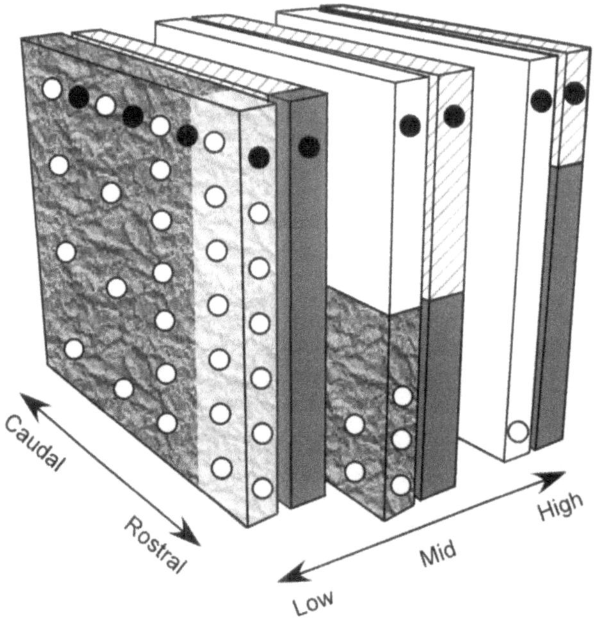

FIG. 4. Scheme of the synaptic domains in the ICC. View of three isofrequency lamina (low, mid, high) in the ICC from a low-frequency perspective. Each lamina is divided into two side-by-side layers. Inputs enter at the base of a layer, course upwards, and terminate where shown. Excitatory inputs are shown as blocks; inhibitory inputs are indicated by circles. Blocks: textured = ipsilateral MSO input (the paler color indicates reduced density); gray = contralateral LSO input; white = contralateral CN input (the hatched white block represents area receiving DCN input). Circles: white = ipsilateral LSO; black = DNLL. (From Oliver (2004), reprinted with permission.)

be noted that the tract-tracing studies involving the DNLL are incomplete; it is likely that this source provides input to more of the ICC than shown (Oliver, personal communication).

In general, the data suggest that an isofrequency lamina is divided into four synaptic domains, which is best observed in the mid frequency lamina. The first two domains, located in dorsal ICC (white background), are both CN types. The left domain is created by the convergence of VCN and DNLL inputs, while the right domain is likely dominated by inputs from the DCN and DNLL. The third domain is an MSO type (located in ventral ICC, shown by a textured background), which receives converging input from the MSO, VCN, and ipsilateral LSO. The final domain (shown in gray) is an LSO type, with input from the LSO, DCN, and VCN. Note, however, that the prevalence of specific domain types changes as a function of frequency; in particular, the MSO type is more common at low frequencies, whereas it is absent at high frequencies.

Two lines of evidence suggest that the four types of functional zones observed in cat ICC are likely to be found in other mammals. First, a similar set of domains was found within a single isofrequency contour in the mustache bat (Ross and Pollak, 1989). These investigators exploited the fact that the ICC in the mustache bat has a greatly enlarged 60 kHz region (Zook *et al.*, 1985) to make small deposits of HRP throughout this area and then to count the numbers of retrogradely labeled cells in the brainstem nuclei. The results suggest that the projection fields of monaural nuclei are widespread within a lamina, whereas the afferents from binaural nuclei, including the MSO, LSO, and DNLL terminate in a more restricted manner. Moreover, the domains of the afferents from LSO and DNLL overlap, but are largely separate from the terminal zone of the MSO.

Second, a comparative study of cytochrome oxidase (CO) activity in the ICC suggests that the pattern of afferent connectivity to the ICC is similar across a number of different species (Cant, 2004). Examples of transverse sections through the middle of the ICC are shown for four species in Fig. 5. In each species, CO activity is relatively high throughout the full extent of the ICC (enclosed by dashed lines), but the labeling is not homogeneous. To highlight this difference in these images, the darkest pixels were selected and then shown against a white background. Clearly, each of these images shows one or more circumscribed patches of CO activity, although the size and orientation of these patches may differ from species to species (e.g., compare the gerbil to the guinea pig). Based on comparisons with results from anterograde tracer studies, the terminals from the MSO and LSO are almost completely confined to the region of high CO activity (gerbil: Cant, 2004; cat: Henkel and Spangler, 1983; Shneiderman and Henkel, 1987; mustache bat: Zook and Casseday, 1987), whereas the terminals from CN afferents can extend past this region.

V. Spectral Processing in Brainstem Nuclei

The nucleotopic organization of the ICC suggests that the VCN, DCN, MSO, and LSO are dominant excitatory inputs to the ICC (reviewed by Oliver, 2004; Oliver and Huerta, 1992). The responses of the principal cells in these nuclei to pure tones as a function of frequency and intensity level (frequency response maps) are shown in Fig. 6. In these plots, a unit's spike count is shown as a bar against stimulus frequency and level. VCN stellate cells (chopper units; Rhode *et al.*, 1983) exhibit type I, III, or I/III properties (Fig. 6A to C: Joris *et al.*, 1994; Shofner and Young, 1985; Winter and Palmer, 1990). Type I units (Fig. 6A) show purely excitatory responses to sound (in the shape of the letter V) and have sufficient spontaneous discharge rate to enable detection of inhibitory sidebands, of which there are none. These maps resemble closely those of their AN fiber

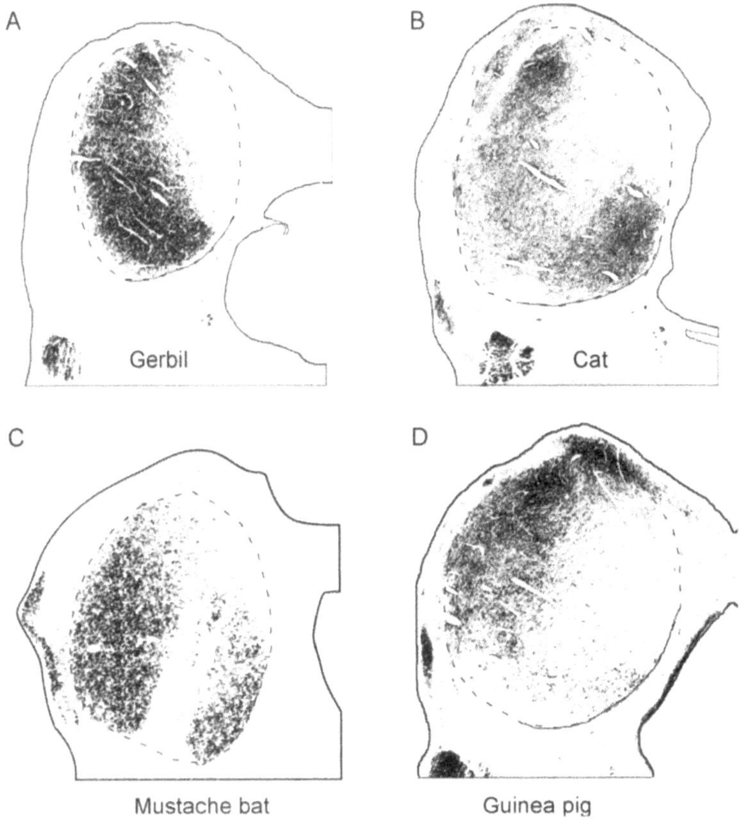

FIG. 5. Patterns of cytochrome oxidase (CO) activity in the ICC of four species. Transverse sections of the middle of the IC of the gerbil (A), cat (B), mustache bat (C), and guinea pig (D). CO activity is relatively high throughout each ICC (enclosed by dashed lines), but not homogeneous; here, the darkest pixels were selected and shown against a white background. (From Cant (2004), reprinted with permission; from Radtke-Schuller, reprinted with permission.)

input (e.g., Kiang *et al.*, 1965). Type III maps (Fig. 6B) have simple V-shaped excitatory areas similar to type I units, but also have inhibitory regions above and/or below the BF. Type I/III units (Fig. 6C) are intermediate between type I and type III units; they have a V-shaped excitatory area, but lack spontaneous activity so inhibitory sidebands cannot be detected.

The responses of principal cells in the DCN are characterized by very complex receptive fields, often exhibiting large areas of inhibition (for review, see Young, 1984). In particular, DCN principal cells show either type III or type IV unit response properties (Fig. 6B and D: Joris, 1998; Young, 1980). DCN type III unit responses are similar to those of the VCN. Type IV maps (Fig. 6D)

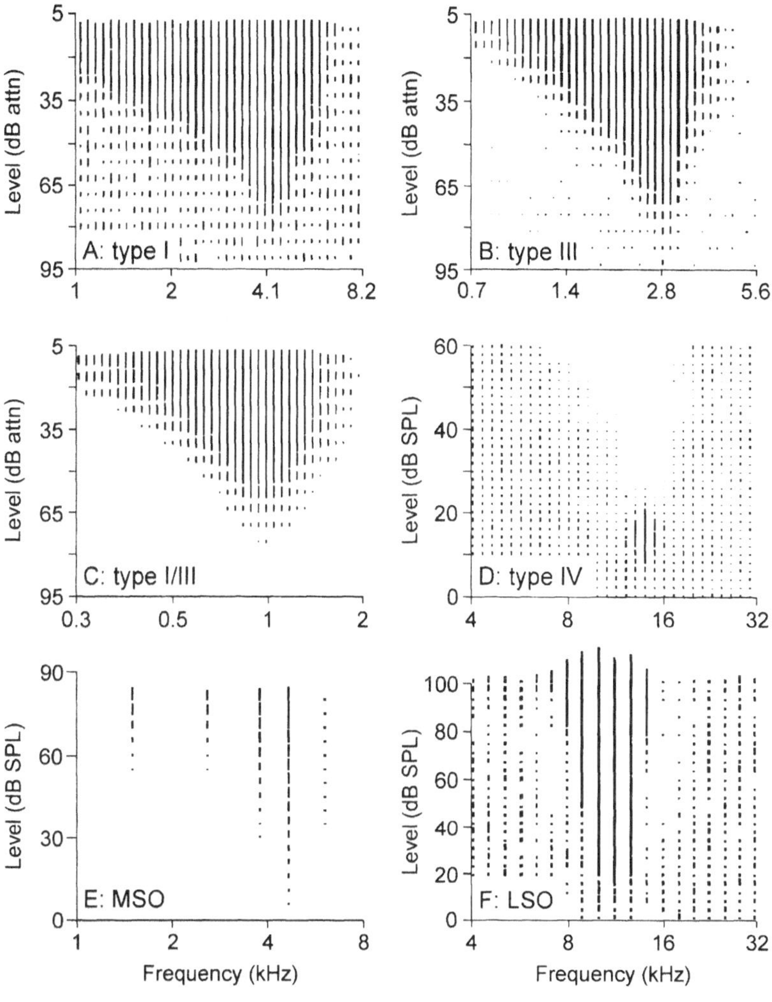

FIG. 6. Spectral processing in the auditory brainstem. Responses to pure tones as a function of frequency and intensity level (frequency response maps) are shown for principal cells in the VCN (A to C), DCN (B, D), MSO (E), and LSO (F), which are four major sources of excitatory input to the ICC. In all panels, spike counts (shown as bars) are plotted against frequency at multiple sound levels. (A to C: modified from Winter and Palmer (1990), reprinted with permission from Elsevier. E: modified from Goldberg and Brown (1969), reprinted with permission from the American Physiological Society. F: modified from Caird and Klinke (1983), reprinted with permission from Springer-Verlag.)

are dominated by inhibition except for an island of excitation at frequencies near the BF and stimulus levels near threshold. Excitatory areas at higher levels and frequencies away from the BF are sometimes observed, but these features are variable (Spirou and Young, 1991).

The response maps of MSO units (Fig. 6E) show simple V-shaped excitatory areas (Goldberg and Brown, 1969). In contrast, the maps of LSO units (Fig. 6F) show very narrow level-tolerant excitatory tuning that is flanked by inhibition (Caird and Klinke, 1983). If the basic response map properties of ICC units are created in this set of four lower-order nuclei and imposed upon collicular cells via excitatory pathways, then one could predict that ICC units would exhibit three general patterns of excitatory tuning: open tuning curves that are V-shaped (Fig. 6A to C and E); level-tolerant curves that remain narrow with level (Fig. 6F); and upper threshold (or closed) tuning curves (Fig. 6D) that respond to a circumscribed range of stimulus levels.

VI. Frequency Response Map Types in the ICC

The frequency response maps of ICC neurons have been obtained in a number of different species (cat: Aitkin *et al.*, 1975; Bock *et al.*, 1972; Ehret and Merzenich, 1988; Ramachandran *et al.*, 1999; Rose *et al.*, 1963; pallid bat: Fuzessery and Hall, 1996; rhesus monkey: Ryan and Miller, 1978; chinchilla: Palombi and Caspary, 1996; Wang *et al.*, 1996; mustache bat: Yang *et al.*, 1992; guinea pig: LeBeau *et al.*, 2001; mouse: Egorova *et al.*, 2001; rat: Hernandez *et al.*, 2005; big brown bat: Casseday and Covey, 1992). Although the nomenclature is often different, ICC neurons can be grouped into three main types based on the patterns of excitation (and inhibition) revealed in pure-tone frequency response maps. Representative data for each unit type from three different species are shown in Figs. 7 and 8.

Figure 7 shows frequency response maps found in the cat (Ramachandran *et al.*, 1999). In these plots, a unit's discharge rate is plotted against frequency at multiple sound levels (labels to the right). Regions where the stimulus-driven activity is consistently above the spontaneous rate (shown as horizontal lines) are defined to be excitatory areas (shown as black fill), whereas regions with activity below the spontaneous rate are inhibitory areas (shown as gray fill). Type V units (top panel) have a V-shaped excitatory area that widens about unit BF (shown as a vertical line) with increasing sound levels. These units do not show inhibitory responses to pure tones. Type I maps (middle panel) show a narrow I-shaped region of excitation that maintains its sharp tuning at high stimulus levels. This level tolerant excitatory area is flanked by inhibitory sidebands. Type O unit maps (bottom panel) are characterized

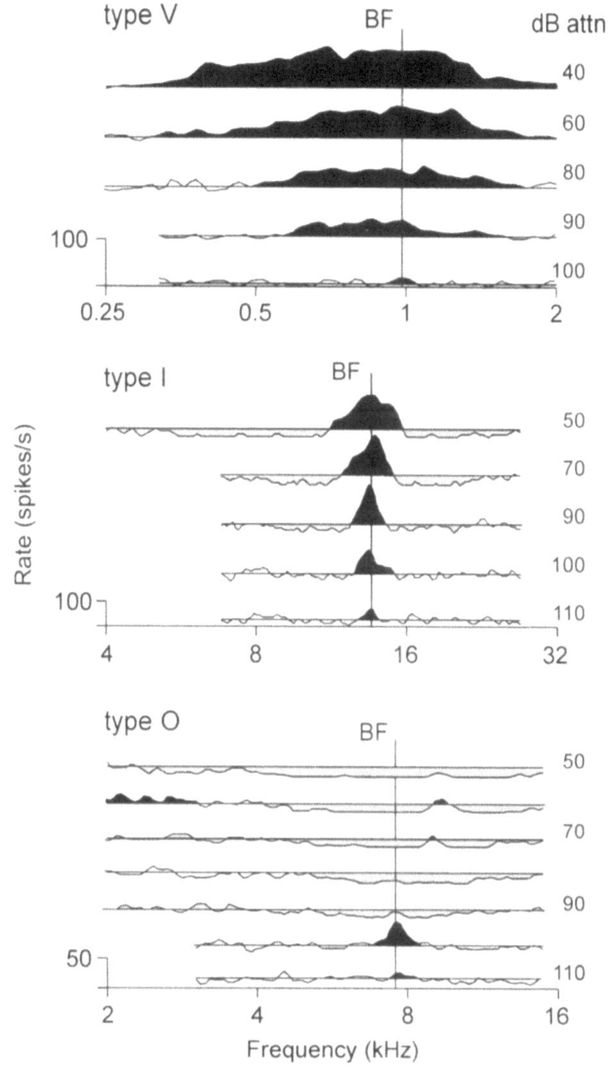

FIG. 7. Three general patterns of excitatory tuning found in the ICC of the cat. Excitatory tuning may be V-shaped (type V, top), remain narrow with level (type I, middle), or be limited to a circumscribed range of stimulus frequencies and levels (type O, bottom). In all panels, stimulus-driven rates are plotted against frequency at multiple sound levels (numerical labels on right). Horizontal lines indicate average spontaneous rate; black (gray) filled areas indicate excitatory (inhibitory) response areas. Sound pressure levels (SPL) are in dB attenuation; absolute SPL varies with the acoustic calibration, but 0 dB attn. is near 100 dB re 20 μPa for tones. Vertical lines indicate BFs. (From Ramachandran *et al.* (1999), reprinted by permission from the American Physiological Society.)

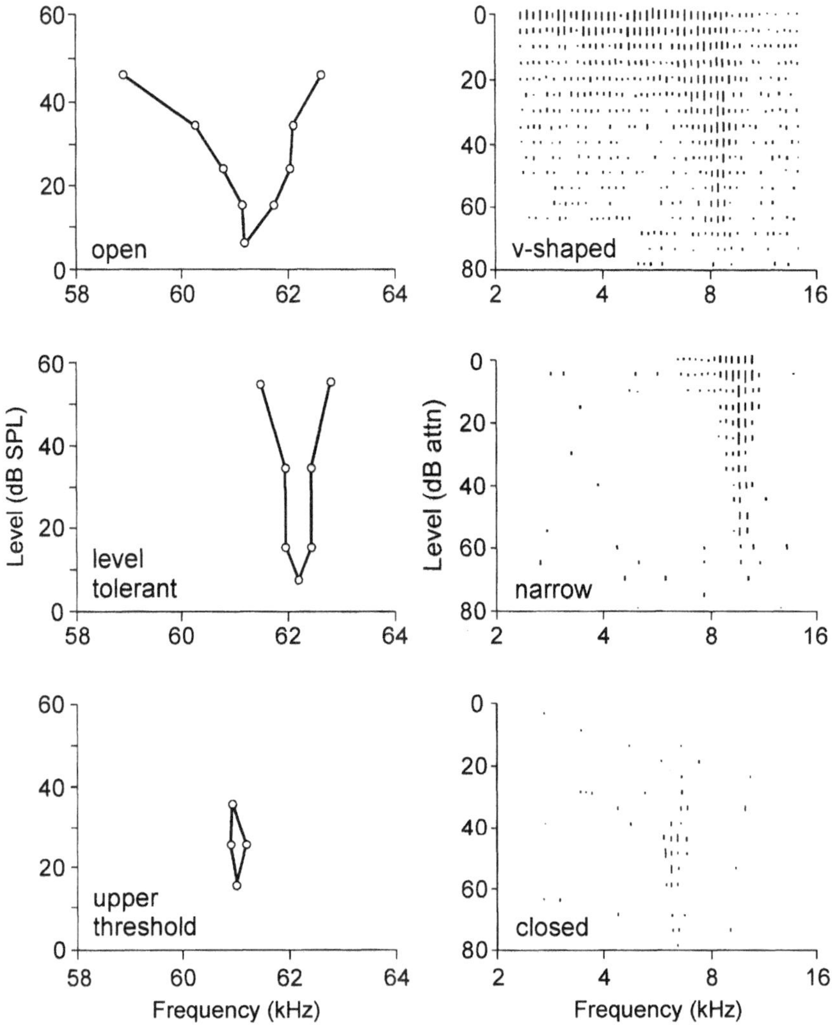

FIG. 8. Patterns of excitatory tuning found in the ICC of mustache bat (left column) and guinea pig (right column); unit names are on plots. Mustache bat data (left column): tuning curves connect the stimulus frequencies and levels (circles) that elicited 10% of the maximum spike count evoked at each level tested. (From Yang *et al.* (1992), redrawn by permission from the American Physiological Society.) Guinea pig data (right column 2): spike counts (shown as bars) are plotted against frequency at multiple sound levels (specified in dB attenuation). (From Le Beau *et al.*, Copyright 2001, by the Society for Neuroscience.)

by an O-shaped island of excitation at BF near threshold, which gives way to inhibition at higher intensities. Excitatory areas at higher levels away from BF are sometimes observed, but these are variable features.

Representative examples of the three major unit types in the mustache bat (Yang *et al.*, 1992) and guinea pig (LeBeau *et al.*, 2001) are shown in columns 1 and 2 of Fig. 8, respectively. The tuning curves in the bat were constructed from frequencies that evoked spike counts equal to 10% of the maximum rate at that sound level. The maps from the guinea pig show spike count (as a bar) against stimulus frequency and level. The similarity of these unit types to those observed in the cat is clear. Similar to their counterparts in the cat, the tuning curves of open units in the bat and V-shaped units in the guinea pig broaden on both sides of BF with increasing sound level, the tuning curves of level tolerant and narrow units show limited expansion at higher intensities, and the tuning curves of upper threshold and closed units have a clearly circumscribed upper limit such that they do not respond to high-intensity tone bursts at their BF.

Despite these basic similarities, the processing of narrowband stimuli in the ICC differs among species in at least two ways. First, the units that produce a type V response in decerebrate cats all have low BFs ($\leq$3 kHz), whereas comparable unit types in other species can be found across the full range of their audible frequencies. Second, the incidence rate of specific unit types shows considerable variation among species (Table I). In particular, type V units are the most prevalent unit type for a number of species (mustache bat, guinea pig, chinchilla, mouse, rat, big brown bat), while type O units constitute a high percentage of the unit types in other species (decerebrate cat, pallid bat, rhesus monkey). These two differences may be related and reflect, at least in part, the fact that different authors used different classification methods. For example, type V unit response maps in the decerebrate cat do not, by definition, show inhibitory sidebands (otherwise they would classified as type I units; Ramachandran *et al.*, 1999), whereas the maps of V-shaped units in other species may exhibit such features (e.g., guinea pig: Le Beau *et al.*, 2001; rat: Hernandez *et al.*, 2005). Also, type V units in cats show monotonic increases in firing rate with sound level, but V-shaped units in other species, e.g., the guinea pig and the rat, may show highly non-monotonic responses that resemble those of type O units in cats. A second explanation for the differences between species is that the different studies used different anesthetics. This is a likely explanation for the differences in unit incidence rates observed within a particular species (e.g., cat and guinea pig). Moreover, it is well known that various anesthetic regimens can reduce or even abolish the amount of inhibition in the response maps of CN units (Evans and Nelson, 1973; Young and Brownell, 1976), which are the precursors directly or indirectly of all response types in the ICC.

TABLE I
INCIDENCE OF ICC UNIT TYPES ACROSS SPECIES

Animal	Preparation	Type V units	Type I units	Type O units
cat [a]	decerebrate	11	32	49
cat [b]	pentobarbital sodium	58	30	12
mustache bat [c]	awake	64	22	14
pallid bat [d]	pentobarbital sodium	20	44	36
guinea pig [e]	urethane	84	6	9
guinea pig [e]	chloralose	67	21	11
rhesus monkey [f]	awake	39	26	35
chinchilla [g]	pentobarbital sodium	55	37	8
mouse [h]	ketamine/xylazine	66	18	10
rat [i]	urethane	69	4	11
big brown bat [j]	methoxyflurane/fentanyl	57	14	20

Table entries are percentages. Units are classified based on general patterns of excitatory tuning: type V maps show a V-shaped excitatory area, type I maps show a restricted I-shaped area of excitation, and type O maps show an O-shaped island of excitation at low stimulus levels. The following equivalences were made:

[a]Ramachandran *et al.,* 1999 (V = V, I = I, O = O);
[b]Ehret and Merzenich, 1988 (V = wide, I = steep, O = closed + tilted);
[c]Yang *et al.,* 1992 (V = open, I = level tolerant, O = upper threshold);
[d]Fuzessery and Hall, 1996 (V = V, I = vertical, O = closed);
[e]Le Beau *et al.,* 2001 (V = V, I = narrow, O = closed + low tilt + high tilt);
[f]Ryan and Miller, 1978 (V = simple, I = W, O = complex);
[g]Wang *et al.*, 1996 (V = open, I = level tolerant, O = upper threshold);
[h]Egorova *et al.,* 2001 (V = class I + class III, I = class II narrow, O = class II closed);
[i]Hernandez *et al.*, 2005 (V = V, I = narrow, O = closed + low tilt + high tilt);
[j]Casseday and Covey, 1992 (V = V, I = narrow, O = closed).

A more intriguing alternative is that the differences between species reflect species-specific variations in the auditory processing of spectral information. For example, it is known that rodent species (gerbils: Davis *et al.*, 1996; mouse: Sawtell *et al.*, 2005) and cats (Young and Brownell, 1976) show important differences in their DCN response patterns. As in the ICC of a lightly anesthetized mouse, closed frequency response maps are relatively rare in the DCN of an unanesthetized mouse. In contrast, closed frequency response maps are common in both the DCN and ICC of the cat. If the DCN is the precursor of most type O units in all mammals, as it is in cat (Davis, 2002; see section entitled "Effects of Blocking the Output of the DCN"), then a decrease in the availability of the closed response areas in DCN would be expected to reduce the number of type O units in the ICC while simultaneously increasing the number of type V units found at all audible frequencies. The role of ethological factors, such as predator/prey relationships, in shaping the functional expression of spectral processing pathways in the auditory system is an area for future study.

VII. How Do the Various Inputs to the ICC Shape the Frequency Response Areas of Units in ICC?

The existence of three physiologically distinct frequency response map types in the ICC suggests that the response types receive dominant excitatory inputs from different sources and are specialized to play complementary roles in the processing of acoustic information. Consistent with this interpretation, the anatomical organization of the ICC predicts that there are four distinct functional domains in the ICC: a VCN type, a DCN type, an MSO type, and an LSO type (see section entitled "Synaptic Domains in the ICC"; reviewed by Oliver, 2004; Oliver and Huerta, 1992). Based on resemblances to response maps in these brainstem nuclei (compare Fig. 6 to 7), the VCN (Joris *et al.*, 1994; Shofner and Young, 1985; Winter and Palmer, 1990) and/or the MSO (Goldberg and Brown, 1969; Guinan *et al.*, 1972) are likely inputs to type V units, the LSO could provide the dominant input to type I units (Caird and Klinke, 1983), and the DCN is a potential source of input to type O units (Spirou and Young, 1991; Young and Brownell, 1976).

On the other hand, neural responses in the ICC are not likely to be simple replicates of those in the brainstem given that the synaptic domains in the ICC are based on combinations of two or more inputs, including at least one from an inhibitory source. Principal cells in the DNLL (Saint-Marie *et al.*, 1997) and neurons intrinsic to the ICC (Oliver and Beckius, 1992) are known sources of GABAergic inhibition in the ICC, whereas projection neurons in the ipsilateral LSO (Glendenning *et al.*, 1992; Oliver *et al.*, 1995; Saint-Marie and Baker, 1990; Saint-Marie *et al.*, 1989) and VNLL (Saint-Marie *et al.*, 1997) are sources of glycinergic inhibition. *In vivo* whole cell patch clamp and intracellular recordings show that most ICC neurons receive synaptic inputs from both excitatory and inhibitory sources and, furthermore, that intrinsic membrane properties can shape a neuron's response to sound (Covey *et al.*, 1996; Kuwada *et al.*, 1997). Thus, the question arises: to what extent do the various excitatory and inhibitory inputs to the ICC shape the spectral response properties of different ICC unit types?

A. Effects of Blocking the Output of the DCN

Principal cells in the CN, MSO, and LSO have unique response properties and are primary excitatory sources of innervation for the ICC. At the present time, only the influence of the DCN on ICC response map types has been tested (Davis, 2002). In these experiments, the frequency response maps of ICC units in the decerebrate cat were recorded before and after reversible inactivation of the output of the DCN, in isolation, by injection of lidocaine into the dorsal acoustic striae (DAS).

An example of the effect of lidocaine on the response of a representative type V unit is shown in Fig. 9A. Each plot shows the unit's discharge rate to tones as function of frequency at a fixed sound level; two levels are shown in the figure, 10 and 40 dB above threshold. The pre-drug (control) response is shown with solid lines while the post-drug (lidocaine) response is shown with heavy dashed lines.

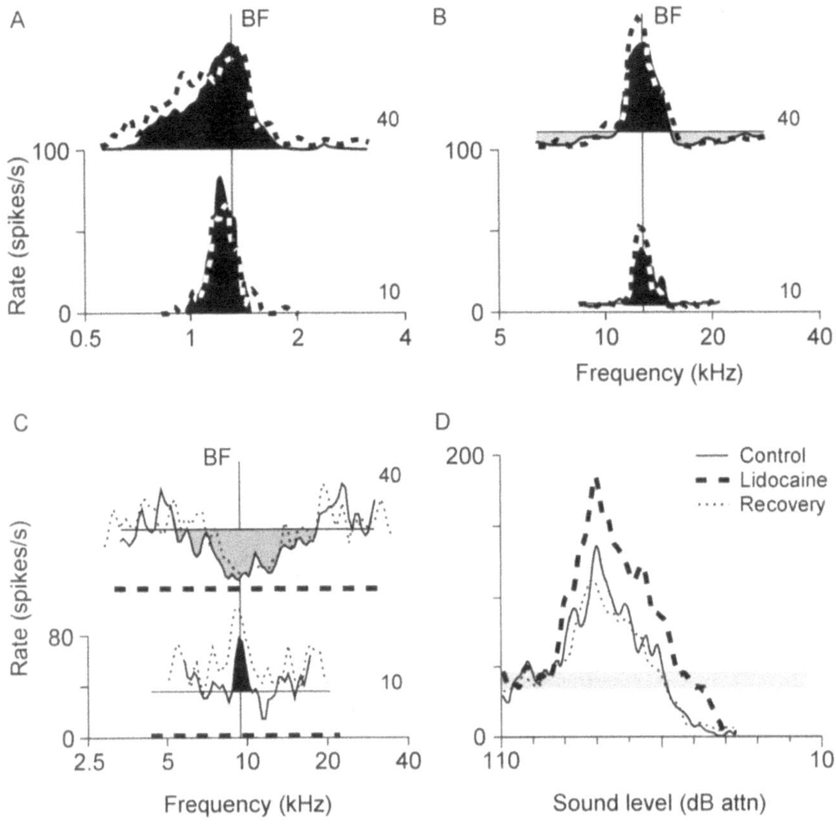

FIG. 9. Effects of reversible inactivation of the output of the DCN on unit types in the ICC of cat. A to C: partial frequency response maps of typical type V (A), type I (B), and type O units (C) before (control = solid lines) and 5 minutes after (dashed lines) injection of lidocaine into the dorsal acoustic striae; the dotted lines indicate recovery curves (where available). Note that the rate of the type O unit (C) was reduced to 0 during the lidocaine injection. Stimulus-evoked discharge rates are plotted against frequency at 10 and 40 dB re threshold (labels at right). Horizontal lines indicate average spontaneous rate; black (gray) filled areas indicate excitatory (inhibitory) response areas under control conditions. The vertical lines indicate BF. (D) BF-tone rate-level functions for an atypical type O unit showing an increase in activity after lidocaine injection. The shaded bar shows the range of spontaneous rate recorded under control conditions. (Panels C and D from Davis, 2002, reprinted by permission from the American Physiological Society.)

To ease recognition of the unit type, regions where the stimulus-driven activity under control conditions is excitatory are black-filled and regions of inhibition are gray-filled. Note that under control conditions, the response map of the type V unit displayed a characteristic V-shaped excitatory area. After the DAS was inactivated, the type V unit showed a small (25%) decrease in stimulus-driven activity at frequencies near BF, but showed increased activity at frequencies remote to BF, particularly on the low-frequency side. Fig. 9B shows an example of the response of a type I unit to DAS inactivation. Compared with the control condition, application of lidocaine resulted in a small (20%) increase in the driven activity at frequencies near BF; otherwise, the map retained a very level-tolerant excitatory area flanked by inhibition.

In contrast to the minor changes in type V and type I units, more dramatic changes in the magnitudes of responses were found in type O units when the DAS was inactivated. The most common response observed in type O units is a marked decrease in rate; an example from this group is shown Fig. 9C. Under control conditions, this unit showed the characteristic low-level island of excitation and inhibition at high levels. When lidocaine was applied to the DAS, the spontaneous rate of the unit decreased from 40 to 0 spikes/s in 2 minutes, and the unit lost entirely its tone-evoked responses. The dotted line shows that the response recovered after about 1 hour. On the other hand, some type O units showed an increase in response magnitude after DAS inactivation. A rate-level curve for one such unit is shown in Fig. 9D. Application of lidocaine produced an increase in stimulus-evoked firing at all stimulus levels. Overall, 85% of affected type O units showed a decrease in activity after the DAS was blocked, while 15% showed a rate increase. Furthermore, all of the former units had low maximal driven rates to tones, whereas the latter units had high rates.

Taken together, these results suggest that the DCN has a variety of affects in ICC. For type V and type I units it is a minimal source of input. For low-rate type O units, however, it is a dominant source of excitation. Consistent with this interpretation, Semple and Aitkin (1980) found that electrical activation of the DAS excited ICC units with highly non-monotonic BF tone rate-level curves. Increased activity in high-rate type O units suggests that, in addition to its excitatory input to the ICC, some pathways from the DCN are capable of producing inhibition. Such effects are likely to involve interneurons, perhaps within the nuclei of the lateral lemniscus or the ICC.

B. Role of Inhibition in Shaping Frequency Response Areas in ICC

The role of inhibition in shaping the frequency response areas in ICC can be assessed directly by combining neuronal recording with iontophoretic application of inhibitory transmitter antagonists. Such experiments have been performed in a variety of species (guinea pig: LeBeau *et al.*, 2001; horseshoe bat: Vater *et al.*, 1992; mustache bat: Yang *et al.*, 1992; pallid bat: Fuzessery and Hall, 1996; cat: Davis, 1999, 2002; chinchilla: Palombi and Caspary, 1996). All of these studies examined the functional role of GABA; only a subset also tested the functional role of glycine (Davis, 1999, 2002; LeBeau *et al.*, 2001; Vater *et al.*, 1992). Representative data on the effects of bicuculline, a $GABA_A$ receptor antagonist, from three species are shown in Figs. 10 and 11.

Figure 10 shows the range of effects of bicuculline on the frequency response types in the decerebrate cat (Davis, 1999, 2002). Response map data for typical type V, I, and O units are shown in Figs. 10A to 10C; the response of an atypical type O unit is shown in Fig. 10D. In all of these plots, the response under control conditions is shown with solid lines and the response during application of bicuculline is shown with heavy dashed lines. The major effects of bicuculline for the entire sample of tested units were increases in the level of spontaneous and maximum stimulus-driven rates (Fig. 10E). On average, spontaneous rates increased by 266% and tone-evoked rates increased by 180%. Application of bicuculline also caused the excitatory areas in all response map types to broaden slightly, particularly at higher stimulus levels (Fig. 10F). For the most part, however, administration of bicuculline did not substantially antagonize existing inhibitory response areas in typical type I (Fig. 10B) and type O units (Fig. 10C); in fact, the inhibition usually appeared stronger during GABAergic blockade because of the higher spontaneous rates. For some type V units (Fig. 10A), the application of bicuculline actually revealed the presence of inhibitory sidebands, almost always on the high frequency side. In contrast to these units, the excitatory areas of some type O units (~20%) substantially expanded under bicuculline into areas that showed inhibition in control conditions (Fig. 10D). All of these units showed high maximum BF tone-driven rates in control conditions and there was a tendency for these units to have low BFs.

The effects of bicuculline on response map types in guinea pig (LeBeau *et al.*, 2001) and mustache bat (Yang *et al.*, 1992) are shown in Figs. 11A and B, respectively. For the guinea pig data (Fig. 11A), the control response map and bicuculline map for a single unit are shown side-by-side. Compared with the control condition, the guinea pig data clearly show that application of bicuculline caused a marked increase in stimulus-evoked firing rate across the whole of the response area. Despite this change in activity, most V-shaped neurons (A1-A2) showed little expansion of their excitatory areas. Furthermore, in the few cases where the maps of V-shaped units showed sideband inhibition, this inhibition was not abolished by bicuculline. In contrast, all units with non–V-shaped response areas, like the narrow (A3-A4) and closed units (A5-A6) shown here, became V-shaped or showed marked expansion in the presence of bicuculline.

The results obtained in the mustache bat are shown in Fig. 11B. In these plots, control tuning curves for each unit are shown with solid lines and post-drug

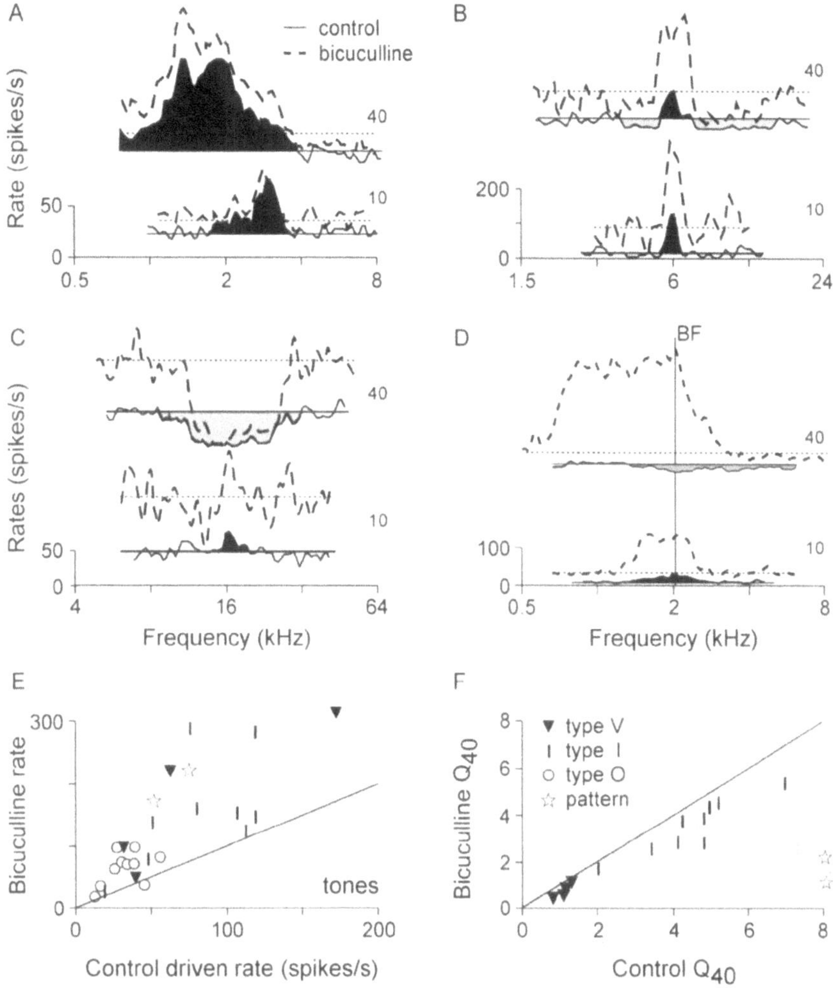

FIG. 10. Effects of bicuculline on the tuning curves of ICC unit types in cat. A to C: examples of type V (A), type I (B), and O unit (C) maps that retain their characteristics features under the influence of bicuculline. D: a type O unit map that shows a pattern change after local application of bicuculline. In panels A to D, responses before the injection of bicuculline (control condition) are shown as solid lines, and responses after the injection are shown as dashed lines. Stimulus driven rates are plotted against frequency at 10 and 40 dB re threshold (labels at right). Horizontal lines indicate average spontaneous rate; black (gray) filled areas indicate excitatory (inhibitory) response areas under control conditions. E: comparison of each unit's maximum driven rate under bicuculline versus control conditions; the diagonal line indicates equal driven rates. Both pattern change units were type O units under control conditions. F: comparison of each unit's Q40 for excitation (BF divided by the excitatory bandwidth 40 dB re threshold) under bicuculline versus control conditions. Type O units are not shown in this plot because these units do not show excitatory responses at this stimulus level under control conditions. (Panels C and D from Davis (2002), reprinted by permission from the American Physiological Society.)

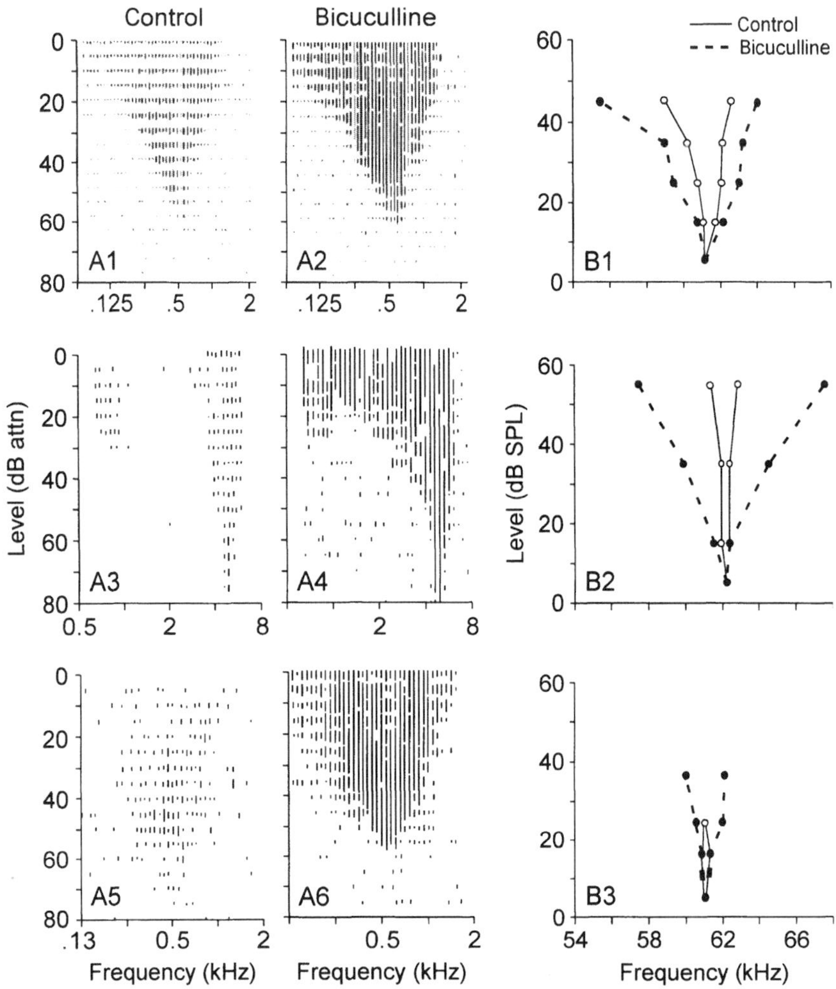

FIG. 11. Effects of bicuculline on the tuning curves of ICC unit types in the guinea pig (A1 to A6) and mustache bat (B1 to B3). A1-A6: frequency response maps of a V-shaped (top), narrow (middle), and closed unit (bottom) in the guinea pig before (left column) and after (right column) application of bicuculline. Spike counts (shown as bars) are plotted against frequency at multiple sound levels (specified in dB attenuation). (From Le Beau *et al.*, Copyright 2001, by the Society for Neuroscience.) (B1–B3) tuning curves for an open (B1), level tolerant (B2), and upper-threshold unit (B3) in mustache bat are shown before (solid lines) and after (dashed lines) an injection of bicuculline. Tuning curves connect the stimulus frequencies and levels (circles) that elicited 10% of the maximum spike count evoked at each level tested. (Redrawn from Yang *et al.* (1992), reprinted by permission from the American Physiological Society.)

curves are shown with dashed lines. One of the key points of this study was that bicuculline caused tuning curves to broaden in approximately 50% of all units. The effects on the different unit types, however, were not uniform. That is, broadening was found in only 30% of units with open tuning curves (B1), whereas it was found in 88% of upper threshold units (B3). The tuning curves of 54% of level tolerant units (B2) broadened after application of bicuculline. In most cases, the expansion of the excitatory areas was most pronounced at higher stimulus levels. As a result, most upper threshold units became V-shaped under the influence of bicuculline.

In general, there is substantial agreement among the results of the various studies that have used pharmacological inactivation to examine the local role of inhibition in the ICC. For example, one similarity is that application of bicuculline causes stimulus-driven firing rates to increase in all response types. This suggests that one common role of GABAergic inhibition is to regulate the overall excitability of neurons. A second similarity among the studies is that bicuculline causes the tuning curves of many units to broaden suggesting that GABAergic inhibition also sharpens tuning. The data suggest that this function is least pronounced in units with simple V-shaped tuning curves, and most pronounced in units with closed tuning curves. Third, the effects of strychnine, when tested, are similar to those of bicuculline in all respects, except that the magnitude of effects is usually smaller.

One clear difference among the studies, however, is the proportion of units with closed excitatory areas that show substantial size and shape changes after application of an inhibitory antagonist, particularly bicuculline. In the cat, only 20% of type O units show a dramatic upward expansion of their excitatory areas (Davis, 2002), whereas in two species of the bat, chinchilla, and guinea pig, the percentage is over 80% (LeBeau *et al.*, 2001; Palombi and Caspary, 1996; Vater *et al.*, 1992; Yang *et al.*, 1992). This discrepancy could reflect a methodological difference. In the cat study, the tonal stimuli were long duration (200 ms) and driven rates were measured over the last 150 ms of the stimulus duration; therefore, these rate estimates represent steady-state rates. In contrast, the other studies used short duration stimuli ($\leq$50 ms) and stimulus-evoked rate estimates included the onset component of the response. If the analysis of the cat data is confined to the first 50 ms of the stimulus, then the percentage of units showing upward expansion of their excitatory area climbs to over 90%, which is comparable to the rates observed in other species. Alternatively, given the differences noted in the section entitled "Frequency Response Map Types in the ICC" between these species (i.e., different distribution of unit types as a function of frequency and unit incidence rates) this discrepancy may reflect a species difference for this role of inhibition in the ICC.

C. Effects of Inactivation of the DNLL

Experiments where inhibitory antagonists are applied directly to the ICC (described in the section entitled "Role of Inhibition in Shaping Frequency Response Areas in ICC") clearly show that both GABA and glycine shape the frequency response areas of ICC neurons (Davis, 1999, 2002; Fuzessery and Hall, 1996; LeBeau *et al.*, 2001; Palombi and Caspary, 1996; Vater *et al.*, 1992; Yang *et al.*, 1992). From such experiments, however, it is not possible to determine the sources of these effects. Principal cells in the DNLL (Saint-Marie *et al.*, 1997) and neurons intrinsic to the ICC (Oliver and Beckius, 1992) are known sources of GABAergic inputs, whereas neurons in the ipsilateral LSO (Glendenning *et al.*, 1992; Oliver *et al.*, 1995; Saint-Marie and Baker, 1990; Saint-Marie *et al.*, 1989) and VNLL (Saint-Marie *et al.*, 1997) are origins of glycinergic effects. To date, only the influence of the DNLL on ICC response map types has been tested (Thornton and Rees, 2001). In these experiments, the frequency response maps of ICC units in anesthetized guinea pigs were recorded before and after reversible inactivation of the output of either the ipsilateral or contralateral DNLL by an injection of kynurenic acid.

The frequency response maps in guinea pig are grouped in two general categories: V-shaped and non–V-shaped, which includes narrow and closed types. None of the V-shaped response areas showed shape changes following inactivation of either DNLL, although changes in overall firing rate (usually decreases) were observed across the whole of the excitatory areas. In contrast, half of the non–V-shaped units showed changes in shape after the DNLL was inactivated. Both expansion and contraction of excitatory response areas were observed. Fig. 12 shows an example of the effect of DNLL inactivation on a narrow unit. Under control conditions (Fig. 12A), this unit showed a level-tolerant excitatory response area with a high-level low-frequency tail. Four minutes after kynurenic acid was injected in the DNLL (Fig. 12B), the driven rate of this unit decreased and the excitatory area contracted. The unit nearly recovered to its control state in approximately one hour (Fig. 12D). Overall, inactivation of the contralateral DNLL resulted in only contraction of non–V-shaped response map areas, whereas after ipsilateral inactivation, both contraction and expansion of narrow units were observed and one closed unit showed expansion.

The difference between the effect of DNLL inactivation on the shape of V-shaped and non–V-shaped ICC units is similar to that observed with the application of bicuculline in the ICC, which also changes only the shape of non–V-shaped units (e.g., Le Beau *et al.*, 2001). Further, expansion of response areas after DNLL inactivation is consistent with a previous finding that the output of the DNLL to the ICC is mainly inhibitory (Shneiderman and Oliver, 1989). Contraction of response areas on inactivation of the DNLL, however, suggests

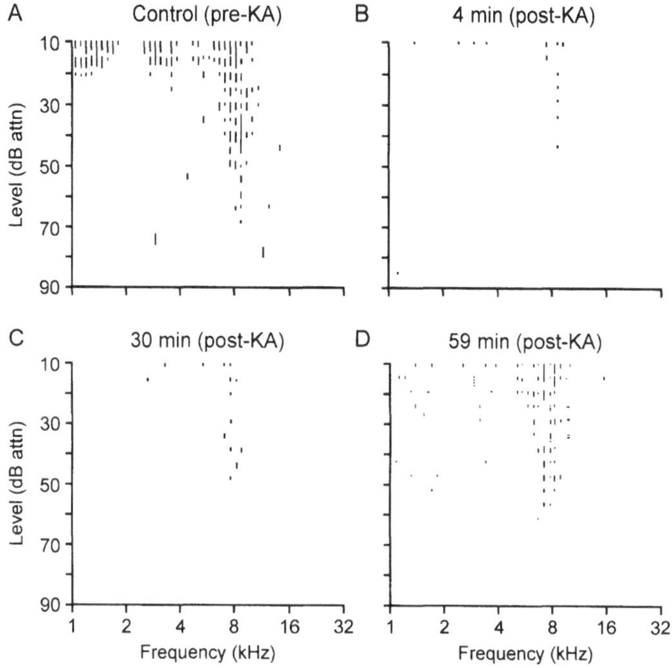

FIG. 12. Influence of the DNLL on the frequency response areas of ICC units in guinea pig. Frequency response map of a narrow unit before (A) and 4, 30, and 59 minutes after (B-D) injection of kynurenic acid into the DNLL. Response areas show spike counts (bars) as a function of stimulus frequency and intensity (specified in dB attenuation). (From Thorton and Rees, 2001, reprinted with permission from Shaker Publishing BV.)

that some pathways from the DNLL are capable of producing inhibition. Such effects could be mediated by the minority of DNLL neurons that are not GABAergic (Saint-Marie *et al.*, 1997) and, therefore, presumably excitatory. Alternatively, there may be multisynaptic pathways from the DNLL to the ICC involving interneurons in the ICC or cross connections between the two DNLLs.

D. Role of the Contralateral IC

In addition to its many ascending inputs, the ICC also receives input from the contralateral IC via a fiber tract called the commissure of the inferior colliculu (CoIC) (Coleman and Clerici, 1987; Malmierca *et al.*, 1995; Saldana and Merchan, 1992). Anatomical evidence suggests that the projection from one IC to the other is excitatory and glutamatergic (Saint-Marie, 1996; Zhang *et al.*,

1998), but physiological evidence from *in vitro* preparations suggests that the projection is both excitatory and inhibitory (Moore *et al.*, 1998; Smith, 1992). The influence of the contralateral IC on response map types in the ICC has been studied by Malmierca *et al.* (2003). In these experiments, the frequency response maps of ICC units in the anesthetized rat were recorded before and after reversible inactivation of the contralateral IC by an injection of kynurenic acid.

The main finding in these experiments is that most units (8 of 12) showed a significant change in the excitatory area of their monaural frequency response maps. Six of these units showed a reduction in the area of their maps, whereas two units showed a significant expansion. An example of the change seen in one of these units is shown in Fig. 13. Compared with the control condition (Fig. 13A), injection of kynurenic acid into the contralateral IC results in a clear decrease in stimulus-driven activity throughout much of the map and a contraction of the excitatory response area (Fig. 13B). The change in this unit is the smallest significant change in the whole sample.

The small sample of units tested precludes detailed analysis of the effects of inactivation on different unit types. However, the observation that block of the commissure leads to contraction or expansion of frequency response areas is consistent with findings that activation of the CoIC elicits both excitatory and inhibitory post-synaptic potentials in ICC neurons (Moore *et al.*, 1998; Smith, 1992). It also demonstrates that one IC can influence the processing of spectral information in the opposite ICC.

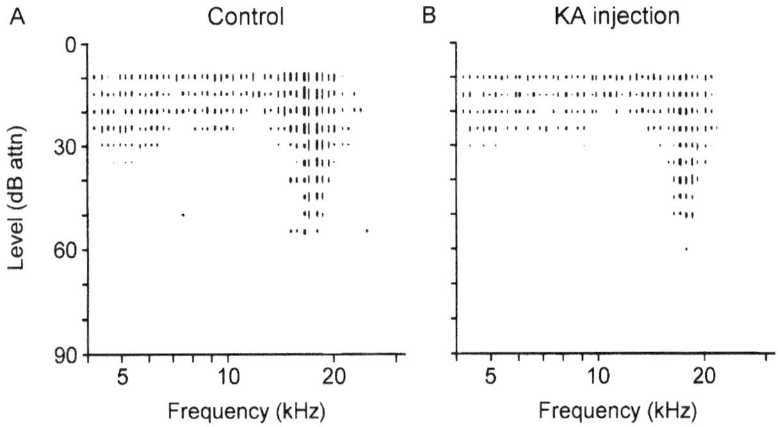

FIG. 13. Effects of blockade of the commissure of the inferior colliculus on the frequency response areas of ICC units in rat. Monaural frequency response maps of a unit before (A) and after (B) injection of kynurenic acid into the contralateral IC. Response areas show spike counts (bars) as a function of stimulus frequency and intensity (specified in dB attenuation). (From Malmierca *et al.*, 2003, reprinted with permission from Springer-Verlag.)

E. Descending Influence of the Auditory Cortex

Anatomic studies suggest that the ICC of most species receives only sparse descending projections from the thalamus and auditory cortex (Adams, 1980; Andersen *et al.*, 1980; Winer *et al.*, 1998; for a review, see Huffman and Henson, 1990; Oliver and Huerta, 1992). Instead, the projections from auditory cortex primarily target the dorsal cortex and paracentral nuclei of the IC. In turn, these subdivisions mainly project to the dorsal and medial nuclei of the medial geniculate body, respectively (Calford and Aitkin, 1983), although some cells in the dorsal cortex also project to the commissural system linking the two IC (Aitkin and Phillips, 1984). Nonetheless, it is clear from physiological studies that electrical stimulation of the auditory cortex can cause both excitatory and inhibitory effects with short latencies in the ICC neurons (Bledsoe *et al.*, 2003; Mitani *et al.*, 1983; Syka and Popelar, 1984).

The influence of the auditory cortex on spectral processing in the ICC has been studied in a variety of species (big brown bat: Gao and Suga, 1998; Jen *et al.*, 1998; Sun *et al.*, 1996; Yan and Suga, 1998; mustache bat: Zhang and Suga, 1997 and 2000; Zhang *et al.*, 1997; mouse: Yan and Ehret, 2001). In some of these experiments, the activity of ICC neurons was recorded before and after the auditory cortex was inactivated by application of lidocaine. In other experiments, the activity of ICC neurons was recorded before and after the cortex was stimulated with either acoustic or electrical stimuli or a combination of both. The results of cortical inactivation and stimulation are complementary. Overall, the results suggest that when the BFs of the cortical area and collicular neuron are matched, stimulation of the cortex does not cause the frequency response curve of the collicular unit to shift in frequency but does sharpen its tuning. In contrast, cortical activation shifts collicular tuning curves when the BFs of the cortical area and collicular neuron are mismatched. Centripedal BF shifts are shifts towards the BF of the cortical area, whereas centrifugal shifts are BF shifts away from the cortical BF. Together, the cortical effect on matched and unmatched neurons is called egocentric selection (Zhang *et al.*, 1997). Importantly, the general shapes of tuning curves do not appear to be altered by this process.

The particular changes associated with the egocentric selection are species-specific. In the mustache bat, cortical neurons mediate a highly focused positive feedback associated with widespread bilateral (symmetric) inhibition (Zhang *et al.*, 1997). Figure 14 shows the effect on two mismatched ICC neurons after the cortex is inactivated with lidocaine. In these figures, the frequency tuning curves under control conditions are shown with open circles, curves during inactivation are shown with filled circles, and the recovery curves are shown with broken lines. Compared to control conditions, inactivation of the cortex shifts these curves towards the BF of the cortical neuron (arrow on abscissa). This suggests that under control conditions, cortical activation evokes centrifugal shifts in

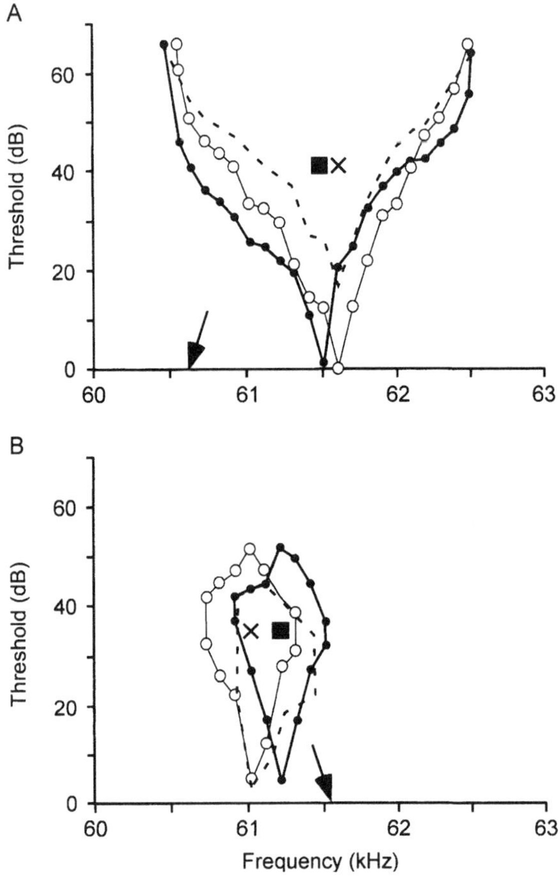

FIG. 14. Corticofugal modulation of frequency response maps in the ICC of mustache bat. Tuning curves of two units before (control, open circles), during (filled circles), and after (recovery, dashed lines) focal cortical inactivation with lidocaine. Tuning curves connect the stimulus frequencies and levels that elicited a firing rate of 4 spikes/s. The BFs of the inactivated cortical neurons are indicated by arrows. Crosses and squares indicate the frequencies with the maximum discharge rates before and after cortical inactivation, respectively. (From Zhang *et al.*, 1997, reprinted with permission from the Nature Publishing Group.)

collicular neurons and thereby enhances the contrast in the neural representation of auditory information. By contrast, egocentric selection is asymmetric in the big brown bat (Yan and Suga, 1998). That is, centrifugal BF shifts were only observed for ICC neurons with BFs below the cortical BF. For neurons above the BF, the collicular neurons showed centripedal shifts. Finally, in mice, shifts in collicular tuning were always toward the cortical BF (centripedal: Yan and Ehret, 2001). Furthermore, comparisons with psychophysical measures suggested that the corticofugal adjustments in mice were related to the critical bandwidths (i.e., to the frequency resolution of the auditory system) in this species.

VIII. On the Roles of the ICC in Processing Spectral Information

Experiments directed towards examining the roles of excitatory and inhibitory inputs in generating the spectral response properties of ICC units reveal that ICC unit types integrate information from a variety of sources. On the surface, the frequency response maps of ICC units in all species exhibit three distinct patterns of excitatory tuning, including V-, I-, and O-shaped, which resemble those in the VCN and MSO, the LSO, and the DCN, respectively. However, it is clear that inactivation of single inputs to the ICC (e.g., the DCN and the DNLL) can affect all three unit types, and sometimes produce counterintuitive results. For example, blockade of the DCN, a putative excitatory source to the ICC, can lead to increased activity in some ICC units, while inactivation of the DNLL, a putative inhibitory source, can lead to decreased activity. These results, therefore, suggest that further refinement of response categories will be necessary before the specific roles of the ICC in processing spectral information are known.

Two general roles of the ICC in spectral processing can be discerned from experiments that have used pharmacological techniques to block inhibition within the ICC. First, inhibitory inputs sharpen the frequency tuning of many neurons in the ICC. In response to blockade of $GABA_A$ receptors with bicuculline, the tuning curves of many ICC units broaden, particularly at high stimulus levels. Non–V-shaped units, in particular, can show a marked expansion of excitatory areas. This finding is in general agreement with the type of sharpening found at higher levels of processing (e.g., the thalamus: Suga *et al.*, 1997; the cortex: Suga and Tsuzuki, 1985), suggesting that frequency tuning sharpens as one moves up the auditory system. One role of such sharpening is to ensure the separation of spectral peaks by establishing critical bands (Ehret and Merzenich, 1988), a necessary condition for the perception and recognition of complex sounds. Second, inhibitory inputs regulate the excitability of ICC units. It is possible that descending inputs from the cortex use such inhibitory effects as attentional filters for gating biologically important information in the ascending auditory system. Such inputs are also known to sharpen or endow ICC units with novel sensitivity to more complex spectral features such as frequency modulations (Fuzessery and Hall, 1996; Koch and Grothe, 1998) and species-specific cells (Klug *et al.*, 2002).

Understanding the functional roles of the separate map types will depend on acquiring three kinds of evidence. First, there is anatomical evidence. If the inputs and outputs of a particular map are known, then it is possible to infer the functional role of the map. For example, if type V units (or some sub-population thereof) receive their dominant input from VCN stellate cells, then it is likely that those type V units would be conveying a rate-based representation of the spectral composition of a complex sound (see Chapter 5 for more information on the response properties of VCN units). If, on the other hand, the dominant input to type V units comes from the MSO, then it is likely that type V units would play an important role in auditory behaviors that required faithful transmission of temporal information such as the encoding of stimulus location based on interaural time differences (ITDs) (Kuwada and Yin, 1983; Yin *et al.*, 1986, 1987). In the case of type I units, if their dominant input is from the LSO, then such units would play a role in encoding stimulus location based on interaural level differences (ILDs) (Boudreau and Tsuchitani, 1968; Caird and Klinke, 1983; Guinan *et al.*, 1972; Tsuchitani and Boudreau, 1969). Finally, type O units, by virtue of input from the DCN, would be sensitive to spectral cues to the location of sounds in space (Imig *et al.*, 2000; Young *et al.*, 1992).

A second source of evidence comes from the physiological response properties of the unit types themselves. By analyzing the response properties of a particular unit type, it may be possible to demonstrate that the units are encoding certain aspects of the stimulus and not others. In cats, for example, it has been shown that only type V units exhibit rate modulations to 1-Hz binaural beat stimuli that attain a maximum (peak) at the same ITD regardless of carrier frequency (Ramachandran and May, 2002). This pattern of response resembles that of MSO units (Batra *et al.*, 1997) suggesting that type V units may indeed encode stimulus location based on ITDs. In contrast, type I units show strong excitatory/inhibitory binaural interactions (Davis *et al.*, 1999), which resemble those of LSO units (Caird and Klinke, 1983; Goldberg and Brown, 1969; Guinan *et al.*, 1972). Therefore, this suggests that type I units are suitable to encode sound location based on processing of ILDs. Finally, type O units alone exhibit frequency-specific excitatory responses to simple analogs of head-related transfer functions (Davis *et al.*, 2003), the filter functions that describe the transformation of a free-field sound to the energy at the eardrum (Wightman and Kistler, 1989). These data are consistent with the hypothesis that type O units are uniquely specialized for processing directionally dependent spectral features.

The third source of evidence is behavioral analysis of the consequences of activating or deactivating a map type. If a behavior is altered when a map is manipulated, then it is likely that the map contributes in generating or controlling that behavior. This sort of experiment is difficult in the ICC because the unit types are intermingled within the nucleus. If, however, the excitatory source is known, then it may be possible to manipulate the source in isolation. For example, lesioning the output pathway of the DCN, the input to low-rate type O units in the ICC of the cat (see section entitled "Effects of Blocking the Output of the DCN"), disrupts the sound localization behaviors of cats, particularly their orientation to sound source elevations (May, 2000; Sutherland *et al.*, 1998).

Overall, the correlation of response types with particular inputs is at an early stage. As our understanding of these relationships grow, so will our understanding of the emergent and specialized forms of spectral processing in the ICC.

Acknowledgments

This work was supported by grants DC00023, DC00979, and DC03758 from the National Institute for Deafness and Other Communications Disorders. Thanks are due to Oleg Lomakin and Yue-Houng Hu for preparation of figures and to colleagues in the University of Rochester Center for Navigation and Communication Sciences for useful discussions of the ideas presented here. The research on decerebrate cats summarized in this chapter was done in collaboration with R. Ramachandran and B. J. May.

References

Adams, J. C. (1980). Crossed and descending projections to the inferior colliculus. *Neurosci. Lett.* **19,** 1–5.

Adams, J. C., and Mugnaini, E. (1984). Dorsal nucleus of the lateral lemniscus: A nucleus of GABAergic projection neurons. *Brain Res. Bull.* **13,** 585–590.

Aitkin, L. (1986). "The Auditory Midbrain: Structure and Function in the Central Auditory Pathway." Humana Press, Clifton, N. J.

Aitkin, L. M., and Phillips, S. C. (1984). Is the inferior colliculus an obligatory relay in the cat auditory system? *Neurosci. Lett.* **44,** 259–264.

Aitkin, L., and Schuck, D. (1985). Low frequency neurons in the lateral central nucleus of the cat inferior colliculus receive their input predominantly from the medial superior olive. *Hear. Res.* **17,** 87–93.

Aitkin, L. M., Webster, W. R., Veale, J. L., and Crosby, D. C. (1975). Inferior colliculus. I. Comparison of response properties of neurons in central, pericentral, and external nuclei of adult cat. *J. Neurophysiol.* **38,** 1196–1207.

Andersen, R. A., Roth, G. L., Aitkin, L. M., and Merzenich, M. M. (1980). The efferent projections of the central nucleus and the pericentral nucleus of the inferior colliculus in the cat. *J. Comp. Neurol.* **194,** 649–662.

Batra, R., Kuwada, S., and Fitzpatrick, D. C. (1997). Sensitivity to interaural temporal disparities of low- and high- frequency neurons in the superior olivary complex. I. Heterogeneity of responses. *J. Neurophysiol.* **78,** 1222–1236.

Bledsoe, S. C., Shore, S. E., and Guitton, M. J. (2003). Spatial representation of corticofugal input in the inferior colliculus: A multicontact silicon probe approach. *Exp. Brain Res.* **153,** 530–542.

Bock, G. R., Webster, W. R., and Aitkin, L. M. (1972). Discharge patterns of single units in inferior colliculus of the alert cat. *J. Neurophysiol.* **35,** 265–277.

Boudreau, J. C., and Tsuchitani, C. (1968). Binaural interaction in the cat superior olive S segment. *J. Neurophysiol.* **31,** 442–454.

Brunso-Bechtold, J. K., Thompson, G. C., and Masterton, R. B. (1981). HRP study of the organization of auditory afferents ascending to central nucleus of inferior colliculus in cat. *J. Comp. Neurol.* **197,** 705–722.

Caird, D., and Klinke, R. (1983). Processing of binaural stimuli by cat superior olivary complex neurons. *Exp. Brain Res.* **52,** 385–399.

Calford, M. B., and Aitkin, L. M. (1983). Ascending projections to the medial geniculate body of the cat: Evidence for multiple, parallel auditory pathways through thalamus. *J. Neurosci.* **3,** 2365–2380.

Cant, N. B. (1992). The Cochlear Nucleus: Neuronal Types and Their Synaptic Organization. *In* "The Mammalian Auditory Pathways: Neuroanatomy" (D. B. Webster, A. N. Popper, and R. R. Fay, Eds.), Vol. 1, pp. 66–116. Springer-Verlag, New York.

Cant, N. B. (2004). Comparative Neuroanatomy of the Mammalian Inferior Colliculus. *Assoc. Res. Otolaryngol. Abstr.* **27,** 259.

Cant, N. B., and Casseday, J. H. (1986). Projections from the anteroventral cochlear nucleus to the lateral and medial superior olivary nuclei. *J. Comp. Neurol.* **247,** 457–476.

Casseday, J. H., and Covey, E. (1987). Central auditory pathways in directional hearing. *In* "Directional Hearing" (W. A. Yost and G. Gourvitch, Eds.), Vol. 1, pp. 109–145. Springer-Verlag, New York.

Casseday, J. H., and Covey, E. (1992). Frequency tuning properties of neurons in the inferior colliculus of an FM bat. *J. Comp. Neurol.* **319,** 34–50.

Coleman, J. R., and Clerici, W. J. (1987). Sources of projections to subdivisions of the inferior colliculus in the rat. *J. Comp. Neurol.* **262,** 215–226.

Covey, E., Kauer, J. A., and Casseday, J. H. (1996). Whole-cell patch-clamp recording reveals subthreshold sound-evoked postsynaptic currents in the inferior colliculus of awake bats. *J. Neurosci.* **16,** 3009–3018.

Davis, K. A. (1999). The basic receptive field properties of neurons in the inferior colliculus of decerebrate cats are rarely created by local inhibitory mechanisms. *Soc. Neurosci. Abstr.* **25,** 667.

Davis, K. A. (2002). Evidence of a functionally segregated pathway from dorsal cochlear nucleus to inferior colliculus. *J. Neurophysiol.* **87,** 1824–1835.

Davis, K. A., Ding, J., Benson, T. E., and Voigt, H. F. (1996). Response properties of units in the dorsal cochlear nucleus of unanesthetized decerebrate gerbil. *J. Neurophysiol.* **75,** 1411–1431.

Davis, K. A., Ramachandran, R., and May, B. J. (1999). Single-unit responses in the inferior colliculus of decerebrate cats. II. Sensitivity to interaural level differences. *J. Neurophysiol.* **82,** 164–175.

Davis, K. A., Ramachandran, R., and May, B. J. (2003). Auditory processing of spectral cues for sound localization in the inferior colliculus. *J. Assoc. Res. Otolaryngol.* **4,** 148–163.

Egorova, M., Ehret, G., Vartanian, I., and Esser, K. H. (2001). Frequency response areas of neurons in the mouse inferior colliculus. I. Threshold and tuning characteristics. *Exp. Brain Res.* **140,** 145–161.

Ehret, G., and Merzenich, M. M. (1988). Complex sound analysis (frequency resolution, filtering and spectral integration) by single units of the inferior colliculus of the cat. *Brain Res.* **472,** 139–163.

Evans, E. F., and Nelson, P. G. (1973). The responses of single neurones in the cochlear nucleus of the cat as a function of their location and the anaesthetic state. *Exp. Brain Res.* **17,** 402–427.

Fuzessery, Z. M., and Hall, J. C. (1996). Role of GABA in shaping frequency tuning and creating FM sweep selectivity in the inferior colliculus. *J. Neurophysiol.* **76,** 1059–1073.

Gao, E., and Suga, N. (1998). Experience-dependent corticofugal adjustment of midbrain frequency map in bat auditory system. *Proc. Natl. Acad. Sci. USA* **95,** 12663–12670.

Glendenning, K. K., Baker, B. N., Hutson, K. A., and Masterton, R. B. (1992). Acoustic chiasm V: Inhibition and excitation in the ipsilateral and contralateral projections of LSO. *J. Comp. Neurol.* **319,** 100–122.

Goldberg, J. M., and Brown, P. B. (1969). Response of binaural neurons of dog superior olivary complex to dichotic tonal stimuli: Some physiological mechanisms of sound localization. *J. Neurophysiol.* **32,** 613–636.

Guinan, J. J., Jr., Norris, B. E., and Guinan, S. S. (1972). Single units in superior olivary complex. II. Location of unit categories and tonotopic organization. *Int. J. of Neurosci.* **4,** 147–166.

Helfert, R. H., and Aschoff, A. (1997). Superior olivary complex and nuclei of the lateral lemniscus. *In* "The Central Auditory System" (G. Ehret and R. Romand, Eds.), Vol., pp. 193–258. Oxford University Press, New York.

Henkel, C. K., and Spangler, K. M. (1983). Organization of the efferent projections of the medial superior olivary nucleus in the cat as revealed by HRP and autoradiographic tracing methods. *J. Comp. Neurol.* **221,** 416–428.

Hernandez, O., Espinosa, N., Perez-Gonzalez, D., and Malmierca, M. S. (2005). The inferior colliculus of the rat: A quantitative analysis of monaural frequency response areas. *Neuroscience* **132,** 203–217.

Huffman, R. F., and Henson, O. W., Jr. (1990). The descending auditory pathway and acousticomotor systems: Connections with the inferior colliculus. *Brain Res. Rev.* **15,** 295–323.

Imig, T. J., Bibikov, N. G., Poirier, P., and Samson, F. K. (2000). Directionality derived from pinna-cue spectral notches in cat dorsal cochlear nucleus. *J. Neurophysiol.* **83,** 907–925.

Irvine, D. R. (1986). The auditory brainstem. *In* "Progress in Sensory Physiology 7" (D. Ottoson, Ed.), Vol. 1, pp. 1–279. Springer-Verlag, Berlin.

Irvine, D. R. (1992). Physiology of the auditory brainstem. *In* "The Mammalian Auditory Pathway: Neurophysiology" (A. N. Popper and R. R. Fay, Eds.), Vol. 1, pp. 153–231. Springer-Verlag, Berlin.

Jen, P. H., Chen, Q. C., and Sun, X. D. (1998). Corticofugal regulation of auditory sensitivity in the bat inferior colliculus. *J. Comp. Physiol. [A]* **183,** 683–697.

Joris, P. X. (1998). Response classes in the dorsal cochlear nucleus and its output tract in the chloralose-anesthetized cat. *J. Neurosci.* **18,** 3955–3966.

Joris, P. X., Smith, P. H., and Yin, T. C. (1994). Enhancement of neural synchronization in the anteroventral cochlear nucleus. II. Responses in the tuning curve tail. *J. Neurophysiol.* **71,** 1037–1051.

Kiang, N. Y. S., Watanabe, T., Thomas, E. C., and Clark, L. F. (1965). *In* "Discharge Patterns of Single Fibers in the Cat's Auditory Nerve." MIT Research Monographs 35, MIT, Cambridge, MA.

Klug, A., Bauer, E. E., Hanson, J. T., Hurley, L., Meitzen, J., and Pollak, G. D. (2002). Response selectivity for species-specific calls in the inferior colliculus of Mexican free-tailed bats is generated by inhibition. *J. Neurophysiol.* **88,** 1941–1954.

Koch, U., and Grothe, B. (1998). GABAergic and glycinergic inhibition sharpens tuning for frequency modulations in the inferior colliculus of the big brown bat. *J. Neurophysiol.* **80,** 71–82.

Kuwada, S., and Yin, T. C. (1983). Binaural interaction in low-frequency neurons in inferior colliculus of the cat. I. Effects of long interaural delays, intensity, and repetition rate on interaural delay function. *J. Neurophysiol.* **50,** 981–999.

Kuwada, S., Batra, R., Yin, T. C., Oliver, D. L., Haberly, L. B., and Stanford, T. R. (1997). Intracellular recordings in response to monaural and binaural stimulation of neurons in the inferior colliculus of the cat. *J. Neurosci.* **17,** 7565–7581.

LeBeau, F. E., Malmierca, M. S., and Rees, A. (2001). Iontophoresis *in vivo* demonstrates a key role for GABA(A) and glycinergic inhibition in shaping frequency response areas in the inferior colliculus of guinea pig. *J. Neurosci.* **21,** 7303–7312.

Loftus, W. C., Bishop, D. C., Saint Marie, R. L., and Oliver, D. L. (2004). Organization of binaural excitatory and inhibitory inputs to the inferior colliculus from the superior olive. *J. Comp. Neurol.* **472,** 330–344.

Maffi, C. L., and Aitkin, L. M. (1987). Differential neural projections to regions of the inferior colliculus of the cat responsive to high frequency sounds. *Hearing Res.* **26,** 211–219.

Malmierca, M. S., Rees, A., Le Beau, F. E., and Bjaalie, J. G. (1995). Laminar organization of frequency-defined local axons within and between the inferior colliculi of the guinea pig. *J. Comp. Neurol.* **357,** 124–144.

Malmierca, M. S., Oliver, D. L., and Merchan, M. A. (1999). Convergence of dorsal and ventral cochlear nuclei input onto frequency-band laminae of the inferior colliculus: A double tracer study in rat and cat. *Assoc. Res. Otolaryngol. Abstr.* **22,** 221.

Malmierca, M. S., Hernandez, O., Falconi, A., Lopez-Poveda, E. A., Merchan, M., and Rees, A. (2003). The commissure of the inferior colliculus shapes frequency response areas in rat: An *in vivo* study using reversible blockade with microinjection of kynurenic acid. *Exp. Brain. Res.* **153,** 522–529.

May, B. J. (2000). Role of the dorsal cochlear nucleus in the sound localization behavior of cats. *Hear. Res.* **148,** 74–87.

Merzenich, M. M., and Reid, M. D. (1974). Representation of the cochlea within the inferior colliculus of the cat. *Brain Res.* **77,** 397–415.

Mitani, A., Shimokouchi, M., and Nomura, S. (1983). Effects of stimulation of the primary auditory cortex upon colliculogeniculate neurons in the inferior colliculus of the cat. *Neurosci. Lett.* **42,** 185–189.

Moore, J. K., and Moore, R. Y. (1987). Glutamic acid decarboxylase-like immunoreactivity in brainstem auditory nuclei of the rat. *J. Comp. Neurol.* **260,** 157–174.

Moore, J. K., Guan, Y. L., and Shi, S. R. (1998). MAP2 expression in developing dendrites of human brainstem auditory neurons. *J. Chem. Neuroanat.* **16,** 1–15.

Morest, D. K., and Oliver, D. L. (1984). The neuronal architecture of the inferior colliculus in the cat: Defining the functional anatomy of the auditory midbrain. *J. Comp. Neurol.* **222,** 209–236.

Oliver, D. L. (1984). Neuron types in the central nucleus of the inferior colliculus that project to the medial geniculate body. *Neuroscience* **11,** 409–424.

Oliver, D. L. (1985). Quantitative analyses of axonal endings in the central nucleus of the inferior colliculus and distribution of 3H-labeling after injections in the dorsal cochlear nucleus. *J. Comp. Neurol.* **237,** 343–359.

Oliver, D. L. (1987). Projections to the inferior colliculus from the anteroventral cochlear nucleus in the cat: Possible substrates for binaural interaction. *J. Comp. Neurol.* **264,** 24–46.

Oliver, D. L. (2004). Specialized synaptic function in the mammalian inferior colliculus. *Assoc. Res. Otolaryngol. Abstr.* **27,** 260.

Oliver, D. L., and Beckius, G. E. (1992). Fine structure of GABA-labeled axonal endings in the inferior colliculus of the cat: Immunocytochemistry on deplasticized ultrathin sections. *Neuroscience* **46,** 455–463.

Oliver, D. L., and Huerta, M. F. (1992). Inferior and superior colliculi. *In* "The Mammalian Auditory Pathways: Neuroanatomy" (D. B. Webster, A. N. Popper, and R. R. Fay, Eds.), Vol., pp. 168–221. Springer-Verlag, New York.

Oliver, D. L., and Morest, D. K. (1984). The central nucleus of the inferior colliculus in the cat. *J. Comp. Neurol.* **222,** 237–264.

Oliver, D. L., and Shneiderman, A. (1991). The anatomy of the inferior colliculus: A cellular basis for integration of monaural and binaural information. *In* "Neurobiology of Hearing" (R. A. Altschuler, R. P. Bobbin, B. M. Clopton, and D. W. Hoffman, Eds.), Vol., pp. 195–222. Raven, New York.

Oliver, D. L., Beckius, G. E., and Shneiderman, A. (1995). Axonal projections from the lateral and medial superior olive to the inferior colliculus of the cat: A study using electron microscopic autoradiography. *J. Comp. Neurol.* **360,** 17–32.

Oliver, D. L., Beckius, G. E., Bishop, D. C., and Kuwada, S. (1997). Simultaneous anterograde labeling of axonal layers from lateral superior olive and dorsal cochlear nucleus in the inferior colliculus of cat. *J. Comp. Neurol.* **382,** 215–229.

Osen, K. K. (1970). Course and termination of the primary afferents in the cochlear nuclei of the cat. An experimental anatomical study. *Arch. Ital. Biol.* **108,** 21–51.

Palombi, P. S., and Caspary, D. M. (1996). GABA inputs control discharge rate primarily within frequency receptive fields of inferior colliculus neurons. *J. Neurophysiol.* **75,** 2211–2219.

Ramachandran, R., and May, B. J. (2002). Functional segregation of ITD sensitivity in the inferior colliculus of decerebrate cats. *J. Neurophysiol.* **88,** 2251–2261.

Ramachandran, R., Davis, K. A., and May, B. J. (1999). Single-unit responses in the inferior colliculus of decerebrate cats. I. Classification based on frequency response maps. *J. Neurophysiol.* **82,** 152–163.

Rhode, W. S., Smith, P. H., and Oertel, D. (1983). Physiological response properties of cells labeled intracellularly with horseradish peroxidase in cat dorsal cochlear nucleus. *J. Comp. Neurol.* **213,** 426–447.

Roberts, R. C., and Ribak, C. E. (1987). GABAergic neurons and axon terminals in the brainstem auditory nuclei of the gerbil. *J. Comp. Neurol.* **258,** 267–280.

Rose, J. E., Greenwood, D. D., Goldberg, J. M., and Hind, J. E. (1963). Some discharge characteristics of single neurons in the inferior colliculus of the cat. I. Tonotopical organization, relation of spike counts to tone intensity and firing pattern of single elements. *J. Neurophysiol.* **26,** 294–320.

Roth, G. L., Aitkin, L. M., Andersen, R. A., and Merzenich, M. M. (1978). Some features of the spatial organization of the central nucleus of the inferior colliculus of the cat. *J. Comp. Neurol.* **182,** 661–680.

Ross, L. S., and Pollak, G. D. (1989). Differential ascending projections to aural regions in the 60 kHz contour of the mustache bat's inferior colliculus. *J. Neurosci.* **9,** 2819–2834.

Ryan, A., and Miller, J. (1978). Single unit responses in the inferior colliculus of the awake and performing rhesus monkey. *Exp. Brain Res.* **32,** 389–407.

Ryugo, D. K., Willard, F. H., and Fekete, D. M. (1981). Differential afferent projections to the inferior colliculus from the cochlear nucleus in the albino mouse. *Brain Res.* **210,** 342–349.

Saint Marie, R. L. (1996). Glutamatergic connections of the auditory midbrain: Selective uptake and axonal transport of D-[3H]aspartate. *J. Comp. Neurol.* **373,** 255–270.

Saint Marie, R. L., and Baker, R. A. (1990). Neurotransmitter-specific uptake and retrograde transport of [3H]glycine from the inferior colliculus by ipsilateral projections of the superior olivary complex and nuclei of the lateral lemniscus. *Brain Res.* **524,** 244–253.

Saint Marie, R. L., Ostapoff, E. M., Morest, D. K., and Wenthold, R. J. (1989). Glycine-immunoreactive projection of the cat lateral superior olive: Possible role in midbrain ear dominance. *J. Comp. Neurol.* **279,** 382–396.

Saint Marie, R. L., Shneiderman, A., and Stanforth, D. A. (1997). Patterns of gamma-aminobutyric acid and glycine immunoreactivities reflect structural and functional differences of the cat lateral lemniscal nuclei. *J. Comp. Neurol.* **389,** 264–276.

Saldana, E., and Merchan, M. A. (1992). Intrinsic and commissural connections of the rat inferior colliculus. *J. Comp. Neurol.* **319,** 417–437.

Sawtell, N., Felix, R., Tzounopoulos, T., and Portfors, C. (2005). Response Properties of Neurons in the Dorsal Cochlear Nucleus of the Unanesthetized Mouse. *Assoc. Res. Otolaryngol. Abstr.* **28,** 239.

Schwartz, I. R. (1992). Superior olive and lateral lemniscus. *In* "The Mammalian Auditory Pathways: Neuroanatomy" (D. B. Webster, A. N. Popper, and R. R. Fay, Eds.), Vol. 1, pp. 117–167. Springer-Verlag, New York.

Semple, M. N., and Aitkin, L. M. (1979). Representation of sound frequency and laterality by units in central nucleus of cat inferior colliculus. *J. Neurophysiol.* **42,** 1626–1639.

Semple, M. N., and Aitkin, L. M. (1980). Physiology of pathway from dorsal cochlear nucleus to inferior colliculus revealed by electrical and auditory stimulation. *Exp. Brain Res.* **41,** 19–28.

Shneiderman, A., and Henkel, C. K. (1987). Banding of lateral superior olivary nucleus afferents in the inferior colliculus: A possible substrate for sensory integration. *J. Comp. Neurol.* **266,** 519–534.

Shneiderman, A., and Oliver, D. L. (1989). EM autoradiographic study of the projections from the dorsal nucleus of the lateral lemniscus: A possible source of inhibitory inputs to the inferior colliculus. *J. Comp. Neurol.* **286,** 28–47.

Shneiderman, A., Oliver, D. L., and Henkel, C. K. (1988). Connections of the dorsal nucleus of the lateral lemniscus: An inhibitory parallel pathway in the ascending auditory system? *J. Comp. Neurol.* **276,** 188–208.

Shneiderman, A., Chase, M. B., Rockwood, J. M., Benson, C. G., and Potashner, S. J. (1993). Evidence for a GABAergic projection from the dorsal nucleus of the lateral lemniscus to the inferior colliculus. *J. Neurochem.* **60,** 72–82.

Shofner, W. P., and Young, E. D. (1985). Excitatory/inhibitory response types in the cochlear nucleus: Relationships to discharge patterns and responses to electrical stimulation of the auditory nerve. *J. Neurophysiol.* **54,** 917–939.

Smith, P. H. (1992). Anatomy and physiology of multipolar cells in the rat inferior collicular cortex using the *in vitro* brain slice technique. *J. Neurosci.* **12,** 3700–3715.

Smith, P. H., Joris, P. X., Carney, L. H., and Yin, T. C. (1991). Projections of physiologically characterized globular bushy cell axons from the cochlear nucleus of the cat. *J. Comp. Neurol.* **304,** 387–407.

Smith, P. H., Joris, P. X., and Yin, T. C. (1993). Projections of physiologically characterized spherical bushy cell axons from the cochlear nucleus of the cat: Evidence for delay lines to the medial superior olive. *J. Comp. Neurol.* **331,** 245–260.

Spirou, G. A., and Young, E. D. (1991). Organization of dorsal cochlear nucleus type IV unit response maps and their relationship to activation by bandlimited noise. *J. Neurophysiol.* **66,** 1750–1768.

Suga, N., and Tsuzuki, K. (1985). Inhibition and level-tolerant frequency tuning in the auditory cortex of the mustached bat. *J. Neurophysiol.* **53,** 1109–1145.

Suga, N., Zhang, Y., and Yan, J. (1997). Sharpening of frequency tuning by inhibition in the thalamic auditory nucleus of the mustached bat. *J. Neurophysiol.* **77,** 2098–2114.

Sun, X., Chen, Q. C., and Jen, P. H. (1996). Corticofugal control of central auditory sensitivity in the big brown bat, Eptesicus fuscus. *Neurosci. Lett.* **212,** 131–134.

Sutherland, D. P., Masterton, R. B., and Glendenning, K. K. (1998). Role of acoustic striae in hearing: Reflexive responses to elevated sound-sources. *Behav. Brain. Res.* **97,** 1–12.

Syka, J., and Popelar, J. (1984). Inferior colliculus in the rat: Neuronal responses to stimulation of the auditory cortex. *Neurosci. Lett.* **51,** 235–240.

Thompson, G. C., Cortez, A. M., and Lam, D. M. (1985). Localization of GABA immunoreactivity in the auditory brainstem of guinea pigs. *Brain. Res.* **339,** 119–122.

Thornton, S. K., and Rees, A. (2001). The dorsal nucleus of the lateral lemniskus shapes auditory responses in the inferior colliculus. *In* "Physiological and Psychological Bases of Auditory Function" (D. J. Breebaart, A. J. M. Houtsma, A. Kohlrausch, V. F. Prijs, and R. Shoonhoven, Eds.), Vol. 12, pp. 298–305. Shaker, The Netherlands.

Tsuchitani, C., and Boudreau, J. C. (1969). Stimulus level of dichotically presented tones and cat superior olive S-segment cell discharge. *J. Acoust. Soc. Am.* **46,** 979–988.

Vater, M., Habbicht, H., Kossl, M., and Grothe, B. (1992). The functional role of GABA and glycine in monaural and binaural processing in the inferior colliculus of horseshoe bats. *J. Comp. Physiol. [A]* **171,** 541–553.

Wang, J., Salvi, R. J., and Powers, N. (1996). Plasticity of response properties of inferior colliculus neurons following acute cochlear damage. *J. Neurophysiol.* **75,** 171–183.

Warr, W. B. (1966). Fiber degeneration following lesions in the anterior ventral cochlear nucleus of the cat. *Exp. Neurol.* **14,** 453–474.

Warr, W. B. (1972). Fiber degeneration following lesions in the multipolar and globular cell areas in the ventral cochlear nucleus of the cat. *Brain. Res.* **40,** 247–270.

Wightman, F. L., and Kistler, D. J. (1989). Headphone simulation of free-field listening. I. Stimulus synthesis. *J. Acoust. Soc. Am.* **85,** 868–878.

Winer, J. A., Larue, D. T., and Pollak, G. D. (1995). GABA and glycine in the central auditory system of the mustache bat: Structural substrates for inhibitory neuronal organization. *J. Comp. Neurol.* **355,** 317–353.

Winer, J. A., Larue, D. T., Diehl, J. J., and Hefti, B. J. (1998). Auditory cortical projections to the cat inferior colliculus. *J. Comp. Neurol.* **400,** 147–174.

Winter, I. M., and Palmer, A. R. (1990). Responses of single units in the anteroventral cochlear nucleus of the guinea pig. *Hearing Res.* **44,** 161–178.

Yan, J., and Ehret, G. (2001). Corticofugal reorganization of the midbrain tonotopic map in mice. *Neuroreport* **12,** 3313–3316.

Yan, W., and Suga, N. (1998). Corticofugal modulation of the midbrain frequency map in the bat auditory system. *Nat. Neurosci.* **1,** 54–58.

Yang, L., Pollak, G. D., and Resler, C. (1992). GABAergic circuits sharpen tuning curves and modify response properties in the mustache bat inferior colliculus. *J. Neurophysiol.* **68,** 1760–1774.

Yin, T. C., Chan, J. C., and Irvine, D. R. (1986). Effects of interaural time delays of noise stimuli on low-frequency cells in the cat's inferior colliculus. I. Responses to wideband noise. *J. Neurophysiol.* **55,** 280–300.

Yin, T. C., Chan, J. C., and Carney, L. H. (1987). Effects of interaural time delays of noise stimuli on low-frequency cells in the cat's inferior colliculus. III. Evidence for cross-correlation. *J. Neurophysiol.* **58,** 562–583.

Young, E. D. (1980). Identification of response properties of ascending axons from dorsal cochlear nucleus. *Brain Res.* **200,** 23–37.

Young, E. D. (1984). Response characteristics of neurones of the cochlear nuclei. *In* "Hearing Science, Recent Advances" (C. I. Berlin, Ed.), Vol. 1, pp. 423–460. College Hill Press, San Diego, CA.

Young, E. D., and Brownell, W. E. (1976). Responses to tones and noise of single cells in dorsal cochlear nucleus of unanesthetized cats. *J. Neurophysiol.* **39,** 282–300.

Young, E. D., Spirou, G. A., Rice, J. J., and Voigt, H. F. (1992). Neural organization and responses to complex stimuli in the dorsal cochlear nucleus. *Philos. Trans. R. Soc. Lond. B. Biol. Sci.* **336,** 407–413.

Zhang, Y., and Suga, N. (1997). Corticofugal amplification of subcortical responses to single tone stimuli in the mustached bat. *J. Neurophysiol.* **78,** 3489–3492.

Zhang, Y., and Suga, N. (2000). Modulation of responses and frequency tuning of thalamic and collicular neurons by cortical activation in mustached bats. *J. Neurophysiol.* **84,** 325–333.

Zhang, Y., Suga, N., and Yan, J. (1997). Corticofugal modulation of frequency processing in bat auditory system. *Nature* **387,** 900–903.

Zhang, D. X., Li, L., Kelly, J. B., and Wu, S. H. (1998). GABAergic projections from the lateral lemniscus to the inferior colliculus of the rat. *Hearing Res.* **117,** 1–12.

Zook, J. M., and Casseday, J. H. (1982). Origin of ascending projections to inferior colliculus in the mustache bat, Pteronotus parnellii. *J. Comp. Neurol.* **207,** 14–28.

Zook, J. M., and Casseday, J. H. (1987). Convergence of ascending pathways at the inferior colliculus of the mustache bat, Pteronotus parnellii. *J. Comp. Neurol.* **261,** 347–361.

Zook, J. M., Winer, J. A., Pollak, G. D., and Bodenhamer, R. D. (1985). Topology of the central nucleus of the mustache bat's inferior colliculus: Correlation of single unit properties and neuronal architecture. *J. Comp. Neurol.* **231,** 530–546.

NEURAL MECHANISMS FOR SPECTRAL ANALYSIS IN THE AUDITORY MIDBRAIN, THALAMUS, AND CORTEX

Monty A. Escabí* and Heather L. Read†

*Department of Electrical Engineering, University of Connecticut
Storrs, Connecticut 06269, USA
†Department of Psychology, The University of Connecticut, Storrs, Connecticut 06269, USA

Frequency organization or cochleotopy is a fundamental aspect of the lemniscal "high fidelity" auditory system. Starting with the cochlear sensory epithelium, a systematic gradient of frequency-tuned neuronal receptors can be observed at all levels of the central auditory system. Distinct spectral processing pathways with unique monaural integration response properties are initially set up in the cochlear

INTERNATIONAL REVIEW OF
NEUROBIOLOGY, VOL. 70
DOI: 10.1016/S0074-7742(05)70007-6

nucleus (CN), and subsequently refined into binaural processing streams in the superior olivary nuclei (SO). At the level of the auditory midbrain and beyond these separate sources of frequency-specific binaural/monaural input intermingle with one another giving rise to new types of modularity and cochleotopy. Here we examine the anatomical organization of neuronal response preferences in the auditory midbrain, thalamaus, and cortex within a framework of functional modularity and bottom up intrinsic connectivity. Spectral preferences are intimately related to the type of binaural preference and sound level dependence and yet we know very little about their basic organization within these centers. An emphasis is placed on reconciling neuronal, anatomic, and physiologic principles underlying spectral analysis and perceptual phenomena, in particular, frequency resolution, loudness perception, and sound localization.

I. Introduction

This chapter examines the principles of spectral analysis and coding in the ascending central auditory system. The chapter is centered around identifying key response properties of single neurons in the lemniscal divisions of the inferior colliculus (ICC), auditory thalamus (MGBv), and primary auditory cortex (AI) that can be related to the underlying anatomical architecture in these stations. Starting with the divergent pathway from the cochlea to the brainstem nuclei, the ICC receives input from the cochlear nucleus (CN), superior olivary nuclei (SO), and the nucleus of the lateral lemniscus (NLL) (Adams, 1979). This highly convergent pattern of innervating input from multiple brainstem nuclei plays a key role in refining the organization of spectral and temporal preferences in the ICC. Within this context, spectral and temporal preferences in the ICC are sequentially modified in the thalamus and cortex leading to distinct differences in the functional organization. This functional and anatomical framework is essential for understanding the distributed code for spectral and temporal features in a sound.

The chapter emphasizes the general principles of coding which serve as a functional and anatomical substrate for spectral analysis in the central auditory system. In particular, we focus on relating details of single neuron activity to the underlying anatomical layout and connectivity at different levels of the lemniscal auditory pathway. Spectral response properties and selectivity are intimately related to binaural preferences and sound level dependence. Throughout we explore the specific relationships between these different response attributes and the underlying local circuitry and global organization of the central auditory system. General properties of single neuron sensitivities are outlined in Sections II to V whereas their relationship to the anatomical circuitry and the underlying functional transformation is described in Sections VI to VIII.

II. Principles of Spectral Analysis in ICC, MGB, and AI

The frequency organization of the cochlear sensory epithelium is conserved in the topographic pattern of neuronal frequency selectivities or cochleotopy. A cochleotopic pattern of frequency response properties is observed throughout the lemniscal auditory pathway including the ICC, MGBv, and AI. Demonstrating the functional consequences associated with specific anatomical patterns of frequency specific projections in the lemniscal pathway requires efficient and meaningful ways of characterizing spectral sensitivities. The receptive field (RF) is the classical method of measuring and quantifying neuronal sensitivities.

Auditory RFs are defined by the rate or time-locked responses of neurons to sound structure including spectral content or frequency, sound level, and modulation rate. Spectral receptive field (SRF) preferences are usually obtained by measuring the response to various simple acoustic parameters, such as frequency and amplitude, as is done when measuring frequency tuning curves (FTCs) or frequency response areas (FRAs). Figure 1 illustrates two AI neuron FTCs generated by varying the frequency and sound pressure level (SPL) of the same transient pure-tone presented to both ears. This is the classic method for determining simple spectral selectivities for auditory neurons; however, many spectral RF properties cannot be assessed with such narrow bandwidth sounds. As will be seen, the measured characteristics of an SRF depend dramatically on a number of factors, such as the type of stimulus used (e.g., broadband versus narrowband), its sound level, dynamic range or contrast, and even the temporal attributes of the sound. Such non-linear properties of the sound sensitivity of collicular, thalamic, and cortical receptive fields make it difficult to relate neuronal activities of one sound to another and present an interpretational challenge in the study of central auditory system function.

III. Sharpness of Tuning, Bandwidth, and Level Dependence

Frequency and absolute sound level are important attributes in complex sounds that are known to influence perception. The spectrum of communication sounds and other behaviorally relevant environmental signals can vary significantly in their frequency content and amplitude level. Frequency content, disparity in frequency composition for the two ears, and absolute sound level can also encode for the location of sound sources. It is the task of the central auditory system to faithfully represent the sound so that different sounds can be discriminated. At the same time, the system must disambiguate frequency and loudness differences that are due to differences in sound location. The first step to

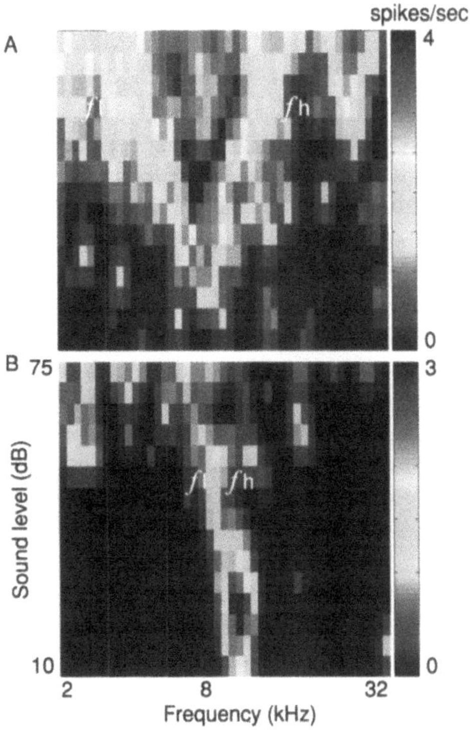

FIG. 1. Two cortical FTCs for pure tone sounds. Tone frequency and amplitude are varied over 5 octaves and 70 dB, respectively. Multi- or single-unit neuronal spike rate histograms are plotted for each frequency and level. The rainbow color spectrum shows lowest to highest spike rates (blue to red). Threshold is the lowest sound level eliciting a response above baseline. Characteristic frequency (CF) is the frequency at threshold. Linear bandwidth is the difference between the high (fh) and lower (fl) cutoff frequency at a given sound level. Examples A and B are broad bandwidth and narrow bandwidth FTCs, respectively. The rate level functions increase monotonically and non-monotonically with sound level for A and B, respectively. Neurons with non-monotonic narrow bandwidth FTCs in cortex are likely azimuth tuned (Clarey *et al.*, 1995; Barone *et al.*, 1996). (See Color Insert.)

understanding how populations of neurons represent frequency content, loudness, and location of sounds is to develop reliable ways of characterizing the responses of individual neurons to these sound cues.

Neurons can be coarsely defined as having either monotonic or non-monotonic sound level dependence for spike rate. Changing the sound level has systematic effects on the response spike rate, temporal precision, and response latency. Monotonic rate-level coding can be demonstrated with a variety of narrowband and broadband sounds. Firing rates increase monotonically with increasing sound level for a large proportion of central auditory neurons (Ehret and Merzenich, 1988a; Heil, 1997c; Heil *et al.*, 1994; Phillips and Irvine, 1981; Phillips *et al.*, 1985; Rees and Palmer, 1988; Rose *et al.*, 1963; Rouiller *et al.*, 1983; Schreiner *et al.*, 1992; Semple and Kitzes, 1985). At the same time, the response latency is reduced and the timing precision improved with increased sound level (Heil, 1997a). While this type of monotonic rate-level coding is characteristic of the auditory nerve (AN) (Evans and Palmer, 1980), coding of sound level in central stations is distinguished by the fact that a significant fraction of neurons are also inhibited by loud sound levels (Ehret and Merzenich, 1988a; Heil *et al.*, 1994; Phillips and Irvine, 1981; Phillips *et al.*, 1985; Rees and Palmer, 1988; Rose *et al.*, 1963; Rouiller *et al.*, 1983; Schreiner *et al.*, 1992). The firing rate of such neurons initially increases with SPL and is subsequently reduced at a very loud SPL, exhibiting a non-monotonic rate-level dependency.

There are generally two conceptual models explaining the distribution of non-monotonic versus monotonic rate level functions between ICC and cortical stations of processing. According to the first, the proportion of non-monotonic responses increases from the ICC to AI (Phillips and Irvine, 1981; Rees and Palmer, 1988; Rouiller *et al.*, 1983; Schreiner *et al.*, 1992), consistent with an increased recruitment of inhibition across this pathway. According to the second model, robust non-monotonic rate-level sensitivities are established early on in CN, in particular within the DCN, and conserved at the level of ICC, MGBv, and AI (Barone *et al.*, 1996; Cant and Benson, 2003; Davis and Young, 2000; Ramachandran *et al.*, 1999). The subtle distinction between these two models is that the first suggests novel inhibitory processes at each level which refine rate-level sensitivities whereas the second suggests that there are "parallel auditory pathways" originating in the CN that pass on and conserve the different rate-level sensitivities observed in ICC, MGBv, and AI (Cant and Benson, 2003). Indeed, the non-monotonic rate level functions observed in the DCN are very similar to those observed for a subset of neurons in the MGBv and AI (Barone *et al.*, 1996). Non-monotonic rate-level dependence could also be created or maintained via ascending inhibitory projections from the sub-collicular nuclei to the colliculus (Oliver and Shneiderman, 1989; Shneiderman and Oliver, 1989; Shneiderman *et al.*, 1988) and again from the colliculus to the MGBv (Peruzzi *et al.*, 1997; Winer *et al.*, 1996). Local inhibitory neurons at each level of processing also could contribute to the level dependence and other SRF properties at each station of the pathway, including early stages such as the CN. Direct correlations between sources of inhibitory input and sound level dependency have been described in the cat DCN (Davis and Young, 2000) and in the bat ICC (Pollak and Park, 1993) but not in the MGBv or AI. Hence, at present both models for how non-monotonic level dependencies are modified centrally are still plausible.

Sound-level-dependent firing rates (monotonicity versus non-monotonicity) and bandwidths often co-vary and these two parameters are derived from the pure-tone FTC or frequency response areas (FRAs). The exact shape of the FRA depends on a number of factors, including the tone-pip duration, rise time, and the time window used to acquire spike activity. FRAs in the ICC, MGBv, and AI typically only account for the excitatory component of the neuronal response. These can vary in their structure from narrowband to broadband excitatory sensitivity and can vary in their shape from simple eighth nerve-like FRA response sensitivity to more intricate FRA shapes (Ehret *et al.*, 2003; Stiebler and Ehret, 1985; Sutter and Schreiner, 1991). Example broadband and narrowband FRAs from cat AI are illustrated in Fig. 1. Because of this diversity in excitatory FRA shapes, FRAs have been categorized according to a number of functional parameters that account for the spectral filtering preferences. These include the frequency with maximal response (best frequency BF), the frequency with the lowest response threshold (characteristic frequency CF), and the sharpness of tuning measured as either a bandwidth in units of octaves or the quality factor (Q, defined as the ratio of CF to bandwidth). A decreased spectral resolution at loud sound levels is observed for all sensory fibers in the cochlea, whereas the filter resolution of ICC, MGBv, and AI neurons and their corresponding level dependence are more heterogeneous (Barone *et al.*, 1996; Langner and Schreiner, 1988; Phillips and Irvine, 1981; Schreiner and Mendelson, 1990).

The type of sound level response dependence is also closely related to other receptive field properties including the spectral sensitivity. In the ICC of decerebrate cats, for instance, monotonic responses have characteristic V-shaped frequency response areas as observed for all VIII AN fibers (Evans and Palmer, 1980). Non-monotonic rate-level response functions, however, are strictly observed for neurons with type I and type O frequency sensitivity (Ramachandran *et al.*, 1999) (see Chapter 7). For thalamic and cortical neurons, sound level-dependent changes in spectral bandwidth also appear to correlate with sound level-dependent changes in firing rate. In the cat, monotonic neurons are often characterized by V-shaped FRAs in which the bandwidth increases with level (Phillips *et al.*, 1985; Schreiner *et al.*, 1992). Neurons with type V-shaped FRAs respond efficiently to both narrowband tones and broadband stimuli (e.g., clicks and noise) and have short latency responses to tones and noise (Phillips *et al.*, 1985). By comparison, non-monotonic neurons often exhibit FRAs with narrowband tuning, typically have longer response latencies, and are inhibited by broadband noise bursts and clicks (Phillips *et al.*, 1985; Ramachandran *et al.*, 2000). Non-monotonic rate level dependence can be demonstrated with either pure-tones or broadband sounds. On the other hand, receptive field properties such as azimuth tuning of non-monotonic neurons can only be observed with broadband sounds (Barone *et al.*, 1996). This dissociation between rate and spectral bandwidth dependence on sound level tuning may be indicative of more than one inhibitory or adaptive mechanism.

IV. Excitatory and Inhibitory Receptive Field Properties

Inhibitory processes are initially established early on in the CN (Davis and Young, 2000), and further refined at more central centers. Feed-forward inhibitory projections from multiple auditory centers play a prominent role in shaping spectral sensitivities. This receptive field shaping could occur via long range tonotopically organized inhibitory projections from the dorsal nucleus of the lateral lemniscus (DNLL) to the ICC (Bajo *et al.*, 1999; Oliver and Shneiderman, 1989; Pollak *et al.*, 2002, 2003; Shneiderman and Oliver, 1989; Shneiderman *et al.*, 1988) and subsequently from the ICC to the MGB (Peruzzi *et al.*, 1997; Winer *et al.*, 1996). The local inhibitory circuitry in various brainstem/midbrain centers also plays a critical role for refining receptive field preferences by sharpening spectral tuning, sound level sensitivities, binaural preferences, or refining neuronal selectivity to specific temporal sound components (Batra and Fitzpatrick, 2002; Casseday *et al.*, 1994; Davis and Young, 2000; Kuwada *et al.*, 1997; Le Beau *et al.*, 1996). Additional inhibitory processes are evident in AI. Cortical inhibition is responsible for adjusting spectral tuning and enhancing temporal preferences (Wehr and Zador, 2003), including directional preferences for FM sweeps (Zhang *et al.*, 2003).

Due to the vast amount of central inhibition in the ascending lemniscal pathway, numerous techniques have been developed to characterize how inhibitory processes contribute to neuronal sensitivities. These include simple two-tone receptive field measurement techniques, the use of bandpass and broadband noise to activate spectral inhibition, and more intricate methods with complex sounds that attempt to quantify spectral and/or temporal aspects of inhibition. While these indirect receptive field mapping techniques generally do not identify the direct source of inhibitory input, they are nonetheless useful for characterizing how spectral preferences are sculpted across multiple levels of processing.

A. SRAs Probed with Narrowband Stimuli

A variety of approaches have been used to quantify neuronal sensitivities to spectral peaks and notches. Measures of two-tone interactions are perhaps the simplest way to measure the combined influence of spectral excitation and inhibition (Brosch and Schreiner, 1997; Calford and Semple, 1995; Egorova *et al.*, 2001; Portfors and Wenstrup, 2002; Shamma *et al.*, 1993). A two-tone frequency response curve can be constructed by presenting a reference tone-pip simultaneously with a probe tone-pip of variable frequency and level. The reference tone-pip is often presented at the neuron's CF and an SPL that evokes a strong excitatory response. The complex interactions between the probe and reference tones may in fact have a number of possible outcomes. Generally speaking, the probe tone may have either an excitatory or inhibitory effect on the response to the reference tone. Excitatory interactions may be either *additive*, i.e., the response to the reference and probe is simply the sum of the responses for both of these components, or *facilitative,* in which case the response to the reference and probe tone is greater in magnitude than the sum of responses resulting from each individual component (Egorova *et al.*, 2001; Portfors and Wenstrup, 2002).

Functional inhibition is encountered whenever the firing rate to the reference and probe tone-pips is less than the firing rate for the reference tone alone. The arrangements of inhibition in two-tone frequency response maps have been used as a marker for categorizing SRF properties according to a number of stereotyped inhibitory interactions (Egorova *et al.*, 2001). Two-tone response profiles can exhibit (1) primary-like response areas with narrow excitatory tuning near threshold and broad low frequency tuning at high sound levels or (2) narrowband excitatory tuning with strong flanking sideband inhibition. Classifications of spectral preference have also been based upon inhibition of spontaneous activity profiles. Inhibition of spontaneous activity in decerebrate cats (Ramachandran *et al.*, 1999) can be used as a marker for spectral integration properties in the ICC that resemble those of the brainstem inputs from the DCN, anterior-ventral cochlear nucleus (AVCN), and superior olivery nuclei (SO). Because of functional similarities in spectral sensitivity between colliculus and brainstem inputs (Davis *et al.*, 1999), Ramachandran and colleagues (Ramachandran and May, 2002; Ramachandran *et al.*, 1999) have argued for an inheritance principle in generating ICC receptive fields (see chapter by Davis). Similar functional relationships may also hold for the construction of thalamic and cortical receptive fields from their brainstem and midbrain inputs although the details of this projection are not so well understood.

B. SRFs Probed with Broadband Sounds: Tone-Noise Interactions

The influences of spectral excitation and inhibition have also been measured with a variety of broadband stimulus paradigms. Ehret and colleagues (Ehret and Merzenich, 1985, 1988b; Ehret and Schreiner, 1997) employed a tone-noise masking paradigm to measure spectral filter resolution in the cat ICC and AI. Their procedure is conceptually analogous to psychophysical paradigms used to measure perceptual filter bandwidths, in which a pure tone is masked by a narrowband noise of variable bandwidth and level. In contrast to pure-tone tuning properties, bandwidth measurements using this technique exhibit level-invariant resolution that closely approaches critical-band masking phenomena and are in agreement with psychophysical measures (see Section IV.D).

C. SRFs Probed with Broadband Sounds: Stationary Spectral Ripples

Spectral ripples are broadband noise with a sinusoidal spectrum that resembles the spectra of communication signals. Spectral ripples have been used to assay excitatory and inhibitory RF properties in AI (Calhoun and Schreiner, 1998; Kowalski *et al.*, 1996; Schreiner and Calhoun, 1994; Versnel and Shamma, 1998). Ripple noises contain peaks and notches in their spectrum at different resolutions and provide a means of systematically manipulating the sound spectral content. The spectral resolution of a peak-notch combination is characterized by the ripple density (Kowalski *et al.*, 1996; Schreiner and Calhoun, 1994) or equivalently the spectral modulation frequency (Miller *et al.*, 2002) (expressed as the number of spectral peaks per octave frequency; cycles/octave). Higher ripple densities correspond to higher spectral resolutions whereas a ripple density of 0 corresponds to a flat spectrum noise. Ripple noise SRF measurements typically exhibit an excitatory component about the BF and often exhibit alternating regions of flanking inhibition above and/or below the BF. Such a receptive field description is conceptually appealing as it parallels the sound spectrum that the neuron prefers (Versnel and Shamma, 1998). This sensitivity could also be expressed as a tuning function with respect to the ripple density parameter, in which case the response is characterized by a response peak at the preferred spectral resolution/ripple density. In the cat and ferret, the inhibitory component of spectral tuning can be asymmetrical such that flanking inhibition appears to be most prominent at frequencies below or above the BF (Kowalski *et al.*, 1996; Schreiner and Calhoun, 1994; Versnel *et al.*, 1995), although in general a variety of arrangements are possible. A related technique using *random spectrum stimuli* (RSS) (see Chapter 6) has revealed similar SRF arrangements in the cat DCN (Yu and Young, 2000) and primate AI (Barbour and Wang, 2003). As for the tone-noise masking studies of Ehret and colleagues (Ehret and Merzenich, 1988b; Ehret and Schreiner, 1997) SRF measurements with broadband ripples and RSS noise are much more level-tolerant than with pure-tone (Barbour and Wang, 2003), which suggests different functional principles underlying the encoding of broadband stimuli.

The prevalence of sideband inhibition observed across many stations in the auditory system has lead to a prevailing view that flanking inhibition about the

CF refines spectral tuning. Two models have been postulated for the arrangement of spectral inhibition that could account for such an observation. On the one hand, inhibition could be confined to flanking regions that extend above and below the central CF region. Such a model would suggest that the inhibitory and excitatory SRF (iSRF and eSRF, respectively) do not fully overlap, but instead are confined to different frequencies. Conversely, inhibition could overlap the entire eSRF and could potentially extend beyond the dominant eSRF region. The arrangement of the excitatory and inhibitory synaptic conductances has been recently measured intracellularly in rat AI and the data support the second model (Wehr and Zador, 2003; Zhang *et al.*, 2003) (Fig. 2). The excitatory and inhibitory SRF of AI neurons are primarily overlapped with similar spectral tuning and high-frequency role-offs, although inhibition is generally temporally delayed relative to the excitation.

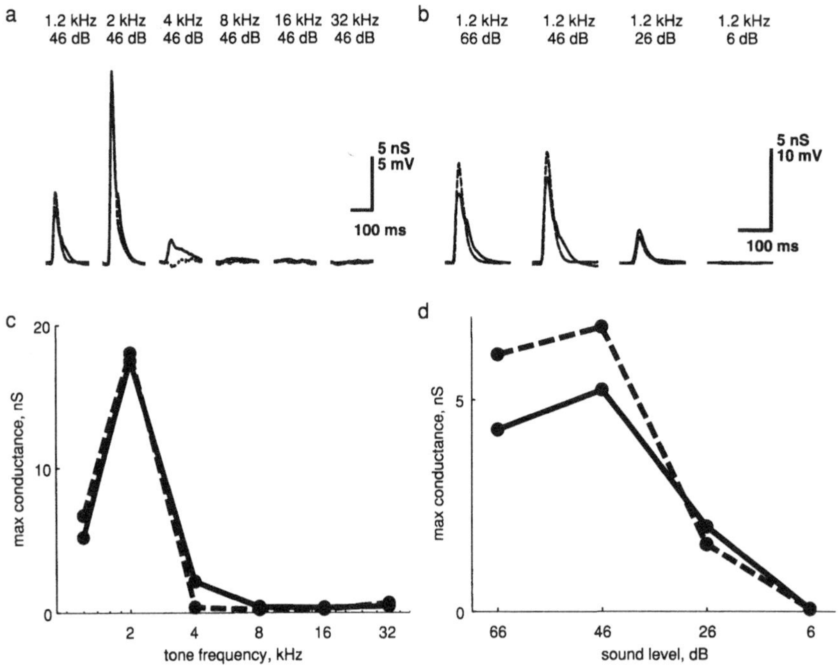

FIG. 2. Excitatory and inhibitory synaptic conductances in the rat auditory cortex are co-tuned for frequency and sound level. (a,b) The excitatory (solid line) and inhibitory (dotted line) synaptic conductances of a neuron with best frequency around 2 kHz show similar timing and amplitude pattern. (c,d) The peak excitatory (solid) and inhibitory (dotted) conductances share similar tuning for frequency (c) and level (d). (Adapted with permission from Wehr and Zador, 2003; Nature Publishing Group).

D. Spectral Resolution and Critical Band Phenomena

Identifying the physiological substrate for perceptual phenomena is a key goal for relating neuronal coding and perception. A number of neuronal mechanisms have been identified in the colliculus, thalamus, and cortex that correlate with critical band spectral integration.

Within the ICC, the frequency gradient of the cochlea is systematically mapped into an array of parallel frequency-band lamina (Merzenich and Reid, 1974; Semple and Aitkin, 1979) (see Section VI.C, Fig. 8. Electrode penetrations along this principal frequency axis of the ICC have demonstrated stepwise 175 μm spacing changes in dominant CF across the lamina. Each lamina encodes for approximately 1/3 octave range of frequencies (Schreiner and Langner, 1997). Such an organization of spectral RF preferences may provide a functional and anatomical substrate for the critical band perceptual resolution. Single neurons in the ICC also exhibit other spectral sensitivities that are consistent with critical band integration. The limiting spectral resolution to pure tones measured in the presence of a noise-masker (Ehret and Merzenich, 1985, 1988b) or to broadband dynamic ripples (Qiu *et al.*, 2003), for instance, mirrors critical band integration.

Critical band resolution phenomena are not exclusive to the ICC. The spectral resolution of the thalamocortical system also approximates critical band integration under a variety of broadband conditions. In cat AI spectral RF measures with both tone-noise masking paradigm (Ehret and Schreiner, 1997) and dynamic ripples (Miller *et al.*, 2002) exhibit spectral sensitivity consistent with 1/3 octave integration. In the awake ferret, Fritz *et al.* (2003) demonstrated that rapid task-related changes in AI receptive fields are confined to approximately 1/3 octave frequency range. This frequency-limited adaptive reorganization of the STRF structure during active learning is consistent with the functional resolution of a subset of thalamocortical projection neurons from the MGBv to AI (Miller *et al.*, 2001; Read *et al.*, 2004). The majority of synaptically connected thalamic and cortical neuron pairs exhibit a high-level spectral alignment within roughly one critical band (Miller *et al.*, 2001). The fact that critical band phenomena along the collicular-cortical pathway invariably require broadband noise conditions might suggest that central inhibition partly underlies these phenomena.

V. Spectral and Temporal Integration

Timing differences in the delay and duration of neuronal excitation and inhibition have a significant impact on the way a neuron responds to a dynamic stimulus. The temporal response pattern of a neuron is not only dependent on the spectral content of a sound but is also affected by the temporal evolution of the stimulus. Neurons in the DCN and beyond are selective for the spectro-temporal content of the sound.

Inhibition shapes spectral selectivity and also refines temporal aspects of the response at the level of the ICC and above (Casseday *et al.*, 1994, 2000; Kuwada *et al.*, 1997; Wehr and Zador, 2003; Zhang *et al.*, 2003). Temporal response patterns can be measured independently of spectral tuning with repetitive clicks, noises, or sinusoidal amplitude modulated tones (SAMs). Neurons can exhibit tuning to the sound modulations, both on the basis of temporal synchronization (phase-locking) or firing rate. In the ICC, neurons that synchronize to the temporal envelope have been reported to exceed several hundred Hertz, whereas tuning on the basis of firing rate has been demonstrated for modulation rates beyond 300 Hz (Langner and Schreiner, 1988; Rees and Møller, 1987; Schreiner and Langner, 1988). Cortical neurons, on the other hand generally lose synchronization at modulation rates beyond 25 Hz (Eggermont, 1999; Liang *et al.*, 2002; Schreiner and Urbas, 1988). Although such findings confirm a clear transformation of temporal preferences beyond the ICC, a similar transformation for spectral features of a sound has not been clearly identified.

A. FM Sweep Selectivity

Measurements of the dynamics of spectral tuning have been performed under a variety of stimulus conditions. The interdependence between spectral and temporal attributes can be quantified with sounds that vary simultaneously along time and frequency. FM tone sweeps are one such stimulus in which the frequency content is continuously incremented or decreased with time. Measurements of FM selectivity have confirmed that spectral tuning is generally not a stationary response property. In the ICC and above, neurons can exhibit a preferred frequency sweep velocity (number of frequency units traversed per unit time) and direction (low to high versus high to low frequency) (Creutzfeldt *et al.*, 1980; Eggermont, 1994; Heil, 1997b; Kowalski *et al.*, 1995; Mendelson *et al.*, 1993; Nelken and Versnel, 2000; Poon *et al.*, 1990; Rees and Møller, 1983; Versnel *et al.*, 1995; Zhang *et al.*, 2003). Consistent with temporal modulation preferences, neurons in the ICC are sensitive to high frequency sweep velocities and this sensitivity is systematically reduced to lower velocities at AI (Creutzfeldt *et al.*, 1980; Poon *et al.*, 1991; Rees and Møller, 1983). This stimulus specificity to the time-varying spectral content is partly attributed to the temporal interplay between neuronal excitation and inhibition (Casseday *et al.*, 1994; Wehr and Zador, 2003; Zhang *et al.*, 2003). A distributed neural representation for FM sweep or modulation rate may exist at the level of AI (Mendelson *et al.*, 1993; Nelken and

Versnel, 2000; Versnel *et al.*, 1995), however, a similar distribution or topography has not been described for ICC or MGBv.

B. Two-Tone Spectro-Temporal Interactions

Spectro-temporal interactions can also be probed with time-varying tone-pip patterns. In the cat AI, the dynamics of two-tone interactions have been measured during a forward-masking paradigm where the response of a delayed probe-tone of variable frequency and level is masked by a reference tone at the CF and at a fixed level (Calford and Semple, 1995; Brosch and Schreiner, 1997). The exact shape of a two-tone FRA is largely variable upon the delay in which the spike activity is measured. Suppression of excitation usually follows the primary excitatory response at the CF. The average time scale of forward inhibition was approximately 150 msec suggesting a relatively slow time course of forward masking in AI. The spectral extent and timing of cortical neurons found in these studies is in close agreement with perceptual measures of forward masking in humans (Jesteadt *et al.*, 1982).

C. RFs Probed with Broadband Sounds—Dynamic Spectral Ripples

An alternative paradigm for measuring the RF preference is to employ a variety of structured sounds that encompass a wide range of acoustic components. Time-varying spectral peaks and notches in the envelope of a sound form important features that are relevant for communication and which are observed in the envelope of natural sounds. Measurements of a neuron's response to a variety of broadband spectrotemporal sound sequences can be used to obtain a more complete description of the STRF properties.

STRFs in the ICC, MGBv, and AI have been measured with a variety of random broadband noise stimuli. Dynamic ripple sounds represent a synthetic class of sounds with sinusoidal spectra and temporal modulations that has been used to systematically probe STRF sensitivities (Depireux *et al.*, 2001; Escabí and Schreiner, 2002; Klein *et al.*, 2000; Kowalski *et al.*, 1996; Miller *et al.*, 2001, 2002; Qiu *et al.*, 2003). Dynamic ripples are a direct extension of static ripple noise but also incorporate time-varying modulations as illustrated in Fig. 3A. The envelope of a dynamic ripple is characterized by a sinusoidal spectrum and sinusoidal temporal envelope with frequency-sweeping spectral peaks. This general class of broadband sounds is theoretically appealing because it offers a complete set of acoustic modulation parameters (spectral and temporal modulation frequencies) that could be used as a building block for constructing an arbitrary complex sound (e.g., vocalizations and speech) (Klein *et al.*, 2000;

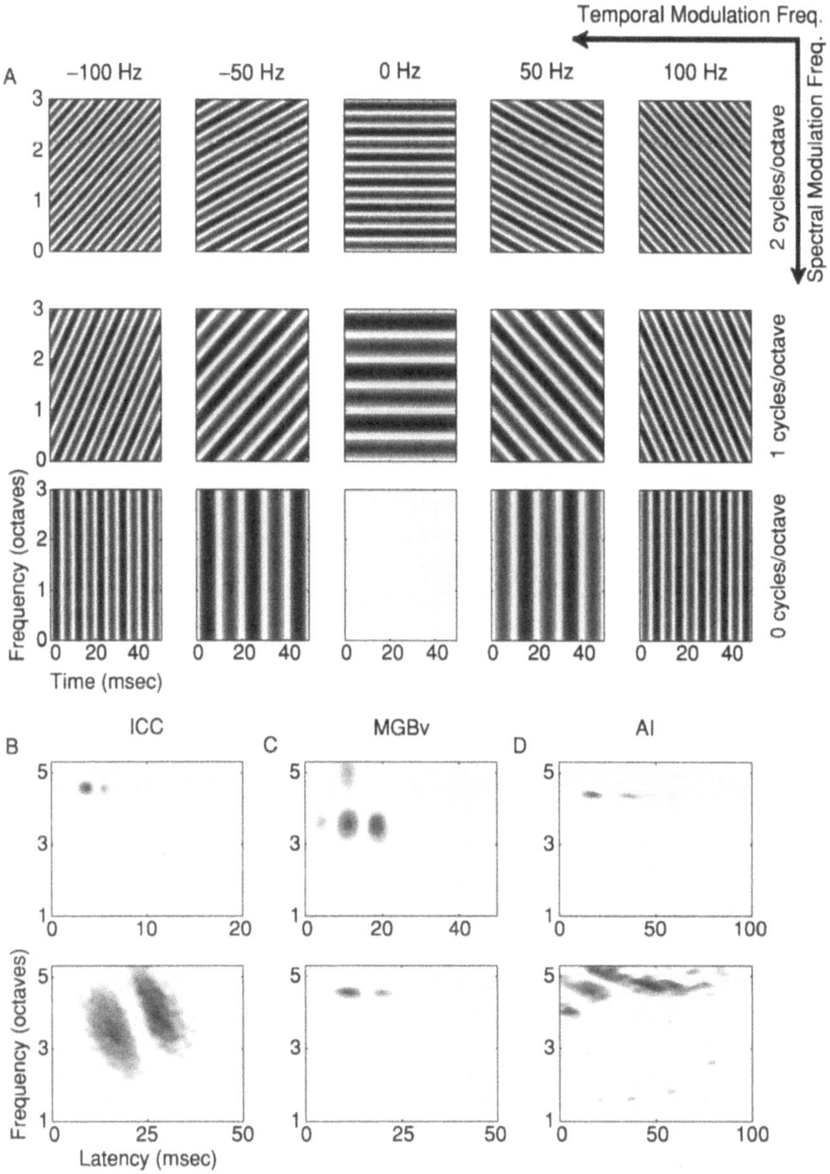

FIG. 3. (A) The spectrotemporal envelope of moving ripple sounds are illustrated as a function of spectral and temporal resolution (SMF and TMF, respectively). Negative SMF produces ripples with upward moving frequency sweeps (top left). Alternately, a high to low frequency transition is observed for ripples with positive SMF (top right). Higher SMF and TMF correspond to finer spectral and temporal resolutions, respectively. Examples of spectrotemporal receptive fields (STRF) from the ICC (B), MGBv (C), and AI (D) were obtained with dynamic ripple sounds. A number of receptive field

Singh and Theunissen, 2003). Although these signals are structurally more complex than simple tone-pips and noise, they are functionally attractive because they offer an intermediate level of complexity that allows for an unbiased assessment of spectral and temporal sensitivities. As illustrated in Fig. 3A the temporal and spectral resolution and ripple sweep direction are controlled by changing the spectral (SMF; units of cycles/octave) and temporal modulation frequencies (TMF; units of Hz). A number of sounds with random spectro-temporal modulations have also been employed for measuring neuronal STRFs. These include random spectrotemporal tone-pips (Blake and Merzenich, 2002; deCharms *et al.*, 1998; Linden *et al.*, 2003; Rutkowski *et al.*, 2002; Theunissen *et al.*, 2000; Valentine and Eggermont, 2004), filtered broadband noise (Escabí and Schreiner, 2002; Keller and Takahashi, 2000; Klein *et al.*, 2000), and natural sounds (Aertsen *et al.*, 1981; Johannesma and Aertsen, 1982; Luczak *et al.*, 2004; Machens *et al.*, 2004; Sen *et al.*, 2001; Theunissen *et al.*, 2000; Yeshurun *et al.*, 1989). In most instances, these are designed so that they retain a flat modulation spectrum and impulsive autocorrelation function within the physiologically relevant range of modulations, thereby satisfying the necessary requirements for estimating STRFs with the reverse correlation technique (de Boer and de Jongh, 1967, 1978; Eggermont *et al.*, 1983; Klein *et al.*, 2000).

The structural arrangement of the STRF can be directly related to the spectrotemporal modulation preferences (Depireux *et al.*, 2001; Escabí and Schreiner, 2002; Klein *et al.*, 2000; Miller *et al.*, 2002; Qiu *et al.*, 2003; Sen *et al.*, 2001) and provides a description of the stimulus features, along time and frequency, which the neuron responds to. Conceptually, the STRF corresponds to the average stimulus pattern that elicited neuronal response. Alternately it could be interpreted as an indirect measure of excitatory and inhibitory RF subregions. Spectrotemporal tuning for neurons in the ICC and beyond is characterized by a variety of arrangements. As for two-tone suppression patterns and for static ripple noise, neurons can exhibit purely excitatory spectral tuning or can exhibit interleaved regions of flanking excitation and inhibition in their STRF (Fig. 3B to D). A variety of temporal arrangements can also be identified. Occasionally, ICC, MGBv, and AI neurons will exhibit a purely excitatory or inhibitory response pattern, but more typically they are characterized by a successive on-off temporal response arrangement. Such temporal patterns could be either independent of the spectral tuning or could exhibit a diagonal time-frequency arrangement suggesting sensitivity to the time-varying frequency content.

---

arrangements are possible: STRF features that are often observed include narrowband versus broadband tuning, presence of sideband inhibition, FM sweep selectivity, temporal inhibition, and/or suppression. In general, a progression from fast to slow timing preferences is observed along the transition from the ICC to AI. (See Color Insert.)

D. STRF Preferences

The relationship between the spectral and temporal components of the STRF can be used to characterize how spectral and temporal aspects of a neuron's integration contribute to a neuron's selectivity. As for direction and motion selectivity in the visual system (Reid *et al.*, 1991), the spectral and temporal components of the integration may interact, leading to selectivity for time-varying components of a sound (e.g., FM sweep velocity and direction). In practice, if temporal RF preferences are independent of spectral preferences, an STRF can be subdivided into a temporal receptive field (TRF) and spectral receptive field (SRF) component from which a neuron's sensitivity could be quantified (Depireux *et al.*, 2001; Qiu *et al.*, 2003; Sen *et al.*, 2001). The SRF or spectral cross-section of the STRF is typically characterized by interleaved patterns of excitation and inhibition and is well approximated by a *Gabor* function (Qiu *et al.*, 2003), i.e., the product of a Gaussian envelope and a sinusoidal carrier. The Gaussian envelope defines the center (BF) and range (bandwidth) over which the neuron integrates spectral information. The sinusoidal carrier accounts for the interleaved spacing of excitation and inhibition and its frequency is closely related to the best ripple density of the neuron (in cycles/octave). A similar decomposition can also be applied to the timing profile of the STRF (Qiu *et al.*, 2003), in which case the Gaussian envelope defines the response duration (Gaussian width) and latency (Gaussian peak) and the sinusoidal component accounts for the best temporal modulation frequency of the neuron (in Hz). In the cat ICC, approximately 80% of the receptive field structure is accounted for with a separable RF model (Qiu *et al.*, 2003), suggesting that spectral and temporal aspects of the integration are largely independent. By comparison, cortical neurons exhibit a greater proportion of inseparable spectrotemporal features (Depireux *et al.*, 2001; Sen *et al.*, 2001) that are consistent with the refinement of directional FM preferences in the cortex (Zhang *et al.*, 2003).

The structural arrangement of spectral and/or temporal excitation and inhibition has also been studied within the context of modulation tuning. The modulation preference of a neuron can be estimated directly from its STRF by Fourier transforming the STRF into a spectro-temporal modulation transfer function (stMTF) (Miller *et al.*, 2002; Qiu *et al.*, 2003; Sen *et al.*, 2001). The stMTF expresses the neurons sensitivity as a function of the sound's temporal (TMF; Hz) and spectral modulation frequency (SMF; cycles/octave or cycles/Hz) (Fig. 4). By considering the arrangement of excitation and inhibition a number of different receptive field preferences can be identified (Miller *et al.*, 2002; Qiu *et al.*, 2003). Temporal *bandpass* tuning occurs whenever the temporal patterning of the STRF has an interleaved arrangement of excitation and inhibition as illustrated in Fig. 4. The temporal MTF exhibits a clear peak at a non-zero best temporal modulation frequency (bTMF). A purely excitatory or inhibitory receptive field

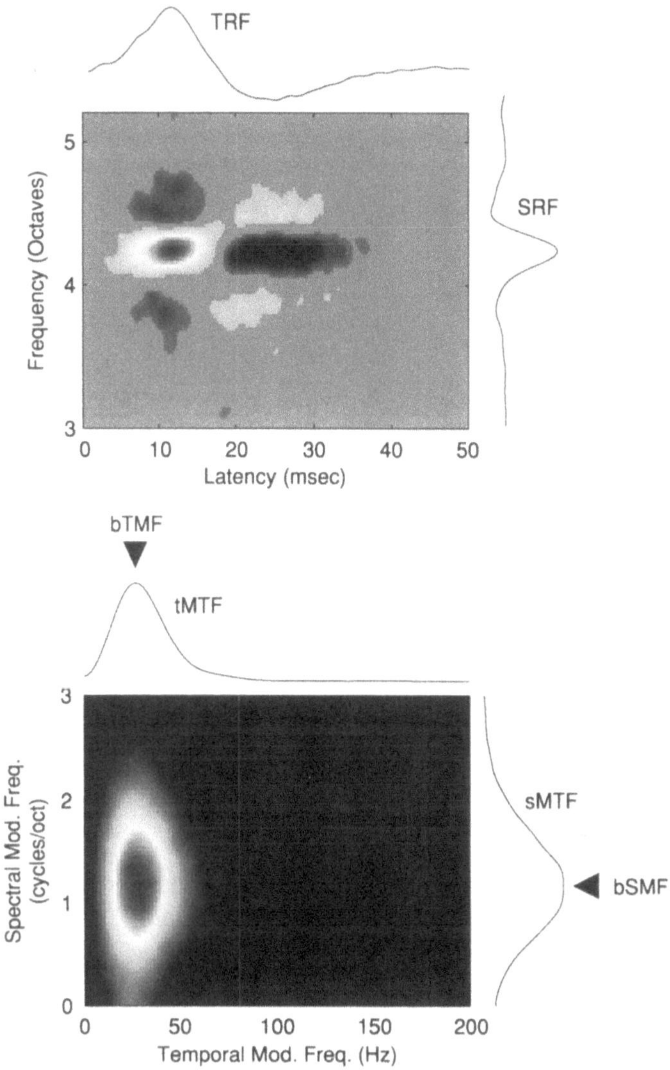

FIG. 4. Spectral and temporal modulation analysis of the STRF preference. A typical ICC receptive field is shown (top) along with its corresponding spectrotemporal modulation transfer function (stMTF) (bottom). The STRF can be decomposed into a temporal receptive field (TRF) and spectral receptive field (SRF) (Qiu *et al.*, 2003). This neuron's TRF is marked by an on-off temporal arrangement whereas its SRF exhibits an off-on-off response pattern. The modulation preference of this neuron can be determined by transforming the STRF to its stMTF, which exhibits a peak response at a preferred spectral and temporal resolution. The spectral and temporal cross sections of the stMTF (to the right and above the stMTF) show a peak about the neurons best temporal modulation (bTMF) and best spectral modulation (bSMF) (denoted by arrows). This neuron responds optimally to 30 Hz temporal modulations and spectral modulations about 1.25 cycles/octave. (See Color Insert.)

temporal pattern, by comparison, will exhibit *lowpass* selectivity. This corresponds to a dominant response at a modulation frequency of 0 Hz. A similar classification scheme can be applied to the spectral sensitivity. Spectral bandpass tuning occurs if the neuron contains flanking sideband inhibition or its stMTF exhibits a best SMF above 0 cycles/octave. Lowpass tuning occurs for a purely excitatory spectral sensitivity, in which case the neuron is sensitive to spectral modulations about 0 cycles/octave (broadband noise).

Consistent with SAM studies, temporal STRF preferences are systematically reduced at each level of the collicular-cortical pathway (Fig. 5A). In the cat, ICC STRFs are characterized by short latency (mean = 10.1 msec; range = 3.5 to 27.4 msec) and short duration (mean = 12.1 msec; range = 1.8 to 82.6 msec) response (Qiu *et al.*, 2003) whereas in the thalamus and cortex the response delays (MGBv mean = 13.2 msec; AI mean = 17.9 msec) are somewhat longer (Miller *et al.*, 2002). Temporal modulation rate measurements from the stMTF reveal that phase-locked modulation rate estimates from STRFs can extend to several hundred hertz in the ICC (Escabí and Schreiner, 2002; Qiu *et al.*, 2003) whereas in thalamus and cortex they are significantly lower (Miller *et al.*, 2002). The mean temporal MTF estimate from the STRF is characterized by bandpass filter profile in all three stations. Furthermore, the best temporal following rate over the population is systematically reduced from 30 Hz at the ICC (observed range = 0 to 255 Hz) to 21.9 Hz and 12.8 Hz in the MGB and AI, respectively (Miller *et al.*, 2002; Qiu *et al.*, 2003). This finding is consistent with temporal modulation studies using SAM and repetitive sequences.

In contrast to temporal preferences, spectral sensitivity in this pathway seems to be much more homogeneous (Fig. 5B). Bandwidth measurements from STRFs reveal a similar range of tuning above the ICC and similar sensitivities to spectral modulations. The average spectral MTF is characterized by a lowpass sensitivity in all three stations (Miller *et al.*, 2002; Qiu *et al.*, 2003) with a peak in the collicular, thalamic, and cortical population MTFs at 0 cycles/octave, suggesting that a large fraction of neurons have little or no sideband inhibition in their STRF, and, therefore, are responsive to broadband stimuli. Furthermore, the range of observed best spectral modulation frequencies (bSMF; observed range = 0 to ~2 cycles/octave) is comparable in all three stations. The limiting spectral resolution estimate from these measurements indicates a maximum resolution of roughly 1/4 octave (1/2 cycle of the maximum observed spectral modulation, ~2 cycles/octave). While the exact nature of these estimates for shaping perceptual resolution still needs further investigation, the limiting STRF resolution in the ICC, MGBv, and AI is surprisingly similar to critical band (1/3 octave) perceptual integration phenomena (Fletcher, 1940; Zwicker *et al.*, 1957).

A unique aspect of the sensory transformation between ICC and thalamocortical receptive field preferences manifests in the combined spectro-temporal population sensitivity. A direct tradeoff in spectral and temporal resolution is seen

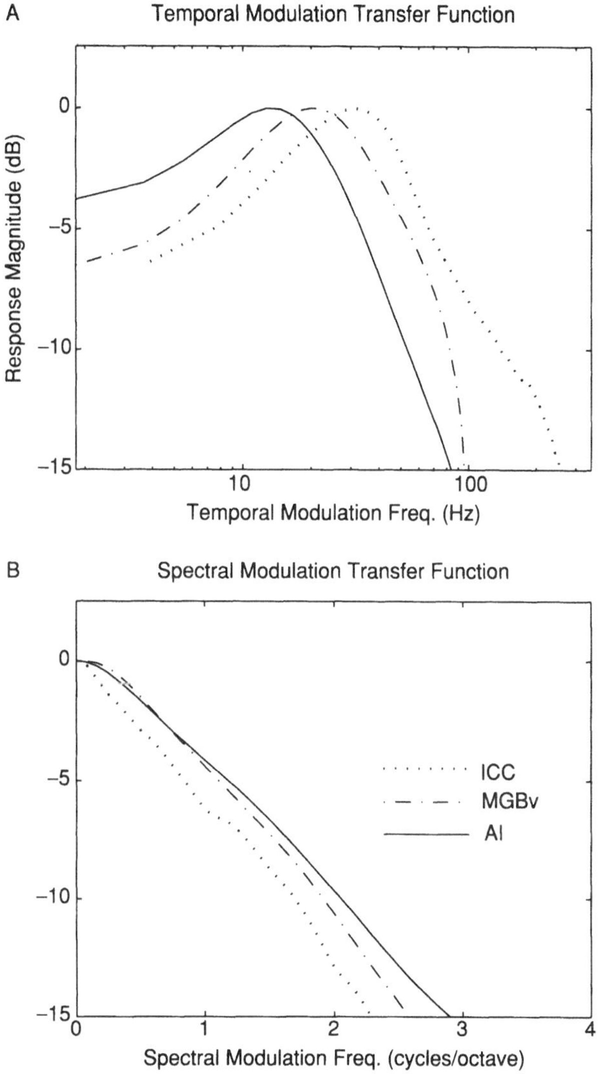

FIG. 5. Population average temporal and spectral modulation transfer functions (tMTF and sMTF, respectively) from the ICC, MGBv, and AI were obtained from the STRF modulation preference (as in Fig. 3). (A) Temporal preferences follow a bandpass selectivity within each station responding optimally to a nonzero temporal modulation. A clear reduction in the preferred/peak temporal modulation rate is observed along the ICC (30 Hz), MGBv (21.9 Hz), and AI (12.8 Hz) pathway. (B) Spectral modulation sensitivity is relatively homogeneous across the three stations, following a low-pass sensitivity with average preferred spectral modulation of 0 cycle/octave (data adapted from Miller *et al.*, 2002 and Qiu *et al.*, 2003).

in the ICC with broadband ripples (Fig. 6) (Qiu *et al.*, 2003) which is not seen using conventional measures of bandwidth or temporal modulation sensitivity. Fast ICC neurons (high bTMF) respond primarily to broad spectral features (low bSMF) while slow integrating neurons are marked by narrow tuning (high bSMF) and strong sideband inhibition. A similar trade-off in spectral and temporal integration resolution has not been observed in thalamic or cortical neurons (Miller *et al.*, 2002). However, there appear to be other forms

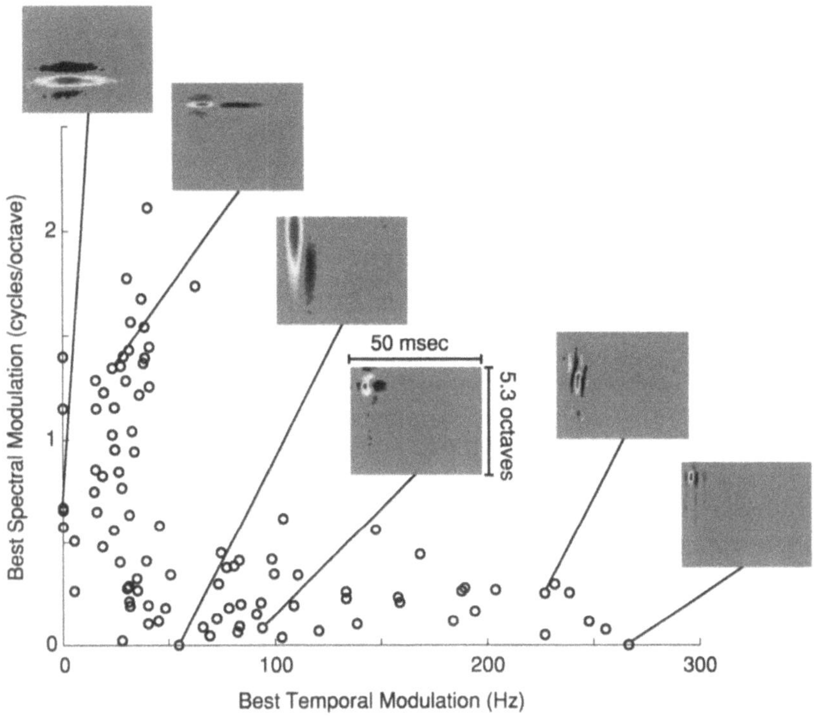

FIG. 6. The best temporal and spectral modulation parameters from ICC neurons exhibit a marked tradeoff in spectral and temporal integration resolution. A similar tradeoff is not observed in the MGBv or AI (Miller *et al.*, 2002; Qiu *et al.*, 2003). Insets show STRFs examples from the ICC at various bSMF and btMF. All STRFs are shown on the same scale (5.3 octave frequency range, ordinate; 50 msec temporal range, abscissa). At fast temporal modulation rates (>50 Hz), ICC neurons are characterized by short temporal duration, fast temporal modulation (Hz), and broad spectral bandwidths. STRFs with high (>50 Hz) modulation rates generally lack spectral sideband inhibition and respond optimally to spectral modulations below 0.5 cycle/octave. Alternately, slow temporal integration neurons (<50 Hz temporal modulation rates) are often characterized by narrow spectral tuning and significant spectral side-band inhibition in the STRF. Neurons with simultaneously high temporal (large bTMF) and spectral resolution (large bSM) are not observed. (See Color Insert.)

of spectro-temporal–dependent properties such as the quadrant non-separability of spectral and temporal components of the STRF for individual neurons (Depireux *et al.*, 2001).

E. Spectral Contrast Sensitivity

Information-bearing components in acoustic signals are often contained within time- or spectral-varying amplitude elements of a complex sound, such as in a peak-notch combination of a vocalization spectrum. The auditory system has an amazing ability to resolve minute differences in amplitude and at the same time is resilient to high levels of background noise and reverberant echoes. It is likely that large peak-to-peak variations observed in natural sounds that are associated with the sound *contrast* may enhance a listener's ability to resolve sounds in noisy environments.

Acoustic contrast can be defined by the peak-to-peak waveform amplitude, waveform standard deviation, or more concisely by the entire amplitude distribution. Thus, contrast is closely related to the sound level and dynamic range. Monotonic and non-monotonic spectral rate-level sensitivities have been described at multiple levels of the auditory system and have been hypothesized as serving as a neuronal correlate for loudness coding. The sound level operating range of a typical auditory neuron (Ehret and Merzenich, 1988a; Evans and Palmer, 1980) is closely matched to the average dynamic range of natural sounds (roughly 30 dB) (Escabí *et al.*, 2003). Thus, it is possible that neurons utilize this operating range not only for level coding but also for encoding spectral and temporal fluctuations associated with the sound contrast. Indeed, AI and ICC neurons can adapt and respond selectively to the dynamic range or contrast of a sound (Barbour and Wang, 2003; Escabí *et al.*, 2003; Kvale and Schreiner, 2004). Neuronal responses to contrast and level are independent suggesting that contrast and loudness coding may in fact involve separate neuronal mechanisms (Escabí *et al.*, 2003).

In most naturally occurring signals, spectral and temporal variations in the instantaneous energy of a sound are normally distributed if defined on a decibel amplitude scale (Attias and Schreiner, 1998b; Escabí *et al.*, 2003). Perceptual studies have demonstrated that loudness perception and discrimination correlate well with the decibel amplitude of a sound but not with its raw sound pressure (Ghitza and Goldstein, 1983; Miller, 1947; Stevens, 1957). Surprisingly, physiological evidence indicates that ICC neurons efficiently utilize decibel fluctuations in natural sounds but are not well suited for encoding linear amplitude differences (Attias and Schreiner, 1998a; Escabí *et al.*, 2003). Furthermore, contrast may serve as a secondary cue for discriminating and categorizing natural signals (e.g., vocalizations and background noise) that have been shown to exhibit different levels of amplitude contrast (Escabí *et al.*, 2003). While such findings suggest that contrast is an important acoustic parameter, further studies are necessary to identify its exact role in various aspects of neuronal coding and perception (e.g., sound source segregation and discrimination).

VI. Organization of Spectral Receptive Field Properties in Neuronal Populations of ICC, MGBv, and AI

In order to appreciate the transformations in spectral receptive field properties at each successive stage of the auditory sensory pathway, it is important to recognize that there are many processes that serve to alter frequency encoding in the peripheral sensory system. There are at least three anatomically and functionally distinct branches of the lemniscal pathway ascending from the periphery to the auditory brainstem. Indeed, some have suggested more (Thompson and Schofield, 2000). These pathways that have very distinct response properties project to adjacent compartments of the same frequency lamina in the ICC (Oliver, 2000; Oliver *et al.*, 1995). ICC is the primary source for ascending projections to the thalamus that in turn projects to AI. The convergence within and between these pathways very likely underlies many of the spectral transformations that occur between the peripheral sensory ganglia and AI.

A. Cochleotopy from the CN to the ICC

The gradient of cochlear hair cell CFs observed in the cat underlies the cat's extensive frequency audibility range and covers over 10 octaves from 45 to 64,000 Hz (Liberman, 1982; Greenwood, 1990). Cochlear frequency sensitivities are represented topographically throughout the lemniscal auditory pathway. The cochleotopic organization differs among the different branches of the brainstem that project to ICC and thalamus. Furthermore, cochleotopy appears to change with processing level. The primary change in cochleotopy with processing level documented to date is a transition from 1/3 octave laminar separation in ICC to 1 octave laminar separation in MGBv (however, see the caveats described in the following text). Such differences in the cochleotopic organization may be associated with distinct spectral processing tasks along the auditory pathway.

Frequency Organization of Sub-Collicular Input to Central IC: The ICC receives convergent input from three cochleotopically organized sub-collicular nuclei (Fig. 7) (Adams, 1979; Bourk *et al.*, 1981; Irvine, 1992; Saint Marie *et al.*, 1999; Spirou *et al.*, 1993). Dorsal cochlear nucleus (DCN), MSO, and LSO are three major branches of the lemniscal auditory pathway projecting to ICC in

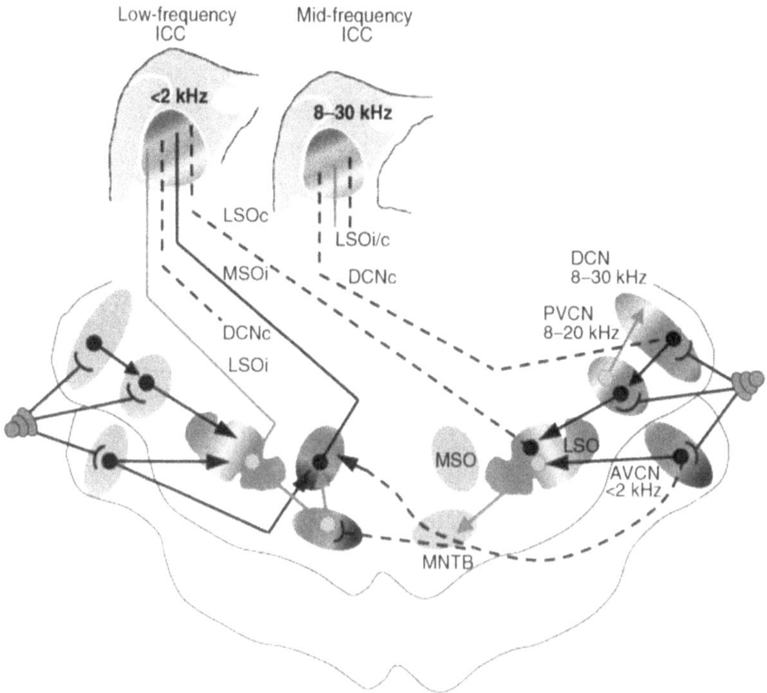

FIG. 7. Major subcollicular pathways projecting to ICC. Each pathway has a separate and in some respects different CF topography. Spiral ganglian input to DCN from two cochlear positions separated by 1 octave is twice the distance for the same octave input in PVCN. Thus, DCN has an expanded area representing mid-frequencies (8 to 30 kHz) whereas PVCN has a compressed area for the same frequencies. DCN inputs go to all iso-frequency laminae in ICC. DCN, MSO, and LSO pathways only project to lower iso-frequency laminae. DCN and LSO pathways project to higher frequency laminae. Direct AVCN/PVCN projections omitted for clarity. DNLL inhibitory projections are not detailed but they cover the same frequency range as LSO. Note that ipsi-hemispheric LSO projections to ICC are inhibitory. This figure is intended to summarize the cochleotopies feeding into ICC and many details have been omitted. Black solid and dotted lines represent excitatory input to ICC. The color scale (blue, green, red, orange) is used to represent the low to high frequency cochleotopic continuum. (See Color Insert.)

mammals.[1] Cochlear sensory ganglion neurons project directly to the ipsi-lateral cochlear nucleus. The CN is functionally and anatomically subdivided into DCN, anterior-ventral (AVCN), and posterior-ventral (PVCN) compartments. The DCN, AVCN, and PVCN have distinct cell populations that are associated with different cochleotopic representations (Bourk *et al.*, 1981; Saint Marie *et al.*, 1999; Spirou *et al.*, 1993).

[1]DCN, MSO, and LSO project to the DNLL which also projects contra-and ipsilateral to ICC.

DCN Pathway: DCN projects heavily to contralateral ICC (dotted line in Fig. 7). The DCN area representing frequencies between 8 and 30 kHz is magnified in size relative to that area representing frequencies below 8 kHz or above 30 kHz (Spirou *et al.*, 1993). A 1/6 octave separation between frequency lamina in DCN corresponds to ~150 μm (Snyder *et al.*, 1997; Spirou *et al.*, 1993). Thus, several frequency laminae in the DCN likely project to a single (1/3 octave) mid-frequency laminae of ICC (Malmierca *et al.*, 1995; Oliver and Morest, 1984; Schreiner and Langner, 1997; Serviere *et al.*, 1984). There are sections of the ICC laminae almost exclusively occupied by DCN terminal fields (Fig. 7) (Oliver, 2000). Furthermore, DCN-like monaural response properties are observed in the ICC and are likely inherited from the ascending synaptic input (Poirier *et al.*, 2003; Ramachandran *et al.*, 1999). In the low iso-frequency laminae of the ICC, DCN input is segregated into a distinct sub-lamina adjacent to MSO/LSO (and DNLL) terminals (Oliver, 2000). Thus, DCN input distributes to appropriate frequency domains but remains relatively segregated within the sub-laminae of ICC.

Functional Distinctions of the DCN Pathway: The DCN pathway clearly plays a role in the spectral integration necessary for processing monaural pinna-based spectral distortions that indicate sound source location relative to the ear and head position (Imig *et al.*, 2000; Oertel and Young, 2004). Spectral properties of local and output projection neurons of the DCN are suited to enhance spectral responses within the middle frequency (8 to 18 kHz) range where pinna distortions in sound spectra are most prominent (Davis and Young, 2000; Imig *et al.*, 2000; Spirou *et al.*, 1999; Young *et al.*, 1992). Response rate for DCN neurons is tuned to spatial location in azimuth and elevation for broad bandwidth sounds; however, the tuning is not observed if probed with narrow bandwidth sounds. Thus, azimuth and elevation tuning in DCN is spectral dependent. A linear relationship between azimuth position, spectral notch, and BF exists for mid-frequency (8 to 18 kHz) projection neurons in the DCN (Imig *et al.*, 2000). Thus, for mid-frequency DCN neurons the cochlear BF reflects spatial location of sound sources! DCN also receives somatosensory input fibers that could carry a neural representation of pinna and head position (Kanold and Young, 2001). Lesions preventing DCN output to higher levels lead to permanent deficits in broad bandwidth spectral-based sound localization tasks (May, 2000).

MSO and LSO Pathways: AVCN and PVCN project directly to the ICC and indirectly to the ICC via the MSO and LSO, respectively (only the MSO/LSO projections are illustrated in Fig. 7) (Adams, 1979). AVCN cochleotopy is complementary to that of the DCN with a greater relative area representing frequencies below 4 kHz (Bourk *et al.*, 1981; Ryugo and May, 1993; Spirou *et al.*, 1993). Cochleotopy in PVCN is approximately one-half the resolution of that in either

DCN or AVCN (Snyder *et al.*, 1997). MSO and LSO have complementary and partially overlapping frequency response property organization (as shown in Fig. 7) (Henkel and Spangler, 1983; Oliver *et al.*, 1995). MSO projects heavily to the low iso-frequency laminae ($\leq$2 kHz) located in dorsal ICC (Fig. 7) (Oliver, 2000; Oliver *et al.*, 1995). MSO and LSO have overlapping terminal field distributions within the low iso-frequency laminae in ICC ($\leq$2 kHz). LSO terminal fields are concentrated in the rostral half whereas MSO terminal fields are concentrated in the caudal half of the laminae (Oliver *et al.*, 1995, 2003). The greatest overlap in MSO/LSO input occurs midway between rostral-caudal borders of the low frequency laminae (labeled "MSO/LSO" in our diagram). LSO projects heavily to the middle and high iso-frequency laminae (=2 kHZ) in ventral ICC. LSO and DCN terminal overlap is greatest for the middle iso-frequency lamina (8 to 18 kHz). However, a sub-laminar segregation is observed for LSO and DCN terminal fields.

Functional Distinctions of the MSO & LSO Pathways: MSO and LSO divisions of the olivary nuclei have some common input from the same cochlear nuclei so a complete functional separation of these pathways is somewhat ambiguous (Fig. 8; Oliver, 2000). However, these sources of converging input have very different functional outcomes as binaural properties in MSO and LSO are quite distinct. MSO neurons process low-frequency sounds and show a distributed code for interaural time (and phase) differences (ITD); whereas LSO neurons process high-frequency sounds and show a distributed representation of interaural level differences (ILD). This fits well with the "duplex theory" of sound localization where low-frequency sounds are localized by timing cues (MSO pathway) and broad bandwidth and high-frequency sounds are located by spectral based ILD cues (LSO pathway) (Middlebrooks and Green, 1991; Populin and Yin, 1998; Pollak *et al.*, 2003; Tollin, 2003). Binaural level differences and spectral shaping of the head-related transfer function are the primary cues for discriminating sound source elevation (Middlebrooks and Green, 1991; Oertel and Young, 2004; Tollin and Yin, 2003; Xu *et al.*, 1999). The latter are likely processed along the DCN and LSO pathways. Thus, MSO and LSO pathways both play prominent roles in processing azimuth while LSO, DCN, and DNLL are important for processing elevation cues. DNLL projections are not shown in Fig. 7. DNLL receives projections from LSO and passes LSO-like response properties to the ICC via long-range inhibitory projections (Pollak *et al.*, 2003). Hence, functionally DNLL can be viewed as part of the LSO pathway. Since the listener and the sound source are often moving in two dimensions (or three) azimuth and elevation cues and the corresponding MSO/LSO pathways likely converge at some level of the auditory pathway in order to generate a cohesive percept.

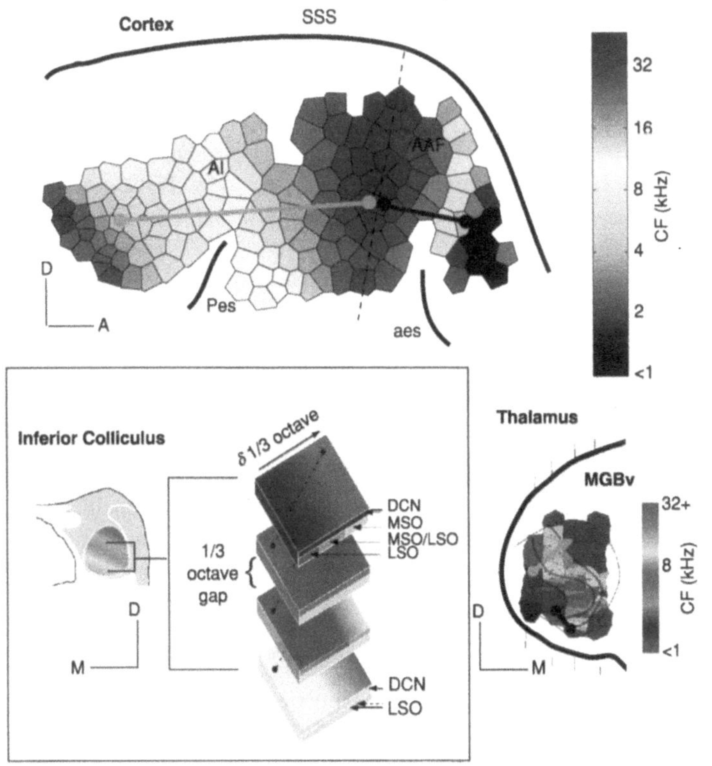

FIG. 8. Multiple cochleotopies in the lemniscal auditory pathway. Rainbow color spectrum indicates CF for each recording site. Blue and red indicate CFs ≤ 1 kHz and ≥32 kHz, respectively. Inset: Three-dimensional organization of CF in the IC. CF change is small for tangential shifts along the "iso-frequency laminae" (dorso-medial to ventro-lateral). CF changes step-wise with 1/3 octave shifts along the dorso-ventral dimension. MSO/LSO and DCN all project to different sub-laminae of the same low CF iso-frequency laminae. Only DCN and LSO project to the high frequency CF (e.g., >16 kHz) laminae. Thalamus: Rainbow color spectrum indicates CF. Recording positions of four recording tracks within MGBv of the cat reconstructed from data of Imig and Morel (1985). Notice that the mm distance for one octave of CF in the inferior part of MGBv is smaller than that of the superior MGBv. Anatomical studies also show two cochleotopies in cat MGBv (Read *et al.*, 2004). Cortex: topographic map of CF showing low-to-high CF gradient for AI and a mirror reversed CF gradient for AAF (Imaizumi *et al.*, 2004). Note that the mm distance for one octave in AI is about twice that of AAF. Apparently the coarse cochleotopy in AAF is inherited from inferior MGBv. (See Color Insert.)

B. Parallel Processing in the ICC

Several groups have suggested that the DCN, MSO, and LSO branches of the auditory pathway provide functionally distinct parallel pathways that project to physiologically distinct cell types and anatomically distinct sub-laminae in the

ICC (Batra *et al.*, 1993; Davis *et al.*, 2002; Ramachandran and May, 2002; Ramachandran *et al.*, 1999). Neurons within the ICC cluster according to monaural and binaural response properties (Semple and Aitkin, 1979). Monaural and binaural properties could segregate in sub-laminae where DCN and olivary projections to the ICC are non-overlapping; however, direct correlations have not been made between physiologic and anatomic sub-laminae (Oliver, 2000). Non-monotonic "O" type ICC neurons are not sensitive to binaural beats and are considered monaural cell types likely inherited from the DCN pathway (Ramachandran and May, 2002). Type "O" cells would include neurons with monaural azimuth position tuning (Poirier *et al.*, 2003). Azimuth tuning in the ICC is shaped by spectral inhibition which is strongest at the cells own pure-tone BF for mid-frequency neurons as is the case for DCN neurons (Davis *et al.*, 2002; Nelken and Young, 1994; Poirier *et al.*, 2003). Spectral and binaural properties including CF of type "V" and type "I" neurons in the ICC are likely inherited from MSO and LSO projections, respectively (Ramachandran and May, 2002; Ramachandran *et al.*, 1999). Type "I" ICC neurons have binaural sensitivities and are sensitive to binaural beats (Ramachandran and May, 2002). Many neurons with type "I" sensitivity are also sensitive to envelope modulations (Batra *et al.*, 1993). Type "I" class would also include binaural azimuth tuned neurons that have ILD sensitivity but lack the spectral inhibitory properties observed for the DCN pathway neurons (Delgutte *et al.*, 1999; Poirier *et al.*, 2003). The ILD sensitivity of binaural azimuth tuned neurons resembles that of LSO neurons (Poirier *et al.*, 2003). A fourth type of neuron is found in the low iso-frequency laminae and has physiologic properties like both MSO and LSO neurons (Batra *et al.*, 1993; Davis *et al.*, 1999; Ramachandran and May, 2002). The ICC neurons with "dual MSO/LSO properties" may be generated in the ICC via convergence of the two input pathways onto single low-frequency laminae or alternatively via intra-laminar ICC stellate neurons (Biebel and Langner, 2002; Oliver and Morest, 1984). However, a subset of the low- and high-frequency dual spectral sensitivities including envelope modulation sensitivities in the ICC are related to monaural cues possibly due to distortions of frequency sensitivities on the cochlea itself (McAlpine, 2004). Thus, in the future we may speak of a "second order" cochleotopic organization that accounts for a subtype of envelope modulation sensitivities in low-frequency neurons that are related to high-frequency cochlear distortions.

Spectral differences in these different acoustic pathways are intimately related to sound location cues including the processing of binaural versus monaural cues. The degree to which these functional streams remain segregated at various levels remains unclear. Many subcollicular properties are conserved in sub-compartments of the ICC (Davis *et al.*, 1999; Ramachandran and May, 2002; Ramachandran *et al.*, 1999). However, it is clear that some physiologic and anatomical convergence occur between the sub-collicular pathways at the level of ICC. Intrinsic inhibitory processes within the ICC can refine binaural and monaural response properties including sharpening rate-level dependence, ITD, and ILD sensitivity (Kuwada *et al.*, 1997; Park and Pollak, 1993; Pollak and Park, 1993). For example, cross-hemispheric inhibitory input from DNLL to ICC neurons converges with LSO excitatory input in ICC to create excitatory-inhibitory responses to binaural sound presentation (Burger and Pollak, 2001). This circuitry conserves and recreates similar binaural excitatory-inhibitory properties observed in LSO and creates de novo an inhibition that appears to allow for temporal adaptation to echoes and accurate sound localization (Pollak *et al.*, 2003). A shift from predominantly monaural azimuth tuning to binaural azimuth tuning at the level of thalamus indicates a large degree of convergence between ICC and thalamic levels of the system (Clarey *et al.*, 1995; Poirier *et al.*, 2003; Samson *et al.*, 2000).

Sub-classes of receptive fields have also been identified in the ICC on the basis of linear and non-linear spectrotemporal integration (Escabí and Schreiner, 2002; Escabí *et al.*, 2003), receptive field resolution, and binaural interactions (Qiu *et al.*, 2003). Approximately 1/3 of colliculus high-frequency neurons exhibit significant binaural STRFs (i.e., for the contralateral and ipsilateral sound). The arrangement of excitatory and inhibitory spectrotemporal interactions in the ICC is well suited for processing binaural disparities in the mid- to high-frequency range (Qiu *et al.*, 2003). Monaural neurons likely arise via projections from the DCN whereas binaural interactions in the mid- to high-frequency stream likely arise via the LSO. The spectrotemporal tuning of ICC receptive fields extends from very narrow bandwidth tuning (~1/3) and slow temporal integration to fast temporal integration and broad bandwidths (Fig. 6). These differences in temporal and spectral resolution could be related to different types of converging inputs from brainstem nuclei (Ramachandran *et al.*, 1999). Alternately, it is plausible that morphological differences in cell class, e.g., stellate versus disk-shaped cells (Oliver and Morest, 1984), or differences in synaptic integration properties (Peruzzi *et al.*, 2000) could contribute to various receptive field classifications. There are presently no studies relating these STRF classifications to the anatomical compartmentalization in ICC or to other conventional receptive field approaches.

C. Three-Dimensional Cochleotopy in ICC

Cochleotopy in the ICC has two axes for frequency organization in three dimensions. Several techniques have confirmed a coarse low-to-high frequency (and CF) organization with dorso-lateral to ventro-medial progression in the ICC (Malmierca *et al.*, 1995; Merzenich and Reid, 1974; Semple and Aitkin,

1979) (Fig. 8). Close examination of recorded CFs along dorso-ventral oriented electrode tracks reveals a ~175 μm frequency laminae where CFs change minimally. A 1/3 octave gap in frequency representation is observed physiologically between adjacent laminae (Schreiner and Langner, 1997) and is consistent with the anatomically defined dorso-medial ventral-lateral laminar organization (Malmierca *et al.*, 1995; Oliver and Morest, 1984). When recording positions are shifted tangentially and recordings are performed along the extent of the frequency laminae (dorso-medial to ventro-lateral) a shallow gradient in CF is observed accounting for the 1/3 octave gap in the dorso-ventral dimension (see δCF axis in Fig. 8). Thus, there is one axis for small changes in CF and another for large changes. These two axes for small versus large changes in CF form a "zig-zag" pattern as illustrated in Fig. 8. A consequence of this pattern is that two very different patterns of CF convergence occur within a radius of ~200 μm. First, disc-shaped neurons with their dendrites restricted to the same iso-frequency laminae interact with neighboring neurons that have very similar CFs. Disc neurons project in parallel to the MGBv and likely conserve the fine resolution cochleotopy of the δCF axis (Fig. 8). Second stellate neurons with dendrites spanning multiple iso-frequency laminae (Oliver and Morest, 1984) are positioned to integrate CFs shifted by =1/3 octave (Biebel and Langner, 2002; Schreiner and Langner, 1997). The dual axes of CF cochleotopy could reflect the formation of new as well as inherited forms of spectral integration in the ICC.

Topographic Spectral Bandwidth Organization of ICC: Spectral bandwidth probed with pure-tones varies topographically across the middle frequency iso-frequency laminae of IC. Bandwidth and CF are spectral properties that can vary independently. When dynamic noise stimuli or pure-tones are used to probe excitatory bandwidths of the STRF, the majority of neurons in ICC respond to a narrow range of frequencies spanning approximately ±1/4 of an octave while a smaller proportion of broadly tuned transient neurons respond over several octaves beyond their CF or BF (Qiu *et al.*, 2003) (see Fig. 6). Such response profiles are consistent with the anatomic restriction of disc cell dendrites to ~175 μm (Oliver and Morest, 1984) and the hypothesis that broad spectral selectivity can be created across laminar integration of stellate neurons. Thus, two neurons displaced in the anterior-posterior dimension of the same iso-frequency laminae can share the same CF but have very different pure-tone bandwidths. Two mirror-symmetric (concentric) gradients for bandwidth extend from the center of an iso-frequency laminae to the respective borders (Ehret *et al.*, 2003; Schreiner and Langner, 1988). Narrow-to-broad bandwidth gradients in the ICC could be generated by a gradient of DCN versus LSO terminal fields to the mid-frequency laminae; however, no such direct correlation has been demonstrated.

D. Cochleotopic Transformations or Heterogeneity in the MGBv

Recently it has been suggested that dramatic changes in cochleotopy occur between the level of ICC and the MGBv division of auditory thalamus (Cetas *et al.*, 2001, 2003). A dramatic difference in cochleotopy could allow for similarly dramatic differences in spectral processing. Unfortunately, this hypothesis is largely derived from comparisons made across species. Namely, McMullen *et al.* recently presented both anatomic and physiologic evidence that discrete 1-octave jumps occur between adjacent iso-frequency laminae in rabbit MGBv. Since the primary input to MGBv is from the ICC, the data may indicate a major transformation in spectral representation between the two stages. Alternatively, the 1-octave cochleotopy in MGBv could be inherited from brainstem projections of the ICC or DCN. Unfortunately, there is little published data supporting a 1-octave MGBv cochleotopy in mammals other than the rabbit. Early physiologic studies of frequency lamina organization in the cat MGBv demonstrated a low-to-high CF gradient with a lateral to medial progression in MGBv (Fig. 8). The cat MGBv iso-frequency laminar organization for one case of Imig and Morel's original physiologic data (Imig and Morel, 1985) is reconstructed using vornoi tessellation and color coding to indicate CF for each recording site. This reconstruction hints at different cochleotopies in central MGBv and inferior parts of MGBv. In central MGBv, many adjacent frequency contours can be physiologically resolved. This high-resolution cochleotopy could be essentially inherited unchanged from the 1/3[2] octave iso-frequency laminae of ICC. In contrast, the inferior aspect of MGBv appears to have a coarse laminar separation resembling that seen in the rabbit MGBv. Unfortunately, it is hard to quantify the cochleotopy differences in the Imig and Morel data because the reconstruction includes only four electrode tracks separated by more than 700 μm.[3] Recently, several anatomic and physiologic distinctions between central and inferior domains of MGBv have been described (Read *et al.*, 2004). First, a difference in cellular morphology and cochleotopy is observed for central versus inferior MGBv. Inferior MGBv has a coarse cochleotopy similar to that described for the rabbit MGBv. Distinct physiologic properties have been attributed to central and inferior MGBv (He, 2002; He and Hu, 2002). The central versus inferior compartments of MGBv defined by cochleotopic differences may correspond to these previously described sub-compartments.

Physiologic Distinctions of MGBv Subdomains within the Iso-Frequency Dimension: The degree of ICC convergence between DCN-, MSO-, and LSO-dominated sub-compartments as they project to MGBv is unknown. Binaural and monaural

[2]1/3 octave lamina have only confirmed for 4 to 20 kHz. (Schreiner and Langher, 1997).

[3]A fine scale reconstruction of the laminar profiles in cat MGBv is more difficult than the rabbit due to the 90 degree rotation in the frequency lamina axis.

response properties are known to segregate along the superior-to-inferior dimension of MGBv. Unfortunately, the methods of characterizing binaural/monaural properties in DCN and ICC are generally different than those used for MGBv neurons. For this reason, it is difficult to describe topographic or organizational differences in cochleotopy in the MGBv with confidence.

Sharpening and Binaural Integration of Azimuth Tuning to Broadband Sounds in MGB: Central MGBv and rostral pole neurons have sharp binaural azimuth tuning sensitivities which are typically not seen at sub-thalamic levels (Barone *et al.*, 1996; Samson *et al.*, 2000). The spectral-dependent properties of MGBv azimuth tuned neurons likely reflects properties inherited from the "DCN dominated" pathway (Imig *et al.*, 2000; Nelken and Young, 1994). Azimuth tuning of MGBv neurons also changes with binaural versus monaural noise presentation (Samson *et al.*, 1993). Binaural frequency disparity tuning is not seen at the level of DCN and likely reflects the convergence of monaural azimuth representations in ICC.

Sharpening and Binaural Integration of ITD Azimuth Tuning in MGB: MGB neurons have sharper ITD azimuth tuning sensitivities than observed for SO and ICC neurons (Fitzpatrick *et al.*, 1997). Sharpening of ITD tuning in MGB could allow for a more efficient neural population coding of this sound parameter. MGB neurons detect ITDs created by either pure-tones or modulated noise. Thus, the spectral dependence of this form of azimuth tuning in MGBv is more like that of the MSO and LSO pathways than the DCN pathway. Unfortunately, there is little information on whether ITD and ILD cues are independently represented at the level of MGBv.

VII. Primary Auditory Cortical Acoustic Feature Representation and Processing

Auditory cortex can be sub-divided according to cochleotopic organization, spectral integration, azimuth tuning properties, and behavioral effects following lesions. Up to four distinct cochleotopically organized regions have been described in mammalian auditory cortex (Schreiner *et al.*, 2000). Examples of differences in cochleotopic representations are described in the following text and illustrated in part in Fig. 8. A direct correlation between the cochleotopy and specializations in spectral or azimuth tuning for a given population of neurons has not been made; however, clear distinctions in these properties characterize the different auditory cortical regions including primary (AI), anterior auditory field (AAF), and posterior auditory field (PAF). Behavioral studies have demonstrated a critical role for both AI and PAF in making fine scale orientations along the azimuth (Adriani *et al.*, 2003; Jenkins and Merzenich, 1984; Malhotra *et al.*, 2004; Smith *et al.*, 2004). The frequency representation for 5 to 20 kHz is expanded in AI versus AAF (Imaizumi *et al.*, 2004). AI neurons are azimuth and elevation tuned to broad bandwidth sounds falling in this range and, therefore, at least a sub-popultation of AI neurons is likely to play a role in representing the pinna distortion cues (Barone *et al.*, 1996; Samson *et al.*, 1994; Schnupp *et al.*, 2001; Xu *et al.*, 1998). Azimuth also is represented coarsely in PAF neuron populations (Barone *et al.*, 1996; Orman and Phillips, 1984). The anterior auditory field, on the other hand, does not appear to be critical for locating sound source in space nor have there been any reports of azimuth or elevation tuning in this area. These dramatic differences in functional organization of auditory cortex can be in part attributed to the fact that each cortical region has a unique dominant thalamic source of input. Accordingly, many of the differences in functional organization between auditory cortices could be inherited from their dominant thalamic input. It is a bit more challenging to say exactly what changes in spectral and sound location tuning occur between the level of thalamus and the corresponding cortical target region. In the following, we describe some of these transformations with an emphasis on what emerges in AI and AAF.

A. Cochleotopy and Spectral Integration in AI and AAF

Most physiologic and anatomic studies of cochleotopic organization in AI suggest a homogenous cochleotopy exists at this level of the auditory pathway. A low-to-high CF gradient (cochleotopy) runs from posterior to anterior AI (Fig. 8). AI receives the bulk of its input from the MGBv which is the thalamic subdivision with the highest resolution cochleotopy. The mm of cortex representing the CF gradient for 2 to 32 kHz domain is ~5 mm. For the 4-octave region between 2 and 32 kHz, 1 octave is represented by ~1 mm in the anterior-posterior dimension (Imaizumi *et al.*, 2004; Merzenich *et al.*, 1975). In contrast, the cochleotopy of AAF is very coarse for much of the same range of frequencies (2 to 32 kHz) (Imaizumi *et al.*, 2004). The coarse cochleotopy in AAF is consistent with the fact that this region derives more of its thalamic input from the inferior part of MGBv that appears to have a similar coarse cochleotopy. The inferior MGBv also projects to parts of AI. However, there is no evidence as of yet that the coarse and fine cochleotopies of inferior and central MGBv persist at the level of AI.

B. Bandwidth and FM Organization in AI

A topographic pattern of organization for level-dependent bandwidth and long-range cortical connections is observed in AI. The range of level-dependent responses and corresponding RF bandwidths is extended and represented topographically in AI as compared to subcollicular stations (Barone *et al.*, 1996; Read *et al.*, 2001). Non-monotonic level dependence is more common and more pronounced in AI than other levels (Barone *et al.*, 1996; Clarey *et al.*, 1995). Neurons with very broad bandwidths covering over five octaves of frequency space are only observed in mature cat AI (Bonham *et al.*, 2004). The developmental dependence of the cortical broadband neurons is distinct from the broad bandwidth tuning seen earlier in development in collicular and sub-collicular neurons. Narrow and broad bandwidth neurons have two extreme operating ranges. Bandwidth is correlated with threshold such that narrow and broad bandwidth neurons have low and high response thresholds respectively (Schreiner *et al.*, 2000). Broad bandwidth neurons have FRAs that increase in frequency range and rate of firing monotonically with sound level whereas narrow bandwidth neurons tend to shut down at the higher sound pressure levels (Phillips *et al.*, 1985; Ramachandran *et al.*, 1999; Schreiner *et al.*, 1992). The extreme differences in operating range may explain why narrow bandwidth domains are not connected to broad bandwidth domains via long-range cortico-cortical connections (Read *et al.*, 2001). In a similar fashion, long-range cortical connections are established between subsets of neurons with similar orientation and ocular dominance tuning in cat primary visual cortex (Gilbert and Wiesel, 1989).

The topographic organization of FM speed tuning and binaural response properties are related to that of the level-dependent bandwidth organization in AI (Heil, 1997b; Kowalski *et al.*, 1995; Mendelson *et al.*, 1993; Versnel *et al.*, 1995; Zhang *et al.*, 2003). Narrow and broad bandwidth neurons correspond in location to those neurons responding to slower versus faster FM modulations, respectively. The functional and anatomic organization related to binaural properties has a different spatial organization than that of bandwidth properties (Middlebrooks *et al.*, 1980). Many of the binaural properties observed in AI are observed within the central narrow bandwidth compartment (Read *et al.*, unpublished). A subset of narrow bandwidth neurons is azimuth tuned for broad bandwidth sounds (Samson *et al.*, 1994). The broad bandwidth neurons in AI may have a more homogeneous monaural or binaural sensitivity since this compartment occupies a relatively smaller area of cortex; however, this has not been demonstrated. There are few if any studies demonstrating topography of ITD, ILD, or broadband spectral-based azimuth tuning of the broad bandwidth neurons. Together these studies would imply that the broad bandwidth neurons are operating at relatively high sound levels and are tuned for faster FM modulations. Conversely, narrow band neurons are operating at relatively low sound levels, are tuned to slow FM modulations, and a subset are likely binaurally tuned to azimuth location. Unfortunately, it is difficult to demonstrate all of these parameters across the $\sim 4 \times 10$ mm area of AI; so no direct comparisons between these feature representations have been made. The value of compartmentalizing and topographically representing neurons with similar response properties is still unknown. An appealing although disputed explanation for a similar compartmentalization in visual cortices is that it maximizes response feature coverage for any given sensory epithelium position (Carreira-Perpinan and Goodhill, 2002, 2004).

VIII. Physiologic Distinctions or Transformations in AI

A. Level-Dependent Broad Bandwidth Azimuth Tuning

A correspondence between level-dependence and sound source location tuning appears for the first time in AI. Azimuth-tuned neurons in MGBv and AI have non-monotonic level-dependent responses to broadband noise; however, the level dependence is greater for AI (Barone *et al.*, 1996; Clarey *et al.*, 1995). At the levels of DCN and MGBv there is no correlation between level dependence and azimuth tuning (Barone *et al.*, 1996). Azimuth tuning also becomes sharper in AI relative to MGBv (Barone *et al.*, 1996; Samson *et al.*, 2000). Binaural azimuth tuning in AI could arise from: (1) contralateral AI callosal input, or (2) convergence of functional sub-domains of MGBv. The anatomical basis or functional repercussions of this change in azimuth tuning are as yet unknown.

B. Spectral Tuning

Two distinct types of spectral transformations occur in the transition from MGBv to AI. Paired single unit recordings between thalamus and AI cortex reveal two forms of thalamocortical CF convergence. AI properties are deemed "inherited" for thalamocortical neuron pairs with excitatory response area CFs that are matched within 0.05 octaves. AI excitatory CFs are deemed "constructive" for thalamocortical neuron pairs with excitatory response area CFs that are mis-matched by as much as 2/3 octave ($\pm 1/3$ octave) (Miller *et al.*, 2001). We have hypothesized that inheritance and constructive frequency convergence arise from central and inferior subdivisions of the MGBv, respectively (Read *et al.*, 2004). However, the limited number of correlated (paired) units prevents a thorough population analysis of the anatomical (MGBv sub-compartment) source for the two physiologic convergence patterns.

C. Binaural Properties

The degree to which left and right ear spectral sensitivities are conserved between thalamic and cortical levels of processing is still somewhat of a controversy. Combined anatomical and physiologic studies indicate a high degree of conservation for many binaural and monaural properties between thalamic neurons and their cortical targets (Cetas *et al.*, 2002; Velenovsky *et al.*, 2003). On the other hand, simultaneous recordings from thalamocortical neuron pairs reveal significant differences in interaural receptive field properties (Miller *et al.*, 2001, 2002). The later studies employed uncorrelated dynamic noise stimuli presented simultaneously to each ear with the same SPL range and variable spectrotemporal content. Differences in the contralateral and ipsilateral STRF structure include CF shifts that may be important for processing binaural spectral disparities in the mid to high frequency range. STRF structure for ipsi versus contra ear inputs can be highly similar, inverted spectrally, or dissimilar in both temporal and spectral character. A similar range of interaural or binaural patterns has been observed for ICC (Qiu *et al.*, 2003). However, interaural pattern for STRFs can be dramatically different for thalamic and cortical neurons that appear synaptically connected based on cross-correlation histogram analysis (Miller *et al.*, 2001, 2002). The latter finding suggests that more than one kind of binaural STRF pattern may converge onto a single layer IV cortical neuron. Thus, at least some cortical neurons do not strictly inherit their binaural pattern. There have been no simultaneous recordings of paired collicular and thalamic neurons to indicate the degree to which binaural properties are transformed between these two levels.

D. Functional Distinctions and Processes Served by AI

Primary auditory cortex (AI) clearly plays a significant role in the spectral integration necessary for sound source location relative to the head (and likely the ear) (Imig *et al.*, 2000; Oertel and Young, 2004). Cats and ferrets with lesions of AI do not orient properly to broad bandwidth noises presented in the contralateral auditory field (Jenkins and Masterton, 1982; Malhotra *et al.*, 2004; Smith *et al.*, 2004). In contrast, lesions of the neighboring anterior auditory cortical fields do not alter orientation to broad bandwidth noises (Malhotra *et al.*, 2004). The orientation deficits seen with AI lesions resemble those observed when DCN projections are interrupted. A subset of neurons in the AI are sharply tuned to sound source location in the azimuth (Barone *et al.*, 1996). A correlation between strongly non-monotonic FRAs and azimuth tuning is observed in the AI but not in the MGBv (Barone *et al.*, 1996; Clarey *et al.*, 1995). Hence, the functional and anatomic compartmentalization of level-dependent narrow bandwidth neurons in AI may be related to pinnae-based spectral integration and azimuth/elevation tuning. However, a correlation between topographic patterns of azimuth/elevation tuning and level-dependent bandwidth has not yet been described in AI. It is also not known if the broad-bandwidth monaural tuned neurons in AI correspond to the neurons with high FM speed selectivity described in the previous text. It is clear that AI has a distributed representation for spectral and temporal integration (e.g., FM speed selectivity); however, it is not clear if this process is independent of binaural processing.

As previously outlined in detail, there are many spectral properties of AI neurons that can be attributed or related to their binaural properties. FM speed selectivities and level-dependent bandwidth properties in AI traditionally have not been related to binaural RFs. It will be interesting to see if there are parallel and independent mechanisms for representing monaural versus binaural spectral (and temporal) features of sounds within AI.

At some point in the auditory pathway it is likely that extensive convergence of the DCN, LSO, and MSO pathways occurs. The striking and perhaps surprising concept to emerge in both physiologic and anatomic descriptions of the past 15 years is that these pathways remain somewhat segregated even at the level of the ICC. Indeed, clear demonstrations of convergence of these pathways on single neurons within the ICC are rare. These different pathways clearly play different roles in sound source localization and orientation and yet most auditory scenes and sounds will clearly activate these three pathways in parallel. It is likely that these processing streams will converge extensively at some point in the lemniscal pathway to allow for unified auditory percepts of sounds. Hence, it will be of interest to see how and if such convergence occurs in the output of ICC to the level of MGB, AI, or at some point later.

IX. Conclusions

Demonstrating the distinct spectral processing streams in the lemniscal auditory pathway has been a formidable task and yet we still do not fully understand their functional role in sound perception. Perhaps the most fundamental task of the cochlea and the lemniscal auditory pathway as a whole is to decompose sounds into frequency channels that can be further analyzed and processed in central stations. Still, numerous aspects of the functional and organizational principles in central auditory stations remain unresolved. At the early brainstem centers, binaural and functionally distinct monaural pathways remain segregated and cochleotopy is established bottom-up via the cochlear projection. From this point of view, the processes served by individual brainstem nuclei can be described as serving a somewhat well-defined task and function, i.e., spectral and intensity analysis in the CN and binaural processing in the SO. Significant differences in cochleotopy are observed at the earliest stages (e.g., PVCN versus AVCN) and perhaps emerge *de novo* at later stages (e.g., MGBv). These heterogeneities in the distributed representation of spectral RFs are explicable in some cases. For example, SRF and cochleotopic differences in MSO versus LSO can be attributed to the role these neural stations play in extracting given spectral features. In contrast, no clear function or statistical properties of sound explain the coarse cochleotopy in PVCN, MGBv, or AAF. It has been suggested that PVCN neurons provide a "wide band" inhibition to DCN neurons (Davis and Young, 2000; Nelken and Young, 1994). Perhaps, a similar wide-band cochleotopy and physiologic property is conferred from PVCN to ICC. It has been difficult to demonstrate a novel process sub-served by the ICC (or MGBv for that matter), including the possibility that the ICC extracts monaural or binaural pitch (Langner, 1997; Langner *et al.*, 2002; McAlpine, 2004). At least three functionally and anatomically distinct sub-collicular pathways converge inputs onto adjacent sub-laminae and compartments of the ICC. Clearly, the recipient ICC neurons are in a position to integrate across these functional domains and yet there is little evidence of a clear functional convergence. We can say with confidence that AI plays a role in representing the contralateral spatial location of sounds, but we cannot say if this is due to convergence between MSO/LSO and DCN pathways in MGBv. In general, future studies will be needed to understand the specific functions or processes served by parallel branches versus subsequent levels of the primary auditory pathway.

It remains a challenge to relate the function of the ICC, MGBv, and AI to the ascending, descending, and local anatomical circuitry. Although some functional properties in the ICC appear to be directly inherited from the brainstem inputs (e.g., spectral integration and binaural type) other aspects of the functional organization are constructed *de novo* via the grid of laminar inputs and its subsequent 1/3 octave cochleotopy and bandwidth topography. Emergent cochleotopies and topographies within the ICC, MGBv, and AI have been described on the basis of bandwidth, sound level, temporal properties, and binaurality. As pointed out in Fig. 7, the dual cochleotopies in ICC, MGBv, and auditory cortex may exist in some form in sub-collicular structures. The specific relationships between these topographies and their roles for acoustic processing and perception still need to be determined. There is sufficient evidence to suggest that simple physiologic properties correlate and may be related to specific acoustic processing tasks such as loudness coding and sound localization. Yet, other high-level functions including sound recognition and source segregation have not been directly related to the physiology or anatomy (but see the chapter by Sinex). The use of sophisticated receptive field approaches with more realistic acoustic stimuli, including natural sounds, should help identify functional and topographic distinctions that cannot be seen under the restricted operating regime afforded by pure-tone and white noise conditions. Although physiologic mapping studies alone are revealing new findings about the organization of the central auditory system, it is much more likely that combined anatomical, behavioral, and physiologic studies will yield key insights to the role subserved by distinct spectral processing streams and cochleotopic organization of the lemniscal pathways.

References

Adams, J. C. (1979). Ascending projections to the inferior colliculus. *J. Comp. Neurol.* **183,** 519–538.

Adriani, M., Maeder, P., Meuli, R., Thiran, A. B., Frischknecht, R., Villemure, J. G., Mayer, J., Annoni, J. M., Bogousslavsky, J., Fornari, E., Thiran, J. P., and Clarke, S. (2003). Sound recognition and localization in man: Specialized cortical networks and effects of acute circumscribed lesions. *Exp. Brain. Res.* **153,** 591–604.

Aertsen, A. M., Olders, J. H., and Johannesma, P. I. (1981). Spectro-temporal receptive fields of auditory neurons in the grassfrog. III. Analysis of the stimulus-event relation for natural stimuli. *Biol. Cybern.* **39,** 195–209.

Attias H., and Schreiner, C (1998a). Coding of naturalistic stimuli by auditory midbrain neurons. *Advances in Neural Information Processing Systems* **10,** 103–109.

Attias H., H., and Schreiner, C. (1998b). Low-order temporal statistics of natural sounds. *Advances in Neural Information Processing Systems* **9,** 27–33.

Bajo, V. M., Merchan, M. A., Malmierca, M. S., Nodal, F. R., and Bjaalie, J. G. (1999). Topographic organization of the dorsal nucleus of the lateral lemniscus in the cat. *J. Comp. Neurol.* **407,** 349–366.

Barbour, D. L., and Wang, X. (2003). Contrast tuning in auditory cortex. *Science* **299,** 1073–1075.

Barone, P., Clarey, J. C., Irons, W. A., and Imig, T. J. (1996). Cortical synthesis of azimuth-sensitive single-unit responses with nonmonotonic level tuning: A thalamocortical comparison in the cat. *J. Neurophysiol.* **75,** 1206–1220.

Batra, R., and Fitzpatrick, D. C. (2002). Monaural and binaural processing in the ventral nucleus of the lateral lemniscus: A major source of inhibition to the inferior colliculus. *Hearing Res.* **168,** 90–97.

Batra, R., Kuwada, S., and Stanford, T. R. (1993). High-frequency neurons in the inferior colliculus that are sensitive to interaural delays of amplitude-modulated tones: Evidence for dual binaural influences. *J. Neurophysiol.* **70,** 64–80.

Biebel, U. W., and Langner, G. (2002). Evidence for interactions across frequency channels in the inferior colliculus of awake chinchilla. *Hearing Res.* **169,** 151–168.

Blake, D. T., and Merzenich, M. M. (2002). Changes of AI receptive fields with sound density. *J. Neurophysiol.* **88,** 3409–3420.

Bonham, B. H., Cheung, S. W., Godey, B., and Schreiner, C. E. (2004). Spatial organization of frequency response areas and rate/level functions in the developing AI. *J. Neurophysiol.* **91,** 841–854.

Bourk, T. R., Mielcarz, J. P., and Norris, B. E. (1981). Tonotopic organization of the anteroventral cochlear nucleus of the cat. *Hearing Res.* **4,** 215–241.

Brosch, M., and Schreiner, C. E. (1997). Time course of forward masking tuning curves in cat primary auditory cortex. *J. Neurophysiol.* **77,** 923–943.

Burger, R. M., and Pollak, G. D. (2001). Reversible inactivation of the dorsal nucleus of the lateral lemniscus reveals its role in the processing of multiple sound sources in the inferior colliculus of bats. *J. Neurosci.* **21,** 4830–4843.

Calford, M. B., and Semple, M. N. (1995). Monaural inhibition in cat auditory cortex. *J. Neurophysiol.* **73,** 1876–1891.

Calhoun, B. M., and Schreiner, C. E. (1998). Spectral envelope coding in cat primary auditory cortex: Linear and non-linear effects of stimulus characteristics. *Eur. J. Neurosci.* **10,** 926–940.

Cant, N. B., and Benson, C. G. (2003). Parallel auditory pathways: Projection patterns of the different neuronal populations in the dorsal and ventral cochlear nuclei. *Brain Res. Bull.* **60,** 457–474.

Carreira-Perpinan, M. A., and Goodhill, G. J. (2002). Are visual cortex maps optimized for coverage? *Neural. Comput.* **14,** 1545–1560.

Carreira-Perpinan, M. A., and Goodhill, G. J. (2004). Influence of lateral connections on the structure of cortical maps. *J. Neurophysiol.* **92,** 2947–2959.

Casseday, J. H., Ehrlich, D., and Covey, E. (1994). Neural tuning for sound duration: Role of inhibitory mechanisms in the inferior colliculus. *Science* **264,** 847–850.

Casseday, J. H., Ehrlich, D., and Covey, E. (2000). Neural measurement of sound duration: Control by excitatory-inhibitory interactions in the inferior colliculus. *J. Neurophysiol.* **84,** 1475–1487.

Cetas, J. S., Price, R. O., Velenovsky, D. S., Sinex, D. G., and McMullen, N. T. (2001). Frequency organization and cellular lamination in the medial geniculate body of the rabbit. *Hearing Res.* **155,** 113–123.

Cetas, J. S., Price, R. O., Crowe, J., Velenovsky, D. S., and McMullen, N. T. (2003). Dendritic orientation and laminar architecture in the rabbit auditory thalamus. *J. Comp. Neurol.* **458,** 307–317.

Cetas, J. S., Price, R. O., Velenovsky, D. S., Crowe, J. J., Sinex, D. G., and McMullen, N. T. (2002). Cell types and response properties of neurons in the ventral division of the medial geniculate body of the rabbit. *J. Comp. Neurol.* **445,** 78–96.

Clarey, J. C., Barone, P., Irons, W. A., Samson, F. K., and Imig, T. J. (1995). Comparison of noise and tone azimuth tuning of neurons in cat primary auditory cortex and medical geniculate body. *J. Neurophysiol.* **74,** 961–980.

Creutzfeldt, O., Hellweg, F. C., and Schreiner, C. (1980). Thalamocortical transformation of responses to complex auditory stimuli. *Exp. Brain Res.* **39,** 87–104.

Davis, K. A., and Young, E. D. (2000). Pharmacological evidence of inhibitory and disinhibitory neuronal circuits in dorsal cochlear nucleus. *J. Neurophysiol.* **83,** 926–940.

Davis, K. A., Ramachandran, R., and May, B. J. (1999). Single-unit responses in the inferior colliculus of decerebrate cats. II. Sensitivity to interaural level differences. *J. Neurophysiol.* **82,** 164–175.

Davis, K. A., Ramachandran, R., and May, B. J. (2003). Auditory processing of spectral cues for sound localization in the inferior colliculus. *J. Assoc. Res. Otolaryngol.* **4,** 148–163.

de Boer, E., and de Jongh, H. R. (1967). Correlation studies applied to the frequency resolution of the cochlea. *J. Aud. Res.* **7,** 209–217.

de Boer, E., and de Jongh, H. R. (1978). On cochlear encoding: Potentialities and limitations of the reverse-correlation technique. *J. Acoust. Soc. Am.* **63,** 115–135.

deCharms, R. C., Blake, D. T., and Merzenich, M. M. (1998). Optimizing sound features for cortical neurons. *Science* **280,** 1439–1443.

Delgutte, B., Joris, P. X., Litovsky, R. Y., and Yin, T. C. (1999). Receptive fields and binaural interactions for virtual-space stimuli in the cat inferior colliculus. *J. Neurophysiol.* **81,** 2833–2851.

Depireux, D. A., Simon, J. Z., Klein, D. J., and Shamma, S. A. (2001). Spectro-temporal response field characterization with dynamic ripples in ferret primary auditory cortex. *J. Neurophysiol.* **85,** 1220–1234.

Eggermont, J. J. (1994). Temporal modulation transfer functions for AM and FM stimuli in cat auditory cortex. Effects of carrier type, modulating waveform and intensity. *Hearing Res.* **74,** 51–66.

Eggermont, J. J. (1999). The magnitude and phase of temporal modulation transfer functions in cat auditory cortex. *J. Neurosci.* **19,** 2780–2788.

Eggermont, J. J., Johannesma, P. M., and Aertsen, A. M. (1983). Reverse-correlation methods in auditory research. *Q. Rev. Biophys.* **16,** 341–414.

Egorova, M., Ehret, G., Vartanian, I., and Esser, K. H. (2001). Frequency response areas of neurons in the mouse inferior colliculus. I. Threshold and tuning characteristics. *Exp. Brain Res.* **140,** 145–161.

Ehret, G., and Merzenich, M. M. (1985). Auditory midbrain responses parallel spectral integration phenomena. *Science* **227,** 1245–1247.

Ehret, G., and Merzenich, M. M. (1988a). Neuronal discharge rate is unsuitable for encoding sound intensity at the inferior-colliculus level. *Hearing Res.* **35,** 1–7.

Ehret, G., and Merzenich, M. M. (1988b). Complex sound analysis (frequency resolution, filtering and spectral integration) by single units of the inferior colliculus of the cat. *Brain Res.* **472,** 139–163.

Ehret, G., and Schreiner, C. E. (1997). Frequency resolution and spectral integration (critical band analysis) in single units of the cat primary auditory cortex. *J. Comp. Physiol. [A]* **181,** 635–650.

Ehret, G., Egorova, M., Hage, S. R., and Muller, B. A. (2003). Spatial map of frequency tuning-curve shapes in the mouse inferior colliculus. *Neuroreport* **14,** 1365–1369.

Escabí, M. A., and Schreiner, C. E. (2002). Nonlinear spectrotemporal sound analysis by neurons in the auditory midbrain. *J. Neurosci.* **22,** 4114–4131.

Escabí, M. A., Miller, L. M., Read, H. L., and Schreiner, C. E. (2003). Naturalistic auditory contrast improves spectrotemporal coding in the cat inferior colliculus. *J. Neurosci.* **23,** 11489–11504.

Evans, E. F., and Palmer, A. R. (1980). Relationship between the dynamic range of cochlear nerve fibers and their spontaneous activity. *Exp. Brain Res.* **40,** 115–118.

Fitzpatrick, D. C., Batra, R., Stanford, T. R., and Kuwada, S. (1997). A neuronal population code for sound localization. *Nature* **388,** 871–874.

Fletcher, H. (1940). Auditory patterns. *Rev. Mod. Phys.* **12,** 47–65.

Fritz, J., Shamma, S., Elhilali, M., and Klein, D. (2003). Rapid task-related plasticity of spectrotemporal receptive fields in primary auditory cortex. *Nat Neurosci* **6,** 1216–1223.

Ghitza, O., and Goldstein, J. L. (1983). JNDs for the spectral envelope parameters in natural speech. *In* "Hearing: Physiological Bases and Psychophysics" (R. Klinke and R. Hartmann, Eds.), pp. 352–359. Springer Verlag, New York.

Gilbert, C. D., and Wiesel, T. N. (1989). Columnar specificity of intrinsic horizontal and corticocortical connections in cat visual cortex. *J. Neurosci.* **9,** 2432–2442.

Greenwood, D. D. (1990). A cochlear frequency-position function for several species—29 years later. *J. Acoust. Soc. Am.* **87,** 2592–2605.

He, J. (2002). OFF responses in the auditory thalamus of the guinea pig. *J. Neurophysiol.* **88,** 2377–2386.

He, J., and Hu, B. (2002). Differential distribution of burst and single-spike responses in auditory thalamus. *J. Neurophysiol.* **88,** 2152–2156.

Heil, P. (1997a). Auditory cortical onset responses revisited. I. First-spike timing. *J. Neurophysiol.* **77,** 2616–2641.

Heil, P. (1997b). Aspects of temporal processing of FM stimuli in primary auditory cortex. *Acta Otolaryngol. Suppl.* **532,** 99–102.

Heil, P. (1997c). Auditory cortical onset responses revisited. II. Response strength. *J. Neurophysiol.* **77,** 2642–2660.

Heil, P., Rajan, R., and Irvine, D. R. F. (1994). Topographic representation of tone intensity along the isofrequency axis of cat primary auditory-cortex. *Hearing Res.* **76,** 188–202.

Henkel, C. K., and Spangler, K. M. (1983). Organization of the efferent projections of the medial superior olivary nucleus in the cat as revealed by HRP and autoradiographic tracing methods. *J. Comp. Neurol.* **221,** 416–428.

Imaizumi, K., Priebe, N. J., Crum, P. A., Bedenbaugh, P. H., Cheung, S. W., and Schreiner, C. E. (2004). Modular functional organization of cat anterior auditory field. *J. Neurophysiol.* **92,** 444–457.

Imig, T. J., and Morel, A. (1985). Tonotopic organization in ventral nucleus of medial geniculate body in the cat. *J. Neurophysiol.* **53,** 309–340.

Imig, T. J., Bibikov, N. G., Poirier, P., and Samson, F. K. (2000). Directionality derived from pinna-cue spectral notches in cat dorsal cochlear nucleus. *J. Neurophysiol.* **83,** 907–925.

Irvine, D. R. (1992). Physiology of the auditory brainstem. *In* "The Mammalian Auditory Pathway: Neurophysiology" (A. N. Popper and R. R. Fay, Eds.), pp. 153–231. Springer, New York.

Jenkins, W. M., and Masterton, R. B. (1982). Sound localization: Effects of unilateral lesions in central auditory system. *J. Neurophysiol.* **47,** 987–1016.

Jenkins, W. M., and Merzenich, M. M. (1984). Role of cat primary auditory cortex for sound-localization behavior. *J. Neurophysiol.* **52,** 819–847.

Jesteadt, W., Bacon, S. P., and Lehman, J. R. (1982). Forward masking as a function of frequency, masker level, and signal delay. *J. Acoust. Soc. Am.* **71,** 950–962.

Johannesma, P., and Aertsen, A. (1982). Statistical and dimensional analysis of the neural representation of the acoustic biotope of the frog. *J. Med. Syst.* **6,** 399–421.

Kanold, P. O., and Young, E. D. (2001). Proprioceptive information from the pinna provides somatosensory input to cat dorsal cochlear nucleus. *J. Neurosci.* **21,** 7848–7858.

Keller, C. H., and Takahashi, T. T. (2000). Representation of temporal features of complex sounds by the discharge patterns of neurons in the owl's inferior colliculus. *J. Neurophysiol.* **84,** 2638–2650.

Klein, D. J., Depireux, D. A., Simon, J. Z., and Shamma, S. A. (2000). Robust spectrotemporal reverse correlation for the auditory system: Optimizing stimulus design. *J. Comput. Neurosci.* **9,** 85–111.

Kowalski, N., Versnel, H., and Shamma, S. A. (1995). Comparison of responses in the anterior and primary auditory fields of the ferret cortex. *J. Neurophysiol.* **73,** 1513–1523.

Kowalski, N., Depireux, D. A., and Shamma, S. A. (1996). Analysis of dynamic spectra in ferret primary auditory cortex. I. Characteristics of single-unit responses to moving ripple spectra. *J. Neurophysiol.* **76,** 3503–3523.

Kuwada, S., Batra, R., Yin, T. C., Oliver, D. L., Haberly, L. B., and Stanford, T. R. (1997). Intracellular recordings in response to monaural and binaural stimulation of neurons in the inferior colliculus of the cat. *J. Neurosci.* **17,** 7565–7581.

Kvale, M. N., and Schreiner, C. E. (2004). Short-term adaptation of auditory receptive fields to dynamic stimuli. *J. Neurophysiol.* **91,** 604–612.

Langner, G. (1997). Neural processing and representation of periodicity pitch. *Acta Otolaryngol. Suppl.* **532,** 68–76.

Langner, G., and Schreiner, C. E. (1988). Periodicity coding in the inferior colliculus of the cat. I. Neuronal mechanisms. *J. Neurophysiol.* **60,** 1799–1822.

Langner, G., Albert, M., and Briede, T. (2002). Temporal and spatial coding of periodicity information in the inferior colliculus of awake chinchilla (Chinchilla laniger). *Hearing Res.* **168,** 110–130.

Le Beau, F. E., Rees, A., and Malmierca, M. S. (1996). Contribution of GABA- and glycine-mediated inhibition to the monaural temporal response properties of neurons in the inferior colliculus. *J. Neurophysiol.* **75,** 902–919.

Liang, L., Lu, T., and Wang, X. (2002). Neural representations of sinusoidal amplitude and frequency modulations in the primary auditory cortex of awake primates. *J. Neurophysiol.* **87,** 2237–2261.

Liberman, M. C. (1982). The cochlear frequency map for the cat: Labeling auditory-nerve fibers of known characteristic frequency. *J. Acoust. Soc. Am.* **72,** 1441–1449.

Linden, J. F., Liu, R. C., Sahani, M., Schreiner, C. E., and Merzenich, M. M. (2003). Spectrotemporal structure of receptive fields in areas AI and AAF of mouse auditory cortex. *J. Neurophysiol.* **90,** 2660–2675.

Luczak, A., Hackett, T. A., Kajikawa, Y., and Laubach, M. (2004). Multivariate receptive field mapping in marmoset auditory cortex. *J. Neurosci. Methods* **136,** 77–85.

Machens, C. K., Wehr, M. S., and Zador, A. M. (2004). Linearity of cortical receptive fields measured with natural sounds. *J. Neurosci.* **24,** 1089–1100.

Malhotra, S., Hall, A. J., and Lomber, S. G. (2004). Cortical control of sound localization in the cat: Unilateral cooling deactivation of 19 cerebral areas. *J. Neurophysiol.* **92,** 1625–1643.

Malmierca, M. S., Rees, A., Le Beau, F. E., and Bjaalie, J. G. (1995). Laminar organization of frequency-defined local axons within and between the inferior colliculi of the guinea pig. *J. Comp. Neurol.* **357,** 124–144.

May, B. J. (2000). Role of the dorsal cochlear nucleus in the sound localization behavior of cats. *Hearing Res.* **148,** 74–87.

McAlpine, D. (2004). Neural sensitivity to periodicity in the inferior colliculus: Evidence for the role of cochlear distortions. *J. Neurophysiol.* **92,** 1295–1311.

Mendelson, J. R., Schreiner, C. E., Sutter, M. L., and Grasse, K. L. (1993). Functional topography of cat primary auditory cortex: Responses to frequency-modulated sweeps. *Exp. Brain Res.* **94,** 65–87.

Merzenich, M. M., and Reid, M. D. (1974). Representation of the cochlea within the inferior colliculus of the cat. *Brain Res.* **77,** 397–415.

Merzenich, M. M., Knight, P. L., and Roth, G. L. (1975). Representation of cochlea within primary auditory cortex in the cat. *J. Neurophysiol.* **38,** 231–249.

Middlebrooks, J. C., and Green, D. M. (1991). Sound localization by human listeners. *Annu. Rev. Psychol.* **42,** 135–159.

Middlebrooks, J. C., Dykes, R. W., and Merzenich, M. M. (1980). Binaural response-specific bands in primary auditory cortex (AI) of the cat: Topographical organization orthogonal to isofrequency contours. *Brain Res.* **181,** 31–48.

Miller, G. A. (1947). Sensitivity to changes in the intensity of white noise and its relation to masking and loudness. *J. Acoust. Soc. Am.* **191,** 609–619.

Miller, L. M., Escabí, M. A., Read, H. L., and Schreiner, C. E. (2001). Functional convergence of response properties in the auditory thalamocortical system. *Neuron* **32,** 151–160.

Miller, L. M., Escabí, M. A., Read, H. L., and Schreiner, C. E. (2002). Spectrotemporal receptive fields in the lemniscal auditory thalamus and cortex. *J. Neurophysiol.* **87,** 516–527.

Nelken, I., and Young, E. D. (1994). Two separate inhibitory mechanisms shape the responses of dorsal cochlear nucleus type IV units to narrowband and wideband stimuli. *J. Neurophysiol.* **71,** 2446–2462.

Nelken, I., and Versnel, H. (2000). Responses to linear and logarithmic frequency-modulated sweeps in ferret primary auditory cortex. *Eur. J. Neurosci.* **12,** 549–562.

Oertel, D., and Young, E. D. (2004). What's a cerebellar circuit doing in the auditory system? *Trends Neurosci.* **27,** 104–110.

Oliver, D. L. (2000). Ascending efferent projections of the superior olivary complex. *Microsc. Res. Tech.* **51,** 355–363.

Oliver, D. L., and Morest, D. K. (1984). The central nucleus of the inferior colliculus in the cat. *J. Comp. Neurol.* **222,** 237–264.

Oliver, D. L., and Shneiderman, A. (1989). An EM study of the dorsal nucleus of the lateral lemniscus: Inhibitory, commissural, synaptic connections between ascending auditory pathways. *J. Neurosci.* **9,** 967–982.

Oliver, D. L., Beckius, G. E., and Shneiderman, A. (1995). Axonal projections from the lateral and medial superior olive to the inferior colliculus of the cat: A study using electron microscopic autoradiography. *J. Comp. Neurol.* **360,** 17–32.

Oliver, D. L., Beckius, G. E., Bishop, D. C., Loftus, W. C., and Batra, R. (2003). Topography of interaural temporal disparity coding in projections of medial superior olive to inferior colliculus. *J. Neurosci.* **23,** 7438–7449.

Orman, S. S., and Phillips, D. P. (1984). Binaural interactions of single neurons in posterior field of cat auditory cortex. *J. Neurophysiol.* **51,** 1028–1039.

Park, T. J., and Pollak, G. D. (1993). GABA shapes sensitivity to interaural intensity disparities in the mustache bat's inferior colliculus: Implications for encoding sound location. *J. Neurosci.* **13,** 2050–2067.

Peruzzi, D., Sivaramakrishnan, S., and Oliver, D. L. (2000). Identification of cell types in brain slices of the inferior colliculus. *NeuroScience* **101,** 403–416.

Peruzzi, D., Bartlett, E., Smith, P. H., and Oliver, D. L. (1997). A monosynaptic GABAergic input from the inferior colliculus to the medial geniculate body in rat. *J. Neurosci.* **17,** 3766–3777.

Phillips, D. P., and Irvine, D. R. (1981). Responses of single neurons in physiologically defined primary auditory cortex (AI) of the cat: Frequency tuning and responses to intensity. *J. Neurophysiol.* **45,** 48–58.

Phillips, D. P., Orman, S. S., Musicant, A. D., and Wilson, G. F. (1985). Neurons in the cat's primary auditory cortex distinguished by their responses to tones and wide-spectrum noise. *Hearing Res.* **18,** 73–86.

Poirier, P., Samson, F. K., and Imig, T. J. (2003). Spectral shape sensitivity contributes to the azimuth tuning of neurons in the cat's inferior colliculus. *J. Neurophysiol.* **89,** 2760–2777.

Pollak, G. D., and Park, T. J. (1993). The effects of GABAergic inhibition on monaural response properties of neurons in the mustache bat's inferior colliculus. *Hearing Res.* **65,** 99–117.

Pollak, G. D., Burger, R. M., and Klug, A. (2003). Dissecting the circuitry of the auditory system. *Trends Neurosci.* **26,** 33–39.

Pollak, G. D., Burger, R. M., Park, T. J., Klug, A., and Bauer, E. E. (2002). Roles of inhibition for transforming binaural properties in the brainstem auditory system. *Hearing Res.* **168,** 60–78.

Poon, P. W., Chen, X., and Hwang, J. C. (1991). Basic determinants for FM responses in the inferior colliculus of rats. *Exp. Brain Res.* **83,** 598–606.

Poon, P. W., Sun, X., Kamada, T., and Jen, P. H. (1990). Frequency and space representation in the inferior colliculus of the FM bat, Eptesicus fuscus. *Exp. Brain Res.* **79,** 83–91.

Populin, L. C., and Yin, T. C. (1998). Behavioral studies of sound localization in the cat. *J. Neurosci.* **18,** 2147–2160.

Portfors, C. V., and Wenstrup, J. J. (2002). Excitatory and facilitatory frequency response areas in the inferior colliculus of the mustached bat. *Hearing Res.* **168,** 131–138.

Qiu, A., Schreiner, C. E., and Escabí, M. A. (2003). Gabor analysis of auditory midbrain receptive fields: Spectro-temporal and binaural composition. *J. Neurophysiol.* **90,** 456–476.

Ramachandran, R., and May, B. J. (2002). Functional segregation of ITD sensitivity in the inferior colliculus of decerebrate cats. *J. Neurophysiol.* **88,** 2251–2261.

Ramachandran, R., Davis, K. A., and May, B. J. (1999). Single-unit responses in the inferior colliculus of decerebrate cats. I. Classification based on frequency response maps. *J. Neurophysiol.* **82,** 152–163.

Ramachandran, R., Davis, K. A., and May, B. J. (2000). Rate representation of tones in noise in the inferior colliculus of decerebrate cats. *J. Assoc. Res. Otolaryngol.* **1,** 144–160.

Read, H. L., Winer, J. A., and Schreiner, C. E. (2001). Modular organization of intrinsic connections associated with spectral tuning in cat auditory cortex. *Proc. Natl. Acad Sci. USA* **98,** 8042–8047.

Read, H. L., Miller, L. M., Escabí, M. A., Schreiner, C. E., and Winer, J. A. (2004). Ventral medial geniculate neucleus contains functionally and anatomically distinct compartments. *Assoc. Res. Otolaryngol. Abs.* **27,** 1457.

Rees, A., and Møller, A. R. (1983). Responses of neurons in the inferior colliculus of the rat to AM and FM tones. *Hearing Res.* **10,** 301–330.

Rees, A., and Møller, A. R. (1987). Stimulus properties influencing the responses of inferior colliculus neurons to amplitude-modulated sounds. *Hearing Res.* **27,** 129–143.

Rees, A., and Palmer, A. R. (1988). Rate-intensity functions and their modification by broadband noise for neurons in the guinea pig inferior colliculus. *J. Acoust. Soc. Am.* **83,** 1488–1498.

Reid, R. C., Soodak, R. E., and Shapley, R. M. (1991). Directional selectivity and spatiotemporal structure of receptive fields of simple cells in cat striate cortex. *J. Neurophysiol.* **66,** 505–529.

Rose, J. E., Greenwood, D. D., Goldberg, J. M., and Hind, J. E. (1963). Some discharge characteristics of single neurons in the inferior colliculus of the cat: I. Tonotopical organization, relation of spike-counts to tone intensity, and firing patterns of single elements. *J. Neurophysiol.* **26,** 294–320.

Rouiller, E., de Ribaupierre, Y., Morel, A., and de Ribaupierre, F. (1983). Intensity functions of single unit responses to tone in the medial geniculate body of cat. *Hearing Res.* **11,** 235–247.

Rutkowski, R. G., Shackleton, T. M., Schnupp, J. W., Wallace, M. N., and Palmer, A. R. (2002). Spectrotemporal receptive field properties of single units in the primary, dorsocaudal and ventrorostral auditory cortex of the guinea pig. *Audiol. Neurootol.* **7,** 214–227.

Ryugo, D. K., and May, S. K. (1993). The projections of intracellularly labeled auditory nerve fibers to the dorsal cochlear nucleus of cats. *J. Comp. Neurol.* **329,** 20–35.

Saint Marie, R. L., Luo, L., and Ryan, A. F. (1999). Spatial representation of frequency in the rat dorsal nucleus of the lateral lemniscus as revealed by acoustically induced c-fos mRNA expression. *Hearing Res.* **128,** 70–74.

Samson, F. K., Clarey, J. C., Barone, P., and Imig, T. J. (1993). Effects of ear plugging on single-unit azimuth sensitivity in cat primary auditory cortex. I. Evidence for monaural directional cues. *J. Neurophysiol.* **70,** 492–511.

Samson, F. K., Barone, P., Clarey, J. C., and Imig, T. J. (1994). Effects of ear plugging on single-unit azimuth sensitivity in cat primary auditory cortex. II. Azimuth tuning dependent upon binaural stimulation. *J. Neurophysiol.* **71,** 2194–2216.

Samson, F. K., Barone, P., Irons, W. A., Clarey, J. C., Poirier, P., and Imig, T. J. (2000). Directionality derived from differential sensitivity to monaural and binaural cues in the cat's medial geniculate body. *J. Neurophysiol.* **84,** 1330–1345.

Schnupp, J. W., Mrsic-Flogel, T. D., and King, A. J. (2001). Linear processing of spatial cues in primary auditory cortex. *Nature* **414,** 200–204.

Schreiner, C. E., and Langner, G. (1988). Periodicity coding in the inferior colliculus of the cat. II. Topographical organization. *J. Neurophysiol.* **60,** 1823–1840.

Schreiner, C. E., and Urbas, J. V. (1988). Representation of amplitude modulation in the auditory cortex of the cat. II. Comparison between cortical fields. *Hearing Res.* **32,** 49–63.

Schreiner, C. E., and Mendelson, J. R. (1990). Functional topography of cat primary auditory cortex: Distribution of integrated excitation. *J. Neurophysiol.* **64,** 1442–1459.

Schreiner, C. E., and Calhoun, B. M. (1994). Spectral envelope coding in the cat primary auditory cortex. *Aud. Neurosci.* **1,** 39–61.

Schreiner, C. E., and Langner, G. (1997). Laminar fine structure of frequency organization in auditory midbrain. *Nature* **388,** 383–386.

Schreiner, C. E., Mendelson, J. R., and Sutter, M. L. (1992). Functional topography of cat primary auditory cortex: Representation of tone intensity. *Exp. Brain Res.* **92,** 105–122.

Schreiner, C. E., Read, H. L., and Sutter, M. L. (2000). Modular organization of frequency integration in primary auditory cortex. *Annu. Rev. Neurosci.* **23,** 501–529.

Semple, M. N., and Aitkin, L. M. (1979). Representation of sound frequency and laterality by units in central nucleus of cat inferior colliculus. *J. Neurophysiol.* **42,** 1626–1639.

Semple, M. N., and Kitzes, L. M. (1985). Single-unit responses in the inferior colliculus: Different consequences of contralateral and ipsilateral auditory stimulation. *J. Neurophysiol.* **53,** 1467–1482.

Sen, K., Theunissen, F. E., and Doupe, A. J. (2001). Feature analysis of natural sounds in the songbird auditory forebrain. *J. Neurophysiol.* **86,** 1445–1458.

Serviere, J., Webster, W. R., and Calford, M. B. (1984). Isofrequency labeling revealed by a combined [14C]-2-deoxyglucose, electrophysiological, and horseradish peroxidase study of the inferior colliculus of the cat. *J. Comp. Neurol.* **228,** 463–477.

Shamma, S. A., Fleshman, J. W., Wiser, P. R., and Versnel, H. (1993). Organization of response areas in ferret primary auditory cortex. *J. Neurophysiol.* **69,** 367–383.

Shneiderman, A., and Oliver, D. L. (1989). EM autoradiographic study of the projections from the dorsal nucleus of the lateral lemniscus: A possible source of inhibitory inputs to the inferior colliculus. *J. Comp. Neurol.* **286,** 28–47.

Shneiderman, A., Oliver, D. L., and Henkel, C. K. (1988). Connections of the dorsal nucleus of the lateral lemniscus: An inhibitory parallel pathway in the ascending auditory system? *J. Comp. Neurol.* **276,** 188–208.

Singh, N. C., and Theunissen, F. E. (2003). Modulation spectra of natural sounds and ethological theories of auditory processing. *J. Acoust. Soc. Am.* **114,** 3394–3411.

Smith, A. L., Parsons, C. H., Lanyon, R. G., Bizley, J. K., Akerman, C. J., Baker, G. E., Dempster, A. C., Thompson, I. D., and King, A. J. (2004). An investigation of the role of auditory cortex in sound localization using muscimol-releasing Elvax. *Eur. J. Neurosci.* **19,** 3059–3072.

Snyder, R. L., Leake, P. A., and Hradek, G. T. (1997). Quantitative analysis of spiral ganglion projections to the cat cochlear nucleus. *J. Comp. Neurol.* **379,** 133–149.

Spirou, G. A., May, B. J., Wright, D. D., and Ryugo, D. K. (1993). Frequency organization of the dorsal cochlear nucleus in cats. *J. Comp. Neurol.* **329,** 36–52.

Spirou, G. A., Davis, K. A., Nelken, I., and Young, E. D. (1999). Spectral integration by type II interneurons in dorsal cochlear nucleus. *J. Neurophysiol.* **82,** 648–663.

Stevens, S. S. (1957). On the psychophysical laws. *Psychol. Rev.* **64,** 153–181.

Stiebler, I., and Ehret, G. (1985). Inferior colliculus of the house mouse. I. A quantitative study of tonotopic organization, frequency representation, and tone-threshold distribution. *J. Comp. Neurol.* **238,** 65–76.

Sutter, M. L., and Schreiner, C. E. (1991). Physiology and topography of neurons with multipeaked tuning curves in cat primary auditory cortex. *J. Neurophysiol.* **65,** 1207–1226.

Theunissen, F. E., Sen, K., and Doupe, A. J. (2000). Spectral-temporal receptive fields of nonlinear auditory neurons obtained using natural sounds. *J. Neurosci.* **20,** 2315–2331.

Thompson, A. M., and Schofield, B. R. (2000). Afferent projections of the superior olivary complex. *Microsc. Res. Tech.* **51,** 330–354.

Tollin, D. J. (2003). The lateral superior olive: A functional role in sound source localization. *Neuroscientist* **9,** 127–143.

Tollin, D. J., and Yin, T. C. (2003). Spectral cues explain illusory elevation effects with stereo sounds in cats. *J. Neurophysiol.* **90,** 525–530.

Valentine, P. A., and Eggermont, J. J. (2004). Stimulus dependence of spectro-temporal receptive fields in cat primary auditory cortex. *Hearing Res.* **196,** 119–133.

Velenovsky, D. S., Cetas, J. S., Price, R. O., Sinex, D. G., and McMullen, N. T. (2003). Functional subregions in primary auditory cortex defined by thalamocortical terminal arbors: An electrophysiological and anterograde labeling study. *J. Neurosci.* **23,** 308–316.

Versnel, H., and Shamma, S. A. (1998). Spectral-ripple representation of steady-state vowels in primary auditory cortex. *J. Acoust. Soc. Am.* **103,** 2502–2514.

Versnel, H., Kowalski, N., and Shamma, S. (1995). Ripple analysis in ferret primary auditory cortex. III. Topographic distribution of ripple response parameters. *Aud. Neurosci.* **1,** 271–285.

Wehr, M., and Zador, A. M. (2003). Balanced inhibition underlies tuning and sharpens spike timing in auditory cortex. *Nature* **426,** 442–446.

Winer, J. A., Saint Marie, R. L., Larue, D. T., and Oliver, D. L. (1996). GABAergic feedforward projections from the inferior colliculus to the medial geniculate body. *Proc. Natl. Acad. Sci. USA* **93,** 8005–8010.

Xu, L., Furukawa, S., and Middlebrooks, J. C. (1998). Sensitivity to sound-source elevation in nontonotopic auditory cortex. *J. Neurophysiol.* **80,** 882–894.

Xu, L., Furukawa, S., and Middlebrooks, J. C. (1999). Auditory cortical responses in the cat to sounds that produce spatial illusions. *Nature* **399,** 688–691.

Yeshurun, Y., Wollberg, Z., and Dyn, N. (1989). Prediction of linear and non-linear responses of MGB neurons by system identification methods. *Bull. Math. Biol.* **51,** 337–346.

Young, E. D., Spirou, G. A., Rice, J. J., and Voigt, H. F. (1992). Neural organization and responses to complex stimuli in the dorsal cochlear nucleus. *Philos. Trans. R. Soc. Lond. B. Biol. Sci.* **336,** 407–413.

Yu, J. J., and Young, E. D. (2000). Linear and nonlinear pathways of spectral information transmission in the cochlear nucleus. *Proc. Natl. Acad. Sci. USA* **97,** 11780–11786.

Zhang, L. I., Tan, A. Y., Schreiner, C. E., and Merzenich, M. M. (2003). Topography and synaptic shaping of direction selectivity in primary auditory cortex. *Nature* **424,** 201–205.

Zwicker, E., Flottorp, G., and Stevens, S. S. (1957). Critical band width in loudness summation. *J. Acoust. Soc. Am.* **29,** 548–557.

SPECTRAL PROCESSING IN THE AUDITORY CORTEX

Mitchell L. Sutter

Center for Neuroscience and Section of Neurobiology, Physiology, and Behavior
University of California Davis
Davis, California 95616, USA

To analyze sounds such as speech, the auditory system must analyze multiple spectral scales in parallel. For example, the spacing between different speech formants requires integration across a wide-frequency scale while analyzing the energy within formants requires a narrower scale. In this chapter, we examine the

DOI: 10.1016/S0074-7742(05)70008-8

0074-7742/05 $35.00

techniques used to probe spectral processing in auditory cortex, and propose a rather simple model for how auditory cortex might be able to achieve this fast multi-spectral scale analysis.

I. Defining Key Issues in Spectral Processing

A. What Is Spectral Processing?

The term spectral processing has different meanings to different people, yet the concept of spectral processing is critical to understanding how we interpret sounds in the world around us. This is because we need some analytical framework to analyze the complex time-varying waveforms of the pressure waves arriving at each ear. When looking for a dictionary definition of spectral, none relate to the concept as applied in the auditory system. While one can use Fourier analysis (Oppenheim 1997; or for more basic introduction to Fourier analysis see Staff, 1995) of sounds to define spectral processing, a much more conceptual approach has come into common parlance when discussing sound processing.

In auditory research most investigators distinguish spectral and temporal processing, even though both refer to aspects of analyzing a single time-varying signal. Unfortunately there is no agreement on a standard definition of spectral processing. Some refer to spectral processing as sound frequency analysis that cannot be achieved by temporal locking of neuronal responses. Therefore, if a neuron phase locks (fires at a specified phase of a sine wave) (see Chapter 1, or also Bear *et al.*, 2001) for a 100-Hz tone, some would refer to this as temporal processing. Complementary to this definition of temporal processing, the definition of spectral processing in this schema is coding that can be represented by place on the cochlea and not by phase locking. This is based on the well-known property that different neurons in the mammalian cochlea are activated by different frequencies of sound waves because of the mechanical structure of the inner ear (see Chapter 1, or Purves *et al.*, 2001). Therefore, because the auditory system distinctly represents higher frequencies with the magnitude of firing rate and lower frequencies by phase locking, this is a relatively natural way for a neuroscientist to divide signals. This sort of a neurocentric view of processing suffers from the problem that the auditory nerve phase locks to much higher frequencies than, for example, auditory cortex so there can be no uniform definition of what frequency defines the border between temporal and spectral processing. For this chapter we will refer to spectral and temporal processing in a way that is more grounded in psychophysics and signal processing than physiology. In this way it is more of a signal- and behavioral-based definition.

The best way to exemplify the definition we will use is by examining how auditory neuroscientists display or graph sounds. Figure 1A shows the time

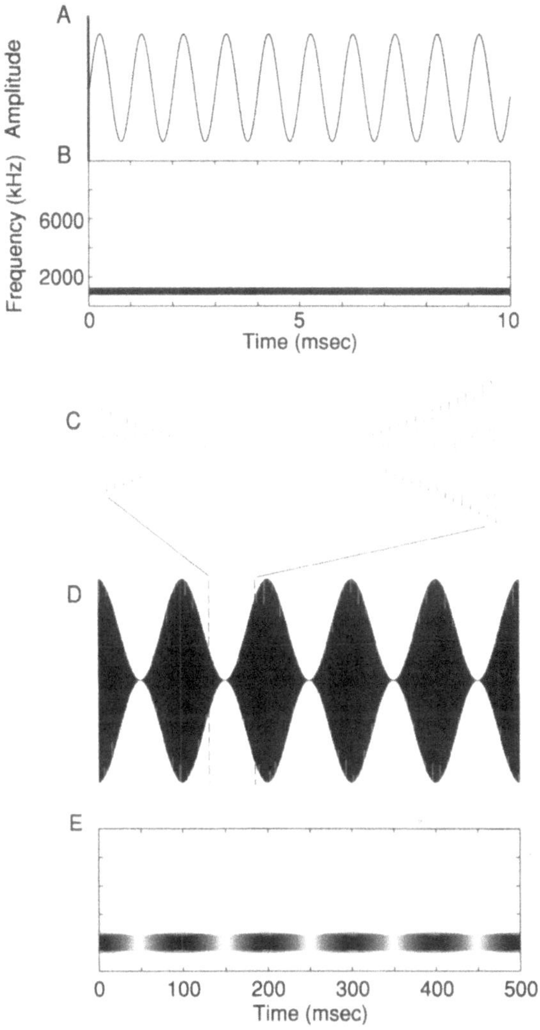

FIG. 1. (A) A sine wave and (B) its spectrogram. The inset on the top right (C) shows the carrier frequency of the AM tone by expanding the time axis for the portion shown with the double-dashed vertical lines. (D) An amplitude-modulated (AM) tone and (E) its spectrogram.

waveform—the amplitude versus time plot—of a sine wave (pure tone). Figure 1B shows a spectrogram—a frequency versus time plot for the same tone. The horizontal line in Fig. 1B represents a sine wave—a sound with the same frequency over time. Notice that although the line should be as narrow as possible at 1 kHz, it is not. This results from necessities in calculating the spectrogram. Auditory neuroscientists often refer to spectral processing as the processing of energy differences along the y-axis of the spectrogram (as opposed to temporal processing which corresponds to analyzing energy changes along the x-axis). This sometimes seems confusing because both spectral and temporal processing involves time-varying features in the amplitude waveform. The distinction is that spectral and temporal processing refers to analyzing sounds at two different time scales; the fast time scale corresponds to spectral processing and the slow one to temporal processing.

This is exemplified better by another kind of stimulus: amplitude modulated (AM) tones. Figure 1D shows the time waveform of an AM tone with 2-kHz carrier and 10-Hz modulation. Because the time scale is so long compared to the 2-kHz carrier, in Fig. 1D all that can be observed is the 10-Hz modulation envelope (the slower modulation in AM which forms an outline of the time waveform), and not the 2-kHz carrier. Therefore, the time waveform is expanded from time = 130 to 190 msec in the top right inset (Fig. 1C). In this expanded view it is easy to see the 2-kHz frequency of the carrier. In the spectrogram (Fig. 1E), both characteristics can be seen in one plot. One can see the horizontal oriented segment at 2 kHz (spectral) turning on and off at a rate of 10 Hz (temporal). The difficulty of defining a border between spectral and temporal processing does not disappear by using spectrograms. This is because, in choosing how to compute the spectrogram, one must choose frequency resolution for the y-axis and the lowest frequency to use. In other words, a border between temporal and spectral still needs to be chosen. Commonly, the cutoff corresponds to approximately 50 to 100 Hz. This border corresponds approximately to where the percept of a sinusoidal AM tone changes from a sine wave that pulsates (roughness) to a complex single tone (Ritsma, 1962; reviews in Joris *et al.*, 2004; Zwicker, 1999), and this cutoff will be used to define the spectral-temporal processing cutoff in this chapter.

B. The Need to Analyze Sounds on Multiple Scales

While pure tones and AM sounds have proved to be useful stimuli for studying the auditory system, a better understanding of the need to separately define spectral processing can be seen by using a natural stimulus. Figure 2 shows the time-amplitude waveform and spectrogram of the author saying the word 'baseball.' The two syllables can be seen in the time amplitude waveform (Fig. 2A) as two large peaks. The consonants from baseball can be seen as the smaller amplitude event. In the spectrogram (Fig. 2B) the long 'a' from the first syllable in baseball can be seen to have three major energy bands, called formants. White arrowheads point to the dark-banded formants on the spectrogram. To discriminate this vowel the auditory system has to analyze the width of each of these bands and the spacing between them. This corresponds to

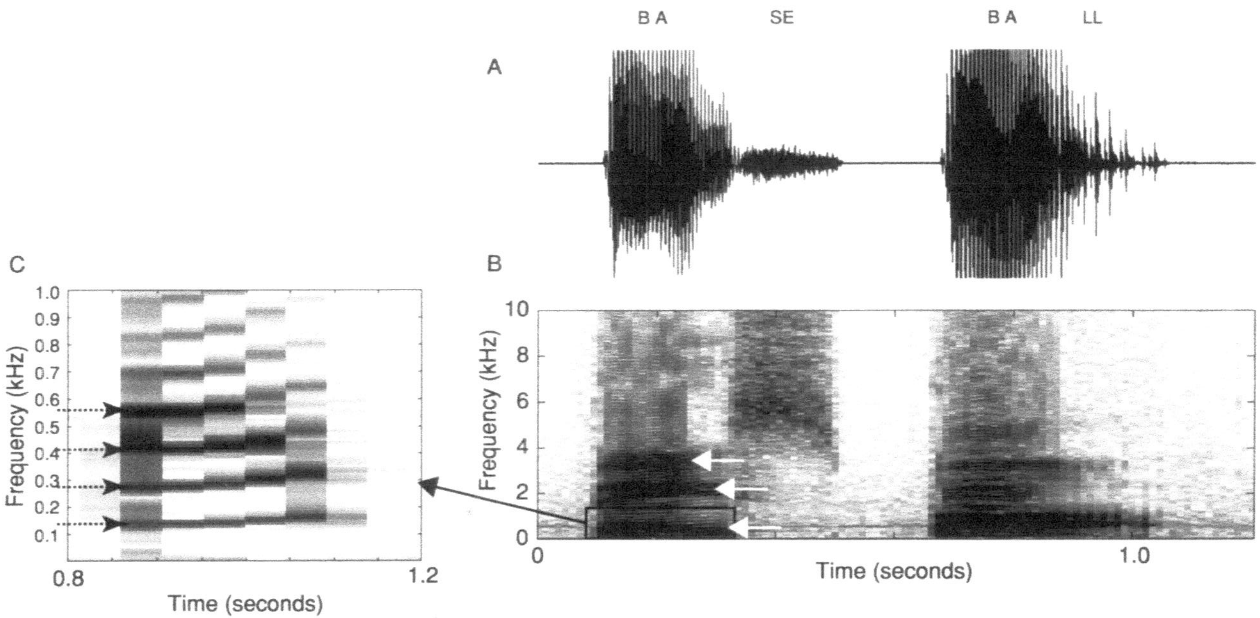

FIG. 2. Multiple spectral scales of information in the word baseball. (A) Time waveform of the author speaking the word baseball. (B) Spectrogram of the spoken word baseball. Note the three white arrows point to the three formants in the long "a" from the first syllable. The box around the lowest arrow distinguishes the part of the spectrogram that is expanded in (C) to show the fundamental frequencies and its harmonics.

analyzing information on a scale of ~1 kHz for the former and across ~4 kHz for the latter. One can see how these properties can be used to help distinguish between the long 'a' in the first syllable and the short 'a' in the second syllable (which has four discernable formants, and the first three have different bandwidths (BWs) and spacing). In addition to analyzing sounds on both these scales, a third frequency scale needs to be analyzed which can be better seen when we narrow the frequency scale for the first formant of the first syllable (Fig. 2C). Here we can see the harmonic structure *within* the formant with energy bands at roughly 140, 280, 420, and 560 Hz (identified by arrowheads at the end of the dashed lines). These bands spaced 140-Hz apart correspond to what is called a fundamental frequency of 140 Hz. Being able to discern the fundamental frequency plays an important role in identifying who is speaking. People with smaller vocal chords have higher fundamental frequencies than those with larger vocal cords. Each person has a fundamental frequency closely corresponding with the perceived pitch of his or her voice and, therefore, fundamental frequency provides an important cue in identifying who is talking. Not only must the auditory system analyze multiple frequency scales, but it must do so quickly. Later in this chapter we present a physiological explanation of how the auditory cortex might analyze and represent multiple frequency scales in parallel.

C. Evidence That the Perception of Broad-Band Stimuli Is Not Easily Predicted by Narrow-Band Stimuli

In perception, there are a range of phenomena indicating that predictions about how we perceive broadband spectrally complex stimuli are not consistent with simple linear filter models that historically have been used in the nervous system. These linear filter models are based on calculating the effective range over which energy in noise can mask one's ability to detect a tone. This has resulted in what many auditory psychophysicists call the critical band or the auditory filter (see Chapters 2 and 3 for more detail).

Many people have tried to explain an array of psychophysical results using linear filter models that stem from critical-band research (Fletcher, 1940; Patterson, 1974, 1976; Zwicker, 1999; Zwicker *et al.*, 1957). The bases of these models are experiments where noise masks tone detection. The critical band is essentially the effective BW for which noise influences the ability to detect a tone. Figure 3 shows two ways in which critical bands are measured. In one experiment band-passed noise maskers, centered in frequency around a tone, are used (Fig. 3A, top left). The ability to detect the tone is then measured as a function of the BW of the masker (Fig. 3A, top right). At first as the width of narrow band maskers is increased, masking is increased and the tone needs to be louder for detection (top, right). However, at some point further increasing

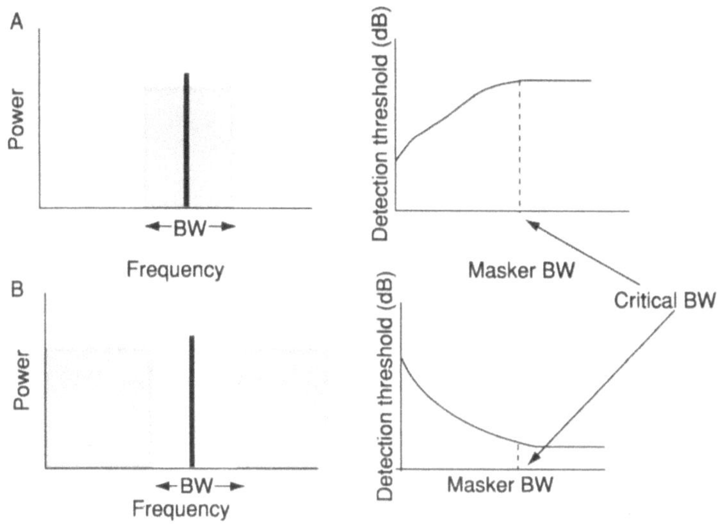

FIG. 3. Schematic of two different types of experiments (A and B) to determine the critical bandwidth. On the left are schematized spectra of the stimuli. On the right are schematic representation of how detection thresholds of the tone change with (A) noise bandwidth and (B) notch bandwidth.

masker BW does not have an increased masking effect; instead the masking effect plateaus and threshold intensity no longer increase (top, right). The cutoff at which masking stops is called the critical bandwidth (BWc). (It is worth noting that values other than the asymptote can be used to define BWc, so different definitions exist; see Chapter 2.) Critical bands can also be measured using notched noise where the notch (lack of energy) is centered around a tone (Fig. 3B). When the notch is very narrow (small BW) the noise is close enough in frequency to the tone to mask it. However, as the notch width is increased (bottom, right) bringing the edges of the notch further in frequency from the tone, it becomes easier to detect the tone so the tone does not need to be as loud to overcome the effect of the noise. The notch width at which the noise no longer affects the ability to detect the tone is another measure of the BWc. Early on, attempts were made to link BWc to individual neurons' sensitivities by claiming that the BW of an auditory nerve (AN) tuning curve at 20 dB above threshold approximated a critical band. There were many problems with this interpretation of critical bands because, depending on the sound pressure level (SPL) chosen, the BW of an AN fiber tuning curve could be nearly any value. This is because the BW of AN fiber tuning curves dramatically increases with increasing intensity. Therefore, since there is nothing particularly special about 20 dB, the choice seems arbitrary. Several properties would be needed to find a neural correlate of critical-band related phenomena, including relative invariance to changes in tone intensity and a dependence on characteristic frequency (CF) consistent with the psychophysical experiments. A more reasonable choice seems to be the responses of neurons from the central nucleus of the inferior colliculus (IC) (Ehret and Merzenich, 1988a,b; Ehret and Schreiner, 1997). In cortex, the focus of this chapter, neurons in ventral primary auditory cortex (A1v) and the central part of A1 show properties more closely approximating critical bands than in the dorsal A1 (A1d) where a wider range of BWs are encountered (Ehret and Schreiner, 1997, 2000).

Often it is presumed that the masking results imply that the critical band defines the effective range of perceptual spectral integration. While these experiments define the limits of energetic masking of noise on tones nicely, they are inherently limited to this type of experiment and are not sufficient for explaining phenomena where frequency integration occurs across critical bands. One example is that critical bands fail to describe how we integrate frequency across the large BW of vowels such as in Fig. 2. In fact, the whole notion of analyzing sounds at multiple spectral scales in parallel indicates that filters of one BW are not sufficient to describe perception, and that a view of neurons simply as filters might limit our ability to understand phenomena where neurons combine information non-linearly across frequency.

There is burgeoning human psychophysical evidence that the ability to discriminate sounds with spectrally complex broad-band structure differs fundamentally from discrimination of simpler, narrow-band sounds. By analogy with the visual system, the processing of a broad spectrum stimulus would, of necessity, entail an integrative stage subsequent to the early stage of frequency decomposition, or 'filtering'; indeed, this is confirmed by human psychophysical data. Human performance indicates that although some perceptual phenomena reflect early frequency filtering (e.g., simple masking and 'critical band' phenomena: Moore, 1995; see Chapter 2), there are conditions that entail more complex stimuli for which peripheral filter models cannot account for performance (e.g., profile analysis: Bernstein and Green, 1987, 1988; Green, 1983, 1986, 1988; O'Connor *et al.*, 2000; Summers and Leek, 1994; auditory scene analysis: Bregman, 1990; O'Connor and Sutter, 2000; comb filtered ripple noise: Yost and Moore, 1987; MDI: Yost and Sheft, 1989; co-modulation masking release: Hall *et al.*, 1984; see Chapter 10). Under these stimulus conditions, perception of frequency components within a sound is influenced by distant frequency components located far beyond the BW of early frequency filters, demonstrating integration across frequency channels. The pattern of these results suggests that the auditory system is so sensitive to global spectral patterns that perception of local frequency features may be altered by certain global configurations. This type of processing where long-range frequency interactions, beyond the range of a classically defined auditory channel, influence perception is called *across-channel* or *global frequency processing* (see Chapter 10). This type of across-channel processing is reminiscent of long-range visual interactions beyond the classical defined receptive field (Allman *et al.*, 1985; Lamme, 1995; Loftus and Sutter, 2001; Sutter *et al.*, 1999).

Few experiments in across-channel processing in non-humans exist (chinchilla: Shofner and Yost, 1997; budgerigar: Amagai *et al.*, 1999; macaque monkey: O'Connor *et al.*, 2000), and knowledge of mechanisms underlying across-channel processing, therefore, is very limited. The data from these animal psychophysical studies are consistent with the notion that animals also process information across frequency channels and share similar mechanisms with humans, albeit possibly with different sensitivities. Before describing why the auditory cortex is a good place to examine spectral processing, we first must define some important ways neuroscientists have measured spectral integration in cortex and how this relates to some basic auditory cortical properties.

II. Different Methods of Measuring Spectral Processing: Approaches Using Tonal Stimuli

A. Linear Versus Non-Linear Spectral Processing

Before discussing the specific methods used to quantify how single neurons process spectral information it is important to define the terms linear and non-linear as applied to auditory analysis. This is important for several reasons. First, a debate has raged about what models should be used to investigate and characterize these systems. The advantage of using linear techniques is that they greatly simplify the approach to the problem and they bring an array of well-developed engineering tools. With these tools, the response to any stimulus can be predicted by characterizing the response to a small subset of stimuli. However, linear tools only work if the processing can be reasonably approximated as linear. It is important to note that just having non-linear components (for example, an action potential threshold) does not preclude using linear tools if the system can be approximated as linear in the range of interest. However, if that approximation is poor or inaccurate, linear tools can give quite misleading answers. So it is important to remember when addressing system linearity, that the question is not whether every element that comprises the ultimate single neurons response is linear but whether the system can be approximated as being linear.

Linear tools do have a disadvantage because their computational power is limited. By introducing non-linearity, even just small ones, the computational power of the system can be improved dramatically. This is particularly important for relating neural processing to perception and psychophysics, where most complex phenomena like illusions and making decisions are non-linear.

If the auditory system were linear with respect to spectral processing, we could predict the response to any complex stimulus based on pure-tone responses. One requirement for linearity is that the response to a combination of tones would be equal to the sum of the responses to each tone individually, or, for those who like equations:

$$R(\text{tone1} + \text{tone2}) = R(\text{tone1}) + R(\text{tone2}) \tag{1}$$

where R(tone1 + tone2) is the response to the two tone combination, and R(tone1) is the response to tone1 by itself, and R(tone2) is the response to tone2 in isolation.

A second rule of linearity is that of scaling which is exemplified by the scaling rule

$$R(A * \text{tone1}) = A * R(\text{tone1}) \tag{2}$$

where A is a scaling factor. So R(A * tone1) is the response to tone1 whose amplitude is multiplied by the scaling factor A, and a * R(tone1) is the response to tone1 multiplied by factor A (notice that in one case the scaling factor A is multiplied by the stimulus and the other by the response). Another simple way of describing this is that for every doubling of a tone's intensity there should be doubling of response magnitude.

The final rule is that of superposition. This means that the response to two tones will preserve timing relationships. Or, in equation form

$$R_{\text{tone1+tone2}}(t) = R_{\text{tone1}}(t) + R_{\text{tone2}}(t) \tag{3}$$

where $R_{\text{tone1+tone2}}(t)$ is the response to the two tone combination as a function of time, and $R_{\text{tone1}}(t)$ is the response to tone1 by itself as a function of time, and $R_{\text{tone2}}(t)$ is the response to tone2 in isolation as a function of time. This equation is mathematically the same as Equation 1, except we are adding responses that are a function of time. Later in this chapter we will show that all of these rules (equations) are violated by the responses of single auditory cortical neurons.

B. The Frequency Response Area (FRA) and Frequency Tuning Curve (FTC)

Spectral processing in auditory cortical neurons has been measured several different ways. In this section we will discuss two commonly used techniques to investigate spectral processing in single-unit responses. The FRA and FTC are two measurements that have been used for many decades on single auditory neurons.

FRAs are three-dimensional plots of a neuron's response versus frequency and intensity (Evans, 1974). While these plots could more appropriately be called frequency/intensity response areas, the less-specific moniker FRA has stuck. The top row of Fig. 4 shows FRAs recorded from six cat primary auditory cortical (A1) neurons. For each FRA, 675 stimuli were used by creating a grid of 45 frequencies by 15 intensities. Each vertical line in the FRA corresponds to the magnitude of responses of the neuron to the stimuli whose frequency is given by the x-axis and intensity by the y-axis. For example, the neuron whose FRA is shown in Fig. 4A responded best to a ~9.5-kHz tone presented at −17 dB (cats are more sensitive than humans to low intensity sounds), responded more weakly to a 10 kHz tone at −10 dB, and did not respond at all to 7-kHz tones. The neuron of Fig. 4B responded best to ~4.5 kHz at ~20 dB, and responded less strongly to nearby frequency intensity combinations. This neuron was not excited by tones >8 kHz.

From the FRA, an FTC can be derived by applying a response criterion to the FRA. Any response greater than the criterion (for excitatory responses) falls within the FTC, any response less than the criterion falls outside of the FTC. The bottom row of Fig. 4 shows the FTCs derived from the FRAs on the top. From the FTCs several important measures of information-bearing parameters (Suga, 1992) related to spectral processing can be calculated. The first is the characteristic frequency (CF): the frequency to which the neuron is most sensitive. This corresponds to the frequency of the lower tip of the tuning curve. The CFs of three neurons are shown in the bottom row of Fig. 4A, B, and D. One can also define the range of frequencies to which the neuron responds as the BW. From the tuning curves BWs are usually defined at some intensity. In the case of the examples of Fig. 4B and D, BW measurements are shown at 10 and 40 db above threshold (BW10 and BW40, respectively) with double-arrowed solid lines inside the tuning curves at intensities shown by dashed horizontal lines spanning all frequencies. BW measurements are commonly expressed in either linear units of frequency (e.g., kHz) or logarithmic units (e.g., octaves). However, another measure derived from the engineering literature is often used: the quality factor, or Q, which is the CF divided by the BW in kHz at a given intensity. This measure was chosen not only because of its use in systems engineering, but also because it was thought it would compensate for the logarithmic representation of frequency in the cochlea. The Q value 10 dB above threshold is called Q_{10} and 40 dB above threshold is called Q_{40}. Arguments have been made about the best way to measure the BW of a neuron. Too often BW is measured only 10 dB above threshold because Q values at low intensities are usually used in filter design for engineering. This is acceptable for designing linear filters because by definition if the filter is linear, measuring the Q at one intensity is sufficient to completely characterize the filter. The problem with applying this idea to neurons is that neurons are not filters and are not linear. Because most discrimination (like speech perception) is done best at more supra-threshold intensities, measuring only Q10 has been misleading (Schreiner and Sutter, 1992; Suga,

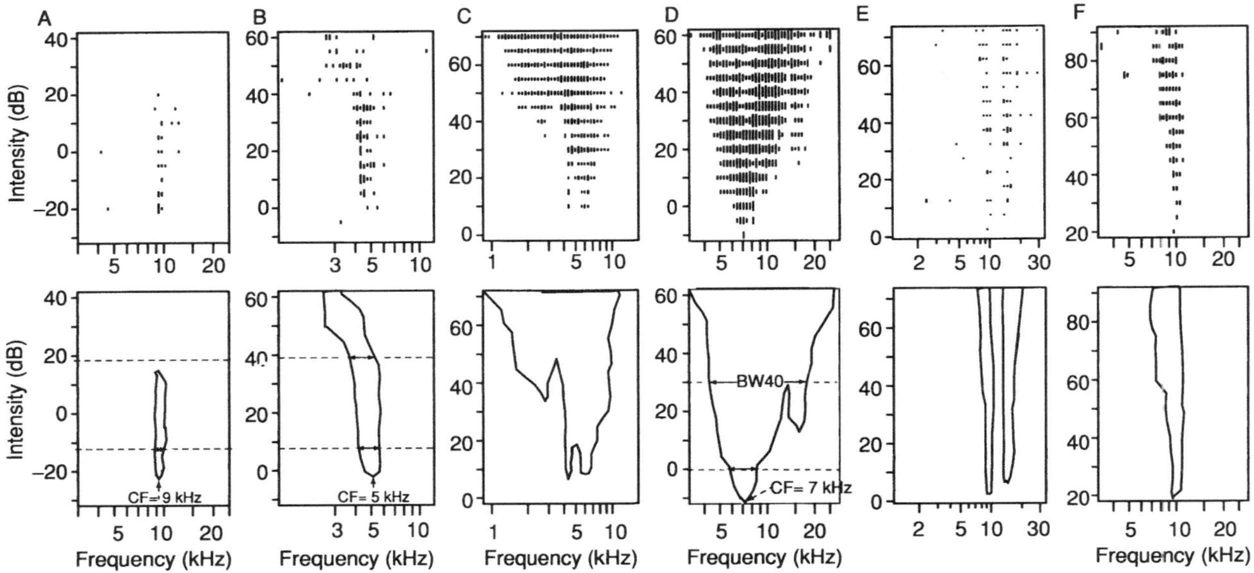

FIG. 4. Frequency response areas (FRAs) and frequency tuning curves (FTCs) for six A1 neurons. The top row shows the FRAs of six neurons (A to F). The x- and y-axis of each FRA indicate the frequencies and intensities of the 675 presented tones. The length of each line is proportional to the number of spikes the neuron fired to the corresponding frequency/intensity combination. For frequency/intensity combinations where there is no line, there was no response. The bottom row shows FTCs calculated from the FRAs directly above them. The areas within the FTCs corresponds to frequency/intensity combinations where the responses exceed a criterion (Sutter and Schreiner, 1991). For A, B, and D the characteristic frequency (CF) is pointed out. For A, B, and D the bandwidth (BW) 10 and 40 dB above thresholds is shown with double arrows on horizontal lines.

1997; Suga and Tsuzuki, 1985) with regard to the effective BW of a neuron when analyzing complex sounds at typical intensities. Therefore, solely defining BW at low intensities is less than ideal. Also it has been pointed out that Q is not a good measure (Calford *et al.*, 1983; Sutter, 2000) because the division by BW makes Q infinite for neurons that are tuned for intensity (non-monotonic). The neuron of Fig. 4A provides a compelling example of this. At 40 dB above threshold this neuron does not respond to tones because it is tuned for intensity (also called non-monotonic neurons) (Brugge *et al.*, 1969; Erulker *et al.*, 1956; Evans and Whitfield, 1964; Greenwood and Maruyama, 1965). Accordingly the Q40 value, which is CF/BW, is 9/0 which is infinite. Therefore, it has been suggested that measuring BW in octaves (rather than Q) and at multiple intensities is better (Sutter, 2000).

Figure 4 also demonstrates why including intensity in these profiles is important. The neurons with circumscribed FTCs (Fig. 4A) comprise about 20% to 30% of cat primary auditory cortical (A1) neurons' FTCs (Barone *et al.*, 1996; Clarey *et al.*, 1994; Phillips and Irvine, 1981; Phillips *et al.*, 1985; Sutter, 2000; Sutter and Schreiner, 1995). In another cat auditory cortical field, the posterior auditory field (PAF), circumscribed FTCs comprise a larger percentage of the total (Heil and Irvine, 1998; Kitzes and Hollrigel, 1996; Loftus and Sutter, 2001; Phillips and Orman, 1984; Phillips *et al.*, 1995; Sutter and Loftus, 2003; Tian and Rauschecker, 1998). In the macaque monkey A1, the percentage of neurons with circumscribed FTCs is similar to cat A1. In the rostral field of macaque auditory cortex (R) higher percentages of this type of FTC can be found (Recanzone *et al.*, 2000). The frequency tuning of neurons with circumscribed FTCs clearly depends on the intensity tested, as evidenced by their inability to be excited by high intensity tones. Therefore, simply looking at a single intensity or collapsing across intensity does not give a complete description of a neuron's spectral response area. The neuron whose FRA and FTC are shown in Fig. 4B shifts its frequency response range towards lower frequencies at higher intensities. At BF the response gets weaker as a function of tone intensity, due to a shifting of responses away from BF to lower frequencies. This type of FTC, which changes the range of responsive frequencies as a function of intensity, has been named 'slant-lower' and comprises ~10% of cat A1 neurons. The neuron whose FRA and FTC are shown in Fig. 4C also changes frequency tuning with intensity. This neuron's BW increases with intensity, with the neuron responding to a larger range of lower frequencies as intensity is increased. The neuron whose FRA and FTC are shown in Fig. 4D increases BW toward higher frequencies with increasing intensity. Figure 4E shows a neuron whose FRA has two sharp distinct response areas separated by a middle frequency region to which the neuron does not respond or is inhibited. This type of FRA is interesting because it indicates the neuron would be good at distinguishing separation between energy bands, like the formants from the long 'a' in the first syllable of 'baseball.' And finally, the neuron whose FRA and FTC are shown in Fig. 4F has a 'U' shape with a similar frequency range of response across intensity. This type of tuning is most common in A1, but still comprises only ~20% of neurons. In a study of tuning curve shape in A1 it was found that the shape of tuning curves varied along multiple continua (Sutter, 2000), and in general a wide range of shapes and complexities of tuning curves were found.

These tuning curve shapes are important because they provide information to indicate the types of complex spectra that cells might prefer. However, many earlier studies of FRAs were performed in barbiturate-anesthetized animals where there was little spontaneous activity (in fact cortical spontaneous activity is larger but not high in awake animals). This makes detecting inhibition difficult. To avoid this obstacle many have modified the technique to present two sounds: one that causes an excitatory response and one that searches frequency/intensity space to see which tones might inhibit the normally excitatory response.

C. Two-Tone FRAs, Inhibitory Response Areas, and Non-Linear Processing

Because it is hard to observe suppressive or inhibitory regions in the FRA using single tones, several methods have been used to circumvent this problem. One is by using multiple tones (Abeles and Goldstein, 1972; Loftus and Sutter, 2001; Shamma and Symmes, 1985; Shamma *et al.*, 1993; Suga, 1965; Sutter and Loftus, 2003; Sutter and Schreiner, 1991a; Sutter *et al.*, 1999). An example of the time waveform, Fourier transform, and spectrogram of a simultaneous two-tone stimulus is shown in Fig. 5. The stimulus is called simultaneous because both tones that devise the complex stimulus turn on and off at the same time. When simultaneous two-tone stimulation is used, and one tone is at a frequency that excites the neuron, inhibitory regions in their FRA are often revealed (Fig. 6). In this example the single tone FRA is shown on the left. The star denotes the excitatory tone used in the two-tone FRA in the middle panel. The middle panel shows how tones with frequencies and intensities on the x- and y-axis modulate the response associated with excitation by a 10-kHz tone at 40 dB. It is easy to see the white areas in the middle plot which show regions where the second tone completely eliminates the response normally associated with the 10 kHz at 40 dB tones. These white areas are the suppressive/inhibitory regions of the FRA. On the right the unfilled outline shows the excitatory FTC derived from the left panel and the gray-filled region shows the suppressive/inhibitory regions derived from the two-tone FRA in the middle panel.

The most commonly observed inhibitory band structure in primary auditory cortex is two flanking bands to a central excitatory area (for examples, see Figs. 6 and 8A), although this comprises only ~35% of A1 neurons (Sutter *et al.*, 1999).

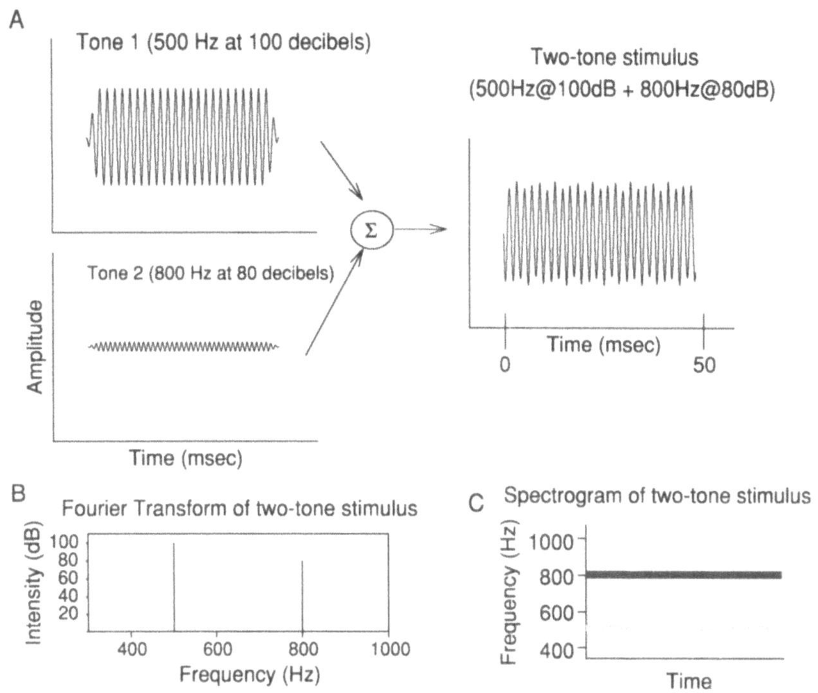

FIG. 5. Simultaneous two-tone stimulus. (A) Shows how two tones are summed to create the two-tone stimulus with the resulting complex two-tone stimulus shown on the right. (B) Shows the Fourier transform of the two-tone stimulus. (C) Shows the spectrogram of the two-tone stimulus.

These can be created by either lateral inhibition or inhibition that is matched in CF, but broader than the excitation. However, there is a large variety of band structures including many neurons with central inhibitory bands (for examples, see Figs. 7A and 8B) or multiple upper and inhibitory lower bands (for examples, Fig. 8C and D). In cat PAF, the simple two-band structure was less common than a more complex structure (Loftus and Sutter, 2001). Later (Section V.D.) we will discuss how the properties of inhibitory band structure vary within A1 and how this relates to analyzing sounds on multiple spectral scales as discussed in the first section.

Some inhibitory areas are intensity tuned, some are not. Intensity-tuned excitatory areas tend to be flanked by non-intensity tuned inhibitory bands (Figs. 7B and 8D). This implies that intensity tuning results from a competition between excitation and inhibition in regions of the FRA where the two overlap (Sutter and Loftus, 2003). Recently, intracellular techniques applied in cats (Ojima and Murakami, 2002) and rats (Kaur *et al.*, 2004; Machens *et al.*, 2004;

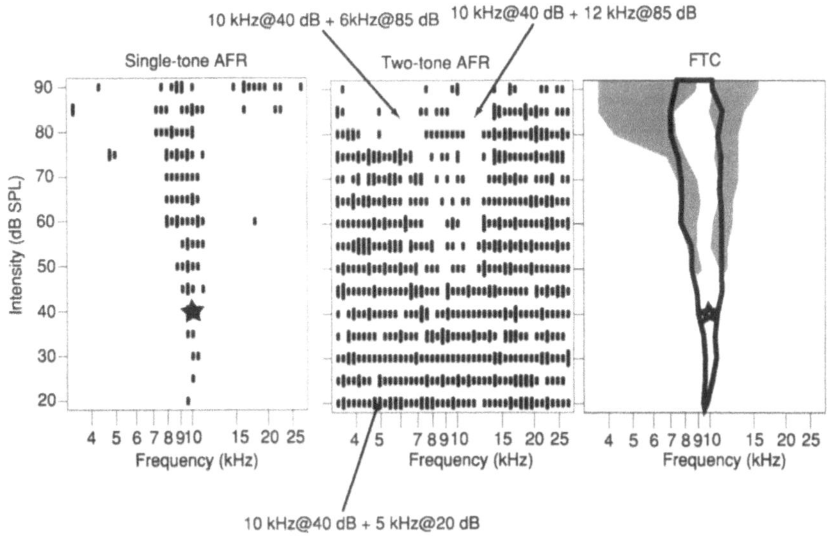

FIG. 6. Method for determining excitatory and inhibitory tuning curves. On the left is the single tone FRA of a neuron derived as in Fig. 4. In the middle is the two-tone FRA where one tone is always fixed at 10 kHz and 40 dB (see star on left and right). Arrows show responses to three different two-tone stimuli. On the right is the FTC. Excitatory areas derived from the single tone FRA are shown in the white outline. Inhibitory areas derived from the two-tone FRA are shown in gray shading. The star shows the excitatory tone used in the two-tone FRA.

Wehr and Zador, 2003) are starting to reveal the intricacies of competition between inhibition and excitation in auditory cortical FRAs.

It is also interesting that when using a forward-masking paradigm, where inhibitory tones precede excitatory tones in time, intensity-tuned excitation is associated with intensity-tuned inhibition (the opposite of when simultaneous two-tones are used) (Calford and Semple, 1995). This implies separate mechanisms of fast-acting inhibition for creating intensity-tuned excitation, and a subsequent persistent inhibition of neurons with similar intensity tuning (Calford and Semple, 1995; Sutter and Loftus, 2003). This also suggests that at higher levels in the auditory system, separate channels are designed for intensity-tuned neurons (Sutter and Loftus, 2003).

These results violate linearity in three important ways. Most inhibitory frequency-intensity ranges observed with two-tone stimulation do not appear to respond to single tones at the same frequency. In Fig. 6 (single-tone, left) it can be seen that 6-kHz tones at 85 dB do not appear to evoke any change of activity. However, when this frequency is presented in combination with a 10-kHz tone at 40 dB (two-tone FRA center), the response normally associated with the 10 kHz

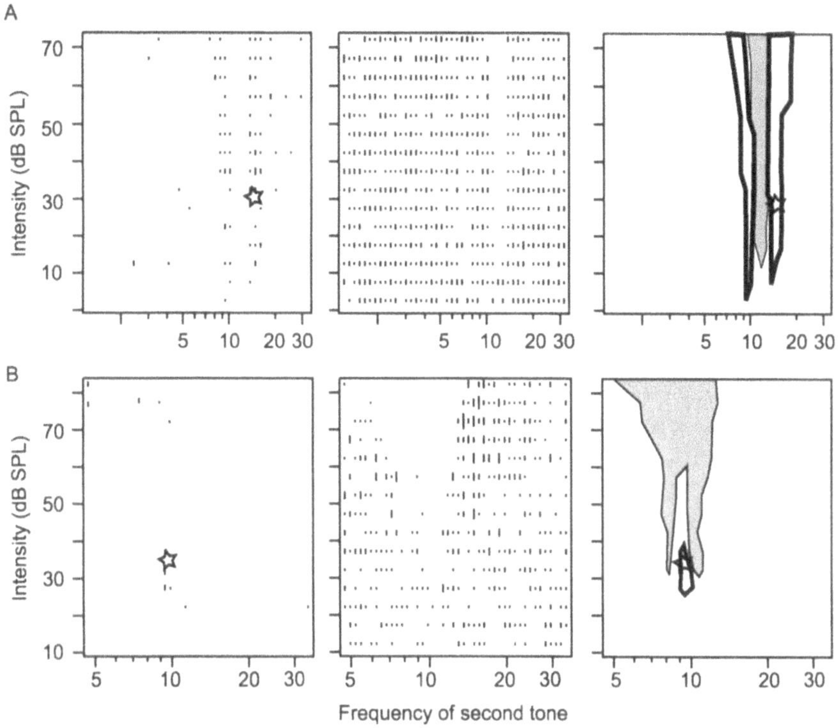

FIG. 7. Single-tone FRAs (left), two-tone FRAs (center), and combined inhibitory/excitatory FTCs (right) for two A1 neurons. The star represents the frequency and intensity used to excite the neurons in the two-tone condition.

tone is eliminated. Clearly, the response to the two-tone stimulus (0 spikes) does not equal the sum of the individual tones (0 spikes + ~1.5 spikes) violating Equation 1.

Intensity-tuned (non-monotonic) areas clearly violate scaling (equation 2). For the neuron in Fig. 4A, doubling the intensity of the sound clearly does not double the neuron's response. This is even more apparent in the two-tone stimulation of Fig. 7B where it is obvious that the neuron is inhibited at high intensities. Finally, many neurons also violate the superposition rule by responding non-linearly to time-varying frequency sounds such as FM sweeps (Erulker *et al.*, 1956), in ways that cannot be predicted by summing tone responses. Overall, studies of the auditory system using simple stimuli such as tones and multi-tone complexes strongly suggest non-linear frequency processing is quite prevalent in auditory cortical responses (Loftus and Sutter, 2001; Nelken *et al.*, 1994a,b; Sutter *et al.*, 1999).

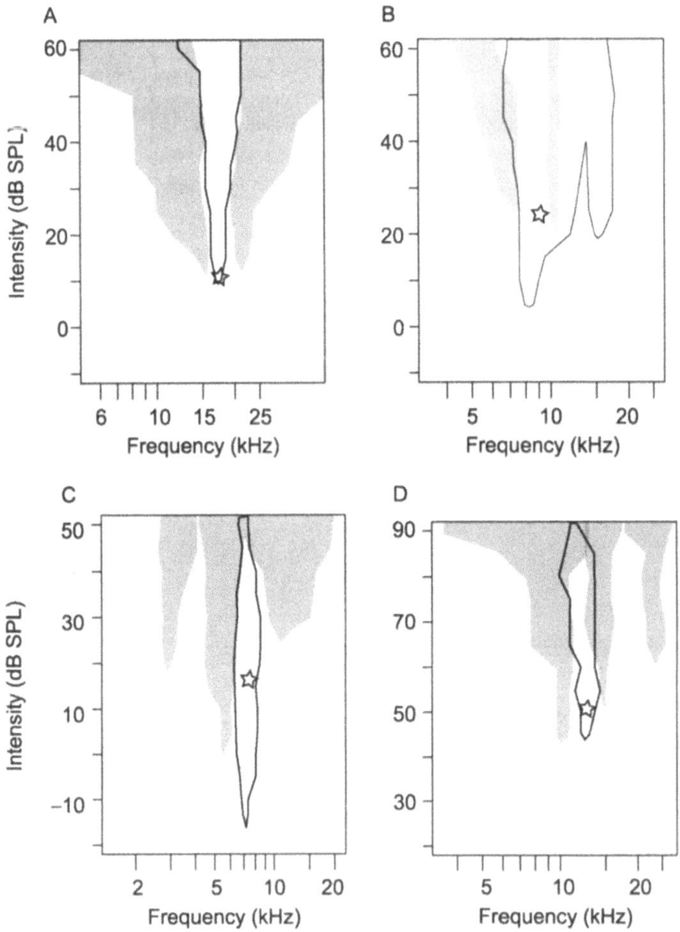

FIG. 8. Combined excitatory (based on single-tone responses) and inhibitory (based on two-tone responses) FTCs for four A1 neurons. Stars represent the frequency and intensity used to excite the neurons in the two-tone condition.

III. Different Methods of Measuring Spectral Processing: Measuring Spectral RFs with Broad-Band Stimuli

A. MEASURING STRFs WITH REVERSE CORRELATION AND TONAL CHORDS

So far we have focused on experiments using narrow-band stimuli such as tones. More recently, techniques using broad-band stimuli have been used to characterize spectral receptive fields (RFs) and spectral temporal receptive fields

(STRFs) (Aertsen and Johannesma, 1981; Eggermont *et al.*, 1983; see also Chapter 7). An advantage of using broad-band stimuli is that if there are interactions across frequency (such as special properties for across-channel processing; see Chapter 3), then broad-band stimuli have energy spanning these potential interactions.

Many methods using broad-band stimuli involve reverse correlation or spike-triggered averaging techniques. The approach applied with spike-triggered averaging is different from that used to derive FRAs with tones. When using tones and multiple tone complexes to obtain FRAs, the investigator chooses a stimulus parameter range and examines how neural responses change as a function of these parameters (e.g., Figs. 6 and 7). In spike-triggered averaging (which conceptually is quite similar to reverse correlation) the investigator uses a long time-varying stimulus and calculates the average stimulus that preceded each spike. Basically, every time a spike is fired, the stimulus (for example, represented by a spectrogram for an STRF) that preceded the spike is recorded and saved. This is repeated for every spike and then the mean stimulus preceding spikes can be determined by averaging all the single-spike associated spectrograms. Figure 9 shows two hypothetical STRFs; lighter shading corresponds to more energy in the stimulus and darker shading corresponds to less energy in the stimulus. The example in Fig. 9A is of a neuron that prefers energy at one frequency (~3 kHz) and less energy to either side (~2.5 and ~3.5 kHz). This is the expected STRF for the most common type of FRA found in A1: neurons with central excitation and two inhibitory sidebands (e.g., Figs. 6 and 8A). The bottom example (Fig. 9B) shows the hypothetical STRF that would be expected for a neuron with two excitatory and two inhibitory bands (e.g., Fig. 8B). This STRF indicates that the average stimulus before each spike had energy at 3 and 5.5 kHz and a lack of energy at 2.5 and 3.8 kHz. In the example of Fig. 8B there is also a slight slant to the STRF that means the average stimulus might have an upward frequency sweep modulation (or it could just mean different latencies for different frequencies).

It is worth noting that in this example the time scale is reversed. Some investigators present STRFs this way, while some do not reverse the time axis. Reversing is done because the time-reversed stimulus represents the predicted optimal stimulus for concentrating spikes into a narrow response-time window. For example, for a neuron that responds with short latency to a 5-kHz tone and with long latency to a 7.5-kHz tone, the optimal stimulus will have the 7.5 kHz tone start before the 5 kHz tone, so that both responses would be superimposed in time.

In its simplest form reverse-correlation/spike-triggered averaging requires that the aggregate over all the presented stimuli contains equal energy at all frequencies. This is called the white noise assumption, since white noise on average contains equal energy at all frequencies. It is important to note that using a true white noise stimulus with near infinite frequency and time resolution,

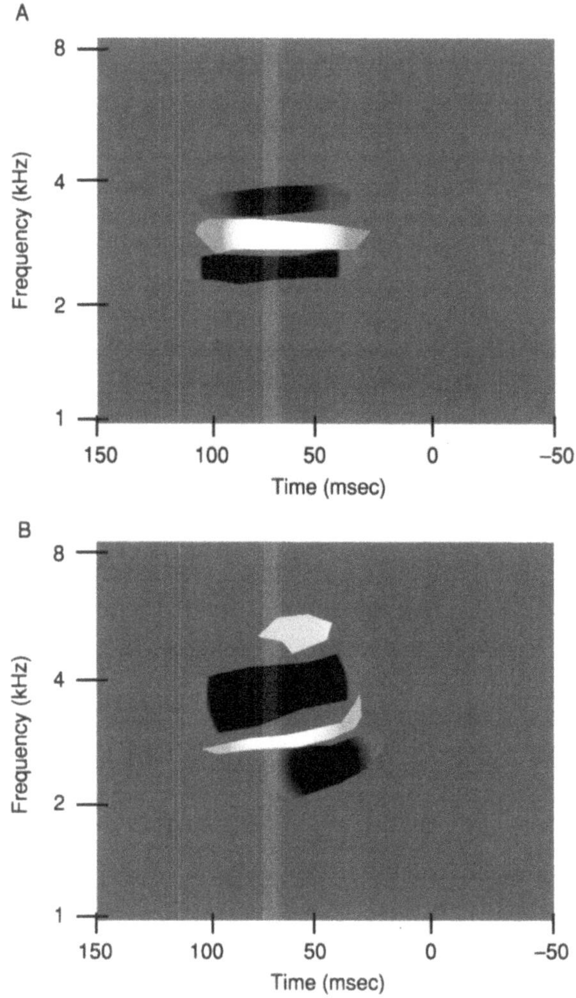

FIG. 9. Schematic STRF of two hypothetical neurons. Light areas represent regions of high stimulus energy preceding action potentials. Dark areas represent regions of low stimulus energy preceding action potentials. Light and dark areas likely correspond to excitatory and inhibitory areas in the receptive field.

which works in some sub-cortical preparations (Hermes *et al.*, 1981; Lewis and Henry, 1994; Lewis and van Dijk, 2004), does not work on auditory cortical neurons (Valentine and Eggermont, 2004), likely because: (1) the time changes in true white noise are too fast for cortical neurons to follow, and (2) strong inhibitory sidebands prevent many cortical neurons from responding.

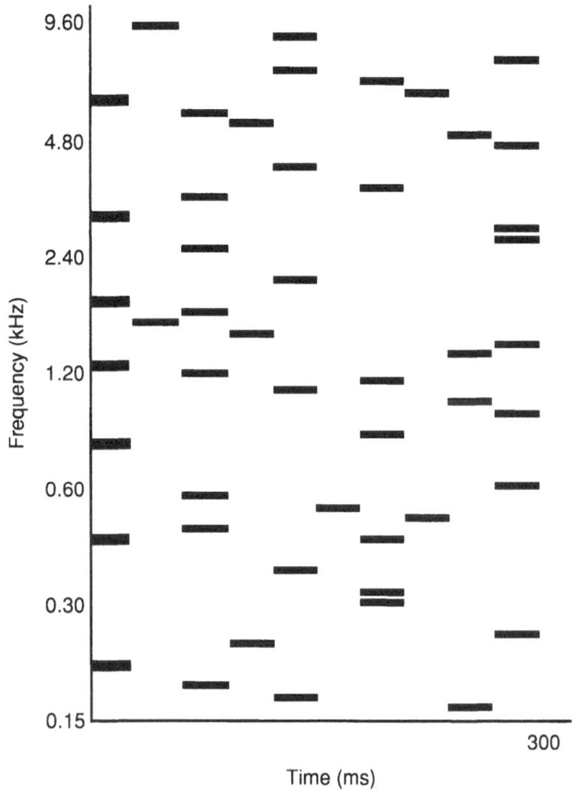

FIG. 10. Schematic spectrogram of 300 ms of a multi-tone chord, similar to those used in deCharms *et al.* (1998).

deCharms and his colleagues worked around this problem by employing time-varying chords to determine cortical STRFs using reverse correlation (deCharms *et al.*, 1998; see also Valentine and Eggermont, 2004). These chords had sparsely spaced, relatively narrow-banded (tone like) frequency components which changed on a relatively slow time scale. Figure 10 shows the spectrogram of a 300-msec segment of a similar type of stimulus. However, to derive an STRF these chord stimuli must be presented for minutes, and over this time the stimuli can be chosen to equally represent all frequencies and thus not violate the white noise assumption. Many of these chord stimuli are not very different from simple two-tone stimuli described in two-tone FRA experiments (except for presentation rate; Valentine and Eggermont, 2004). For example, in deCharms *et al.* (1998), tones were spaced on average at about one per octave, which is not very different than using pure tone and two-tone stimuli to define the STRF, since on average only one or two tones are in the RF at a time. However, conceptually this work was important because it demonstrated the practicality of using reverse correlation to determine cortical STRFs. More recent studies (Blake and Merzenich, 2002) have shown that STRF properties profoundly depend on the density of presented tones. This suggests across-frequency interactions are being recorded in the STRFs.

B. Reverse Correlation and Non-Linear Across Frequency Interactions

So far the description of reverse correlation and spike-triggered averaging has focused on characterization of the spectro-temporal receptive field. There is evidence that when reverse correlation with broad-band stimuli is used to determine a neuron's STRF, non-linear interactions across frequency can be detected (Blake and Merzenich, 2002). This might seem counterintuitive to some who have heard that reverse correlation derived STRFs are inherently linear. This paradox exists because the reverse correlation technique can do two slightly different things. First, it can be used to determine which frequencies excite and which inhibit the neuron; that is, it can determine the general shape of the inhibitory and excitatory regions of the spectral receptive field (SRF). Second, the calculated average stimulus preceding each spike can be used to predict a neurons response magnitude (spikes) to arbitrary stimuli. It is in this latter predictive capacity that reverse-correlation based STRFs are limited by assumptions of non-linearity.

This apparent dichotomy between an inherently linear predictor being able to detect non-linear frequency interactions might be best explained by an example. In bats there are auditory cortical neurons that only respond to specific combinations of two tones, but not to either tone individually (Suga *et al.*, 1979). These neurons are called combination-selective CF-CF neurons. Suppose we record from a combination-selective CF-CF neuron, and the neuron only responds when 60- and 90-kHz tones are presented together, but not to 60- or 90-kHz tones presented by themselves. Single-tone FRAs would yield no responses. Recording FRAs with tone-pairs would reveal the non-linear interaction of selectivity for the two-tone combination. Interestingly using reverse correlation to derive STRFs with chords should reveal response areas in the RF at 60 and 90 kHz. This is because when sounds in the stimulus set contain energy both at 60 and 90 kHz the neuron will usually respond. Therefore, the average stimulus preceding has energy at 60 and 90 kHz. This technique does predict that the neuron will respond best to 60 and 90 kHz presented together and would show the correct general RF shape of two response frequency ranges: one at 60 and one at 90 kHz.

So if the reverse correlation technique is detecting regions of the RF that result from non-linear frequency interactions, why is it said to be limited by being an inherently linear method? It is because the technique does not correctly quantify the non-linear interaction between the two response areas. This means that the STRF will not yield accurate spike count predictions in this and most non-linear cases. In fact the STRF under this condition predicts that the neuron would respond to 60 or 90 kHz in isolation (which it does not). So for this type of non-linearity, the STRF accurately portrays the general shape of the inhibitory and excitatory regions of the STRF, but fails to be useful in predicting the response magnitude of the neurons to arbitrary stimuli.

C. Reverse Correlation with Broad-Band and Natural Stimuli

Reverse correlation has been used to derive STRFs with more complex stimuli. For example, several groups have used sine spectral profiles (also called ripples) to acquire STRFs (Depireux *et al.*, 2001; Fritz *et al.*, 2003; Klein *et al.*, 2000; Kowalski *et al.*, 1996; Miller *et al.*, 2002). Sine spectral profiles are sounds with sinusoidally modulated power spectra, usually with log frequency. These stimuli consequently are related to the cochlea roughly the same way a sine wave grating is related to the retina (Fig. 11). Sine spectral profiles are particularly interesting because the ability to detect these spectral modulations depends on across-channel processing, possibly through global frequency normalization (O'Connor *et al.*, 2000). Essentially, linear filter models (Section I.C., Chapter 2) are poor predictors of human and macaque ability to determine differences between these broad-band spectrally modulated stimuli and un-modulated white noise. Therefore, comparing STRFs collected with sine spectral profiles to tonal produced STRFs would be interesting.

Using reverse correlation to obtain STRFs with natural stimuli has recently been achieved in Field L of songbird forebrain. While Field L of birds technically is not cortex, it shares many anatomical and physiological properties with A1

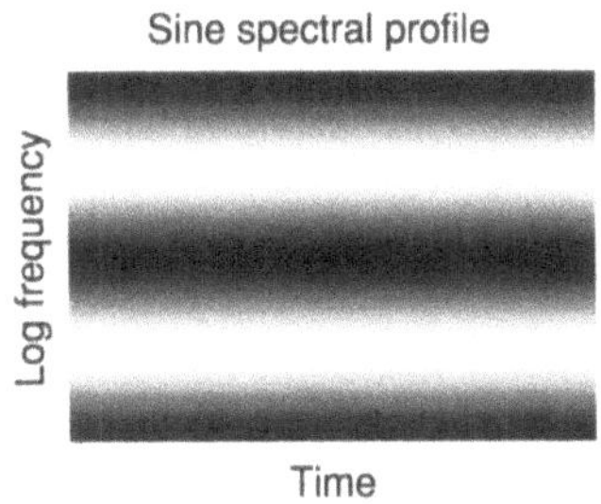

FIG. 11. Spectrogram of sine spectral profile.

(Heil and Scheich, 1991a,b, 1992b; Heil *et al.*, 1992a; Mendelson *et al.*, 1993, 1997; Schreiner and Sutter, 1992). Using reverse correlation with natural stimuli is not trivial because natural sounds can (and usually do) violate the white noise assumption (Theunissen *et al.*, 2000). This is problematic because it would cause a bias towards showing structure in STRFs that looks like the statistical tendencies in natural sounds. For example, suppose a neuron responded equally to all frequencies of sound. The true spectral RF of this neuron would be flat with equal responses for all frequencies. However, if natural sounds were used to derive the RF through reverse correlation, and the average sound presented to generate the STRF had a lot more energy at low than higher frequencies, which most natural sounds like vocalizations do (Attias and Schreiner, 1997; Dusenbery, 1992; Nelken *et al.*, 1999; Singh and Theunissen, 2003), then the STRF would inaccurately show that the neuron prefers lower frequencies. To adjust for this, STRFs must be corrected for the statistics of the stimulus set used to derive them (Theunissen *et al.*, 2000). In simple terms this procedure is very similar to dividing the STRF by the average stimulus presented over all trials, so any frequency (or frequency-time interaction) occurring frequently in the sounds is not artificially inflated in the STRF. Using this technique, Theunissen *et al.* (2000) found that tone- and vocalization-derived reverse correlation STRFs could be substantially different, again implying non-linear across-frequency interactions. Interestingly, a recent study that adapted the use of cross-correlation to measure STRFs of intracellular sub-threshold membrane potentials in rat auditory cortex (Machens *et al.*, 2004) found that differences between tone- and vocalization-derived STRFs could not be explained by simple non-linearities (such as threshholding). They concluded that membrane potential non-linearity likely resulted from non-linear interactions across sound frequencies. The bulk of evidence seems to point to the fact that across-frequency non-linear interactions play an important role in auditory cortical processing of sounds.

D. Reverse Correlation and Time-Frequency Separability

From STRFs important spectral and temporal interactions—time-frequency separability—can be determined. In very simple terms, if the time and frequency components of the STRF are separable, then the entire STRF can be described by multiplying a single time function by a single frequency function. Therefore, for example, if there are slanting components in the STRF suggestive of FM sensitivity, separability is likely to be violated. In A1 responses of mammals (Depireux *et al.*, 2001; Linden *et al.*, 2003) and Field L of birds (Sen *et al.*, 2001) there is a mixture of separable and inseparable responses. This suggests that some cortical responses can be quite complex with regard to how neurons integrate across frequency and time, while others are considerably less complex.

E. Limitations of Reverse Correlation

In summary, there are both advantages and limitations of using reverse correlation to determine STRFs. The advantages, as stated previously, are that this technique allows the use of broad-band and natural sounds to determine the STRF. As such, this technique can detect across-frequency interactions in the STRF which single-tone derived spectral RFs cannot. This technique also has the advantage of identifying inhibition that cannot be detected with single tones. This inhibition is highly likely to be non-linear; Blake and Merzenich (2002) found the predictive power of STRFs for responses to novel stimuli (by making linear predictions) is particularly poor for neurons showing pronounced inhibitory regions.

There are also some weaknesses. Magnitudes within reverse-correlation derived STRFs (for example, gray shading in Fig. 9) do not allow for good predictions of responses to arbitrary stimuli, because the linear assumptions about these magnitudes cause non-linear interactions to be inappropriately quantified. This leads to these STRFs having poor predictive powers for response magnitudes to complex stimuli. Therefore, while STRFs can detect complex interactions, they do not necessarily quantify them well. Reverse correlation also can give grossly inaccurate estimates if neurons have intensity-tuned (non-monotonic) circumscribed response areas. Most of these neurons are inhibited by higher intensity tones at the best excitatory frequency (e.g., Fig. 7B) (Loftus and Sutter, 2001; Sutter and Loftus, 2003; Sutter *et al.*, 1999). For example, for the neuron in Fig. 7B, a sound with energy at 9.5 kHz and 30 dB will cause 9.5 kHz to be associated (reverse-correlated) with spiking activity. However, a sound with energy at 9.5 kHz and 70 dB which inhibits this neuron would be associated with a lack of spikes and actually cause the spike-triggered stimulus average to reduce the 9.5 kHz component because when high energy at 9.5 kHz is presented the neuron's response would decrease to below the mean. These influences would cancel to some degree, and the fact that this neuron prefers 9.5 kHz, but only at low intensities, would not be detected or would be grossly miscalculated. This would be particularly severe for a neuron with a circumscribed response area at one frequency band and a monotonically increasing function at another frequency band. Such FRA structure can be seen for type IV units in the dorsal cochlear nucleus of the brainstem (Young and Brownell, 1976) and the reverse correlation technique indeed works poorly if at all for this type of neuron (Nelken *et al.*, 1997). While neurons with all the properties of type IV units have not been reported in auditory cortex, neurons with non-monotonic and monotonic areas in the spectral RF have been shown in A1 and PAF of cats (Loftus and Sutter, 2001; Sutter and Schreiner, 1991a). Similar arguments can be made for why reverse correlation will yield poor estimates for neurons with slanted FRAs (e.g., see the FRA of Fig. 4B). The reverse correlation technique also would work poorly for units that have sustained responses that are not tightly time linked to the fine structure of stimuli (Sadagopan and Wang, 2004) and for offset responding units.

F. Random Spectral Shape Stimuli and Adaptive Optimization

Some reverse-correlation–related problems are not present in another broadband based technique that applies random spectral shape (RSS) stimuli to derive spectral RF estimates (Yu and Young, 2000; see also Chapter 5). An RSS consists of many (often densely) packed tonal elements. When all are the same amplitude, the density of tones is high, and the frequency range is broad, then that particular RSS approximates white noise (Fig. 12A). The technique involves taking this stimulus and randomly changing each tone element's amplitude within a specified amplitude range (this is where the random in RSS comes from). Two examples of such random spectral stimuli are shown in Fig. 12B and C. A large number of RSS stimuli are presented to a neuron. With enough of these stimuli, the random fluctuations in each individual RSS average out so the mean stimulus contains equal energy at all frequencies. The next step is basically to multiply each response by the vector that defines the RSS for that stimulus presentation (the intensity of each frequency component) and to divide by the total energy. By doing this, an estimate of the spectral RF is obtained since the largest responses are obtained for RSSs with more energy at preferred frequencies, barring circumscribed response areas. Several interesting properties of auditory cortical neurons have been revealed with this technique. One is that A1 neurons show frequency tuning that is more level-tolerant than tuning obtained with pure tones (Barbour and Wang, 2003a). The other is that not all cells like high spectral contrast, rather many prefer lower contrast (Barbour and Wang, 2003b). The third is that if linear assumptions are made to predict responses to novel RSS stimuli, the predictability is not very good (Barbour and Wang, 2003a). This, once again, indicates strong across-frequency non-linear interactions in A1 neurons.

Another method is adaptive stimulus optimization (Nelken *et al.*, 1994; O'Connor *et al.*, 2005). With this technique, stimuli are presented to the neuron, and successive stimuli are modified depending on the neuron's response. In this way the stimulus can be changed to better drive the neuron to maximize its response. Using this technique we found that the preferred stimulus for many neurons converged to the same general form for two independent optimization runs starting at different initial conditions. This suggests that the optimization is converging on a preferred stimulus form, likely reflecting the neuron's spectral RF. We also found some consistent structure in the preferred spectra of the neurons, with most neurons preferring a peak of energy surrounded by low

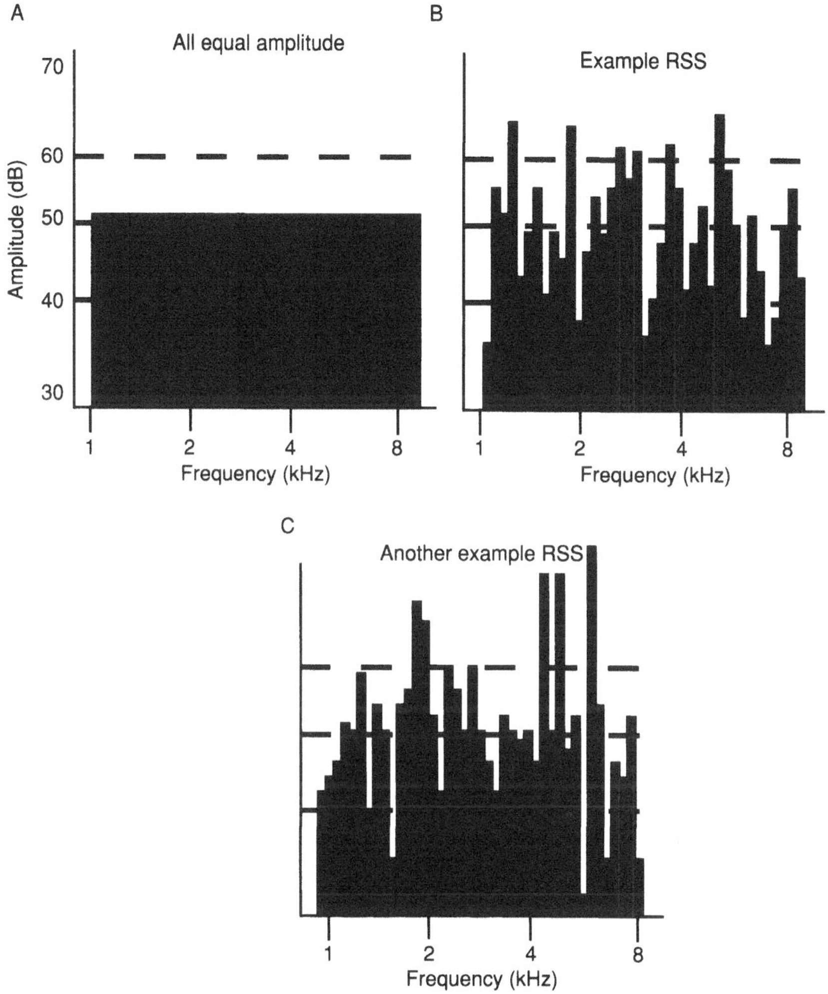

FIG. 12. (A) Schematic amplitude spectrum of flat spectral profile. (B, C) Schematic amplitude spectra of two random spectral shape (RSS) stimuli.

energy at neighboring frequencies, or a trough surrounded by high energy at neighboring frequencies. This suggests a spectral RF exhibiting either an excitatory center and an inhibitory surround (or inhibitory center and excitatory surround). The inhibitory and excitatory BWs obtained with this method were consistent with a multiple spectral resolution analysis in the auditory cortex. Both RSS-based and adaptive techniques are potentially powerful, because non-linear rules can be used to search the space for spectral stimuli eliciting strong or near optimal responses (Yu and Young, 2000).

G. A Summary of Spectral Estimation of RFs with Broad-Band Stimuli

While there have been many studies using different types of broadband stimuli, differences in the details of the studies have produced often contradictory results. However, some generalities about spectral processing can be found. The first is that estimates of spectral receptive fields and response magnitude estimations made with broad-band stimuli are not always the same as those derived with narrow-band sound such as tones. For example, spectral RFs derived with tones are often different from spectral RFs derived with vocalizations (Theunissen *et al.*, 2000), RSS stimuli (Barbour and Wang, 2003a), or with chords, or even tones at different presentations rates (Valentine and Eggermont, 2004). Spectral RF shape is also dependent on the density of tones in 20- to 400-tone complexes (Blake and Merzenich, 2002), and spectral RFs can change with the average modulation depth of RSSs (Barbour and Wang, 2002). Also, the ability to use spectral RFs to predict responses to sounds is worse when the properties of the second sound are different than those used to derive the STRF. For example, using STRFs derived with birdsong stimuli to predict responses to synthetic sounds works poorly, as does using low-density (few tones per octave) spectral profiles to predict responses to high-density spectral profiles. While these across-stimulus type comparisons are very poor, even within-type comparisons are not very good (e.g., RSSs with the same spectral density; Barbour and Wang, 2003a). In aggregate these data demonstrate relatively strong across frequency non-linear integration in auditory cortical responses.

The results suggesting non-linear integration across frequencies are quite relevant to across-channel processing and are particularly interesting with respect to psychophysical results with sine spectral profiles (SSPs). Psychophysical performance with spectral profiles depends on the density of tones within the profile (Green, 1983, 1988). The ability to detect spectral modulation in SSPs is worse than one would predict if the subject were just to observe the output of selective single critical bands or even single auditory nerve fibers (O'Connor *et al.*, 2000). This implies what is called across-channel interference, and might result from some kind of global normalization across frequency. It is notable that broad-band derived STRFs do not match narrow-band derived ones, suggesting a similarity to across channel psychophysics. Blake and Merzenich (2002) found that the predictive power of STRFs for responses to novel stimuli (by making linear predictions) is particularly poor for neurons showing pronounced inhibitory regions. This suggests that inhibition from outside the classically derived (pure-tone–derived) RF might contribute to the across-channel interference observed with SSPs.

IV. Spectral Integration and the Responses of Auditory Cortical Neurons

A. Why Examine the Auditory Cortex?

In order to determine the neural substrates of auditory perceptual performance, the appropriate neural structures must be targeted. Physiological and lesion studies of the ascending auditory pathway suggest that the auditory cortex is an appropriate place to search for neural correlates of spectral integration. Destruction of auditory cortical areas leads to a wide range of deficits in processing vocal communications, frequency analysis, signal detection, and sound localization, indicating that the cortex is necessary for normal spectral processing (vocalization: Coslett *et al.*, 1984; Graham *et al.*, 1980; Heffner and Heffner, 1986a,b; Michel *et al.*, 1980; frequency analysis: Riquimaroux *et al.*, 1992; signal detection: Maruyama and Kanno, 1961; Heffner and Heffner, 1990).

B. Spectral Integration, RF Properties, and Bandwidth Tuning

Spectral processing in the cortex probably involves both decomposition and integration. There is recent evidence of progressive frequency decomposition along the classical lemniscal pathway (central nucleus of the inferior colliculus [ICC], ventral medial geniculate body [MGBv], and ventral primary auditory cortex [A1v]). Prior to this, there had been some controversy as to whether or not frequency tuning sharpens in the ascending lemniscal auditory pathway. Some laboratories found no difference or slight broadening (Calford *et al.*, 1983) and others found a sharpening (Katsuki *et al.*, 1959) with ascension through the lemniscal auditory system. However, recent findings by Suga (1997) in bats suggest that much of these differences are due to how sharpness of tuning changes with sound intensity. It appears that in comparison to AN fibers, neurons at successive stations in the pathway, on average, have increasingly sharper tuning at 40 to 50 dB above threshold. At 10 to 20 dB above threshold such successive sharpening is not observed. This results in more intensity invariant BW (U-shaped tuning curves, for example, Fig. 8C) at higher levels in at least bat A1 (Suga, 1997). We have performed a meta-analysis of the cat literature and found results consistent with Suga's interpretation in bats (2005). This suggests that central neurons play a role in spectral decomposition. However, many neurons at high levels of the lemniscal system, particularly the cortex, have receptive field characteristics congruent with frequency integration. Such neurons exhibit multiple excitatory and inhibitory regions, very broad inhibition, and spectrally complex non-linear tonal interactions (Abeles and Goldstein, 1972; Kadia and Wang, 2003; Loftus and Sutter, 2001; Oonishi and Katsuki, 1965; Suga and Manabe, 1982; Sutter and Schreiner, 1991a; Sutter *et al.*, 1999). Some neurons have spectral RFs with an off-center, on-surround organization (Kadia and Wang, 2003; Sutter and Schreiner, 1991a) that seems well-suited to analyze spectral cues used for vertical sound localization (Butler *et al.*, 1990; Clarey *et al.*, 1994; Middlebrooks and Green, 1991; Musicant and Butler, 1984; Rice *et al.*, 1992; Samson *et al.*, 1993; Xu *et al.*, 1999; Young *et al.*, 1996). Many of these and other cortical neurons in the lemniscal auditory cortex have complex spectral receptive fields with up to six distinct excitatory and/or inhibitory sub-regions; in cats, these are concentrated primarily in the PAF (Loftus and Sutter, 2001) and the dorsal region of A1 (Sutter *et al.*, 1999). Such multipart receptive fields would seem well-suited for analyzing spectrally complex signals rather than performing frequency decomposition. Additionally, the non-lemniscal pathway (the external nucleus of the IC, the medial and dorsal divisions of the MGB, and some of non-primary auditory cortex) contains a high proportion of neurons with broad frequency tuning and complex properties and may be also be a likely candidate for integrative processes, as opposed to filtering properties (Aitkin *et al.*, 1975; Calford, 1983; Calford *et al.*, 1983; Lennartz and Weinberger, 1992; Olsen and Suga, 1991; Phillips and Irvine, 1979; Recanzone *et al.*, 2000; Shamma and Symmes, 1985; Wenstrup, 1999).

While A1 is normally associated with the lemniscal pathway and some secondary areas with the non-lemniscal pathway most auditory cortical areas can represent both lemniscal and non-lemniscal channels, owing to the varying input of each pathway to different auditory cortical fields. It has recently been hypothesized that two areas of monkey auditory cortex, the lateral belt (L) and caudomedial field (CM), are preferentially involved in spectral integration and sound localization, respectively (Hackett *et al.*, 1998a,b, 1999a,b; Kaas and Hackett, 1998, 1999; Kaas *et al.*, 1999; Rauschecker, 1997, 1998; Rauschecker *et al.*, 1997; Romanski *et al.*, 1999). These areas are posited to be part of a cortical 'what' pathway involved in object identification and a cortical 'where' pathway involved in sound localization. This hypothesis leads to specific predictions about spectral processing in these two pathways that have only begun to be tested (it should be noted that sound localization requires spectral processing but with different, but to some degree overlapping, properties than required for object identification). One such interesting property with regard to spectral analysis is that neurons in the object identification pathway have a high degree of tuning for sound BW which appears to form a topographic map along the cortical surface. The significance of this BW tuning and a possible BW map, with respect to multiple frequency resolution analysis, will be discussed in the concluding Section V.E. of this chapter.

C. The Relationship of Auditory Cortical Responses to Across Channel Processing

Some perceptual phenomena, such as co-modulation masking release, automatic perceptual grouping (streaming; see Chapter 10), and even simple modulation detection cannot be explained by critical band models (see also Section I.C.). Despite the interest in across-channel processing, very few studies have been performed on the responses of auditory neurons to the actual stimuli used in across-channel psychophysics. In general, when auditory cortical physiology has been compared to specific psychophysical experiments the results have been valuable (Mickey and Middlebrooks, 2001; Middlebrooks, 2003; Xu *et al.*, 2000). This is particularly true with experiments that have examined the relationship of cortical activity to across-channel psychophysics (e.g., CMR: Nelken *et al.*, 1999; auditory scene analysis: Fishman *et al.*, 2001a, 2004; Nelken, 2004; Petkov *et al.*, 2003a,b; Sugita, 1997; gap detection: Eggermont, 2000; consonance and dissonance: Fishman *et al.*, 2001b; Zwicker tones: Norena and Eggermont, 2003). Most of the results support the view that auditory cortical responses are often inconsistent with simple linear filter models and are likely involved in across-channel processing (Bar-Yosef *et al.*, 2002; Ehret and Schreiner, 1997). Therefore, there are remaining questions to be asked about across-channel processing, and it seems likely that much can be revealed if future studies are specifically designed to tackle these questions. In the following sections a hypothesis that might be tested by such studies will be presented.

V. The Functional Organization of Auditory Cortex for Spectral Processing

A. Many Different Auditory Cortical Areas

The auditory cortical areas of cats, macaque monkeys, and mustached bats are shown in Fig. 13. In general, in mammals, there are one to several auditory cortical fields that are tonotopically organized with cells that respond with a short latency ~10 msec after stimulus onset (Imaizumi *et al.*, 2004). It is important to note that the tonotopy is most obvious by looking at the tips of tuning curves (CF, Fig. 4) and when using a multiple unit recording (Schreiner and Sutter, 1992; Sutter and Schreiner, 1991a). When higher intensity tones are used the tonotopy degrades dramatically (Phillips *et al.*, 1994). In the bottom panel of Fig. 13, these tonotopic fields are shown in gray or black with schematized isofrequency lines drawn in white with typical CFs (in kHz) drawn next to the lines. One of these fields is always called primary auditory cortex (A1) (black for the expanded panels on the bottom of Fig. 13). In macaque monkeys, A1 is buried in a sulcus so it

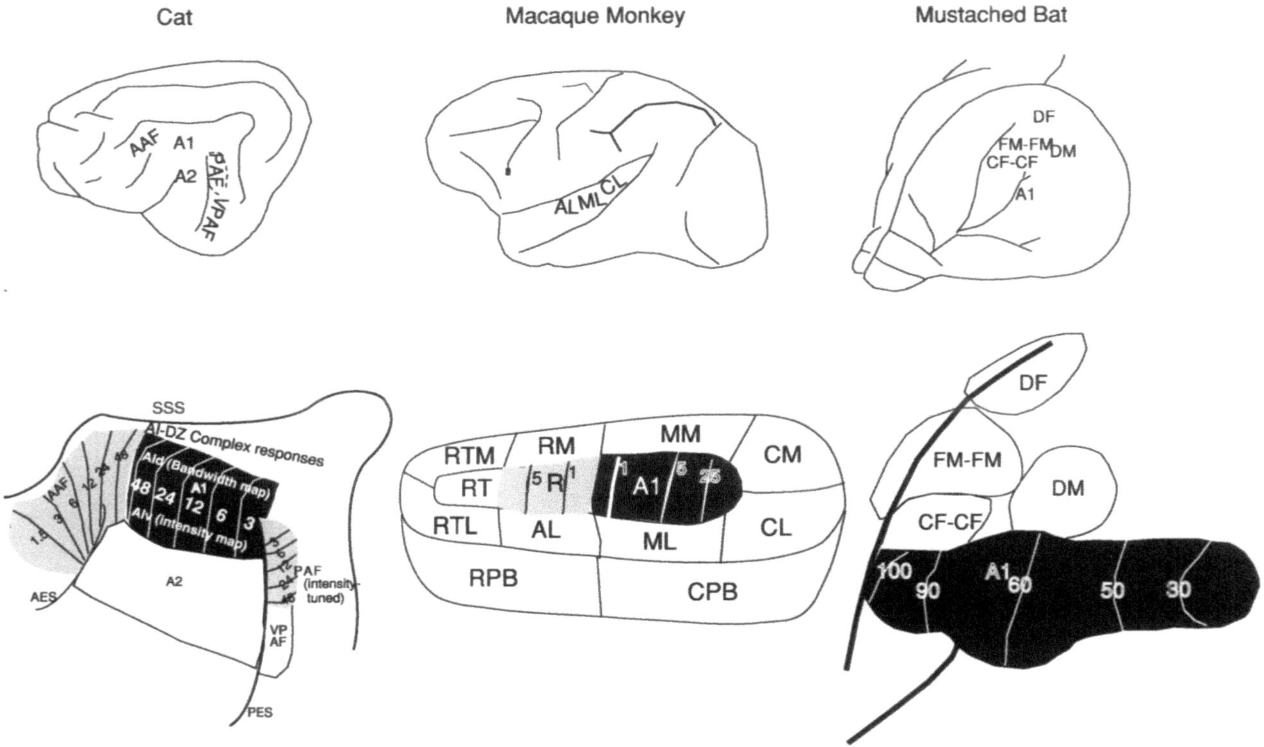

FIG. 13. Schematic representations of auditory cortex for cats, macaque monkeys, and mustached bats. The top row shows lateral views of the entire brains. The bottom row zooms in on auditory cortex. Tonotopic A1 is shaded black and other tonotopic areas are shaded gray. Schematized typical isofrequency lines are shown in tonotopic areas. It should be noted that there is a tremendous amount of individual variability in these isofrequency lines. Areas outlined with white fill are not tonotopic (except the CF-CF area of mustached bats that has a dual tonotopic map).

cannot be seen in the gross brain on the top panel. The response properties of A1 are remarkably similar across species. Surrounding the tonotopically organized fields are other non-primary fields that are not tonotopically organized (shown in white in the bottom panels of Fig. 13), and which show marked differences across species. Neurons in these areas tend to respond with longer latencies and higher thresholds to pure tones than neurons in tonotopic areas. They also tend to respond more weakly to pure tones than neurons in the tonotopic areas, but respond to a broader range of frequencies. The tonotopic areas receive the majority of their input from the ventral division of the medial geniculate body (MGBv) of the thalamus which is also generally sharply tuned and tonotopically organized. The non-tonotopic areas generally receive much of their input from the non-tonotopic dorsal and medial divisions of the MGB (MGBd and MGBm) which have broadly tuned neurons. These non-primary auditory cortical areas also receive and send a large number of inputs from and to other auditory cortical areas including A1 (Lee *et al.*, 2004), but the degree to which higher cortical areas are primarily driven from parallel thalamic inputs versus a serial cortical hierarchy might be species dependent (Rauschecker, 1997, 1998; Rauschecker *et al.*, 1997; Rouiller *et al.*, 1991). Cats appear to have non-primary auditory cortical areas whose responses are dominated by thalamic inputs (e.g., Kitzes and Hollrigel, 1996), whereas responses in macaque non-primary auditory cortex appear to be more dependent on intra-cortical connections (e.g., Rauschecker *et al.*, 1997).

The differences between broadly and narrowly tuned cortical areas is suggestive of different scales for spectral integration. Before delving into the potential functional organization for spectral processing in cats and monkeys (which will be revisited at the end of this chapter), we will examine some compelling cases for a functional organization for spectral processing in echo-locating mustached bat auditory cortex.

B. Functional Organization of Non-Linear Facilitation in Bat Auditory Cortex

One of the most studied animals is the echo-locating mustached bat (reviewed in Suga, 1988), as well as other kinds of bats (Chen and Jen, 2000; Dear *et al.*, 1993b; Galazyuk and Feng, 1997; Jen and Zhang, 1999; Jen *et al.*, 2002; Razak *et al.*, 1999). Some have claimed that these bats are 'specialists' and, therefore, are not an appropriate model for understanding hearing in mammals. Two facts contradict this view. First, bats are no more specialists than the very well-studied cat model. Cats also have their unique usage of hearing. Second, and more importantly, many of the general properties of cortical organization found in bats have also been found in cats and other mammals.

Mustached bats echo-locate by emitting an ultrasonic chirp and measuring the properties of the echoes. Bats compare the chirps and echoes by analyzing auditory responses to combined chirp-echo features. For example, the frequency difference between the emitted chirp and the received echo gives information about the velocity of the target being tracked by the bat because the relative velocity of a target object causes a Doppler shift of the sound wave. Doppler shift is the property that also makes the frequency of a train's whistle increase as it approaches. The same principle applies to the bat and a target it is echo-locating. If the bat is approaching the object, the echo contains higher frequencies than those the bat emits. The bat appears to calculate this using neurons that selectively respond to tone combinations. Neurons in the CF-CF brain areas (Fig. 13) do not respond to pure tones, but respond to tone pairs with specific frequency combinations corresponding to the natural range of encountered Doppler shifts (Suga, 1988). Therefore, the responses to the complex tone combinations cannot be predicted from the response to individual tones. This combination selectivity is a prominent form of non-linearity observed in many auditory systems. There is a clear functional organization to this combination selectivity. In fact, in the CF-CF area, there is a topographic map of target velocity. This is achieved by arranging dual tonotopic maps for the frequency combinations.

While Doppler-shifted frequency analysis provides an example of cortical maps for non-linear spectral integration, it is interesting that neurons with non-linear frequency integration properties in bats are not limited to Doppler-shifted frequency analysis. Neurons in different parts of the mustached bat's auditory cortex—even those areas previously associated with selectivity for echo-location signals—have been also found to non-linearly process species-specific vocalizations with lower frequencies than the echo-location chirp (Kanwal *et al.*, 1999; Medvedev and Kanwal, 2004; Medvedev *et al.*, 2002; Ohlemiller *et al.*, 1996). It should be noted that this non-linear facilitation is not limited to bats, as cat auditory cortical neurons have shown non-linear facilitation in response to tone sequences (Brosch and Schreiner, 2000; McKenna *et al.*, 1989). Finally, single-unit response properties in big brown bats have led to theories of wavelet (analyzing on multiple frequency scales) and processing of spectral interference patterns for target range (Dear *et al.*, 1993a; Neretti *et al.*, 2003; Sanderson and Simmons, 2002).

C. Organization of Cat Primary Auditory Cortex to Perform Multi-Resolution Spectral Analysis

Re-examining Fig. 2 and the question of how the brain can analyze complex sounds like the word 'baseball,' let us determine how the brain might do this. Several recent papers have noted the importance of multi-resolution spectral processing required to analyze some speech sounds (Nelken *et al.*, 2003; Schreiner *et al.*, 2000). In this section we will propose how primary auditory cortex might contribute to this analysis. First, multi-spectral analysis appears to happen rapidly, by processing information about multiple frequency bands in parallel. One way the brain could do this would be if different neurons were to analyze different BWs. In cats, a candidate for a topographical representation of this type of processing has been found in A1. A single-unit BW map has been found in dorsal A1 (A1d) (Schreiner *et al.*, 1992). In this map the part of A1d towards the dorsal-ventral center of A1 has the most sharply tuned single units (Fig. 14). This would be well suited to analyze the very narrow energy bands. Progressing more dorsally, in A1d, neurons become more broadly tuned. In the 0.4 to 0.8 octave range, such neurons would be well suited to analyze and estimate the BW of the individual formants (white arrowheads in Fig. 2). Finally, as one approaches the dorsal extreme of A1 and an area called the dorsal zone (DZ) (He and Hashikawa, 1998; Middlebrooks and Zook, 1983; Ojima and He, 1997; Ojima and Takayanagi, 2004), multi-peaked units emerge (Sutter and Schreiner, 1991a). These neurons' tuning curves display two distinct peaks (Fig. 4E) with a lack of excitation and usually inhibition between the peaks (Abeles and Goldstein, 1972; Oonishi and Katsuki, 1965; Suga and Tsuzuki, 1985; Sutter and Schreiner, 1991a). These units also show evidence of integrating energy across bands as their latency shortens when energy is presented in the two bands. These multi-peaked units from dorsal A1 would be well suited for calculating the frequency separation between formants. Anatomy within dorsal A1 also supports a special

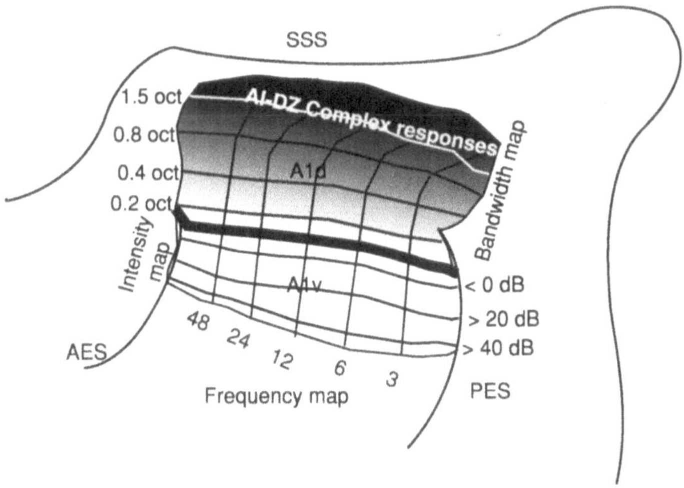

FIG. 14. Schematic of cat A1 showing the bandwidth and intensity maps. For the bandwidth maps, the darker area corresponds to neurons that analyze over a larger bandwidth.

role for A1d in spectral integration (Ojima and Takayanagi, 2004; Read *et al.*, 2001). Therefore, for a complex stimulus like the vowel in Fig. 2, different regions along the dorso-ventral extent of A1 could analyze harmonics of the fundamental, formant BW, and formant separation in parallel. This provides a way to understand how the cortex, in parallel, can analyze multiple spectral BWs within a complex stimulus simultaneously.

In the ventral part of A1 (A1v), almost all single neurons are sharply tuned for frequency (Schreiner and Sutter, 1992). However, a single-unit map of intensity can be found in A1v (Sutter and Schreiner, 1995). This map could find the amplitudes of the peaks in complex spectra. However, this map is not a typical map. For lower-intensity stimuli, a region of intensity-tuned cells can be found near the center of A1 on the ventral side (e.g., <0 dB in Fig. 6). These neurons only respond to low-intensity sounds and have been implicated in the detection and retrieval of very low energy bands within a complex sound (Sutter and Schreiner, 1995). With more ventral progression, neurons have higher thresholds and fewer intensity-tuned neurons can be found. Thresholds progressively become higher with more ventral locations heading towards A2. Therefore, in this region outside of the detection area, as sounds get louder, more ventral regions of A1 will be recruited to become active. In this way the ventral part of A1 can detect the amount of energy within narrow bands of a complex stimulus.

In conclusion, in cat A1d there is a topographical arrangement of neurons, well-suited to perform a multi-resolution spectral analysis. When taken in combination with the topographical representation of intensity in A1v, a simple scheme for analyzing complex spectral sounds such as vowels is present. While owl monkeys show some evidence for a similar scheme (Recanzone *et al.*, 1999), this type of organization has yet to be looked for in other animals, such as macaque monkeys and mustached bats. However, at the end of this chapter (V.E.) we will introduce a novel hypothesis about how a similar organization for multi-resolution analysis might exist across primates. But first, we will examine the relationship of cortical spectral analysis to non-linear and inhibitory response properties.

D. Inhibitory Frequency Tuning Is Consistent with the Organization of Cat Primary Auditory Cortex to Perform Multi-Resolution Spectral Analysis

Not only do excitatory BW maps support an ability of A1d to perform analysis of complex spectral patterns by analyzing multiple spectral scales in parallel, but inhibitory response areas in A1 also support this view. Most A1v neurons are sharply tuned for frequency and prefer narrow-band stimuli. The most common form of inhibitory band structure in A1v is two lateral flanking bands around a central excitatory area. Dorsal A1 in the previous section was proposed to have a BW map for multi-resolution analysis with the most dorsal parts of A1 analyzing the widest spectral scale and perhaps the most complex spectra. In dorsal A1 the inhibitory band structure tends to be more varied and complex than in A1v, and this complexity increases progressively with more dorsal recordings (Sutter *et al.*, 1999). These results are consistent with more dorsal locations in A1 processing more complex stimulus patterns and a wider BW (Schreiner *et al.*, 1992; Sutter and Schreiner, 1991b). It also is consistent with the notion that these neurons are likely involved in psychophysical across-channel processing (e.g., see Chapters 10 and Section I.C. of this chapter). Therefore, it appears that both excitatory and inhibitory areas in cat A1 are constructed to perform multi-resolution analysis of complex sounds.

E. Organization of Auditory Cortex to Perform Multi-Resolution Spectral Analysis

In the previous sections, we have made a case for how cat primary auditory cortex might be organized to analyze multiple spectral scales. This would allow for rapid analysis both within and across formants, as well as fundamental frequency analysis in the example using the word "baseball" in Fig. 2. While this topographical map of the cortex with different parts of the cortex simultaneously processing and integrating across different BWs has not been investigated as thoroughly in other species, some similarities have been found in owl monkeys (Recanzone *et al.*, 1999). More importantly, in macaques there appears to be a gradual progression of neurons preferring increased BW with progression from A1 towards and into more lateral fields (Rauschecker and Tian, 2004; Rauschecker *et al.*, 1995; Tian and Rauschecker, 2004). This BW map could serve similarly to the multi-resolution BW map reported in cats. If this is so, it represents an attractively different way to examine the BW selectivity in the cortex of primates. Such a result points to the idea that analyzing sounds with multiple BW scales in parallel might be a general organizational property of mammalian auditory cortex.

References

Abeles, M., and Goldstein, M. H., Jr. (1972). Responses of single units in cat primary auditory cortex of the cat to tones and tone pairs. *Brain Res.* **42,** 337–352.

Aertsen, A. M., and Johannesma, P. I. (1981). The spectro-temporal receptive field. A functional characteristic of auditory neurons. *Biol. Cybern.* **42,** 133–143.

Aitkin, L. M., Webster, W. R., Veale, J. L., and Crosby, D. C. (1975). Inferior colliculus. I. Comparison of response properties of neurons in central, pericentral, and external nuclei of adult cat. *J. Neurophysiol.* **38,** 1196–1207.

Allman, J., Miezin, F., and McGuinness, E. (1985). Stimulus specific responses from beyond the classical receptive field: Neurophysiological mechanisms for local-global comparisons in visual neurons. *Ann. Rev. Neurosci.* **8,** 407–430.

Amagai, S., Dooling, R. J., Shamma, S., Kidd, T. L., and Lohr, B. (1999). Detection of modulation in spectral envelopes and linear-rippled noises by budgerigars (Melopsittacus undulatus). *J. Acoustic. Soc. Am.* **105,** 2029–2035.

Attias, H., and Schreiner, C. E. (1997). Temporal low-order statistics of natural sounds. *Adv. Neural. Inf. Process.* **9,** 27–33.

Barbour, D. L., and Wang, X. (2003a). Auditory cortical responses elicited in awake primates by random spectrum stimuli. *J. Neurosci.* **23,** 7194–7206.

Barbour, D. L., and Wang, X. (2003b). Contrast tuning in auditory cortex. *Science* **299,** 1073–1075.

Barbour, D. L., and Wang, X. (2002). Temporal coherence sensitivity in auditory cortex. *J. Neurophysiol.* **88,** 2684–2699.

Barone, P., Clarey, J. C., Irons, W. A., and Imig, T. J. (1996). Cortical synthesis of azimuth-sensitive single-unit responses with nonmonotonic level tuning: A thalamocortical comparison in the cat. *J. Neurophysiol.* **75,** 1206–1220.

Bar-Yosef, O., Rotman, Y., and Nelken, I. (2002). Responses of neurons in cat primary auditory cortex to bird chirps: Effects of temporal and spectral context. *J. Neurosci.* **22,** 8619–8632.

Bear, M. F., Conners, B. W., and Paradiso, M. A. (2001). "Neuroscience Exploring the Brain." Lippincott, Williams, and Wilkins, Baltimore, MD.

Bernstein, L. R., and Green, D. M. (1988). Detection of changes in spectral shape: Uniform vs. non-uniform background spectra. *Hearing Res.* **34,** 157–165.

Bernstein, L. R., and Green, D. M. (1987). Detection of simple and complex changes of spectral shape. *J. Acoustic. Soc. Am.* **82,** 1587–1592.

Blake, D. T., and Merzenich, M. M. (2002). Changes of AI receptive fields with sound density. *J. Neurophysiol.* **88,** 3409–3420.

Bregman, A. S. (1990). "Auditory Scene Analysis." MIT Press, Cambridge, MA.

Brosch, M., and Schreiner, C. E. (2000). Sequence sensitivity of neurons in cat primary auditory cortex. *Cereb. Cortex.* **10,** 1155–1167.

Brugge, J. F., Dubrovsky, N. A., Aitkin, L. M., and Anderson, D. J. (1969). Sensitivity of single neurons in auditory cortex of cat to binaural tonal stimulation; effects of varying interaural time and intensity. *J. Neurophysiol.* **32,** 1005–1024.

Butler, R. A., Humanski, R. A., and Musicant, A. D. (1990). Binaural and monaural localization of sound in two-dimensional space. *Perception* **19,** 241–256.

Calford, M. B. (1983). The parcellation of the medial geniculate body of the cat defined by auditory response properties of single units. *J. Neurosci.* **3,** 2350–2364.

Calford, M. B., and Semple, M. N. (1995). Monaural inhibition in cat auditory cortex. *J. Neurophysiol.* **73,** 1876–1891.

Calford, M. B., Webster, W. R., and Semple, M. M. (1983). Measurement of frequency selectivity of single neurons in the central auditory pathway. *Hearing Res.* **11,** 395–401.

Chen, Q. C., and Jen, P. H. S. (2000). Bicuculline application affects discharge patterns, rate-intensity functions, and frequency tuning characteristics of bat auditory cortical neurons. *Hearing Res.* **150,** 161–174.

Clarey, J. C., Barone, P., and Imig, T. J. (1994). Functional organization of sound direction and sound pressure level in primary auditory cortex of the cat. *J. Neurophysiol.* **72,** 2383–2405.

Coslett, H. B., Brashear, H. R., and Heilman, K. M. (1984). Pure word deafness after bilateral primary auditory cortex infarcts. *Neurology* **34,** 347–352.

Dear, S. P., Fritz, J., Haresign, T., Ferragamo, M., and Simmons, J. A. (1993b). Tonotopic and functional organization in the auditory cortex of the big brown bat, *Eptesicus Fuscas*. *J. Neurophysiol.* **70,** 1988–2009.

Dear, S. P., Simmons, J. A., and Fritz, J. (1993a). A possible neuronal basis for representation of acoustic scenes in auditory cortex of the big brown bat. *Nature* **364,** 620–623.

deCharms, R. C., Blake, D. T., and Merzenich, M. M. (1998). Optimizing sound features for cortical neurons. *Science* **280,** 1439–1443.

Depireux, D. A., Simon, J. Z., Klein, D. J., and Shamma, S. A. (2001). Spectro-temporal response field characterization with dynamic ripples in ferret primary auditory cortex. *J. Neurophysiol.* **85,** 1220–1234.

Dusenbery, D. B. (1992). "Sensory Ecology: How Organisms Acquire and Respond to Information." W. H. Freeman, New York, NY.

Eggermont, J. J. (2000). Neural responses in primary auditory cortex mimic psychophysical, across-frequency-channel, gap-detection thresholds. *J. Neurophysiol.* **84,** 1453–1463.

Eggermont, J. J., Johannesma, P. I., and Aertsen, A. (1983). Reverse correlation methods in auditory research. *Quart. Rev. Biophys.* **16,** 341–414.

Ehret, G., and Merzenich, M. M. (1988b). Complex sound analysis (frequency resolution, filtering and spectral) by single units of the inferior colliculus of the cat. *Brain Res.* **472,** 139–163.

Ehret, G., and Merzenich, M. M. (1988a). Neuronal discharge rate is unsuitable for encoding sound intensity at the inferior-colliculus level. *Hearing Res.* **35,** 1–7.

Ehret, G., and Schreiner, C. E. (1997). Frequency resolution and spectral integration (critical band analysis) in single units of the cat primary auditory cortex. *J. Compar. Physiol. Sens. Neural Behavior Physiol.* **181,** 635–650.

Ehret, G., and Schreiner, C. E. (2000). Regional variations of noise-induced changes in operating range in cat AI. *Hearing Res.* **141,** 107–116.

Erulker, S. D., Rose, J. E., and Davies, P. W. (1956). Single unit activity in the auditory cortex of the cat. *Johns Hopkins Hospital Bulletin* **39,** 55–86.

Evans, E. F. (1974). Auditory frequency selectivity and the cochlear nerve. *In* "Facts and Models in Hearing" (E. Zwicker and E. Terhardt, Eds.), pp. 118–129. Springer-Verlag, Berlin.

Evans, E. F., and Whitfield, I. C. (1964). Classification of unit responses of the auditory cortex of the unanesthetized and unrestrained cat. *J. Physiol.* **171,** 476–493.

Fishman, Y. I., Reser, D. H., Arezzo, J. C., and Steinschneider, M. (2001a). Neural correlates of auditory stream segregation in primary auditory cortex of the awake monkey. *Hearing Res.* **151,** 167–187.

Fishman, Y. I., Volkov, I. O., Noh, M. D., Garell, P. C., Bakken, H., Arezzo, J. C., Howard, M. A., and Steinschneider, M. (2001b). Consonance and dissonance of musical chords: Neural correlates in auditory cortex of monkeys and humans. *J. Neurophysiol.* **86,** 2761–2788.

Fishman, Y. I., Arezzo, J. C., and Steinschneider, M. (2004). Auditory stream segregation in monkey auditory cortex: Effects of frequency separation, presentation rate, and tone duration. *J. Acoust. Soc. Am.* **116,** 1656–1670.

Fletcher, H. (1940). Auditory patterns. *Reviews of Modern Physics*.

Fritz, J., Shamma, S., Elhilali, M., and Klein, D. (2003). Rapid task-related plasticity of spectrotemporal receptive fields in primary auditory cortex. *Nat. Neurosci.* **6,** 1216–1223.

Galazyuk, A. V., and Feng, A. S. (1997). Encoding of sound duration by neurons in the auditory cortex of the little brown bat, Myotis lucifugus. *J. Comp. Physiol. [A]* **180,** 301–311.

Graham, J., Greenwood, R., and Lecky, B. (1980). Cortical deafness: A case report and review of the literature. *J. Neurol. Sci.* **48,** 35–49.

Green, D. M. (1986). 'Frequency' and the detection of spectral shape change. *In* "Auditory Frequency Selectivity" (B. Moore, Ed.), pp. 351–359. Plenum Press, New York, NY.

Green, D. M. (1983). "Profile Analysis: A Different View of Auditory Intensity Discrimination." *Am. Psychol.* **38**.

Green, D. M. (1988). "Profile Analysis: Auditory Intensity Discrimination." Oxford University Press, New York, NY.

Greenwood, D. D., and Maruyama, N. (1965). Excitatory and inhibitory response areas of auditory neurons in the cochlear nucleus. *J. Neurophysiol.* **28,** 863–892.

Hackett, T. A., Stepniewska, I., and Kaas, J. H. (1999a). Callosal connections of the parabelt auditory cortex in macaque monkeys. *Eur. J. Neurosci.* **11,** 856–866.

Hackett, T. A., Stepniewska, I., and Kaas, J. H. (1999b). Prefrontal connections of the parabelt auditory cortex in macaque monkeys. *Brain Res.* **817,** 45–58.

Hackett, T. A., Stepniewska, I., and Kaas, J. H. (1998a). Subdivisions of auditory cortex and ipsilateral cortical connections of the parabelt auditory cortex in macaque monkeys. *J. Comp. Neurol.* **394,** 475–495.

Hackett, T. A., Stepniewska, I., and Kaas, J. H. (1998b). Thalamocortical connections of the parabelt auditory cortex in macaque monkeys. *J. Comp. Neurol.* **400,** 271–286.

Hall, J. W., Haggard, M. P., and Fernandes, M. A. (1984). Detection in noise by spectro-temporal pattern analysis. *J. Acoustic. Soc. Am.* **76,** 50–56.

He, J., and Hashikawa, T. (1998). Connections of the dorsal zone of cat auditory cortex. *J. Comp. Neurol.* **400,** 334–348.

Heffner, H. E., and Heffner, R. S. (1986b). Effect of unilateral and bilateral auditory cortex lesions on the discrimination of vocalizations by Japanese macaques. *J. Neurophysiol.* **56,** 683–701.

Heffner, H. E., and Heffner, R. S. (1986a). Hearing loss in Japanese macaques following bilateral auditory cortex lesions. *J. Neurophysiol.* **55,** 256–271.

Heffner, H. E., and Heffner, R. S. (1990). Effect of bilateral auditory cortex lesions on absolute thresholds in Japanese macaques. *J. Neurophysiol.* **64,** 191–205.

Heil, P., and Irvine, D. R. (1998). The posterior field P of cat auditory cortex: Coding of envelope transients. *Cerebral Cortex.* **8,** 125–141.

Heil, P., Rajan, R., and Irvine, D. R. F. (1992a). Sensitivity of neurons in cat primary auditory cortex to tones and frequency modulated stimuli: Organization of response properties along the 'isofrequency' dimension. *Hearing Res.* **63,** 134–156.

Heil, P., and Scheich, H. (1991a). Functional organization of the avian auditory cortex analogue. I. Topographic representation of isointensity bandwidth. *Brain Res.* **539,** 110–120.

Heil, P., and Scheich, H. (1991b). Functional organization of the avian auditory cortex analogue. II. Topographic distribution of latency. *Brain Res.* **539,** 121–125.

Heil, P., and Scheich, H. (1992b). Spatial representation of frequency-modulated signals in the tonotopically organized auditory cortex analogue of the chick. *J. Comp. Neurol.* **322,** 548–565.

Hermes, D. J., Aertsen, A. M., Johannesma, P. I., and Eggermont, J. J. (1981). Spectro-temporal characteristics of single units in the auditory midbrain of the lightly anaesthetised grass frog (Rana temporaria L) investigated with noise stimuli. *Hearing Res.* **5,** 147–178.

Imaizumi, K., Priebe, N. J., Crum, P. A., Bedenbaugh, P. H., Cheung, S. W., and Schreiner, C. E. (2004). Modular functional organization of cat anterior auditory field. *J. Neurophysiol.* **92,** 444–457.

Jen, P. H., Chen, Q. C., and Wu, F. J. (2002). Interaction between excitation and inhibition affects frequency tuning curve, response size and latency of neurons in the auditory cortex of the big brown bat, Eptesicus fuscus. *Hearing Res.* **174,** 281–289.

Jen, P. H., and Zhang, J. P. (1999). Corticofugal regulation of excitatory and inhibitory frequency tuning curves of bat inferior collicular neurons. *Brain Res.* **841,** 184–188.

Joris, P. X., Schreiner, C. E., and Rees, A. (2004). Neural processing of amplitude-modulated sounds. *Physiol. Rev.* **84,** 541–577.

Kaas, J. H., and Hackett, T. A. (1998). Subdivisions of auditory cortex and levels of processing in primates. *Audiol. Neuro-Otolo.* **3,** 73–85.

Kaas, J. H., and Hackett, T. A. (1999). 'What' and 'where' processing in auditory cortex [news; comment]. *Nat. Neurosci.* **2,** 1045–1047.

Kaas, J. H., Hackett, T. A., and Tramo, M. J. (1999). Auditory processing in primate cerebral cortex. *Cur. Opin. Neurobiol.* **9,** 164–170.

Kadia, S. C., and Wang, X. (2003). Spectral integration in A1 of awake primates: Neurons with single- and multipeaked tuning characteristics. *J. Neurophysiol.* **89,** 1603–1622.

Kanwal, J. S., Fitzpatrick, D. C., and Suga, N. (1999). Facilitatory and inhibitory frequency tuning of combination-sensitive neurons in the primary auditory cortex of mustached bats. *J. Neurophysiol.* **82,** 2327–2345.

Kaur, S., Lazar, R., and Metherate, R. (2004). Intracortical pathways determine breadth of subthreshold frequency receptive fields in primary auditory cortex. *J. Neurophysiol.* **91,** 2551–2567.

Katsuki, Y., Wantanabe, T., and Maruyama, N. (1959). Activity of auditory neurons in upper levels of brain of cat. *J. Neurophysiol.* **22,** 343–359.

Kitzes, L. M., and Hollrigel, G. S. (1996). Response properties of units in the posterior auditory field deprived of input from the ipsilateral primary auditory cortex. *Hearing Res.* **100,** 120–130.

Klein, D. J., Depireux, D. A., Simon, J. Z., and Shamma, S. A. (2000). Robust spectrotemporal reverse correlation for the auditory system: Optimizing stimulus design. *J. Comput. Neurosci.* **9,** 85–111.

Kowalski, N., Depireux, D. A., and Shamma, S. A. (1996). Analysis of dynamic spectra in ferret primary auditory cortex. II. Prediction of unit responses to arbitrary dynamic spectra. *J. Neurophysiol.* **76,** 3524–3534.

Lamme, V. A. (1995). The neurophysiology of figure-ground segregation in primary visual cortex. *J. Neurosci.* **15,** 1605–1615.

Lee, C. C., Imaizumi, K., Schreiner, C. E., and Winer, J. A. (2004). Concurrent tonotopic processing streams in auditory cortex. *Cereb. Cortex.* **14,** 441–451.

Lennartz, R. C., and Weinberger, N. M. (1992). Frequency selectivity is related to temporal processing in parallel thalamocortical auditory pathways. *Brain Res.* **583,** 81–92.

Lewis, E. R., and Henry, K. R. (1994). Dynamic changes in tuning in the gerbil cochlea. *Hearing Res.* **79,** 183–189.

Lewis, E. R., and van Dijk, P. (2004). New variation on the derivation of spectro-temporal receptive fields for primary auditory afferent axons. *Hear. Res.* **189,** 120–136.

Linden, J. F., Liu, R. C., Sahani, M., Schreiner, C. E., and Merzenich, M. M. (2003). Spectrotemporal structure of receptive fields in areas AI and AAF of mouse auditory cortex. *J. Neurophysiol.* **90,** 2660–2675.

Loftus, W. C., and Sutter, M. L. (2001). Spectrotemporal organization of excitatory and inhibitory receptive fields of cat posterior auditory field neurons. *J. Neurophysiol. (Bethesda)* **86,** 475–491.

Machens, C. K., Wehr, M. S., and Zador, A. M. (2004). Linearity of cortical receptive fields measured with natural sounds. *J. Neurosci.* **24,** 1089–1100.

Maruyama, N., and Kanno, Y. (1961). Experimental study of functional compensation of bilateral removal of auditory cortex in cats. *J. Neurophysiol.* **24,** 193–202.

McKenna, T., Weinberger, N. M., and Diamond, D. M. (1989). Responses of single auditory cortical neurons to tone sequences. *Brain Res.* **481,** 142–153.

Medvedev, A. V., Chiao, F., and Kanwal, J. S. (2002). Modeling complex tone perception: Grouping harmonics with combination-sensitive neurons. *Biol. Cybern.* **86,** 497–505.

Medvedev, A. V., and Kanwal, J. S. (2004). Local field potentials and spiking activity in the primary auditory cortex in response to social calls. *J. Neurophysiol.* **92,** 52–65.

Mendelson, J. R., Schreiner, C. E., and Sutter, M. L. (1997). Functional topography of cat primary auditory cortex: Response latencies. *J. Compar. Physiol. Sens. Neural Behavior Physiol.* **181,** 615–633.

Mendelson, J. R., Schreiner, C. E., Sutter, M. L., and Grasse, K. L. (1993). Functional topography of cat primary auditory cortex - responses to frequency-modulated sweeps. *Exp. Brain Res.* **94,** 65–87.

Michel, F., Peronnet, F., and Schott, B. (1980). A case of cortical deafness: Clinical and electrophysiological data. *Brain Lang.* **10,** 367–377.

Mickey, B. J., and Middlebrooks, J. C. (2001). Responses of auditory cortical neurons to pairs of sounds: Correlates of fusion and localization. *J. Neurophysiol.* **86,** 1333–1350.

Middlebrooks, J. C. (2003). Hearing phantoms. Focus on "Spectral Cues Explain Illusory Elevation Effects with Stereo Sounds in Cats." *J. Neurophysiol.* **90,** 1.

Middlebrooks, J. C., and Green, D. (1991). Sound localization by human listeners. *Ann. Rev. Psych.* **42,** 135–159.

Middlebrooks, J. C., and Zook, J. M. (1983). Intrinsic organization of the cat's medial geniculate body identified by projections to binaural response-specific bands in the primary auditory cortex. *J. Neurosci.* **3,** 203–224.

Miller, L. M., Escabi, M. A., Read, H. L., and Schreiner, C. E. (2002). Spectrotemporal receptive fields in the lemniscal auditory thalamus and cortex. *J. Neurophysiol.* **87,** 516–527.

Moore, B. C. J. (1995). Frequency Analysis and Masking. *In* "Hearing" (B. C. J. Moore, Ed.), 2nd ed. pp. 161–206. Academic Press, San Diego, CA.

Musicant, A. D., and Butler, R. A. (1984). The influence of pinnae-based spectral cues on sound localization. *J. Acoust. Soc. Am.* **75,** 1195–1200.

Nelken, I. (2004). Processing of complex stimuli and natural scenes in the auditory cortex. *Curr. Opin. Neurobiol.* **14,** 474–480.

Nelken, I., Fishbach, A., Las, L., Ulanovsky, N., and Farkas, D. (2003). Primary auditory cortex of cats: Feature detection or something else? *Biol. Cybern.* **89,** 397–406.

Nelken, I., Kim, P. J., and Young, E. D. (1997). Linear and nonlinear spectral integration in type IV neurons of the dorsal cochlear nucleus. II. Predicting responses with the use of nonlinear models. *J. Neurophysiol.* **78,** 800–811.

Nelken, I., Prut, Y., Vaadia, E., and Abeles, M. (1994a). In search of the best stimulus: An optimization procedure for finding efficient stimuli in the cat auditory cortex. *Hear. Res.* **72,** 237–253.

Nelken, I., Prut, Y., Vaadia, E., and Abeles, M. (1994b). Population responses to multifrequency sounds in the cat auditory cortex: Four-tone complexes. *Hear. Res.* **72,** 223–236.

Nelken, I., Rotman, Y., and Bar Yosef, O. (1999). Responses of auditory-cortex neurons to structural features of natural sounds. *Nature* **397,** 154–157.

Neretti, N., Sanderson, M. I., Intrator, N., and Simmons, J. A. (2003). Time-frequency model for echo-delay resolution in wideband biosonar. *J. Acoust. Soc. Am.* **113,** 2137–2145.

Norena, A. J., and Eggermont, J. J. (2003). Neural correlates of an auditory afterimage in primary auditory cortex. *J. Assoc. Res. Otolaryngol.* **4,** 312–328.

O'Connor, K. N., Barruel, P., and Sutter, M. L. (2000). Global processing of spectrally complex sounds in macaques (Macaca mullata) and humans. *J. Compar. Physiol. Sens. Neural Behavior Physiol.* **186,** 903–912.

O'Connor, K. N., Petkov, C. I., and Sutter, M. L. (2004). Stimulus optimization for auditory cortical neurons. *Soc. Neurosci. Abstr.*

O'Connor, K. N., Petkor, C. I.., and Sutter, M. L. (2005). Adaptive stimulus optimization for auditory cortical neurons. *J. Neurophysiol.* in press.

O'Connor, K. N., and Sutter, M. L. (2000). Global spectral and location effects in auditory perceptual grouping. *J. Cog. Neurosci.* **12,** 1–13.

Ohlemiller, K. K., Kanwal, J. S., and Suga, N. (1996). Facilitative responses to species-specific calls in cortical FM-FM neurons of the mustached bat. *Neuroreport.* **7,** 1749–1755.

Ojima, H., and He, J. F. (1997). Cortical convergence originating from domains representing different frequencies in the cat AI. *Acta. Otolaryngol. Suppl.* **532,** 126–128.

Ojima, H., and Murakami, K. (2002). Intracellular characterization of suppressive responses in supragranular pyramidal neurons of cat primary auditory cortex *in vivo. Cereb. Cortex.* **12,** 1079–1091.

Ojima, H., and Takayanagi, M. (2004). Cortical convergence from different frequency domains in the cat primary auditory cortex. *Neuroscience* **126,** 203–212.

Olsen, J. F., and Suga, N. (1991). Combination-sensitive neurons in the medial geniculate body of the mustached bat: Encoding of target range information. *J. Neurophysiol.* **65,** 1275–1296.

Oonishi, S., and Katsuki, Y. (1965). Functional organization and integrative mechanism of the auditory cortex of the cat. *Jap. J. Physiol.* **15,** 342–365.

Oppenheim, A. (1997). "Signals and Systems." Prentice Hall, Upper Saddle River, NJ.

Patterson, R. D. (1974). Auditory filter shape. *J. Acoust. Soc. Am.* **55,** 802–809.

Patterson, R. D. (1976). Auditory filter shapes derived with noise stimuli. *J. Acoust. Soc. Am.* **51,** 640–654.

Petkov, C. I., O'Connor, K. N., and Sutter, M. L. (2003a). Illusory sound perception in macaque monkeys. *J. Neurosci.* **23,** 9155–9161.

Petkov, C. I., O'Connor, K. N., and Sutter, M. L. (2003b). Scene analysis in macaque primary auditory cortical (A1) neurons: Perceptual fill-in and masking. *Soc. Neurosci. Abstr.*. .

Phillips, D. P., and Irvine, D. R. (1979). Acoustic input to single neurons in pulvinar-posterior complex of cat thalamus. *J. Neurophysiol.* **42,** 123–136.

Phillips, D. P., and Irvine, D. R. (1981). Responses of single neurons in physiologically defined primary auditory cortex (AI) of the cat: Frequency tuning and responses to intensity. *J. Neurophysiol.* **45,** 48–58.

Phillips, D. P., and Orman, S. S. (1984). Responses of single neurons in posterior field of cat auditory cortex to tonal stimulation. *J. Neurophysiol.* **51,** 147–163.

Phillips, D. P., Orman, S. S., Musicant, A. D., and Wilson, G. F. (1985). Neurons in the cat's primary auditory cortex distinguished by their response to tones and wide-spectrum noise. *Hearing Res.* **18,** 73–86.

Phillips, D. P., Semple, M. N., Calford, M. B., and Kitzes, L. M. (1994). Level-dependent representation of stimulus frequency in cat primary auditory cortex. *Exp. Brain Res.* **102,** 210–226.

Phillips, D. P., Semple, M. N., and Kitzes, L. M. (1995). Factors shaping the tone level sensitivity of single neurons in posterior field of cat auditory cortex. *J. Neurophysiol.* **73,** 674–686.

Purves, D., Augustine, G. J., FItzpatrick, D., Katz, L. C., La Mantia, A., McNamara, J. O., and Williams, S. M. (2001). "Neuroscience." Sinauer Associates, Inc., Sunderland, MA.

Rauschecker, J. P. (1998). Parallel processing in the auditory cortex of primates. *Audiol. Neuro-Otol.* **3,** 86–103.

Rauschecker, J. P. (1997). Processing of complex sounds in the auditory cortex of cat, monkey, and man. *Acta. Otolaryngol. Suppl.* **532,** 34–38.

Rauschecker, J. P., and Tian, B. (2004). Processing of band-passed noise in the lateral auditory belt cortex of the rhesus monkey. *J. Neurophysiol.* **91,** 2578–2589.

Rauschecker, J. P., Tian, B., and Hauser, M. (1995). Processing of complex sounds in the macaque nonprimary auditory cortex. *Science* **268,** 111–114.

Rauschecker, J. P., Tian, B., Pons, T., and Mishkin, M. (1997). Serial and parallel processing in rhesus monkey auditory cortex. *J. Comp. Neurol.* **382,** 89–103.

Razak, K. A., Fuzessery, Z. M., and Lohuis, T. D. (1999). Single cortical neurons serve both echolocation and passive sound localization. *J. Neurophysiol.* **81,** 1438–1442.

Read, H. L., Winer, J. A., and Schreiner, C. E. (2001). Modular organization of intrinsic connections associated with spectral tuning in cat auditory cortex. *Proc. Natl. Acad. Sci. USA* **98,** 8042–8047.

Recanzone, G. H., Gaurd, D. C., and Pahn, M. L. (2000). Frequency and intensity response properties of single neurons in the auditory cortex of the behaving macaque monkey. *J. Neurophysiol.*

Recanzone, G. H., Schreiner, C. E., Sutter, M. L., Beitel, R. E., and Merzenich, M. M. (1999). Functional organization of spectral receptive fields in the primary auditory cortex of the owl monkey. *J. Comp. Neurol.* **415,** 460–481.

Rice, J. J., May, B. J., Spirou, G. A., and Young, E. D. (1992). Pinna-based spectral cues for sound localization in cat. *Hear. Res.* **58,** 132–152.

Riquimaroux, H., Gaioni, S. J., and Suga, N. (1992). Inactivation of the DSCF area of the auditory cortex with muscimol disrupts frequency discrimination in the mustached bat. *J. Neurophysiol.* **68,** 1613–1623.

Ritsma, R. J. (1962). Existence region of the tonal residue. Part I. *J. Acoustic. Soc. Am.* **34,** 1224–1229.

Romanski, L. M., Tian, B., Fritz, J., Mishkin, M., Goldman-Rakic, P. S., and Rauschecker, J. P. (1999). Dual streams of auditory afferents target multiple domains in the primate prefrontal cortex [see comments]. *Nat. Neurosci.* **2,** 1131–1136.

Rouiller, E. M., Simm, G. M., Villa, A. E., de Ribaupierre, Y., and de Ribaupierre, F. (1991). Auditory corticocortical interconnections in the cat: Evidence for parallel and hierarchical arrangement of the auditory cortical areas. *Exp. Brain Res.* **86,** 483–505.

Sadagopan, S., and Wang, X. (2004). Temporal response parameters affect accuracy and predictive power of linear STRF methods. *Soc. Neurosci. Abstr.*

Samson, F. K., Clarey, J. C., Barone, P., and Imig, T. J. (1993). Effects of ear plugging on single-unit azimuth sensitivity in cat primary auditory cortex. I. Evidence for monaural directional cues. *J. Neurophysiol.* **70,** 492–511.

Sanderson, M. I., and Simmons, J. A. (2002). Selectivity for echo spectral interference and delay in the auditory cortex of the big brown bat Eptesicus fuscus. *J. Neurophysiol.* **87,** 2823–2834.

Schreiner, C. E., Mendelson, J. R., and Sutter, M. L. (1992). Functional topography of cat primary auditory cortex: Representation of tone intensity. *Exp. Brain Research.* **92,** 105–122.

Schreiner, C. E., Read, H. L., and Sutter, M. L. (2000). Modular organization of frequency integration in primary auditory cortex. *Annu. Rev. Neurosci.* **23,** 501–529.

Schreiner, C. E., and Sutter, M. L. (1992). Topography of excitatory bandwidth in cat primary auditory cortex: Single-neuron versus multiple-neuron recordings. *J. Neurophysiol.* **68,** 1487–1502.

Sen, K., Theunissen, F. E., and Doupe, A. J. (2001). Feature analysis of natural sounds in the songbird auditory forebrain. *J. Neurophysiol.* **86,** 1445–1458.

Shamma, S. A., Fleshman, J. W., Wiser, P. R., and Versnel, H. (1993). Organization of response areas in ferret primary auditory cortex. *J. Neurophysiol.* **69,** 367–383.

Shamma, S. A., and Symmes, D. (1985). Patterns of inhibition in auditory cortical cells in awake squirrel monkeys. *Hearing Res.* **19,** 1–13.

Shofner, W. P., and Yost, W. A. (1997). Discrimination of rippled-spectrum noise from flat-spectrum noise by chinchillas: Evidence for a spectral dominance region. *Hearing Res.* **110,** 15–24.

Singh, N. C., and Theunissen, F. E. (2003). Modulation spectra of natural sounds and ethological theories of auditory processing. *J. Acoust. Soc. Am.* **114,** 3394–3411.

Staff, T. Co L. (1995). Who Is Fourier?: "A Mathematical Adventure Language Research Foundation." Tokyo, Japan.

Suga, N. (1965). Analysis of frequency-modulated sounds by auditory neurones of echo-locating bats. *J. Physiol.* **179,** 26–53.

Suga, N. (1988). Auditory combination sensitivity and speech processing: Complex-sound processing by combination sensitive neurons. *In* "Auditory Function: Neurobiological Bases of Hearing" (G. W. Edelman, W. E. Gall, and W. M. Cowan, Eds.), pp. 670–720. John Wiley and Sons, NY.

Suga, N. (1992). Philosophy and stimulus design for neuroethology of complex-sound processing. *Philos. Trans. R. Soc. Lond. B. Biol. Sci.* **336,** 423–428.

Suga, N. (1997). Tribute to Yasuji Katsuki's major findings: Sharpening of frequency tuning in the central auditory system. *Acta. Oto-Laryngologica. Suppl.* **532,** 9–12.

Suga, N., and Manabe, T. (1982). Neural basis of amplitude-spectrum representation in auditory cortex of the mustached bat. *J. Neurophysiol.* **47,** 225–255.

Suga, N., O'Neill, W. E., and Manabe, T. (1979). Harmonic-sensitive neurons in the auditory cortex of the mustache bat. *Science* **203,** 270–274.

Suga, N., and Tsuzuki, K. (1985). Inhibition and level-tolerant frequency tuning in the auditory cortex of the mustached bat. *J. Neurophysiol.* **53,** 1109–1145.

Sugita, Y. (1997). Neuronal correlates of auditory induction in the cat cortex. *Neuroreport.* **8,** 1155–1159.

Summers, V., and Leek, M. R. (1994). The internal representation of spectral contrast in hearing-impaired listeners. *J. Acoustic. Soc. Am.* **95,** 3518–3528.

Sutter, M. L. (2000). Shapes and level tolerances of frequency tuning curves in primary auditory cortex: Quantitative measures and population codes. *J. Neurophysiol.* **84,** 1012–1025.

Sutter, M. L., and Loftus, W. C. (2003). Excitatory and inhibitory intensity tuning in auditory cortex: Evidence for multiple inhibitory mechanisms. *J. Neurophysiol.* **90,** 2629–2647.

Sutter, M. L., and Schreiner, C. E. (1991a). Physiology and topography of neurons with multipeaked tuning curves in cat primary auditory cortex. *J. Neurophysiol.* **65,** 1207–1226.

Sutter, M. L., and Schreiner, C. E. (1991b). Physiology and topography of neurons with multi-peaked tuning curves in cat primary auditory cortex. *J. Neurophysiol.* **65,** 1207–1226.

Sutter, M. L., and Schreiner, C. E. (1995). Topography of intensity tuning in cat primary auditory cortex: Single-neuron versus multiple-neuron recordings. *J. Neurophysiol.* **73,** 190–204.

Sutter, M. L., Schreiner, C. E., McLean, M., O'Connor, K. N., and Loftus, W. C. (1999). Organization of inhibitory frequency receptive fields in cat primary auditory cortex. *J. Neurophysiol.* **82,** 2358–2371.

Theunissen, F. E., Sen, K., and Doupe, A. J. (2000). Spectral-temporal receptive fields of nonlinear auditory neurons obtained using natural sounds. *J. Neurosci.* **20,** 2315–2331.

Tian, B., and Rauschecker, J. P. (1998). Processing of frequency-modulated sounds in the cat's posterior auditory field. *J. Neurophysiol.* **79,** 2629–2642.

Tian, B., and Rauschecker, J. P. (2004). Processing of frequency-modulated sounds in the lateral auditory belt cortex of the rhesus monkey. *J. Neurophysiol.* **92,** 2993–3013.

Valentine, P. A., and Eggermont, J. J. (2004). Stimulus dependence of spectro-temporal receptive fields in cat primary auditory cortex. *Hearing Res.* **196,** 119–133.

Wehr, M., and Zador, A. M. (2003). Balanced inhibition underlies tuning and sharpens spike timing in auditory cortex. *Nature* **426,** 442–446.

Wenstrup, J. J. (1999). Frequency organization and responses to complex sounds in the medial geniculate body of the mustached bat. *J. Neurophysiol.* **82,** 2528–2544.

Xu, L., Furukawa, S., and Middlebrooks, J. C. (1999). Auditory cortical responses in the cat to sounds that produce spatial illusions. *Nature* **399,** 688–691.

Xu, L., Furukawa, S., and Middlebrooks, J. C. (2000). Cortical mechanisms for auditory spatial illusions. *Acta. Otolaryngol.* **120,** 263–266.

Yost, W. A., and Moore, M. J. (1987). Temporal changes in a complex spectral profile. *J. Acoustic. Soc. Am.* **81,** 1896–1905.

Yost, W. A., and Sheft, S. (1989). Across-critical-band processing of amplitude-modulated tones. *J. Acoustic. Soc. Am.* **85,** 848–857.

Young, E. D., and Brownell, W. E. (1976). Responses to tones and noise of single cells in dorsal cochlear nucleus of unanesthetized cats. *J. Neurophysiol.* **39,** 282–300.

Young, E. D., Rice, J. J., and Tong, S. C. (1996). Effects of pinna position on head-related transfer functions in the cat. *J. Acoust. Soc. Am.* **99,** 3064–3076.

Yu, J. J., and Young, E. D. (2000). Linear and nonlinear pathways of spectral information transmission in the cochlear nucleus. *Proc. Natl. Acad. Sci. USA* **97,** 11780–11786.

Zwicker, E., and Fastl, H. (1999). "Psychoacoustics Berlin," Springer.

Zwicker, E., Flottrop, G., and Stevens, S. S. (1957). Critical band width in loudness summation. *J. Acoust. Soc. Am.* **29,** 548–557.

PROCESSING OF DYNAMIC SPECTRAL PROPERTIES OF SOUNDS

Adrian Rees* and Manuel S. Malmierca[†]

*School of Neurology, Neurobiology, and Psychiatry, University of Newcastle
Newcastle upon Tyne NE2 4HH, United Kingdom
[†]Laboratory for the Neurobiology of Hearing, the Institute of Neuroscience of 'Castilla y Leon'
and Faculty of Medicine, University of Salamanca,
Salamanca 37007, Spain

Dynamic changes in frequency, whether as modulations of single tones or sets of harmonics, are important information-bearing elements of natural sounds, and distinguishing attributes in the vocal repertoires of many species. In this chapter we review the ways in which neurons at different levels of the auditory pathway respond to frequency modulations (FMs). We emphasize how responses to FM are elaborated at successive levels of the auditory pathway, and attempt to identify trends in the ways that dynamic changes in frequency are represented. Specificity for parameters like the direction of frequency change is apparent at all levels above the auditory nerve (AN). However, at more central locations there is evidence for great specificity of response, topographic mapping of FM parameters, and differences in response properties between cortical subdivisions. Nevertheless, many questions remain; there is considerable variation between datasets, and it is striking that many neurons respond best to rates of frequency change that are unlikely to occur in natural sounds.

INTERNATIONAL REVIEW OF
NEUROBIOLOGY, VOL. 70
DOI: 10.1016/S0074-7742(05)70009-X

0074-7742/05 $35.00

I. Introduction

Dynamic changes in spectral content, where the frequency of one or more components of the sound varies with time, are important for sound perception. In many biologically significant sounds, these components distinguish one token from another. For example, in speech, upward or downward changes in frequency occurring over a few tens of milliseconds characterize the formants of consonants and diphthongs. On a longer time scale, changes in the fundamental frequency of the voice carry suprasegmental or prosodic information in many languages. These changes in stress and intonation signal not only the speaker's mood, but also important semantic content. For example, the change in pitch at the end of a sentence can signal a question, statement, or qualification, depending on the direction or pattern of the change. In contrast, pitch changes in tonal languages, like Mandarin Chinese, can distinguish totally different word meanings. Dynamic changes in frequency are not restricted to human vocalizations. Con-specific sounds in other species contain a wide range of glides or periodic changes in frequency, e.g., monkeys (Ghazanfar and Hauser, 2001; Winter *et al.*, 1966), cats (Brown *et al.*, 1978), guinea pigs (Suta *et al.*, 2003), rats (Kaltwasser, 1990), bats (Suga, 1988), and birds (Greenwald, 1968).

A second role often postulated for FMs is in auditory streaming or grouping. One of the most intriguing aspects of auditory processing, particularly when compared with machine-based sound recognition systems, is the capacity to separate and identify concurrent sound sources (see the chapter by Donal G. Sinex), particularly when those sources contain frequency components in common. The recovery of independent sources depends on a process that identifies components belonging to the same source and enables them to be perceived as a single stream (Bregman, 1990). Several cues contribute to this process, and it is likely that their exploitation in auditory processing depends on their saliency in a particular context. One possible cue is the coherence of FM across frequency components. If a group of harmonics is played together, one perceives the sound as a whole and the individual components are difficult to resolve. However, if selected harmonics are modulated in frequency they can be clearly heard to stand out from the rest (Bregman, 1990; McAdams, 1989). Musicians have long exploited this phenomenon by using vibrato to give a vocal or solo instrumental line greater prominence against the accompaniment.

Formal psychophysical studies suggest that coherent FM appears to be relatively unimportant for grouping vowels or even inharmonic sounds (Carlyon, 1991; Culling and Summerfield, 1995; McAdams, 1989). Nevertheless, FM may be important when listeners are required to identify whole sentences under unfavorable conditions, as in the presence of competing sounds. Zeng *et al.* (2005) extracted amplitude modulation (AM) and FM from different frequency bands of sentences and then created stimuli in which these components modulated pure tones or noise bursts. In quiet, AM alone was sufficient for good recognition, but in the presence of a competing sentence subjects performed significantly better when both AM and FM were present. This suggests that when the task is sufficiently demanding, FM does contribute to the separation of auditory streams.

The importance of FM as a distinguishing feature in animal sounds and speech suggests that the analysis of spectral modulations is a prerequisite for auditory perception. In this chapter we review experiments that have reported how single neurons at different levels of the auditory pathway respond to FM, and discuss what is known about the extraction and representation of this information. Many of the stimuli used in these studies are abstractions of real vocalizations, being either periodic FMs with a single carrier and a simple modulation waveform, such as a sinusoid, or a frequency sweep. Such stimuli have the advantage of being more tractable because they allow the experimenter greater control over their parameters. However, it is important to remember they are considerably simpler than most of the natural sounds they seek to emulate.

This review focuses on data from non-specialized mammals. This may seem surprising given that FM is, an important component in the echolocation calls of many bat species and the subject of intensive investigation in these animals. However, the success of these investigations has resulted in a body of literature too voluminous to do justice to here, and, in any case, it has received excellent treatment elsewhere (e.g., Suga, 1988; O'Neill, 1995; Covey and Casseday, 1999). Our comments on processing in bats, therefore, are mainly restricted to points of comparison, or to aspects of bat hearing not specifically concerned with echo-location.

II. Auditory Nerve Fibers: Responses to FM Predicted by Pure-Tone Responses

In general, AN fibers respond to frequency-modulated sounds in a manner that is predictable from their responses to pure tones (Britt and Starr, 1976; Sinex and Geisler, 1981). In a study of neurons in the cochlear nucleus (CN) of cat, Britt and Starr (1976) described the responses of a small number of AN fibers to triangular frequency modulations. The period histograms of such responses show a burst of spikes for each frequency transition that crosses the fiber's response area. All but a few fibers had symmetrical responses to the upward and downward sweeps of the FM at the four modulation rates tested. The remainder showed some degree of asymmetry at the highest sweep rate, which the authors ascribed to adaptation. The interpretation of the data was complicated by the manner in which they were collected. Responses were measured by the total spike count to the stimulus, but the duration of the stimulus was constant for all modulation rates and thus the number of sweeps presented increased with modulation rate. Furthermore, with a triangular waveform it is not possible to separate the rate of change of frequency from the repetition rate of the stimulus.

Sinex and Geisler (1981), also recording in cat, avoided these confounds by using trapezoidal modulation. Fibers fired impulses while the sweep was within the frequency response area (FRA) (Fig. 1). Histograms of firing over the course of the sweep showed that the occurrence of spikes became more restricted in time as the rate of frequency change increased. In general, the average firing rate increased with sweep rate up to the maximum tested (usually 16 or 160 kHz/s, depending on the fiber's best frequency [BF]). Except at the highest sweep rates,

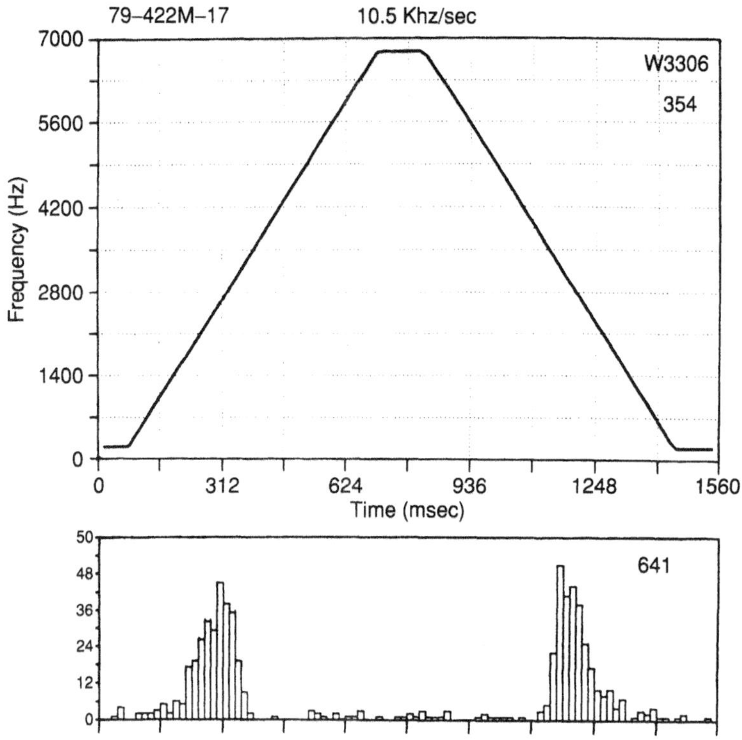

FIG. 1. Instantaneous frequency plot of a trapezoidally modulated sweep tone (upper panel), and the period histogram (lower panel) of the discharges it elicited in a single cochlear nerve fiber in cat. Twenty-one repetitions of the 10 kHz/s sweep were presented at 80 dB SPL. The responses to upward and downward sweeps were approximately mirror images of each other, consistent with the response depending on the instantaneous frequency of the sweep and the fiber's frequency selectivity. The asymmetry was attributable to the interaction between the fiber's frequency response and firing rate adaptation. (Adapted from Sinex and Geisler, 1981.)

the average rates for the upward and downward sweeps were the same. In cases where small asymmetries occurred, the descending sweep usually showed the stronger response. Firing rates also increased with the sound level at which the sweeps were presented, consistent with the broadening of the frequency response of the fibers at higher sound levels.

An important strength of this study was the authors' careful comparison of the fibers' response areas derived using frequency sweeps with those to constant tones. This analysis suggested that the increased average rate with sweep rate was explained by adaptation since it occurred most noticeably at rates where the time for the sweep to cross the response area was less than the time for adaptation. Similarly, adaptation also explained the small asymmetries in the responses to upward and downward sweeps. At high sweep rates the larger response occurred for the direction (usually downward) with the smallest difference between the edge of the response area and the most effective frequency for driving the fiber. Together, adaptation and the fiber's FRA also predicted the greater separation of the sweep-tone response areas as sound level increased (Fig. 2).

In summary, the responses of cochlear nerve fibers are well predicted by the fibers' responses to static tones, and neural adaptation seems adequate to account for most of the features in the data. Importantly, the cochlear nerve appears not to display any substantive difference for the responses to frequency sweeps in different directions.

III. Cochlear Nuclei: The Appearance of Directional Asymmetry for FM

A. Responses to Frequency Sweeps

As one might expect given the diversity of response properties among the neurons of the CN (for reviews, see Rhode and Greenberg, 1992; Smith and Spirou, 2002) there is also diversity in the responses of these neurons to FMs. Unfortunately, most studies that used frequency sweeps to study the responses of neurons in the CN predate the detailed physiological and morphological classifications of these neurons. In some cases, even the location of the units within the CN complex was not known.

Despite these difficulties of interpretation, there is a striking difference between the responses of neurons in the CN and those of AN fibers. In contrast to the responses of nerve fibers, CN responses to FM are less easily predicted from their responses to static tones. The emergence of selectivity for the direction of frequency change and more complex response patterns, particularly at high rates of frequency change, are notable differences (Britt and Starr, 1976; Erulkar *et al.*, 1968). Erulkar *et al.* (1968) described three categories of response: mirror

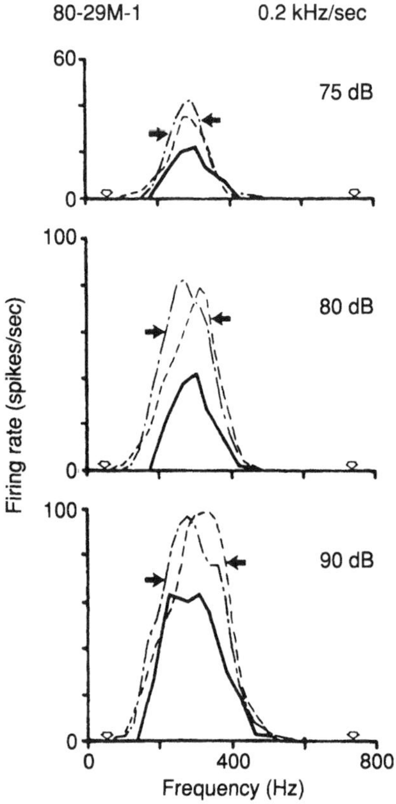

FIG. 2. Isointensity frequency responses of a cochlear nerve fiber in cat for pure tones (3-s duration, solid lines) and FM sweeps (dashed lines), measured at three different sound pressure levels (parameter). The arrow next to each sweep plot indicates the direction of the sweep. The frequency-sweep response functions were higher than the pure tone response because the fiber was less adapted during the sweep. Also as a result of adaptation, the sweep functions diverged from the pure tone function as the sound level increased (see text). Fiber CF, 340 Hz; sweep speed, 0.2 kHz/s, and FM range 50 to 700 Hz (open arrows). (Adapted from Sinex and Geisler, 1981.)

symmetry, translation symmetry, and asymmetric. A total of 70% of responses in their sample were asymmetric. The firing patterns of these units departed markedly from the single burst of spikes elicited by each sweep that is typical of AN fibers and other CN units. It is impressive that intracellular recording in these pioneering studies identified inhibition as a component in the generation of these complex firing patterns (Britt and Starr, 1976; Erulkar *et al.*, 1968).

Several studies in cat and rat emphasize the importance of sweep rate as a determinant of the response to FM sweeps in the CN (Britt and Starr, 1976;

Erulkar *et al.*, 1968; Møller, 1969, 1971, 1974b). Most agree that at low sweep-rates the responses of the neurons are more readily predicted from the static tone response area than when rapid frequency changes are employed (Møller, 1969). In rat, Møller (1974b) reported that rapid changes in frequency (up to 20 MHz/s) restricted the response to progressively narrower frequency range close to the neurons characteristic frequency (CF), but with little change in the average firing rate, consequently the peak height of the response increased. Responses tended to become more asymmetrical as sweep rate increased and in general they were larger for downward sweeps (Møller, 1971, 1974b). Tuned sweep rate response functions, with preferences for rates between 400 kHz/s and 10 MHz/s, were apparent in many cases, particularly at higher stimulus levels (Fig. 3). Møller presented evidence indicating that these effects could not be explained by adaptation alone, but depended on the interaction of inhibition and excitation. Another factor will also influence the response at such high rates. As sweep rate increases, the stimulus spends so little time within the excitatory response area that refractoriness becomes an increasingly important determinant of the number of

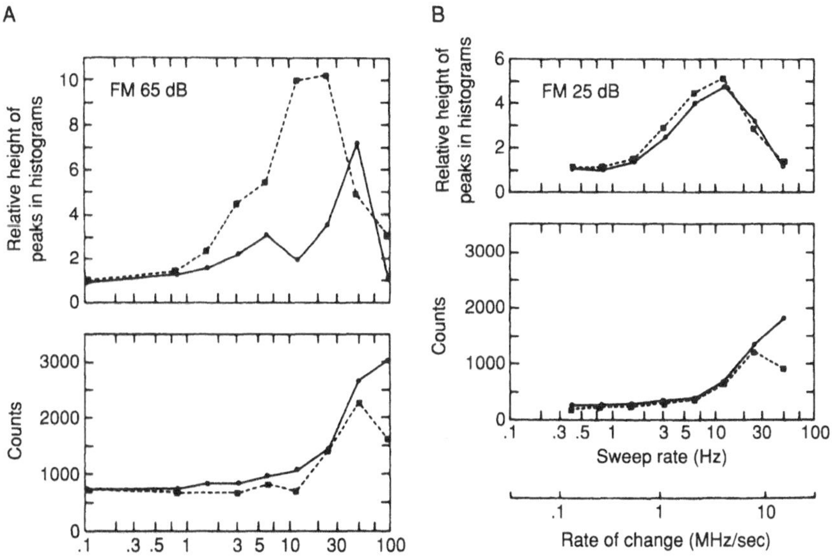

FIG. 3. Responses of a neuron in the CN of rat to frequency sweeps at two different stimulus levels: A = 65 dB SPL and B = 25dB SPL. The upper panel in each plot shows the height of the period histograms and the lower panel the total spike counts for upward (solid line) and downward (dashed line) sweeps. On the abscissa, "sweep rate" represents the repetition rate of the sweep, and "rate of change" the equivalent rate of change of frequency. The histogram-height measure shows bandpass tuning to the rate of change of frequency, and at the higher sound level the response to the downward sweep was bigger than that to the upward sweep. (Adapted from Møller, 1974b.)

impulses the neuron fires. The influence of such effects may not be obviously when the number of sweeps presented increases as the rate of change of frequency is increased (see Sinex and Geisler, 1981, for discussion).

Cochlear function also has a bearing on the responses to frequency sweeps. For high rates of frequency change, the temporal pattern of activation generated by an array of AN fibers with different CFs converging on a neuron in the CN will depend on sweep direction. Owing to the accumulating delay of the traveling wave as it moves from base to apex in the cochlea, a sweep that rises in frequency (initiating a peak displacement that shifts from apex to base on the basilar membrane (BM) should produce a more synchronized input than an equivalent downward frequency sweep, since the latter heightens the dispersive effect of cochlear travel time (Shore *et al.*, 1987). Response latency measurements of neurons in the ventral CN to such sweeps are broadly in line with these predictions (Shore *et al.*, 1987).

Two studies have correlated the responses of CN neurons to FM with their pure-tone classification (Britt and Starr, 1976; Rhode and Smith, 1986). Britt and Starr (1976) found that primary-like units resemble AN fibers and show a symmetrical response to frequency sweeps, whereas build-up, onset, and pauser units showed a greater propensity for asymmetrical or unidirectional responses. None of the onset and pauser units showed purely symmetrical responses. Pauser units were often characterized by inhibitory responses that depended on the sweep rate. In a much larger sample of neurons in the anteroventral cochlear nuclei (AVCN) and posteroventral cochlear nuclei (PVCN), where subcategories of unit type were more precisely defined, onset inhibitory (O_I) units showed the strongest preference for sweep direction. Most units preferred a low to high frequency change. Onset locker (O_L) units with low spontaneous activity (<2/s) showed a lesser degree of directional preference, whereas their counterparts with high spontaneous activities showed no preference for sweep direction (Rhode and Smith, 1986). Evidence from intracellular labeling suggests that O_I and O_L responses derive from octopus cells in the PVCN (Rhode *et al.*, 1983; Rouiller and Ryugo, 1984). These neurons receive inputs from a wide range of frequencies because their dendrites are orientated across the incoming array of AN fibers. Models incorporating such patterns of input suggested how the systematic activation in response to a frequency sweep (e.g., Fernald, 1971; Rall, 1962, 1964) could explain directional selectivity. However, such an arrangement seems not to be the only explanation since not all of these onset units are likely to be octopus cells (Palmer *et al.*, 1996). In the dorsal CN only the onset graded (O_G) units showed directional selectivity (Rhode and Greenberg, 1992).

B. Responses to Periodic FM

Responses to triangular periodic modulations were previously discussed in the context of frequency sweeps, but fewer studies have examined the responses of CN neurons with sinusoidal frequency modulation. Møller, who pioneered the first delineations of modulation transfer functions for AM and FM in the CN, found that neurons responded to sinusoidal frequency modulations with a response synchronized to the modulation envelope (Møller, 1972a,b). Best modulation frequencies were typically between 100 to 500 Hz, but synchronization persists up to ~800 Hz (Møller, 1972a). Interestingly, the response patterns and modulation transfer functions for FM were similar to those for AM at least for relatively small frequency excursions (Møller, 1972a, 1974a).

C. Overview of CN

Our understanding of how neurons in the CN respond to frequency sweeps, with respect both to response patterns associated with particular neuron types, and the mechanisms that underlie these differences, is far from clear. A problem in interpreting the data is the difficulty of comparing the stimuli used in different studies, with their marked differences in sweep waveform, sweep rate, and repetition rate. Similarly, the studies use a wide range of modulation depths from 1/8 octave around CF (Erulkar *et al.*, 1968) to several octaves (Møller, 1974b). Thus, in some cases the sweep was completely contained within the neuron's frequency response area, while for others it would have begun and ended well outside its boundaries.

In comparison with these studies in non-specialized animals, it is perhaps surprising that neurons selective for sweep direction are not found in the CN of echo locating bats whose calls contain FM sweeps (Suga, 1965). In these animals sweep direction appears to be a feature that is extracted at more central levels of the pathway (Suga, 1969, 1988).

Our understanding of how neurons in the CN respond to frequency sweeps, and the mechanisms that underlie their responses is still incomplete. With the more extensive information about the functional organization of the CN now available, these questions are ripe for revisiting.

IV. Higher Brainstem and Midbrain Processing of FM

Little is known about how neurons in nuclei between the CN and inferior colliculus (IC) respond to FM, and the only available data are from bats (Huffman *et al.*, 1998). Neurons in the nuclei of the lateral lemniscus in the big brown bat show selectivity for downward going FM sweeps similar to those contained in the biosonar calls of this species. Some neurons grouped in (nucleus of the lateral lemniscus) the ventral NLL responded selectively to downward frequency sweeps when the stimulus crossed the high frequency boundary of the excitatory response area. Neurons in the dorsal division of the NLL responded to smaller depths of FM when the frequency excursion was restricted entirely within excitatory response area of the neuron (Huffman *et al.*, 1998).

A. Responses to FM Sweeps in the IC

Frequency sweeps have proved popular for studying the IC in non-specialized mammals (Clopton and Winfield, 1974; Felsheim and Ostwald, 1996; Hage and Ehret, 2003; Poon *et al.*, 1991; Vartanian, 1974) and in bats (e.g., Fuzessery, 1994; Gordon and O'Neill, 1998; Pollak *et al.*, 1978; Suga, 1969). In bats, special mechanisms emerge at this level for the detection of FM. Neurons in the ICs of these species show a high degree of selectivity to downward sweeps that are characteristic of echolocation calls, and some fail to respond to pure tones, noise bursts, or clicks (e.g., Pollak *et al.*, 1978; Suga, 1969).

Evidence that this selectivity is a specialization for echolocation is apparent from experiments on the pallid bat (*Antrozous p. pallidus*), a species that hunts by combining echolocation of high frequency sounds with passive listening to low frequencies. In the ventromedial IC, which represents high frequencies, approximately one-half of the neurons were found to be selective for downward going FM sweeps and about one-third responded exclusively to these stimuli. In contrast, in the dorsolateral, low-frequency region very few neurons with directional selectivity for FM were found (Fuzessery, 1994).

The extent to which FM-specific neurons occur in the IC of non-specialized species is controversial.

In rat, 25% to 35% of units across two studies were defined as "FM specialized" because they responded to linear FM sweeps but not to pure tones (but see the following text). A further ~45% responded to both classes of stimuli (mixed type) while the remainder were defined as FM insensitive (Poon *et al.*, 1991, 1992). Most of the FM specialists had bandpass transfer functions for FM sweep rate, and on average preferred higher rates of frequency change, with a mean of 50 octaves/s compared with 20 octaves/s for the mixed group. The FM specialized cells also responded better to broader frequency sweeps, and the majority (>80%) were described as unidirectional. While these data indicate the existence of neurons with FM specificity in the IC of the rat, it is important to note that in this context, not responding to a pure tone, means that the neurons failed to respond to *continuous* pure tones (for discussion see Poon *et al.*, 1992). Since it cannot be stated that the "FM specialized" neurons responded to FM but not to pure tone *bursts* (the more usual meaning of the expression, "pure-tone"), the FM specificity described in rat may not be as complete as in bats.

In marked contrast, another study in rat found no FM selective units (pure tone responses were defined with tone bursts in this study) and only weak directional selectivity (Felsheim and Ostwald, 1996). On the other hand, selectivity for sweep speed often showed bandpass tuning as in Poon *et al.* (1992), with most units responding best to sweeps between 10 to 400 octaves/s. The use of exponential rather than linear sweeps can only partially explain the differences between these studies. Felsheim and Ostwald (1996) argued that apparent FM directional selectivity could be observed if the positions of the upward and downward FM sweeps were not positioned symmetrically with respect to the unit's frequency response area.

Two studies have suggested that certain aspects of FM responses correlate with the cytoarchitecture and structure of the IC (Hage and Ehret, 2003; Poon *et al.*, 1992). A particularly informative approach adopted by Poon *et al.* (1992) was to label recorded units by intracellular injection of horseradish peroxidase. Morphological analyses showed that "FM specialized" cells were multipolar with larger, more extensively branched, dendritic fields and higher densities of dendritic spines than other neurons (Fig. 4). Neurons responding to FM were found in all three main subdivisions of the IC, but a higher proportion of FM specialized cells were located in the external nucleus. The smallest proportion of FM specialized neurons was in the dorsal cortex. It is not clear, therefore, whether FM response properties depended solely on cellular morphology or whether location within the IC was also a determinant. However, across all cell types responding to FM, the best FM range was highly correlated with the degree to which cell's dendrites extended perpendicularly with respect to the orientation of the frequency band laminae (Fig. 5) (Poon *et al.*, 1992).

Evidence for a topographic organization of FM responses within frequency band laminae of the IC has been reported in mouse (Hage and Ehret, 2003). The proportion of cells that responded to high sweep speeds increased with distance from the center of the lamina, as did the firing rate to high, and to a lesser extent medium, sweep speeds. This study also reported a spatial segregation of directional selectivity. About one-half the sample showed directional selectively for at least one of the three sweep speeds tested (300 kHz/s, 600 kHz/s, and 3 MHz/s). A total of 60% of the directionally selective neurons preferred downward sweeps and these were concentrated in the center of the central nucleus, with neurons preferring upward sweeps distributed on either side, or possibly in a concentric arrangement around them (Hage and Ehret, 2003).

These distribution patterns correlated with those for the units' frequency response area type, sharpness of tuning, and temporal response pattern. The authors proposed that the common feature underlying all these distributions was a gradient of inhibition from the edge to the center of the lamina (Ehret *et al.*, 2003; Hage and Ehret, 2003).

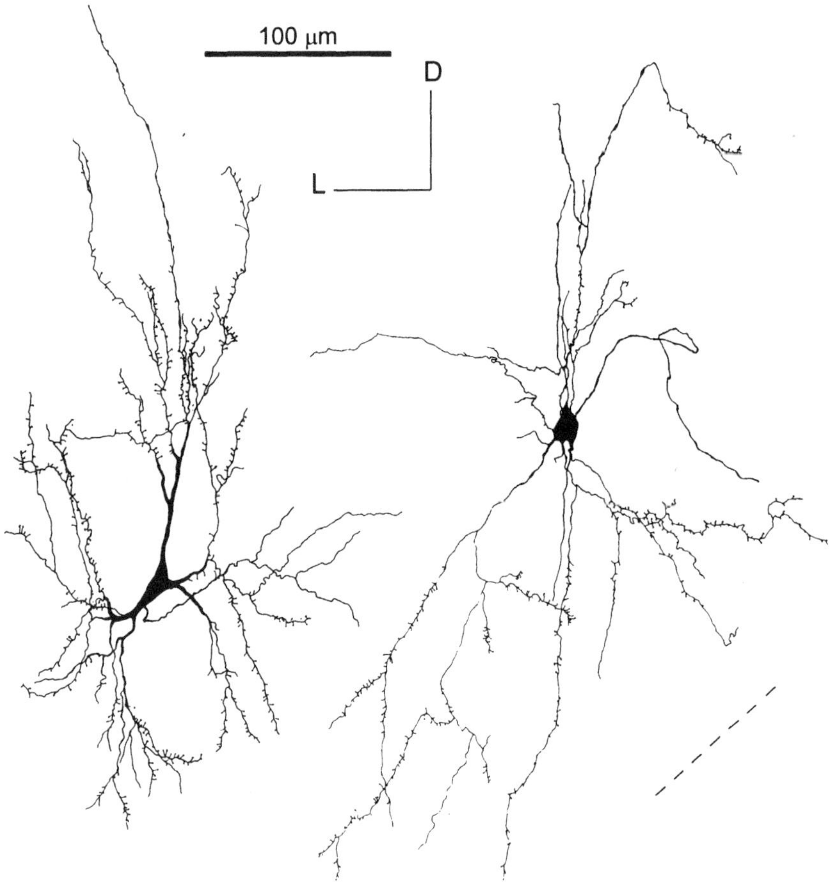

FIG. 4. Camera lucida drawings of two "FM-specialized" neurons (see text for definition) in the IC of rat. These neurons had more extensive dendritic arborizations, a higher density of dendritic spines, and dendrites that extended further along the axis perpendicular to the orientation of the frequency-band laminae (dashed line) than other neuronal types. D = dorsal; L = lateral. (Adapted from Poon *et al.*, © 1992 with kind permission of Springer Science and Business Media.)

B. Does FM Selectivity Depend on Inhibition?

Spectral side-band inhibition has long been hypothesized to play a role in determining directional and rate specificity to FM at several levels of the auditory pathway, including the IC. The hypothesis predicts that an FM-selective unit will fail to respond if the frequency trajectory of the stimulus crosses an inhibitory region prior to entering the excitatory region of its frequency response area. Several variants on this theme have been proposed to accommodate the complexities that

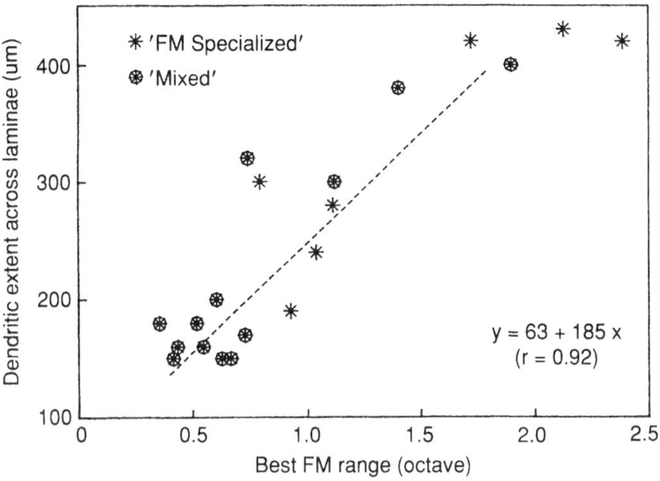

FIG. 5. Correlation, for neurons in the IC of rat, between the FM range that elicited a maximal response, and the extent to which their dendrites spanned the axis perpendicular to the orientation of the frequency-band laminae. See text for definition of "FM-specialized" and "mixed" neuronal types. (Adapted from Poon *et al.*, © 1992 with kind permission of Springer Science and Business Media.)

always feature in biology, in particular the finding that such FM specialized neurons may not respond to pure tones (for review see Fuzessery and Hall, 1996).

The postulated role of inhibition as a component in generating directional selectivity in the IC is open to test by iontophoretic application of antagonists to the putative inhibitory neurotransmitters. Thus far, the hypothesis has only been tested in bat. In the pallid bat, blockade of GABAergic inhibition reduced directional selectivity for FM sweeps in all cells, but in some cases without reducing their selectivity for FM over tones or noise (Fuzessery and Hall, 1996). Moreover, the directional specificity observed in some neurons was not consistent with the organization of the inhibitory sidebands in the neuron's frequency response area. These findings suggest that while inhibition is involved in generating directional specificity it may not operate via frequency-dependent inhibition. Alternatively, directional selectivity in the IC could be explained without the need to invoke inhibitory mechanisms at all if it reflects the results of inhibitory processing in more peripheral nuclei of the pathway.

C. Responses to Sinusoidal Frequency Modulations

Neurons in the IC respond to sinusoidal FM with an approximately sinusoidal modulation of their neural discharge (cat: Nelson *et al.*, 1966; bat: Schuller, 1979; rat: Rees and Møller, 1983; Casseday *et al.*, 1997). Some show a high

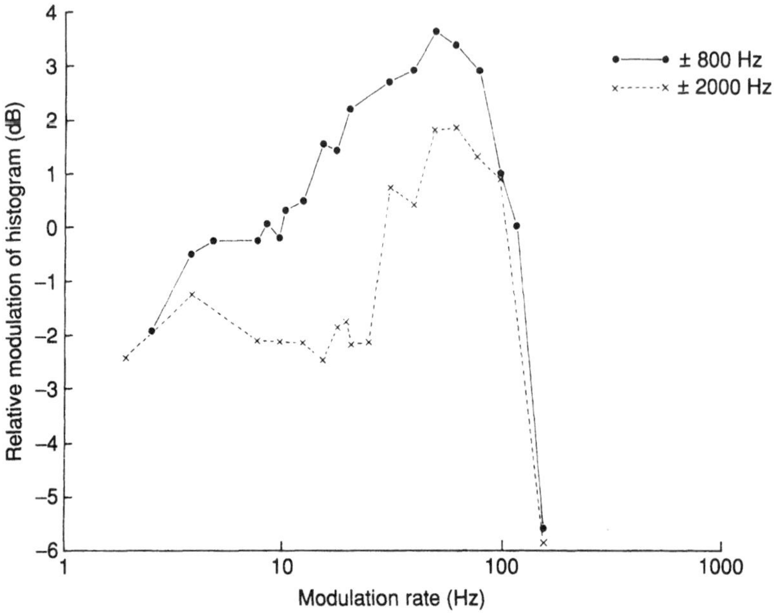

FIG. 6. Modulation transfer function for a single neuron in the IC of rat to sinusoidal FM at two different modulation depths. Carrier frequency, 4.5 kHz, presented at 20 dB above the unit's pure-tone threshold. (Adapted from Rees and Møller, 1983.)

degree of sensitivity to FM, and a substantial modulation of the response was observed with modulation depths of <10% of the carrier frequency (Rees and Møller, 1983). As with AM, neurons in rat IC have lower best modulation frequencies (≤100 Hz) for sinusoidal FM than in the CN (Fig. 6) (Møller, 1972a, 1974a). However, as reported for the CN, the range of best modulation frequencies for FM and AM was similar in IC (Rees and Møller, 1983). A correspondence between FM and AM responses has also been reported for neurons in IC of rats using pseudorandom AM or FM exponential envelopes (Poon and Chiu, 2000).

V. Responses to FM in Thalamus and Cortex

Responses to FM in the thalamus of non-specialized mammals have received little attention. Creutzfeldt *et al.* (1980) reported that, compared with neurons in the cortex, fewer MGB neurons showed directional selectivity in their responses to sinusoidal FM. Studies using frequency sweeps to stimulate units in the ventral division of the MGB found that about one-half showed a preference for sweep direction, and of those 60% responded best to upward sweeps (Lui and Mendelson, 2003). Selectivity for sweep speed was similar to that observed in some cortical fields (30 to 800 kHz/s) and directional selectivity was often dependent on sweep speed.

The most complete evidence for functional specialization in the encoding of FM in the auditory cortex has been obtained from echo-locating bats. In these animals distinct cortical regions exist for encoding components of the emitted sound and the returning echo. One area, the FM-FM, responds to particular combinations of the FM components in the pulse and echo when they are separated by an optimal delay. Such neurons provide information about the range of the target. For further information about these mechanisms, the reader is directed to excellent reviews published elsewhere (O'Neill, 1995; Suga, 1988).

Studies of neuronal responses to FM in the auditory cortex of non-specialized mammals can be grouped into those that have used relatively small frequency excursions and rates of frequency change (Phillips *et al.*, 1985; Whitfield and Evans, 1965), and the majority in which much broader and faster frequency sweeps have been used (Heil *et al.*, 1992a; Mendelson *et al.*, 1993; Nelken and Versnel, 2000; Shamma *et al.*, 1993; Tian and Rauschecker, 1994). Another important aspect of cortical processing is the extent to which responses to FM differ between different subdivisions of the auditory cortex.

A. Responses to FM in Primary Auditory Cortex (AI)

Whitfield and Evans (1965) were the first to report evidence of neurons showing FM specificity in the auditory cortex of unanesthetized cat. They presented sinusoidal and sweep tones in which frequency varied by around 10% of the carrier, and observed responses to FM that were not predictable from the responses to static pure tones. These included neurons that responded to FM positioned outside the pure tone response area and others that responded to FM in the absence of any response to a steady pure tone. Directional specificity was also a conspicuous property of such units and in some cases this depended on the position of the FM sweep relative to the response area. In contrast, other units responded in a manner that was generally predictable from their pure tone response areas, particularly when low (<5 Hz) modulation frequencies were employed (Whitfield and Evans, 1965). As will be discussed in later text, responses to FM but not pure tones have not been as evident in more recent studies and might reflect the effects of anesthetic.

Phillips *et al.* (1985) used 2 kHz frequency sweeps (around 10% to 20% depth depending on the carrier) to study neurons in cat AI under barbiturate anesthesia. They concluded that the directional specificity of a neuron in response to ramps placed at different frequencies relative to the response area depended on the relationship between the direction of frequency change, the neuron's best frequency, and the slope of the function relating frequency to spike-count. Units responded best when the ramp was centered on the steepest slope of the function and the direction of frequency change was orientated towards BF. A sweep moving towards BF generated a response, while a sweep centered on the same frequency, but changing in the opposite direction, produced no response. Phillips *et al.* (1985) argued that because latency of firing decreases when frequency changes in the direction of BF, sweeps directed towards BF generated temporal summation that excited the neuron. However, short-term adaptation could also be a contributing factor. Interestingly, while these units responded to narrow sweeps traveling in either direction, some units responded with a pronounced unidirectional selectivity when stimulated with a much broader sweep.

Similar directionally selective responses to broad ramps with high rates of frequency change have been confirmed by other studies in cat in both single unit (Mendelson and Cynader, 1985; Watanabe, 1972) and multi-unit studies (Eggermont, 1994; Heil *et al.*, 1992a). The number of directionally selective units varies between studies (Mendelson and Cynader, 1985, 65%; Heil *et al.*, 1992a, 30%) despite their use of identical criteria for determining directional selectivity. This difference may be accounted for by the comparison of single- and multi-unit data. However, both studies concur that about two-thirds of directionally selective units preferred downward directed sweeps. In contrast, data from rat AI showed a greater propensity for upward sweeps, indicating that species differences exist (Gaese and Ostwald, 1995; Ricketts *et al.*, 1998). In these studies 65% to 75% of directionally selective units preferred upward sweeps; in one case the data were obtained using sinusoidal frequency modulation (Gaese and Ostwald, 1995), but the other used linear FM sweeps with rates similar to those used in cat (Ricketts *et al.*, 1998). Gaese and Ostwald (1995) also pointed out a similarity between responses to sinusoidal FM and AM in the cortex, as has been noted in the midbrain (see the previous text).

As discussed earlier, the relationship between the excitatory and inhibitory regions of the neuron's response area has been evoked to account for directional selectivity. Contrary to the expectations of this hypothesis, Heil *et al.* (1992a) failed to find any differences in directional selectivity when responses to narrow or broad sweeps were compared, or for slow versus fast rates of change of frequency. However, they did not routinely map inhibitory response areas.

Many responses showed sensitivity to the rate of frequency change as well as its direction. Functions describing sensitivity to rate had a range of different shapes, including bandpass and low pass as well as more complex functions

(Fig. 7). However, the most common (approximately 60%) were the predominantly ascending type (high pass) followed by bandpass (~20%, Heil *et al.*, 1992a). In general, neurons in AI responded best to high rates of frequency change (800 kHz/s, Mendelson and Cynader, 1985; ~1000 kHz/s, Heil *et al.*, 1992a), but both studies found a wide distribution, with some units, particularly those with bandpass rate of change of frequency (RCF) functions, giving best responses at rates of change as low as 4 kHz/s (Heil *et al.*, 1992a).

Heil *et al.* (1992a), like Phillips *et al.* (1985), found that firing occurred when the frequency sweep was modulated towards CF rather than away from it. Detailed analysis of the relationship between the frequency sweep and the cluster's frequency response area revealed that a response was initiated when the instantaneous frequency of the sweep reached an "effective instantaneous frequency" (F_i). The effective F_i always occurred before the sweep reached the BF and, therefore, was lower or higher than BF for ascending and descending sweeps respectively. Effective F_i was dependent on the slope of the response area and shifted progressively further away from BF as the level of the stimulus was raised. This dependency in cortex of the effective F_i on the response area is similar to that observed in AN fibers (Sinex and Geisler, 1981).

Since the topographic representation of the stimulus domain is a general feature of cortical organization, a key question is whether FM properties are topographically represented in auditory cortex in mammals. Heil *et al.* (1992b) reported that sweep speed parameters, e.g., best RCF and type of RCF function, showed a statistically significant tendency to cluster at sites in a line of penetrations running along an isofrequency strip. But there was no obvious consistency in the ordering of these positions with respect to the cortical surface. Furthermore, there was no evidence of order with respect to directional preference (Fig. 8).

Mendelson *et al.* (1993) used exponential sweeps to categorize multi-unit clusters on the basis of their response to three sweeps speeds: slow, medium, and fast (10, 33, and 110 octaves/s, respectively). This analysis revealed a trend for sites distributed dorsal to ventral across AI to prefer fast-medium-slow and medium speeds, respectively. At the most ventral positions another region preferring fast sweeps was evident, but it was not certain whether these recordings were in AI or AII. Directional preference was also topographically distributed, with downward sweeps giving stronger responses in the center and upward sweeps preferred at the dorsal and ventral borders of AI (Fig. 9).

Methodological differences between the studies of Heil *et al.* (1992b) and Mendelson *et al.* (1993), might account for these disparate findings namely, sampling density, statistical methods, linear versus exponential sweeps, the number of sweeps speeds tested, and the value of the directional selectivity index adopted as the criterion for selectivity (0.33 vs. 0.15, respectively).

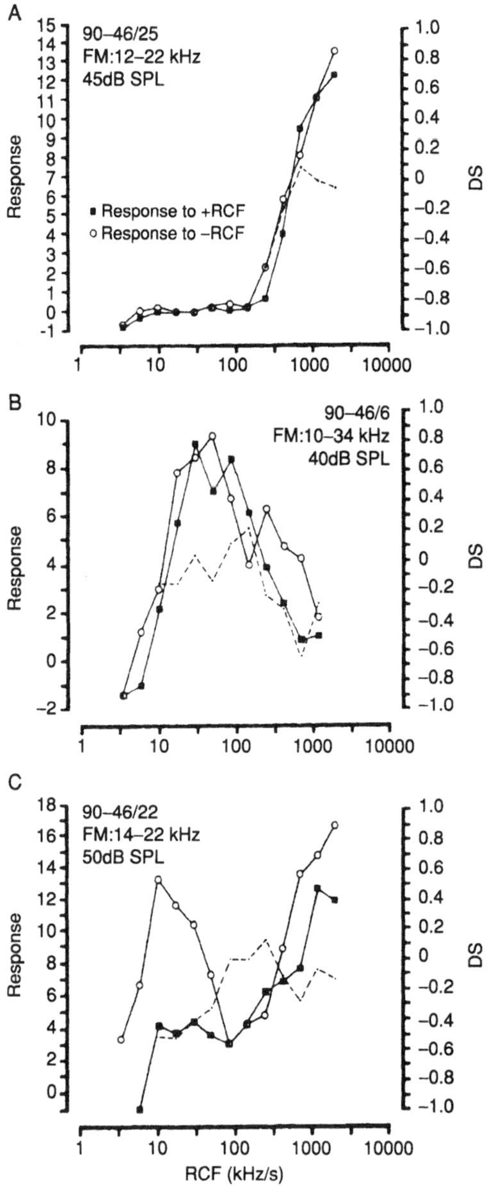

FIG. 7. Responses of three multi-unit clusters (A to C) in the primary auditory cortex of cat elicited by linear FM sweeps plotted as a function of rate of change of frequency (RCF). "Response" (left ordinate and solid lines): number of spikes elicited to each presentation of the stimulus. Solid squares and open circles show responses to upward and downward sweeps, respectively. DS (directional sensitivity, right ordinate, and dashed line): index of directional sensitivity. Positive and

Mendelson *et al.* (1993) found that clusters preferring the slow speed were in general more narrowly tuned and those preferring fast sweeps responded well to broadband transients. Thus, the maps for these properties overlap (Schreiner and Mendelson, 1990). Interestingly, Heil *et al.* (1992b) reported similar trends in frequency tuning with the most sharply tuned units located in the central region of the isofrequency strip, but this did not correlate with the preferred sweep speed.

Evidence for a somewhat different topographic representation of preferred sweep direction has been reported in ferret AI (Kowalski *et al.*, 1995; Shamma *et al.*, 1993). In recordings from multi-unit clusters and single units, responses showing a preference to upward sweeps were found to dominate the caudal region of an isofrequency lamina, and downward sweeps were found to dominate the rostral region (Shamma *et al.*, 1993). Between ~25% to 35% of the responses had a directional index consistent with the response to one direction giving more than twice the response to the other (index $>\pm$ 0.3). Sites in the center of the lamina responded equally well to sweeps in either direction.

Underlying this organization was a weak, but significant, correlation between the observed directional preferences and the asymmetry of inhibition in a neuron's frequency response area mapped with pairs of tones (Kowalski *et al.*, 1995; Shamma *et al.*, 1993). Thus, sites which preferred upward sweeps had stronger inhibition above BF and vice versa. Inhibition had a tendency to be symmetrically organized around BF in the center of the isofrequency band, and excitatory response areas were narrower in this region than on either side. Although this trend in bandwidth across the lamina is in keeping with that reported in cat (Schreiner and Mendelson, 1990), in contrast to findings in cat, there appears to be no clear relationship between preferred sweep rate and response area bandwidth, or between sweep rate and directional preference (Kowalski *et al.*, 1995; Shamma *et al.*, 1993).

In the Shamma *et al.* (1993) study, it was not clear whether a single band with the described directional preferences existed across a single isofrequency lamina, or whether in some cases the pattern repeated. They stated that reversals of asymmetry were observed at the edges of some bands, but these could not be investigated in detail.

Yet another pattern of organization has been reported in rat AI (Zhang *et al.*, 2003). They found that directional sensitivity co-varied with BF, with low- and high-BF neurons preferring upward and downward sweeps, respectively. In keeping with this, asymmetry of the response areas was dependent on BF:

negative values denote a preference for upward and downward sweeps, respectively. A value of ±0.3 represents a response with twice as many spikes to one sweep direction relative to the other. Over one-half the neurons had high pass response functions (A), while 20% were bandpass (B). Other neurons showed more complex functions, e.g., (C). (Adapted from Heil *et al.*, 1992a.)

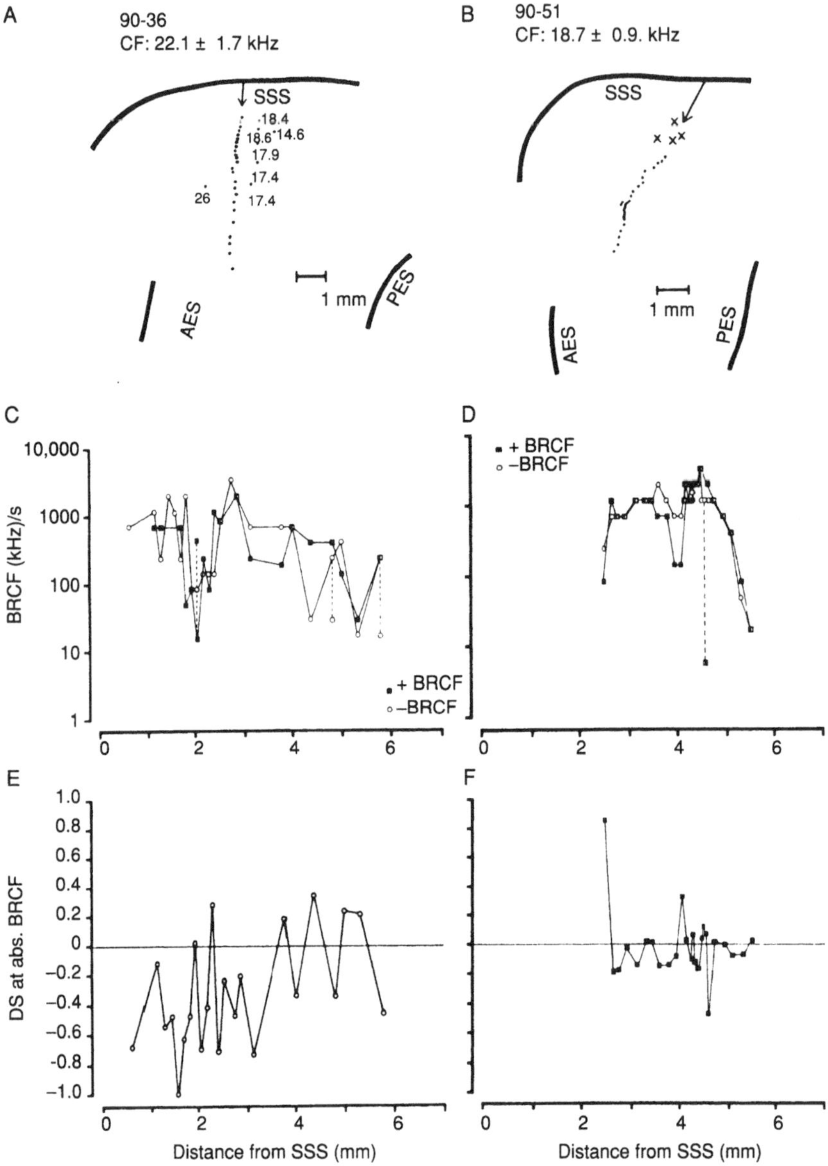

FIG. 8. Surface maps (A, B) of the primary auditory cortex in two cats showing lines of electrode penetrations (dots) made parallel with the isofrequency contours. The CF value on the plots indicates the mean CF of the multi-unit clusters encountered at the unlabelled recording sites and its standard deviation. C and D, values of best rate of change of frequency (BRCF), and E and F, values of directional sensitivity (DS, see Fig. 7) at BRCF, for the corresponding penetrations in A and B. The bases of the arrows in A and B denote the origin of the abscissas in C to F. In C and D, solid squares

inhibition tended to be most apparent below BF in high-BF neurons, and above BF in low-BF neurons. Further insights into these mechanisms were teased out by characterization of the intracellular currents in these neurons. Directional selectivity was evident in the excitatory synaptic input to the neuron, implying that some element of directional selectivity was generated prior to the cortex. However, peak inhibitory currents (i.e., of cortical origin) were more delayed relative to the excitatory current when the sweep moved in the preferred direction, and would thus be less effective in suppressing the excitation. Interestingly, the frequency tuning of the excitatory and inhibitory currents was almost identical, but both showed an asymmetry: low BF neurons showed stronger currents on their low frequency side and high BF neurons on their high frequency side. Thus, for the preferred direction the neuron receives a strong excitation followed by a strong but delayed inhibition. In the non-preferred direction, excitation is initially weak and subject to inhibition as it becomes stronger (Zhang *et al.*, 2003). These data suggest that in rats both spectral and temporal mechanisms contribute to directional selectivity.

As the foregoing discussion indicates, although there is evidence for the topographic organization of FM sweep parameters across AI, fundamental differences exist between studies in the same and different species. In cat, mapping of directional preference, where it occurs, correlates with sweep speed, while in ferret the correlation is with asymmetry of inhibitory sidebands. A correlation with sideband asymmetry is also apparent in rat but, in contrast to ferret, asymmetry and, therefore, directional preference is strongly correlated with BF in the rat.

It is difficult to determine to what extent such contrasts are true species differences or due to stimulus or other methodological differences. One important difference between studies is the choice of linear or exponential/log FM sweeps. The only study to address this issue specifically was in ferret AI where the responses to different linear and log sweeps were compared (Nelken and Versnel, 2000). These ranged between 200 to 2000 kHz/s over 10 kHz for the linear sweeps and between 30 to 300 octaves/s over four octaves for the log sweeps. The response property that remained constant across sweep types was sensitivity to sweep speed. With both sweep types, most multi-units preferred slower speeds, and over one-half responded best to speeds below 400 kHz/s linear sweeps. In contrast, directional preferences depended markedly on sweep type, although all

---

and open circles show data for upward and downward sweeps, respectively. Although over short distances some penetrations shared similar BRCFs, sharp transitions also occurred, and consistent patterns were not seen between animals. The lack of order is particularly evident for DS values. SSS = superior, AES = anterior, and PES = posterior, ectosylvian sulci, respectively. Numbers next to dots denote CF values of penetrations used to verify the tonotopic gradient; crosses show sites were no response to a tone was obtained. (Adapted from Heil *et al.*, 1992b.)

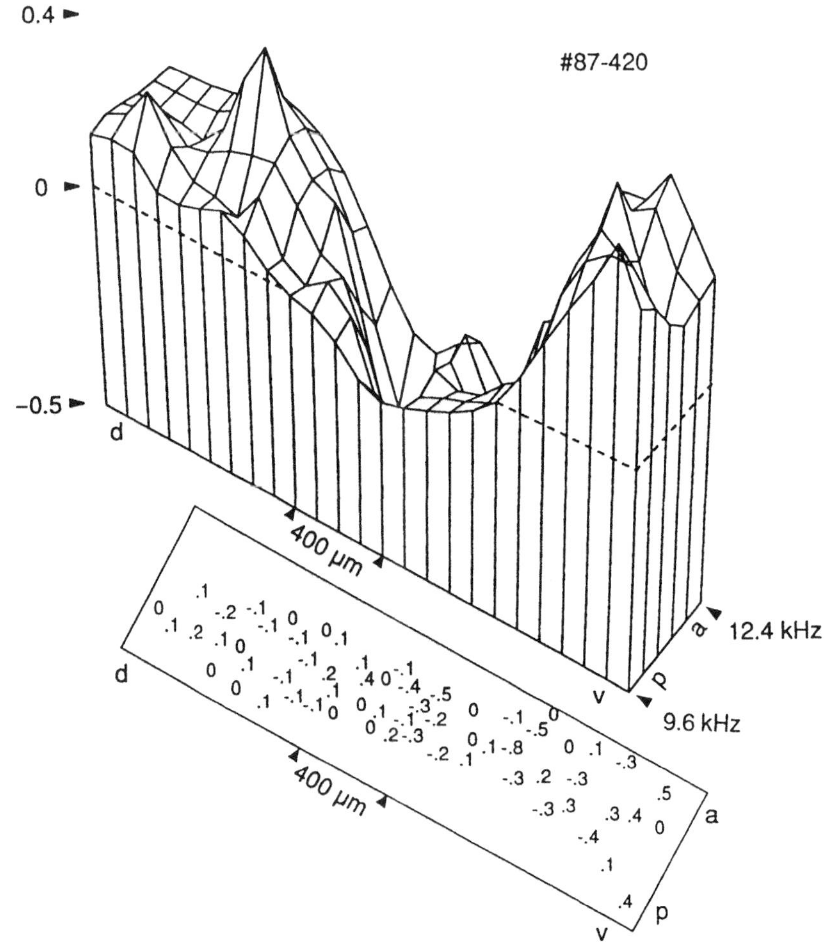

FIG. 9. Topography of preferred FM sweep direction in primary auditory cortex (AI) of cat. The pseudo three-dimensional plot runs parallel with the isofrequency contours along the dorso-ventral (d-v) axis. The arrowheads and values on the anterior-posterior (a-p) axis indicate the approximate positions and frequencies of the isofrequency contours. The elevation of the surface indicates the directional preference of the recorded multiunit clusters. Positive values (above the dashed line) indicate a preference for upward sweeps, and negative values (below the dashed line) a preference for downward sweeps (see Fig. 7 and text). The numbers in the rectangle below the plot are the values of the directional selectivity index used to produce the interpolated three-dimensional plot. Upward sweeps were preferred along the dorsal and ventral edges of the cortex and downward sweeps in the center. (Adapted from Mendelson *et al.*, © 1993 with kind permission of Springer Science and Business Media.)

sweep types revealed a response bias for up sweeps. Nelken and Versnel (2000) argued that this might reflect the preference for sweep speed. In log sweeps, upward sweeps begin with a slow rate of change of frequency that gets faster, while in downward sweeps the pattern is reversed. The correlation between inhibitory sideband asymmetry and directional preference discussed in the previous text (Kowalski *et al.*, 1995; Shamma *et al.*, 1993) was also only found with one type of log sweep in which the low-high transition occurred first.

When they compared the topographic plots of data gathered with linear or log sweeps, Nelken and Versnel (2000) noted greater consistency between the clustering of data for sweep speed than for directional selectivity. In some cases values for particular areas of cortex were reversed for different paradigms and in others no clustering was observed. Where clustering of similar data points occurred, the authors emphasized that this did not take the form of a regular progression.

Finally, in this discussion of responses in AI it should be acknowledged that most studies used monaural stimulation to the ear contralateral to the recorded cortex, but responses to FM may be different for monaural and binaural conditions. Some cells in AI are not directionally selective with monaural stimulation, but are with binaural stimulation (Mendelson and Grasse, 1992).

B. Responses to FM in the Anterior Auditory Field (AAF)

Responses to FM sweeps have been examined in areas beyond AI in cat and in ferret to determine whether the nature of FM responses differs between cortical areas (Kowalski *et al.*, 1995; Tian and Rauschecker, 1994). Both studies report that the bandwidths of neurons in AAF are on average wider than those in AI. Using logarithmic FM sweeps, Kowalski *et al.* (1995) found in ferret that neurons with wider bandwidths tended to prefer faster sweep rates. The mean preferred sweep speed did not differ between single neurons in AAF and AI, but the standard deviation of the data was greater for AI. A weak but significant correlation was found between the asymmetry of response area inhibition and the neuron's directional selectivity, as these authors previously noted in AI.

Recordings in cat AAF with linear FM sweeps (Tian and Rauschecker, 1994) found that most units responded to sweep rate with a high pass or bandpass function. About 60% of the sample responded best to fast sweeps >200 kHz/s, and the range of sweeps speeds and the distribution of units within these groups was reasonably similar to that reported in AI (Fig. 10) (Heil *et al.*, 1992a). However, a higher percentage of bandpass tuned units was obtained when units that had bandpass functions for only one sweep direction were included. The directional preferences of neurons in AAF were in marked contrast to those

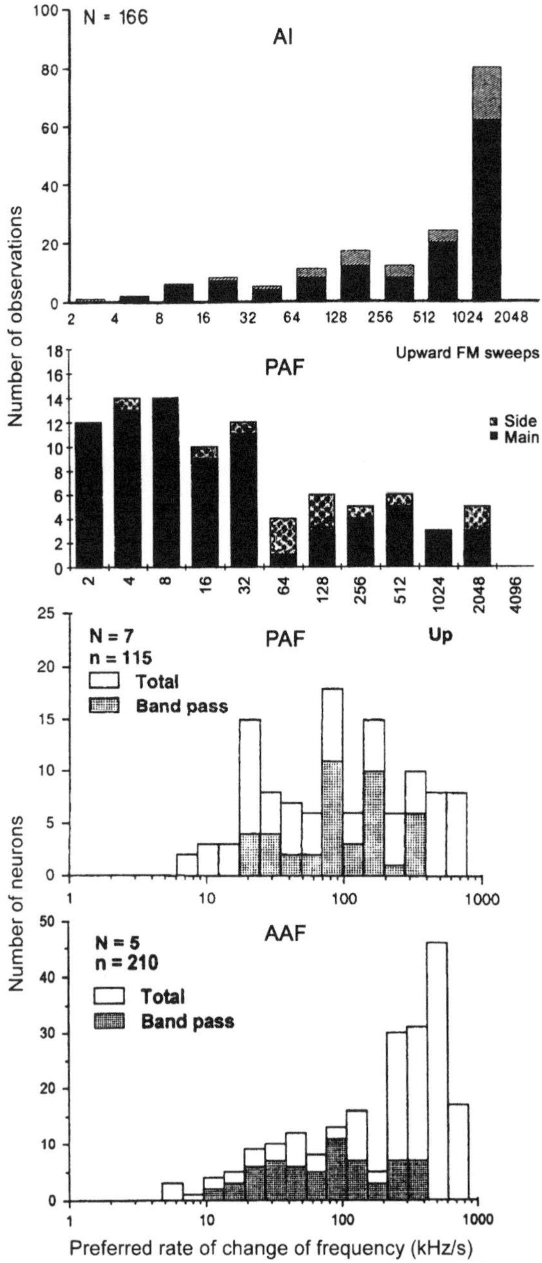

FIG. 10. Distributions of preferred rate of change of frequency for neurons and multi-unit clusters in different regions of the cat's auditory cortex. Data represent maxima in rate of change of frequency (RCF) functions (c.f. Fig. 7) obtained with linear sweeps. Upper two panels, data from primary

of AI neurons. About 70% of units showed selectivity for sweep direction, and of these 60% preferred upward sweeps, in contrast to the downward sweep preference reported in cat AI (Heil *et al.*, 1992a; Mendelson *et al.*, 1993).

C. Responses to FM in the Posterior Auditory Field (PAF)

Whereas neurons in AI and AAF appear to prefer sweeps >200 kHz/s, neurons in PAF of cats respond best to lower rates of change. Heil and Irvine (1998) reported a median value around 10 kHz/s (Fig. 10). The dominant response function for sweep speed was low pass followed by bandpass, again in contrast to the high pass functions common in AI. The response of PAF neurons to sound level also correlates with their response to FM. Many neurons that responded best to low sweep speeds had non-monotonic rate level functions that are characteristic for neurons in this area (Phillips *et al.*, 1995). Such neurons also responded best to low rates of amplitude change (Schreiner and Urbas, 1988).

In a second study in cats, Tian and Rauschecker (1998) noted two particular features of the response of neurons in PAF to frequency sweeps. First, for the majority of neurons, the instantaneous frequency that initiated a response was dependent on the sweep speed, in contrast to what pertains in AI. This failure to predict the response to FM on the basis of the pure-tone response suggests that neurons in PAF are more non-linear in their properties. Second, the majority of neurons showed bandpass functions for sweep speed, at least for one sweep direction, with low pass the next most common. This distribution of response functions is broadly similar to that reported by Heil and Irvine (1998) taking into account the difference in criteria used to define a bandpass response (Fig. 10). However, the range of best sweep speeds (median ~100 kHz/s) reported by Tian and Rauschecker (1998) is higher than that reported by Heil and Irvine (1998). Of the directionally selective units recorded, the proportion preferring upward sweeps was greater than that preferring downward sweeps, but compared with AAF fewer units were directionally selective. Furthermore, directional selectivity in PAF was maintained across all sweep rates, in contrast to AAF.

---

auditory cortex (AI) (adapted from Heil *et al.*, 1992a) and posterior auditory field (PAF) (adapted from Heil and Irvine, 1998). Observations made from recordings of multi-unit clusters. Data from side peaks of RCF functions are indicated separately. N = number of recorded clusters. Lower two panels, data from PAF (adapted from Tian and Rauschecker, 1998) and anterior auditory field (AAF) (adapted from Tian and Rauschecker, 1994). Recordings made from single neurons. Data from neurons with bandpass RCF functions are indicated separately. N = number of animals, n = number of units.

D. Responses to FM in Lateral Belt Areas of Auditory Cortex of Primates

In the rhesus monkey, a striking difference is apparent when comparing the different areas of the lateral belt region of auditory cortex, in which neurons tended to give stronger responses to FM stimuli than to pure tones (Tian and Rauschecker, 2004). Neurons in the anterolateral area (AL) preferred rates of change between about 25 to 50 kHz/s, neurons in caudolateral area (CL) 160 kHz/s, with those in the mediolateral area (ML) responding across the range. Furthermore, the majority of neurons in AL had low-pass functions for sweep rate, while those in ML and CL were bandpass. Tian and Rauschecker (2004) argue that the preference for slower sweeps in AL makes these units more appropriate for encoding the frequency changes present in con-specific vocalizations (Ghazanfar and Hauser, 2001), and that the differences between lateral belt areas are consistent with the notion of ventrally direct "what" and dorsally directed "where" processing streams in the macaque cortex (Rauschecker and Tian, 2000).

E. Effects of Cortical Lesions on the Discrimination of Frequency Sweeps

Evidence from cortical lesions suggests auditory cortex a role for in discriminating the direction of frequency change. In such studies it is important to be aware of the presence of cues other than the change in frequency; for example, the animal might use the initial or final frequency of the stimulus rather than its dynamic component.

Cats with large auditory cortical lesions could still discriminate two sweeps differing in direction at levels above chance for frequency sweeps of only 10% of the carrier, but their performance was impaired and showed higher numbers of false positives (Kelly and Whitfield, 1971). This was the case even when carrier frequency was varied from sweep to sweep to eliminate absolute frequency as a cue. However, it might be argued that the animals responded to the difference in frequency between the beginning and end of the sweeps since the paradigm involved detection of the interval in which repeating sweeps changed direction from up to down. Deficits have also been reported following cortical lesions in gerbil (Ohl *et al.*, 1999).

In primates cortical ablation had a more dramatic effect although it cannot be excluded that this reflects the use of a paradigm different to that used by Kelly and Whitfield (1971). Macaques with ablation of auditory cortex failed to discriminate steady tones from sweeps when the frequency of the pure tone varied from trial to trial over the range of frequencies played in the sweep (Harrington *et al.*, 2001; Heffner, 2005). Furthermore, lesions that selectively destroyed rostral or core regions of the auditory cortex resulted in a failure to discriminate frequency sweeps from pure tones. Lesions targeting more posterior regions had no effect, thus lending support to a segregation of function for the processing of dynamic spectra within the cortical fields (Heffner, 2005).

VI. Conclusions

In this chapter we have shown that the responses of neurons in the auditory pathway to frequency sweeps and sinusoidal FM depend on several parameters. Surprisingly, several features in the responses to sweeps observed at the cortical level have already emerged at the lowest levels of the auditory pathway. The relationship between the effective instantaneous frequency of the FM sweep and the slope of the frequency response area is apparent in the responses of cochlear nerve fibers, and selectivity for sweep direction is present in the responses of some neurons in the CN. However, there are quantitative differences in the responses between brainstem and cortical areas, and, in rat at least, synaptic activity in the cortex enhances directional selectivity.

A finding that is conspicuous across different auditory centers is the similarity between responses to AM and FM. It is apparent both in neuronal discharge patterns and in quantitative analyses such as modulation transfer functions. This correspondence is perhaps not surprising since a frequency sweeping across a neuron's response area can be viewed as a change in amplitude. Psychophysical evidence supporting this concept has been reported (Saberi and Hafter, 1995).

Evidence has been presented that FM response properties of neurons are topographically represented in different regions of auditory cortex, but the precise nature of the distribution requires further clarification. Evidence supports both a model in which response properties change gradually across the cortical surface and an organization in which discrete clusters of neurons share similar properties. Such organizational differences occur both within and across species. The situation is further complicated because some aspects of the response to FM depend on the nature of the stimulus, e.g., whether the frequency trajectory is linear or exponential, and the dependence of directional selectivity on sweep speed. This lack of invariance in the responses to sweeps makes it difficult to conceive that individual neurons are selectively tuned to specific sweep properties.

A particularly striking and puzzling finding to emerge from the studies discussed is that neurons at all levels of the pathway respond best to rates of frequency change that are considerably faster than the frequency sweeps in many animal vocalizations. In human speech the rate of change of frequency in consonant transitions is about 10 to 50 kHz/s (e.g., in consonants a frequency change of about 1 kHz in 100 ms). While it can reasonably be argued that speech did not influence the evolution of cats and ferrets, rather similar values have also been reported for the vocalizations of these species and their prey (for discussion see Heil *et al.*, 1992a). Thus, the relevance of many of these responses for the perception of sounds encountered in nature is difficult to discern. Only areas PAF in cat and AL in the lateral belt in macaque are dominated by neurons selective for rates of change of frequency found in sounds likely to be of interest to these animals.

Dedication

This review is dedicated to two physiologists who made pioneering contributions to the hearing of FM in sounds and whose interests in modulation inspired my own. To Aage R. Møller (M. F. Jonsson Professor, University of Texas at Dallas) on his retirement as founder editor of *Hearing Research*, and to the memory of my tutor, the late Roy Kay (formerly University Lecturer in Physiology, and Fellow of Keble College, Oxford). To borrow R. H. K.'s favorite dedication, "Homage!" (A. R.).

Acknowledgments

We are grateful for financial support in the writing of this review from the Wellcome Trust (066348; A.R., M.S.M.), Spanish DGES: (BFI-2003-09147-02-01), and JCYL-UE (SA040.04) M.S.M.

References

Bregman, A. S. (1990). "Auditory scene analysis: The perceptual organization of sound." Bradford Books, MIT Press, Cambridge, MA.

Britt, R., and Starr, A. (1976). Synaptic events and discharge patterns of cochlear nucleus cells. Ii frequency-modulated tones. *J. Neurophysiol.* **39,** 179–194.

Brown, K. A., Buchwald, J. S., Johnson, J. R., and Mikolich, D. J. (1978). Vocalization in the cat and kitten. *Dev. Psychobiol.* **11,** 559–570.

Carlyon, R. P. (1991). Discriminating between coherent and incoherent frequency-modulation of complex tones. *J. Acoust. Soc. Am.* **89,** 329–340.

Casseday, J. H., Covey, E., and Grothe, B. (1997). Neural selectivity and tuning for sinusoidal frequency modulations in the inferior colliculus of the big brown bat, eptesicus fuscus. *J. Neurophysiol.* **77,** 1595–1605.

Clopton, B. M., and Winfield, J. A. (1974). Unit responses in the inferior colliculus of rat to temporal auditory patterns of tone sweeps and noise bursts. *Exp. Neurol.* **42,** 532–540.

Covey, E., and Casseday, J. H. (1999). Timing in the auditory system of the bat. *Annu. Rev. Physiol.* **61,** 457–476.

Creutzfeldt, O., Hellweg, F. C., and Schreiner, C. (1980). Thalamocortical transformation of responses to complex auditory-stimuli. *Exp. Brain Res.* **39,** 87–104.

Culling, J. F., and Summerfield, Q. (1995). The role of frequency-modulation in the perceptual segregation of concurrent vowels. *J. Acoust. Soc. Am.* **98,** 837–846.

Eggermont, J. J. (1994). Temporal-modulation transfer-functions for AM and FM stimuli in cat auditory-cortex—effects of carrier type, modulating wave-form and intensity. *Hear. Res.* **74,** 51–66.

Ehret, G., Egorova, M., Hage, S. R., and Muller, B. A. (2003). Spatial map of frequency tuning-curve shapes in the mouse inferior colliculus. *Neuroreport* **14,** 1365–1369.

Erulkar, S. D., Butler, R. A., and Gerstein, G. L. (1968). Excitation and inhibition in cochlear nucleus. II. Frequency-modulated tones. *J. Neurophysiol.* **31,** 537–548.

Felsheim, C., and Ostwald, J. (1996). Responses to exponential frequency modulations in the rat inferior colliculus. *Hear. Res.* **98,** 137–151.

Fernald, R. D. (1971). Neuron model with spatially distributed synaptic input. *Biophys. J.* **11,** 323–340.

Fuzessery, Z. M. (1994). Response selectivity for multiple dimensions of frequency sweeps in the pallid bat inferior colliculus. *J. Neurophysiol.* **72,** 1061–1079.

Fuzessery, Z. M., and Hall, J. C. (1996). Role of GABA in shaping frequency tuning and creating fm sweep selectivity in the inferior colliculus. *J. Neurophysiol.* **76,** 1059–1073.

Gaese, B. H., and Ostwald, J. (1995). Temporal coding of amplitude and frequency-modulation in the rat auditory-cortex. *Eur. J. Neurosci.* **7,** 438–450.

Ghazanfar, A. A., and Hauser, M. D. (2001). The auditory behaviour of primates: A neuroethological perspective. *Curr. Opin. Neurobiol.* **11,** 712–720.

Gordon, M., and O'Neill, W. E. (1998). Temporal processing across frequency channels by FM selective auditory neurons can account for FM rate selectivity. *Hear. Res.* **122,** 97–108.

Greenwald, C. H. (1968). "Birdsong: Acoustics and physiology." Smithsonian Institution Press, Washington, DC.

Hage, S. R., and Ehret, G. (2003). Mapping responses to frequency sweeps and tones in the inferior colliculus of house mice. *Eur. J. Neurosci.* **18,** 2301–2312.

Harrington, I. A., Heffner, R. S., and Heffner, H. E. (2001). An investigation of sensory deficits underlying the aphasia-like behavior of macaques with auditory cortex lesions. *Neuroreport* **12,** 1217–1221.

Heffner, H. E. (2005). The neurobehavioural study of auditory cortex. *In* "The auditory cortex: A synthesis of human and animal research" (R. König, P. Heil, E. Budinger, and H. Scheich, Eds.), pp. 111–126. Lawrence Erlbaum Associates, Mahwah, NJ.

Heil, P., Rajan, R., and Irvine, D. R. F. (1992a). Sensitivity of neurons in cat primary auditory-cortex to tones and frequency-modulated stimuli.1. Effects of variation of stimulus parameters. *Hear. Res.* **63,** 108–134.

Heil, P., Rajan, R., and Irvine, D. R. F. (1992b). Sensitivity of neurons in cat primary auditory-cortex to tones and frequency-modulated stimuli .2. Organization of response properties along the isofrequency dimension. *Hear. Res.* **63,** 135–156.

Heil, P., and Irvine, D. R. F. (1998). Functional specialization in auditory cortex: Responses to frequency-modulated stimuli in the cat's posterior auditory field. *J. Neurophysiol.* **79,** 3041–3059.

Huffman, R. F., Argeles, P. C., and Covey, E. (1998). Processing of sinusoidally frequency modulated signals in the nuclei of the lateral lemniscus of the big brown bat, eptesicus fuscus. *Hear. Res.* **126,** 161–180.

Kaltwasser, M. T. (1990). Acoustic signaling in the black rat (Rattus rattus). *J. Compar. Psychol.* **104,** 227–232.

Kelly, J. B., and Whitfield, I. C. (1971). Effects of auditory cortical lesions on discriminations of rising and falling frequency-modulated tones. *J. Neurophysiol.* **34,** 802–816.

Kowalski, N., Versnel, H., and Shamma, S. A. (1995). Comparison of responses in the anterior and primary auditory fields of the ferret cortex. *J. Neurophysiol.* **73,** 1513–1523.

Lui, B., and Mendelson, J. R. (2003). Frequency modulated sweep responses in the medial geniculate nucleus. *Exp. Brain Res.* **153,** 550–553.

McAdams, S. (1989). Segregation of concurrent sounds.1. Effects of frequency-modulation coherence. *J. Acoust. Soc. Am.* **86,** 2148–2159.

Mendelson, J. R., and Cynader, M. S. (1985). Sensitivity of cat primary auditory-cortex (AI) neurons to the direction and rate of frequency-modulation. *Brain Res.* **327,** 331–335.

Mendelson, J. R., and Grasse, K. L. (1992). A comparison of monaural and binaural responses to frequency modulated (FM) sweeps in cat primary auditory-cortex. *Exp. Brain Res.* **91,** 435–454.

Mendelson, J. R., Schreiner, C. E., Sutter, M. L., and Grasse, K. L. (1993). Functional topography of cat primary auditory-cortex—responses to frequency-modulated sweeps. *Exp. Brain Res.* **94,** 65–87.

Møller, A. R. (1969). Unit responses in the cochlear nucleus of the rat to sweep tones. *Acta. Physiologica. Scandinavica* **76,** 503–512.

Møller, A. R. (1971). Unit responses in the rat cochlear nucleus to tones of rapidly varying frequency and amplitude. *Acta. Physiologica. Scandinavica* **81,** 540–556.

Møller, A. R. (1972a). Coding of amplitude and frequency modulated sounds in the cochlear nucleus of the rat. *Acta. Physiologica. Scandinavica* **86,** 223–238.

Møller, A. R. (1972b). Coding of sounds in lower levels of the auditory system. *Q. Rev. Biophys.* **5,** 59–155.

Møller, A. R. (1974a). Responses of units in the cochlear nucleus to sinusoidally amplitude-modulated tones. *Exp. Neurol.* **45,** 105–117.

Møller, A. R. (1974b). Coding of sounds with rapidly varying spectrum in the cochlear nucleus. *J. Acoust. Soc. Am.* **55,** 631–640.

Nelken, I., and Versnel, H. (2000). Responses to linear and logarithmic frequency-modulated sweeps in ferret primary auditory cortex. *Eur. J. Neurosci.* **12,** 549–562.

Nelson, P. G., Erulkar, S. D., and Bryan, J. S. (1966). Response of units in the inferior colliculus to time-varying acoustic stimuli. *J. Neurophysiol.* **29,** 834–860.

Ohl, F. W., Wetzel, W., Wagner, T., Rech, A., and Scheich, H. (1999). Bilateral ablation of auditory cortex in mongolian gerbil affects discrimination of frequency modulated tones but not of pure tones. *Learn Mem.* **6,** 347–362.

O'Neill, W. E. (1995). The bat auditory cortex. *In* "Hearing by bats" (A. N. Popper and R. R. Fay, Eds.), pp. 416–498. Springer-Verlag, New York, NY.

Palmer, A. R., Jiang, D., and Marshall, D. H. (1996). Responses of ventral cochlear nucleus onset and chopper units as a function of signal bandwidth. *J. Neurophysiol.* **75,** 780–794.

Phillips, D. P., Mendelson, J. R., Cynader, M. S., and Douglas, R. M. (1985). Responses of single neurons in cat auditory-cortex to time-varying stimuli—frequency-modulated tones of narrow excursion. *Exp. Brain Res.* **58,** 443–454.

Phillips, D. P., Semple, M. N., and Kitzes, L. M. (1995). Factors shaping the tone level sensitivity of single neurons in posterior field of cat auditory-cortex. *J. Neurophysiol.* **73,** 674–686.

Pollak, G. D., Marsh, D. S., Bodenhamer, R. D., and Souther, A. (1978). A single unit analysis of inferior colliculus in unanesthetized bats: Response patterns and spike-count functions generated by constant-frequency and frequency-modulated sounds. *J. Neurophysiol.* **41,** 677–691.

Poon, P. W. F., Chen, X., and Hwang, J. C. (1991). Basic determinants for FM responses in the inferior colliculus of rats. *Exp. Brain Res.* **83,** 598–606.

Poon, P. W. F., Chen, X. Y., and Cheung, Y. M. (1992). Differences in FM response correlate with morphology of neurons in the rat inferior colliculus. *Exp. Brain Res.* **91,** 94–104.

Poon, P. W. F., and Chiu, T. W. (2000). Similarities of FM and AM receptive space of single units at the auditory midbrain. *Biosystems* **58,** 229–237.

Rall, W. (1962). Theory of physiological properties of dendrites. *Ann. N Y Acad. Sci.* **96,** 1071–1092.

Rall, W. (1964). Theoretical significance of dendritic trees for neuronal input-output relations. *In* "Neural theory and modeling" (R. F. Reiss, Ed.), pp. 73–97. Stanford University Press, Stanford.

Rauschecker, J. P., and Tian, B. (2000). Mechanisms and streams for processing of "what" and "where" in auditory cortex. *Proc. Natl. Acad. Sci. USA* **97,** 11800–11806.

Rees, A., and Møller, A. R. (1983). Responses of neurons in the inferior colliculus of the rat to AM and FM tones. *Hear. Res.* **10,** 301–330.

Rhode, W. S., Oertel, D., and Smith, P. H. (1983). Physiological-response properties of cells labeled intracellularly with horseradish-peroxidase in cat ventral cochlear nucleus. *J. Comp. Neurol.* **213,** 448–463.

Rhode, W. S., and Smith, P. H. (1986). Encoding timing and intensity in the ventral cochlear nucleus of the cat. *J. Neurophysiol.* **56,** 261–286.

Rhode, W. S., and Greenberg, S. (1992). Physiology of the cochlear nuclei. *In* "The mammalian auditory pathway: Neurophysiology" (A. N. Popper and R. R. Fay, Eds.), pp. 94–152. Springer-Verlag, New York.

Ricketts, C., Mendelson, J. R., Anand, B., and English, R. (1998). Responses to time-varying stimuli in rat auditory cortex. *Hear. Res.* **123,** 27–30.

Rouiller, E. M., and Ryugo, D. K. (1984). Intracellular marking of physiologically characterized cells in the ventral cochlear nucleus of the cat. *J. Comp. Neurol.* **225,** 167–186.

Saberi, K., and Hafter, E. R. (1995). A common neural code for frequency-modulated and amplitude-modulated sounds. *Nature* **374,** 537–539.

Schreiner, C. E., and Urbas, J. V. (1988). Representation of amplitude-modulation in the auditory-cortex of the cat.2. Comparison between cortical fields. *Hear. Res.* **32,** 49–64.

Schreiner, C. E., and Mendelson, J. R. (1990). Functional topography of cat primary auditory-cortex -distribution of integrated excitation. *J. Neurophysiol.* **64,** 1442–1459.

Schuller, G. (1979). Coding of small sinusoidal frequency and amplitude modulations in the inferior colliculus of the "cf-fm" bat, *Rhinolophus ferrumequinum. Exp. Brain Res.* **34,** 117–132.

Shamma, S. A., Fleshman, J. W., Wiser, P. R., and Versnel, H. (1993). Organization of response areas in ferret primary auditory-cortex. *J. Neurophysiol.* **69,** 367–383.

Shore, S. E., Clopton, B. M., and Au, Y. N. (1987). Unit responses in ventral cochlear nucleus reflect cochlear coding of rapid frequency sweeps. *J. Acoust. Soc. Am.* **82,** 471–478.

Sinex, D. G., and Geisler, C. D. (1981). Auditory-nerve fiber responses to frequency-modulated tones. *Hear. Res.* **4,** 127–148.

Smith, P. H., and Spirou, G. A. (2002). From cochlea to cortex and back. *In* "Integrative functions in the mammalian auditory pathway" (D. Oertel, A. N. Popper, and R. R. Fay, Eds.), pp. 6–71. Springer-Verlag, New York.

Suga, N. (1965). Analysis of frequency-modulated sounds by auditory neurons of echo-locating bats. *J. Physiol.* **179,** 26–53.

Suga, N. (1969). Classification of inferior collicular neurons of bats in terms of responses to pure tones, fm sounds and noise bursts. *J. Physiol.* **200,** 555–574.

Suga, N. (1988). Auditory neuroethology and speech processing: Complex-sound processing by combination-sensitive neurons. *In* "Auditory function: Neurobiological bases of hearing" (G. M. Edelman, W. E. Gall, and W. M. Cowan, Eds.), pp. 679–720. Wiley, New York.

Suta, D., Kvasnak, E., Popelar, J., and Syka, J. (2003). Representation of species-specific vocalizations in the inferior colliculus of the guinea pig. *J. Neurophysiol.* **90,** 3794–3808.

Tian, B., and Rauschecker, J. P. (1994). Processing of frequency-modulated sounds in the cat's anterior auditory field. *J. Neurophysiol.* **71,** 1959–1975.

Tian, B., and Rauschecker, J. P. (1998). Processing of frequency-modulated sounds in the cat's posterior auditory field. *J. Neurophysiol.* **79,** 2629–2642.

Tian, B., and Rauschecker, J. P. (2004). Processing of frequency-modulated sounds in the lateral auditory belt cortex of the rhesus monkey. *J. Neurophysiol.* **92,** 2993–3013.

Vartanian, I. (1974). On mechanisms of specialized reactions of central auditory neurons to frequency-modulated sounds. *Acoustica* **31,** 305–310.

Watanabe, T. (1972). Fundamental study of the neural mechanism in cats subserving the feature extraction process of complex sounds. *Jpn. J. Physiol.* **22,** 569–583.

Whitfield, I. C., and Evans, E. F. (1965). Responses of auditory cortical neurons to stimuli of changing frequency. *J. Neurophysiol.* **28,** 655–672.

Winter, P., Ploog, D., and Latta, J. (1966). Vocal repertoire of the squirrel monkey (Saimiri sciureus), its analysis and significance. *Exp. Brain Res.* **1,** 359–384.

Zeng, F. G., Nie, K., Stickney, G. S., Kong, Y. Y., Vongphoe, M., Bhargave, A., Weit, C. G., and Cao, K. (2005). Speech recognition with amplitude and frequency modulations. *Proc. Natl. Acad. Sci. USA* **102,** 2293–2298.

Zhang, L. I., Tan, A. Y. Y., Schreiner, C. E., and Merzenich, M. M. (2003). Topography and synaptic shaping of direction selectivity in primary auditory cortex. *Nature* **424,** 201–205.

REPRESENTATIONS OF SPECTRAL CODING IN THE HUMAN BRAIN

Deborah A. Hall

MRC Institute of Hearing Research
Nottingham NG7 2RD, United Kingdom

Modern neuroscience has a range of tools at its disposal for physiological measurements of human brain function. The techniques described in this chapter have presented certain challenges for acquiring, analyzing, and interpreting data on central auditory function, yet the results have enabled auditory neuroscientists to characterize patterns of response sensitivity to different acoustical features with reasonably fine precision. To build a general scheme of the functional anatomy of the human auditory cortex, patterns of brain activity have been described by referring to the underlying morphological structure of the auditory cortex in the superior temporal region. However, the fine-detailed arrangement of the functional patterns can vary markedly across individual brains, so that generalized conclusions establish guidelines rather than rules. So far, human brain mapping methods have demonstrated that the auditory cortex is most concerned with the analysis of sounds that comprise multiple, dynamically varying component frequencies. Sounds that have a complex spectrotemporal pattern most obviously include speech, music, and environmental sounds. While being somewhat simpler, laboratory-generated sounds, such as tones that are sinusoidally modulated in frequency, represent fundamental elements of these natural sounds yet still

INTERNATIONAL REVIEW OF
NEUROBIOLOGY, VOL. 70
DOI: 10.1016/S0074-7742(05)70010-6

0074-7742/05 $35.00

maintain some of their essential acoustic complexity. This chapter is concerned with reporting recent key findings from neuroimaging studies using both natural and laboratory-generated sounds. Sections will review the functional neuroanatomy of processing single-frequency tones, wide-spectrum signals, pitch, spectral cues for sound localization, speech, and music. The results reveal important insights into how the human auditory cortex deals with spectral information and some comments will be made on the direction of future research.

I. Introduction

A large component of my research program over the past seven years at the MRC Institute of Hearing Research has been concerned with examining the functional anatomy of the human auditory cortex. What acoustic features are important for the human auditory cortex? Can subdivisions of the human auditory cortex be defined on the basis of differential sensitivity to these acoustical features and what is its spatial distribution? Can the functional organization be described as modular? Whatever type of sound is presented to a listener, a large proportion of the auditory cortex is activated (Hall *et al.*, 2003). Indeed, carbon-labeled deoxyglucose mapping in primates has revealed that passive listening can engage not only the entire auditory cortex, but also parts of the inferior parietal cortex, prefrontal cortex, frontal pole, and the upper back of the superior temporal sulcus (Poremba *et al.*, 2003). While these other regions are probably involved in attentional and goal-directed aims of the listening task, auditory cortical activity is more likely to represent the analysis of sound properties. Many neuroimaging techniques have measured feature-specific responses, either by contrasting pairs of stimuli that differ only in one acoustic feature or by systematically varying the value of that feature across sound conditions.

Five families of brain mapping technique are available for assessing human sensation, perception, and cognition; positron emission tomography (PET), functional magnetic resonance imaging (fMRI), electroencephalography (EEG), magnetoencephalography (MEG), and lesion analysis. All of these techniques have made valuable contributions to a global overview of spectral coding mechanisms in the auditory cortex. Each technique has its own peculiar strengths and weaknesses, with the selection of the most suitable one being driven predominantly by the nature of the scientific question. For determining where in the human auditory cortex are those neuronal populations that are most sensitive to the spectral properties of sound, the most successful technique has undoubtedly been fMRI because it is the most spatially precise. The first section in this chapter introduces the basic principles of human brain mapping and makes some general statements about spatial resolution, since no technique matches the single-unit precision of invasive animal electrophysiological recordings.

II. Techniques and Concerns About Spatial Resolution

A. PET

PET is the name given to a general method for measuring tissue metabolism by detecting the local tissue concentration of biologically important compounds that have been labeled with positron-emitting isotopes. Measuring metabolic activity in the brain takes advantage of the local coupling between neuronal activity and blood flow and commonly uses an isotope of oxygen incorporated into water ($H_2{}^{15}O$) that is injected into the blood stream and quickly perfuses the brain parenchyma. The concentration of the isotope builds up over a period of about 30 seconds while the subject is participating in an experimental task. Safety guidelines on the radiation dosage typically restrict the experiment to a small number of scans, thus permitting the presentation of only a limited number of sound conditions. Although the functionally related blood flow signal is local, it is not possible to determine how precisely it is localized to the cortical grey matter since more distant, large diameter blood vessels can also display detectable changes. Spatial specificity is further reduced during analysis by spatially smoothing the image data (up to 18 mm) and averaging across different brains to improve the signal-to-noise ratio. For example, Fig. 1A shows the diffuse group activation pattern for a 1-kHz tone relative to a silent baseline, presented to the right ear (Mirz *et al.*, 1999). In the 20-year history of the availability of PET for human auditory brain mapping, the lack of relevant publications indicates that it has probably failed to provide reliable spatial distinctions between responses to different sound attributes. Of all of the methods, poor spatial and temporal resolution limit the other benefits of using PET to localize functional responses within the auditory cortex (see Johnsrude *et al.*, 2002, for a review) and the use of PET for examining central auditory function in healthy individuals has been overtaken by fMRI.

B. fMRI

fMRI is just one of a family of MRI techniques developed from nuclear spectroscopy; a field which relies on the resonant properties of different atomic nuclei. A charged nucleus has a specific energy level related to a property called spin, and the spinning-charged particle creates its own small electromagnetic field. When nuclei are placed in an external magnetic field, for example when a

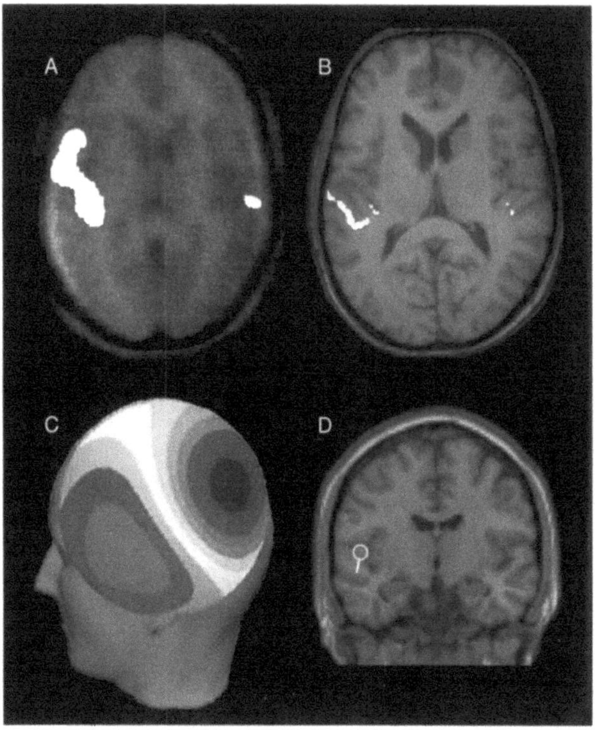

FIG. 1. Examples of the types of spatial maps that represent the neural activation pattern for a single frequency tone. (A) PET data for five listeners showing activation for a train of 1-kHz tone pulses relative to a silent baseline presented to the right ear (Mirz et al., 1999); note the greater contralateral activation. (B) fMRI data for a single listener for a train of 800-Hz tone pulses relative to a silent baseline (unpublished data from the author, 2004). (C) The scalp distribution of the M100 field for a single 1-kHz tone burst (Lütkenhöner and Steinsträter, 1998). (D) A schematic example of a fixed dipole source estimate in the left hemisphere from an auditory-evoked field. (A) and (C) reproduced with kind permission from the publishers.

patient lies in the MRI scanner, they tend to align in the direction of the main magnetic field. The orientation of the net magnetization can be flipped perpendicularly by applying a short burst of radiofrequency (RF) energy at the same resonant frequency as the atomic nuclei. When the RF pulse ceases, the nuclei lose energy by emitting their own RF signal and this is measured by a conductive field coil placed around the object—the head, for example—being imaged. This measurement is reconstructed to obtain a three-dimensional grey-scale brain image. The rate of the return to equilibrium is determined by the magnetic state of the surrounding tissue. fMRI capitalizes on the fact that deoxygenated hemoglobin affects this magnetic environment, with the rate of decay being more rapid the greater the neighboring concentration of deoxyhemoglobin. Given that the metabolic demands of neuronal activity determine the oxygenation state of the blood in localized regions of the brain, the MR signal provides a convenient and spatially-specific measure of neuronal activity. The fMRI signal is derived from the behavior of hydrogen nuclei which are single, positively charged protons. Hydrogen (in the form of H_2O) is abundant in living tissue. The resulting marker of neural activity is referred to as blood-oxygenation-level-dependent (BOLD) contrast. In reality, the relationship between neuronal activity and the BOLD signal is highly complex since the contrast is based on combined changes in relative blood flow, volume, and oxygenation. However, it is generally accepted that the BOLD signal has a reasonably direct correlation with population synaptic activity (including inhibitory and excitatory activity) and a more variable correlation with spiking activity (Arthurs and Boniface, 2002). Nevertheless, the values of the BOLD signal are relative not absolute; one value can be interpreted only with respect to another acquired within the same experiment.

The time course of the BOLD response to any brief stimulus event can be defined by three temporal components: a short negative dip, a longer positive phase, and a final negative component that includes a protracted return to baseline (Fig. 2). Statistical maps of the subtracted brain activity are typically

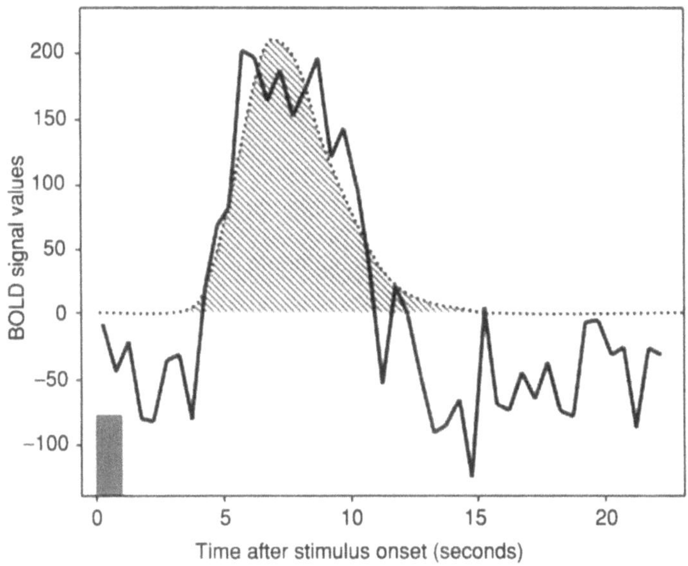

FIG. 2. The three phases of the BOLD response shown by the time-averaged response (solid line) of a typical voxel to a brief 900-ms stimulus (grey bar). In this example, two negative phases occur 0 to 4 s and 15 to 22.5 s after stimulus onset. The dotted line represents the model estimate of the positive phase (4 to 15 s post-onset) used for generating statistical maps of activation. This graph is adapted from Hall *et al.* (2002a).

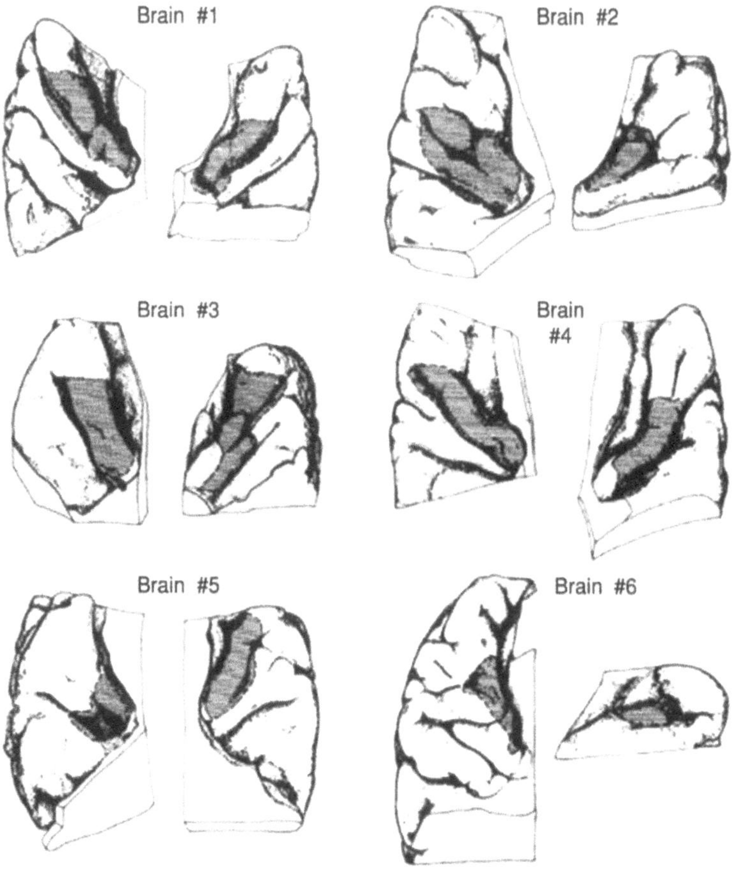

FIG. 3. Drawings of Heschl's gyrus and the surrounding superior temporal plane in six postmortem brains. Grey shading illustrates how the topography of the primary region varies in its association with the two sulci that lie anterior (FTS) and posterior (HS) to the gyrus. Reproduced from Rademacher *et al.* (2001) with permission.

generated from the second, moderately long-lasting, high magnitude, positive component that starts about 300 ms after the stimulus onset. This phase probably reflects an increase in the oxyhemoglobin (and concomitant decrease in the deoxyhemoglobin) concentration that is due to the surge in arterial blood volume and flow following neural activity. However, the precise relationship between the phases of the BOLD response and neural activity is still debated (Arthurs and Boniface, 2002; Heeger and Ress, 2002; Logothetis and Wandell, 2004). The spatial resolution of the functional images is generally about 3 mm^3. The true spatial precision of the activation maps generated from the positive BOLD phase may be somewhat poorer than this because a greater proportion of this signal component derives from changes in larger draining veins than it does from smaller venules and capillaries which are more closely located to the site of neuronal activity (Hall *et al.*, 2002a). Despite these caveats, the hemodynamic signal is sufficiently co-localized to the site of population activity to be informative and the major emphasis of fMRI (and PET) research has been in measuring the spatial extent, magnitude, and topographical distribution of the signals.

The huge rise in the popularity of BOLD fMRI is a testimony to its benefits for functional brain mapping. First, since the signal contrast has an intrinsic origin, fMRI presents a significantly lower safety risk than PET. Hence, the image acquisition capacity is limited more by the hardware and software configurations of the magnetic resonance scanner and the tolerance of the subject than by the level of personal risk. Second, fMRI studies can typically run into many hundreds of image acquisitions, enabling more sophisticated experimental designs with greater numbers of different sound conditions and valid statistical inferences to be made from individual data rather than group averages. Without the necessity for inter-subject averaging, much less spatial smoothing needs to be imposed on the data (e.g., 4 mm). For example, Fig. 1B shows individual data that were acquired at a resolution of 1.6 mm and smoothed by 2 mm (unpublished data from the author, 2004).

Despite these advantages, the static and time-varying electromagnetic fields of the magnetic resonance scanner pose unique technical challenges for conducting auditory research within this hostile environment. Concerted efforts by a number of research groups have helped to overcome some of the problems by designing high-quality magnetic resonance-compatible stimulus delivery equipment (Baumgart *et al.*, 1998) and by reducing the effects of intense (132 dB(A-weighted); Foster *et al.*, 2000) scanner acoustic noise (de Zwart *et al.*, 2002; Hennel *et al.*, 1999; Mansfield *et al.*, 1995). Typically, auditory fMRI strategies combine a passive method of ear protection with an experimental protocol that carefully controls the timing between stimulus presentation and image acquisition (see Amaro *et al.*, 2002, for a review). Despite these available solutions, the number of laboratories with a dedicated research program in auditory fMRI appears not to have risen in line with the exponential increase in the total number of fMRI facilities worldwide, possibly due to the additional cost implications of magnetic resonance-compatible, high-quality sound presentation equipment.

C. EEG AND MEG

Any measure that relies on changes in the vascular system (indirectly related to neuronal activity) is not going to reflect the temporal accuracy needed to observe the time-based coding of stimulus properties. While PET and fMRI best provide knowledge about the spatial location of many sensory processes, complementary time-dependent measures (EEG and MEG) are used to examine highly transient activity and temporal patterns in the signal. The basic premise of EEG and MEG is that summated and synchronized activity within a neuronal population elicits variations in the electrical and magnetic fields that can be remotely recorded on the surface of the scalp (illustrated in Fig. 1C). The columnar organization of the cortical tissue matches many of the physical constraints that determine whether neural activity is detectable by surface sensors, although synchronous spiking activity in many hundreds of thousands of neurons is required for an electromagnetic deflection to be measurable (Nunez, 1981). The measured electric potentials arise from both tangential and radial (with respect to the scalp) components of neural activity, while the corresponding magnetic field changes reflect only the tangential component. Consequently, some source-current components that are measured using EEG are not detectable by MEG (Huotilainen *et al.*, 1998).

Two different types of response are related to a stimulus event; evoked activity is time-locked to the stimulus however, induced activity is not. Time-locked responses have been used widely to explore sensory processes. An evoked potential is generally defined in terms of its neural source or the psychological process that it manifests. The amplitude, latency, and distribution of a component can be largely dependent on the physical properties of the stimulus that evoked it, while other components depend more on the subject's interaction with the stimulus. Different components of both auditory-evoked middle latency (10 to 60 ms after stimulus onset) and late responses (70 to 300 ms after stimulus onset) have been attributed to specific computational processes. While middle latency deflections tend to reflect the physical attributes of the acoustic signal—hence, stimulus processing—the later responses are more affected by the goals of the listening task—hence, the attentional and integrative aspects of auditory processing.

A number of modeling approaches have been developed which enable the neural sources to be inferred directly from the scalp measurement at a specific time period after stimulus onset. For example, electrical source analysis breaks down the signal waveform according to a summation of activity across a number of sources that can each be modeled as equivalent dipoles (Scherg, 1990). Fig. 1D illustrates the type of result obtained using a single dipole model. Analyses can be conducted using either fixed or moving dipole estimates (Lütkenhöner, 2003). However, no modeling strategy is assumption-free and no model generates a unique solution for the distribution of the scalp activity. Even within a narrow time window, the scalp distribution of both electrical and magnetic signals is difficult to interpret because multiple fields are superimposed at any given recording location and the topography is influenced by the orientation of the underlying generators with respect to both the cortical surface and scalp recording sites. Wide variations across subjects in the pattern of cortical folding means that neuronal populations in an equivalent anatomical region can generate sources with different orientations and depths to the cortical surface (Lütkenhöner *et al.*, 2003). In addition to this, electrical signals are spatially very diffuse, because the head, particularly the cerebrospinal fluid, is a highly conductive medium. The strengths of non-invasive EEG and MEG methods lie in their temporal, not their spatial, properties. Indeed, early studies estimating the location of auditory-evoked components have been criticized for their analytical and interpretive simplicity (Lütkenhöner, 2003; Lütkenhöner *et al.*, 2003).

Intracerebral electrophysiology can provide a direct and accurately localized demonstration of the electrical response to an input sound because the recording site is very close to the source of the neural activity. This technique is invasive and has been used for human studies only in exceptional circumstances. However, advances in medical procedures over the last decade now enable auditory neuroscientists to take advantage of a standard presurgical evaluation procedure for patients with severe epilepsy in which depth electrodes are chronically implanted onto the superior temporal gyrus (STG) of the non-dominant hemisphere to ascertain the generator of their seizures (Celesia, 1976; Godey *et al.*, 2001; Liegeois-Chauvel *et al.*, 1991, 1994). Intracerebral recordings provide direct and accurate information on the localization of a component's generator and so they been used to validate surface recordings.

D. Lesion Analysis

Neuropsychology is a many-centuries-old discipline that examines the abilities of patients suffering from known cortical lesions and makes inferences about the relations between brain organization and mental functioning. Lesion analysis was responsible for much of our early understanding of the neuroanatomical organization of language. In 1861, Paul Broca first argued for there being a language center in the left posterior frontal lobe, based on the postmortem examination of a patient, known as 'Tan,' who lost the ability to produce speech. Not long after Broca's localization of language production, Karl Wernicke in 1874 was able to pinpoint a center for language comprehension in the left posterior temporal lobe.

Using the lesion analysis approach, the neuroanatomical organization of hearing has been more difficult to ascertain than spoken language. Stroke or head injury often results in unilateral, not bilateral, damage and so selective disorders of auditory processing that occur in the absence of additional problems of attention or making a behavioral response, are unusual. For example, deafness due to disorders of the brain stem are rare, since complete damage to the auditory pathway also carries a risk of affecting neighboring key motor and sensory tracts and nuclei. Nevertheless, some cases of multiple sclerosis (that damages the brainstem) have been causally linked to specific auditory deficits in time-domain processing, affecting the discrimination of interaural time difference

(Van der Poel *et al.*, 1988), changes in frequency (Quine *et al.*, 1984), and changes in spectrotemporal sound patterns (Jones *et al.*, 2002). Lesions affecting higher centers of the auditory brain have also been documented. For example, auditory agnosia is a neurological disorder of complex sound perception that occurs despite normal, or nearly normal, pure-tone detection thresholds and brainstem auditory-evoked potentials. Auditory agnosia is defined by an impairment in perceiving sound patterns along the dimensions of frequency, time, amplitude, or space. Three separate subtypes have been described in which patients have difficulties in perceiving words, music, and environmental sounds, respectively (see Griffiths *et al.*, 1999, for a review). Auditory agnosias follow a broad distribution of lesion sites involving temporal, frontal, and parietal lobes, but damage to the primary and/or non-primary auditory cortex on the STG appears to be critical to the disorder and a majority of cases report bilateral infarcts to this region. The lack of systematic psychophysical investigation is probably the single greatest obstacle to providing a unified account of the human neuropsychological literature on disorders of auditory perception. However, if detailed psychophysical testing is appropriately conducted, the neuropsychological approach provides an invaluable tool for identifying which parts of the auditory brain are necessary for a specific hearing process.

III. Variability in Structure-Function Relations

The rise in the popularity of functional neuroimaging has nurtured an associated interest in acquiring high quality *in vivo* images of human brain structure. In particular, the advent of neuroimaging techniques, such as fMRI and PET, which provide spatially accurate measures of the functional responses to both sensory and cognitive events, demand high-resolution anatomical images for interpreting the spatial location of that functional activity. A traditional strategy for establishing a comprehensive brain atlas is based upon linking specific sensory or cognitive functions to their gyral or sulcal location in the human brain because, for many years, it was understood that these macroscopic anatomical landmarks had an important physiological relevance. This section explains the limitations on the confidence with which one can associate brain function with a particular anatomical structure, first, because the macroscopic anatomical landmarks have little functional relevance, and second, because there is high variability across individual brains. Smoothing and averaging data across individuals reduces some of these concerns, but at the cost of losing some of the richness of the dataset.

Most of the human auditory cortex lies within a deep fold of the cortical surface that forms the boundary between the temporal and frontal lobes. The auditory cortex itself occupies a region called the supratemporal plane, the upper surface of the STG. In the human brain, there is considerable anatomical variability in the position, size, and shape of cortical structures; this is particularly true for the STG. Nevertheless, certain topographical components are consistently present. One of the most distinctive visible landmarks is an anterior-most transverse gyrus that cuts obliquely across the supratemporal plane in an antero-lateral to postero-medial direction (Steinmetz *et al.*, 1989). This gyrus was first identified in 1855 by an Austrian anatomist, Richard Heschl, and is named after him: the transverse gyrus of Heschl (HG). HG is bounded by the medial margin of the temporal lobe, adjacent to the insula, and by the lateral lip of the Sylvian fissure. Its anterior border is defined by the first transverse sulcus (FTS) and its posterior border is defined by Heschl's sulcus (HS). While a single gyrus is most common (70% of hemispheres) (Rademacher *et al.*, 2001, see Fig. 3), HG can be intersected by intermediate sulci and duplications of up to five HG have been reported in postmortem brains (Campain and Minckler, 1976). The large undulating cortical surface extending behind HG to the end of the Sylvian fissure is known as the planum temporale (PT). PT is another well-known macroscopic anatomical feature. The anterior border of PT is marked by HS, or behind the first intermediate sulcus in cases where HG is duplicated. Its medial border occurs where PT fades into the insula and its lateral border extends to the convexity of the STG.

Obviously, any scheme that ascribes a specific functional role to specific macroscopic anatomical landmarks relies on a consistent one-to-one mapping between structure and function. Recent data challenge this view. The cortex displays a high degree of tissue differentiation and connectivity which, when stained appropriately, can be viewed under the microscope and quantified. Animal studies demonstrate that microscopic anatomical landmarks, such as patterns of cell staining across the cortical layers, are more tightly coupled with functional specificity and input-output relations than are macroscopic features, such as gyri (Kosaki *et al.*, 1997; Morel *et al.*, 1993). Hence, architectural characteristics are now providing a reliable and independent marker for functionally specific regions. While this strategy is suitable for establishing the principles of structure-function variability, there is no opportunity to match individual results with corresponding functional data in humans because the required histological staining can only be performed on postmortem tissue. Functional activation often varies in its spatial distribution from one individual to another, and we might surmise that this is related to the supporting microscopic anatomical variability (see Fig. 3).

It is beyond the scope of this chapter to present a thorough examination of all of the evidence for anatomical variability in the auditory cortex (see Hackett, 2003, for a comparative review). For many years, FTS and HS were taken to represent the anterior and posterior borders, respectively, of the primary auditory field (Leonard *et al.*, 1998). In terms of its microscopic structure, the primary auditory field is characterized by densely packed small granule cells in layer IV and thick layers V and VI. Surrounding the primary region are non-primary (belt and parabelt) fields, distinguished by a reduction in the granularity and cell density in the outer layers and an increase in the size of the pyramidal cells in layer III. This architecture can be demonstrated using a silver stain for highlighting cell bodies and the human data have been compared to the core (primary) belt and parabelt (non-primary) divisions of the primate auditory cortex. Using architectonic criteria to identify the primary auditory field in each of 27 human brains, Rademacher *et al.* (2001) has shown that the size and location of the primary region varies somewhat independently from both FTS and HS. Likewise, in those brains containing duplications of HG, the intermediate sulcus does not reliably represent the posterior border of the primary auditory field. Rademacher concludes that HG serves only as an approximate visible marker for the location of the primary field, since the sulci do not define its borders. PT is another part of the auditory cortex that shows little correspondence between its macroscopic and microscopic structure. First, there has been some ambiguity about the macroscopic anatomical criteria by which the borders of PT have been defined (Westbury *et al.*, 1999). The ascending and descending branches that form a '>' shape at the end of the Sylvian fissure are generally taken to mark the posterior boundary of PT. Yet for some brains, it is difficult to distinguish a strongly curved PT, terminating in a short '>,' from a flat PT terminating in a relatively long ascending branch. Since right PT tends to bend more sharply than the left, errors are more easily made in quantifying right PT. Westbury *et al.* (1999) defined a set of operational rules for identifying the posterior border of PT based on the observation that the posterior wall behind the '>' always bends forward while any ascending slope of PT does not. When PT is defined in this way, the significant hemispheric differences in volume, that have previously been described (Galaburda *et al.*, 1978), disappear. The region defined as PT has no clear visible macroscopic subdivisions, yet this does not mean it represents a single functional field. Even the early anatomists recognized systematic variations in the cellular distribution across human PT (Brodmann, 1909) and, on the evidence of different histological staining profiles, at least three subdivisions have recently been identified in the anterior part of PT alone: the lateral area, posterolateral area, and supratemporal area (Rivier and Clarke, 1997; Wallace *et al.*, 2002). The distribution of markers for neural activity (parvalbumin and acetylcholinesterase) delineate a separate narrow region in the base and walls of HS which could represent an additional primary field because of its characteristic layer IV. The range of individual differences in the location of these multiple auditory regions behind HG is currently uncertain because studies have only been conducted on small samples (two brains, Rivier and Clarke, 1997; and six brains, Wallace *et al.*, 2002).

The message for the neuroscience community is that individual differences in the pattern of auditory activation might arise either from anatomical variability or from individual differences in sensory processes. Hence, it is misleading to draw a generalized conclusion from data that have been averaged over the whole sample using conventional methods (Hall *et al.*, 2003; Lütkenhöner *et al.*, 2003). A preferable approach would determine activity patterns for individual subjects, then average those functional results in a more anatomically informed way. In the future, it should be possible to make statements about the structural-functional correlations by making *in vivo* measures of brain microstructure, as well as function. Quantitative investigations using postmortem brain tissue have already demonstrated that significant changes in transmitter receptor density coincide well with traditional architectonically defined borders (Zilles *et al.*, 2002). Research is already underway to compare *in vitro* and *in vivo* measures acquired using PET. The results from a single subject confirm that of one these receptor binding sites ($GABA_A$) fits to the *in vitro* receptor autoradiographic data.

IV. Single Frequency Tones

Von Békésy (1947) developed the place theory of hearing by revealing that sinusoids produce a traveling wave along the cochlear partition that maximally displaced the basilar membrane at a specific place corresponding to the CF of the sound. In simple terms, the cochlea acts by separating sound frequencies according to the location along the membrane where the *hair cells* are stimulated. High-frequency sound information is represented within those nerve fibers near the base of the cochlea and lower-frequency information towards the apex (shown schematically in Fig. 4A). The cochleotopic representation persists throughout the auditory pathway and multiple complete representations of the audible frequency range are present (Kosaki *et al.*, 1997; Merzenich and Brugge, 1973). In the primary cortex, neurons tuned to the same sound frequency are co-localized in a band across the surface, with an orderly progression of frequency tuning across adjacent bands. It is generally observed that frequency tuning is sharper in the primary auditory fields than in the surrounding non-primary cortical regions.

The mapping of frequency sensitivity (tonotopy) has for many years been the dominant electrophysiological method for defining fields across the mammalian auditory cortex (Kosaki *et al.*, 1997; Merzenich and Brugge, 1973; Morel *et al.*, 1993). At the boundary where one field abuts another, the direction of the frequency progression reverses or the tonotopic axis clearly breaks down. The resulting activity pattern provides a distinctive marker for defining the functional boundaries between adjacent auditory fields. For example, within the primate

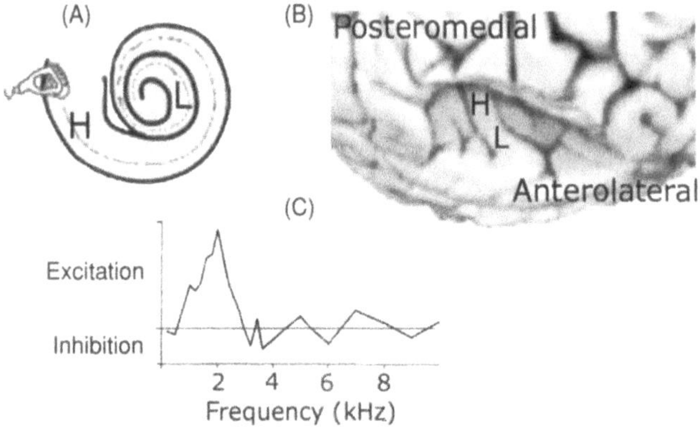

FIG. 4. Place specificity of frequency coding in the human cochlea (A) and on Heschl's gyrus (B). The frequency tuning curve for a unit in the primary cortex (C) displays a single-peak best-frequency (2000 Hz) with regions of surrounding inhibition. Reproduced from Howard *et al.* (1996) with permission.

primary region, two tonotopically organized areas (the first primary field and the rostral primary field) can be separated at their shared low-frequency border. Electrophysiological recordings from primate cortex confirm that frequency gradients are present in many auditory fields, but they have not been fully investigated across all known auditory fields (Kaas *et al.*, 1999; Merzenich and Brugge, 1973; Morel *et al.*, 1993). Establishing precise correspondences across species is difficult, but the current working hypothesis is that the human and non-human primate auditory cortices share a common organizational framework that supports both parallel and hierarchical processing (Kaas and Hackett, 1998). A detailed mapping of the tonotopic layout across the human auditory cortex would represent one of the many important steps towards assessing the validity of this model. This section describes the considerable effort that has been put into demonstrating the spatial organization of frequency sensitivity in human auditory cortex.

Human intracerebral measures of auditory-evoked potentials support the view that HG contains several functional subdivisions. For example, early evoked potentials (13 to 26 ms), indicative of the primary auditory field, are restricted to the posteromedial part of HG, while latencies increase to the range 50 to 75 ms as the location of the recording site is moved towards the anterolateral direction (Liegeois-Chauvel *et al.*, 1991, 1994). Howard *et al.* (1996) reported measurements of the firing activity pattern evoked by pure tones in electrode recording sites within the posteromedial portion of HG (Fig. 4B). Frequency tuning curves were calculated from the peristimulus time histograms by using a measure of the number of action potentials per second in the period 0 to 100 ms after stimulus onset divided by the background firing rate. Many units in the posteromedial portion of HG exhibited sharply tuned excitatory best-frequency responses to pure tone stimuli and inhibitory responses to tones that were just off best-frequency (Fig. 4C). The best frequencies of the three recording sites showed a systematic progression in sensitivity from high (3360 Hz) to low (1480 Hz) in a medial to lateral direction. The frequency-selective characteristics in posteromedial HG are consistent with experimental animal microelectrode studies in primary regions of the auditory cortex (e.g., Phillips and Cynader, 1985).

Non-invasive methods for human brain mapping lend themselves more readily to measuring frequency responses. The results from PET have been slightly disappointing because they have not provided clear evidence for a tonotopic gradient, but they have at least confirmed the presence of a dominant frequency separation across HG. For example, an early study reported a spatial separation of the peak responses to 500-Hz and 4-kHz tones along HG, with the 500-Hz tone response being more anterior and lateral (Lauter *et al.*, 1985). Almost 15 years later, a study by Lockwood *et al.* (1999) demonstrated the same basic displacement in the anterolateral direction between responses to 500-Hz and 4-kHz tones.

Source estimates derived from electromagnetic recordings of the late transient deflection (the N100 potential or M100 field) have also exposed a dominant sensitivity to frequency (Pantev *et al.*, 1988). Evidence for tonotopy has been provided by other categories of auditory-evoked fields, for example, the sustained field (Huotilainen *et al.*, 1995; Pantev *et al.*, 1994), the steady-state response (Pantev *et al.*, 1996; Romani *et al.*, 1982), and peak of the middle-latency response, 'Pam' (Pantev *et al.*, 1995). However, the most common marker for frequency sensitivity remains the 100-ms late transient deflection, possibly because it is present in all normally-hearing listeners and is easy to record. The conventional approach to modeling the dipole source of the late transient deflection has assumed a single prominent frequency-dependent generator within the superior temporal region. Coupled with the presentation of multiple sound frequencies, this modeling approach suggests that frequency-sensitivity is linearly arranged on the supratemporal plane in a gradient according to log(frequency). Typically, the axis of the gradient is in the medial-to-lateral orientation, with the dipole source being located closer to the lateral surface as sound frequency decreases (Pantev *et al.*, 1988, 1995; Verkindt *et al.*, 1995). However, not all of the available data support this interpretation. For example, some studies fail to find any frequency dependence of the N100/M100 dipole location, but instead report that non-spatial factors such as latency are important (Lütkenhöner *et al.*, 2003; Roberts and Poeppel, 1996; Stufflebeam *et al.*, 1998). Evidence is gathering to indicate that the electromagnetic fields used to map frequency sensitivity are more spatially and temporally complex than can be accounted for by many of the current modeling strategies. In particular, it has been reported that the estimated dipole source for the M100 is not fixed in a single position, but clearly moves by about 1 cm when analyzed across a narrow time window centered around the peak of the deflection (Lütkenhöner and Steinsträter, 1998; Rogers *et al.*, 1990). The spatial shift in neural activity might either represent the spreading of the activation across cortical columns or the sequential activation across several fixed, independent sites. Even when the data are modeled using a moving trajectory approach, the different analysis strategies for source estimation can still give rise to uncertainty about the precise location of the dipole source that is in the order of several mm (Lütkenhöner and Steinsträter, 1998).

A number of other, more serious, problems for conventional dipole modeling of tonotopy have been discussed in the literature and call into question the precision of many of the conclusions from electromagnetic recordings. These are discussed at some length by Lütkenhöner and Steinsträter (1998) and Lütkenhöner *et al.* (2003), but are summarized here. First, there is an incorrect assumption often made by those who use the current dipole model that the contribution of the contralateral auditory cortex can be ignored. Second, the complexity of the M100 phenomenon indicates that it is unlikely to represent a single auditory generator. Supporting this view, evidence from intracerebral recordings indicates that the M100 probably arises from several non-primary auditory areas on the lateral parts of HG and PT (Godey *et al.*, 2001). Third, the high degree of intersubject variability ensures that grand average fails to reflect any group trend, because individual trends are typically oriented in different directions. One account for the high intersubject variability follows from the claim that the 100-ms deflection arises from multiple cortical generators, only some of which might be tonotopically organized. In this case, the detection of a tonotopic organization would be highly dependent upon those contributing fields being aligned in a manner that is favorable for their detection by EEG and MEG techniques (Lütkenhöner, 2003). Despite this skepticism, high-precision dipole source analysis methods that take into account the temporal complexity of the dipole source, and do so for a single listener, provide some confidence that the M100 reliably reflects a tonotopically organized source in the auditory cortex (Lütkenhöner *et al.*, 2003), that probably includes a field within PT (Lütkenhöner and Steinstrater, 1998).

Given that fMRI provides good spatial resolution and the analysis requires few *a priori* modeling assumptions, it is an ideal tool for exploring the spatial distribution of the frequency-dependent responses across the human auditory cortex (see Hall, 2004, for a review). Nevertheless, early reports were probably much influenced by the views at the time and fMRI results tended to find the same depth-sensitive change in the location of the response peak from low to high frequencies that had been claimed from other sorts of imaging data. For example, Bilecen *et al.* (1998) reached this conclusion by comparing the topographical distribution of voxels significantly activated by 500-Hz and 4-kHz tones relative to a resting baseline, after discarding those voxels activated by both frequencies. A single medial-lateral frequency axis certainly dominates the pattern of frequency-dependent activation, illustrated here in Fig. 5 using data from our laboratory (unpublished data from the author, 2004). More recent fMRI studies have sought to determine the spatial arrangement of multiple tonotopic maps that, based on the primate electrophysiological data, are expected to exist in the human auditory cortex (Formisano *et al.*, 2003; Schönwiesner *et al.*, 2002; Talavage *et al.*, 2000, 2004). Perhaps an equally important contribution to progress is that these researchers have avoided averaging data across individuals, preferring to report common patterns using the more qualitative method of visual comparison. In those studies where characterization of the frequency sensitivity in non-primary auditory cortex has been of particular interest, dynamic band-limited signals have been presented in preference to pure sinusoids because they generate more robust activations. For example, Talavage *et al.* (2000, 2004) favored the use of amplitude-modulated signals and Schönwiesner *et al.* (2002) favored the use of sine tones that were frequency modulated across a narrow bandwidth rather than their steady-state equivalents. Talavage *et al.* (2000) contrasted pairs of low (below 66 Hz) and high (above 2490 Hz) frequency stimuli of moderate intensity and sufficient spectral separation to produce spatially resolvable differences in activation (low > high and high > low) across the putative tonotopic

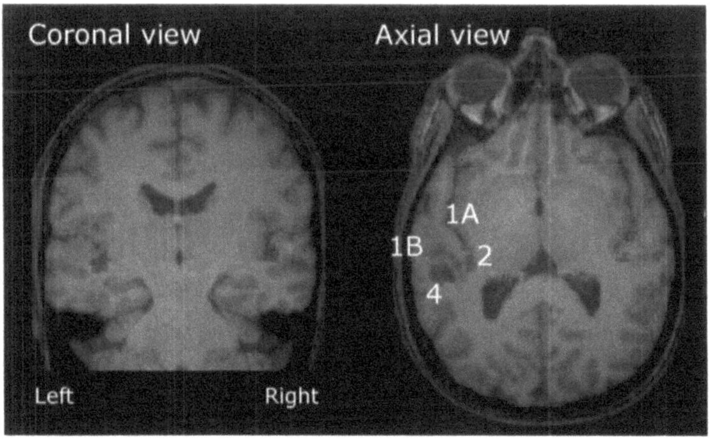

FIG. 5. Coronal and axial views of frequency-sensitivity in a single listener. Red regions represent activity that was significantly greater for stimuli with 800-Hz center frequency than for 4.5 kHz center frequency ($p < 0.001$). Blue represents the significant converse comparison. The coronal view illustrates the dominant medial-lateral frequency axis that has been described by electromagnetic studies. Numerical labels identify the multiple frequency-sensitive sites around HG as defined by Schönwiesner *et al.* (2002). The source is unpublished data from the author. (See Color Insert.)

fields. In summary, they defined eight frequency-sensitive sites on HG and the surrounding superior temporal region. Each site was repeatedly identified in a majority of the eight imaging sessions and had a fairly consistent position relative to the other frequency-sensitive sites and to visible structural landmarks. Hence, the sites were defined by a numerical label (1–8). The spatial arrangement of the frequency-sensitive foci is illustrated in Fig. 6, for one hemisphere. The first four numbered sites were located in the immediate vicinity of the medial two-thirds of HG, putatively the primary auditory cortex. Region (1) was a low-frequency sensitive site on the superior aspect of HG, two-thirds of the length from the medial end of the gyrus, region (2) was a high-frequency sensitive focus on or near the anteromedial aspect of HG, region (3) was a high-frequency focus located on or just posterior to HG, possibly in HS, and region (4) was a high-frequency sensitive site on or near the most posteromedial part of HG. Although not providing any direct indication of tonotopic organization of architectonical areas, Talavage *et al.* argued that the first three frequency-sensitive sites are at least consistent with the locations of the low- and high-frequency end points of two tonotopic primary fields that are predicted by the model of the primate auditory cortex; the two fields cross the primary auditory cortex, they share a common low-frequency border, and the direction of one frequency axis is a mirror-image

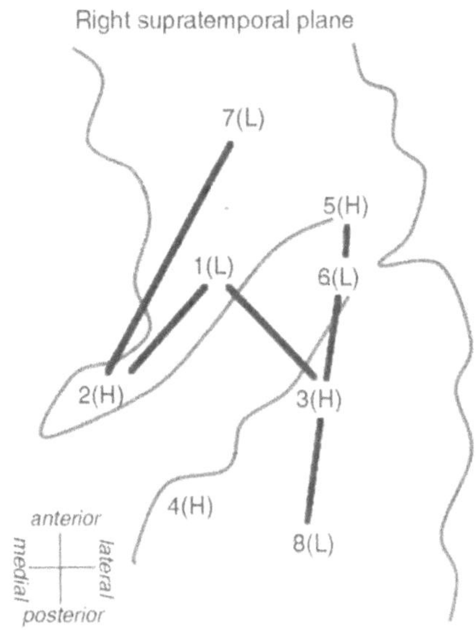

FIG. 6. Surface of the auditory cortex showing an outline of Heschl's gyrus. The location of the eight numbered frequency-sensitive foci identified by Talavage *et al.* (2000, 2004) are overlaid onto the cortex. The letter in parentheses denotes whether the region is sensitive to low (L) or high (H) frequencies.

reversal of the other. The tonotopy argument was strongly criticized in a paper by Schönwiesner *et al.* (2002). Schönwiesner *et al.* reported low (below 500 Hz) and high (above 4000 Hz) frequency-sensitive foci that broadly co-localized with those reported previously by Talavage *et al.* (2000)—a small difference being that what corresponded to region (1) was further subdivided into two distinct low-frequency sensitive sites; region (1a) overlapping with the anterior aspect of HG and region (1b) overlapping with the posterior aspect of HG. Again, these low and high frequency-sensitive regions are generally replicable and are represented for a single listener in Fig. 5 (unpublished data from the author, 2004). Schönwiesner *et al.* argued that it was wrong to associate these foci with specific tonotopic fields because pairs of low- and high-frequency foci could not clearly be attributed to specific frequency axes nor to anatomically defined fields. Furthermore, Schönwiesner *et al.* did not observe the predicted gradual decrease in frequency-response amplitude at locations away from the best-frequency peak, but instead observed rather complex response profiles. The alternative interpretation favored by Schönwiesner *et al.* is that regions of frequency sensitivity reflect distinct cortical areas that each prefer different acoustic features that are associated with a limited bandwidth signal. An example from our laboratory is the cortical representation of interaural correlation (IAC). IAC is a binaural characteristic that carries information about the source of the sound and is largely confined to low-frequency channels. Human fMRI results have revealed the presence of an IAC-sensitive region in the same lateral portion of HG that independent studies have shown to respond preferentially to low-frequency signals (Budd *et al.*, 2003).

Direct evidence for tonotopy has recently emerged from two fMRI studies that mapped a progression of activation as a smooth function of tone frequency, in HG (Formisano *et al.*, 2003) and more widely across the supratemporal plane (Talavage *et al.*, 2004). Frequency sensitivity in HG was studied at 1 mm^3 spatial resolution using a novel ultra-high field technology to improve the signal-to-noise ratio (SNR) (Formisano *et al.*, 2003). Two linear frequency progressions were mapped across HG by computing the locations of the best response to pure sinusoids presented at a number of discrete frequencies (0.3, 0.5, 0.8, 1, 2, and 3 kHz). Flattened cortical maps of best frequency revealed two mirror-symmetric gradients (high-to-low and low-to-high) traveling along the depth of HG from an anterolateral point to the posteromedial extremity in each of the six listeners. In general, the amplitude of the response monotonically decreased as the difference between pure tone frequency and best frequency increased suggesting that a high SNR might be required to characterize some of the more subtle changes in the response away from best frequency. Imaging the auditory cortex by using a receiver coil placed close to the scalp is another way to achieve a high SNR and this was the method used by Talavage *et al.* (2004) to generate flattened tonotopic maps for HG and beyond. Fig. 6 illustrates the six frequency progressions that were revealed by measuring best frequency responses to a signal that slowly swept across frequencies between 0.1 and 8 kHz; these frequency progressions were consistently identified in all seven listeners. Again, two mirror-symmetric maps that shared a low-frequency border were reported to cross HG, traveling from the anterior to the posterior sulci. The remaining four gradients occurred in non-primary regions of the auditory cortex extended across the supratemporal plane towards the lateral convexity of the STG: one gradient anterior to HG, one in the lateral portion of HG, and two posterior to HG. Although there are difficulties in associating each frequency progression with a particular anatomically-defined field, Talavage *et al.* nevertheless suggest relationships between the location of their functional maps and various architectonic parcellation schemes (Galaburda and Sanides, 1980; Rivier and Clarke, 1997; Wallace *et al.*, 2002).

Although more evidence will be required before a clear consensus is established, these recent studies have made important advances in characterizing the functional anatomy of human tonotopy, especially in the primary auditory region on the medial two-thirds of HG. In order to validate homologies between primate and human tonotopic organization future studies should acknowledge several key methodological differences, including the sound presentation level (SPL) employed, effects of anesthesia, and the frequency range. First, it is important to realize that the response pattern of a given neuron is not rigidly fixed, but can change as a function of other acoustic features, including SPL. Human tonotopic maps have, in the main, been created using acoustic signals presented at moderate to high sound levels (e.g., 42 to 50 dB above threshold; Talavage *et al.*, 2000), while microelectrode recordings in animals identify the location of the best response at the near-threshold level of acoustic stimulation. The cortical area responsive to a suprathreshold stimulus cannot be predicted from the threshold map. The spatial pattern of auditory cortical activity evoked by narrowband signals depends on sound level in complex ways, since populations of frequency-specific units include those with monotonic and non-monotonic rate/level functions (Phillips *et al.*, 1994). At high sound levels, units with non-monotonic rate/level functions respond poorly while the remaining active units are widely spread across the cortex giving a pattern of suprathreshold responses that bears little relationship to the best frequency map at threshold. Since one expects human auditory cortical neurons to also exhibit monotonic and non-monotonic response properties, the reported spatial distribution of frequency sensitivity may not be replicated at near-threshold sound levels. Second, the concepts of tonotopic organization that form the basis for interpreting human neuroimaging data are derived from microelectrode recordings in the deeply anesthetized animal which reveal primary auditory neurons with narrowly tuned responses to frequency (Merzenich and Brugge, 1973). Neuronal responses measured from rat primary auditory cortex suggest that the representation of sound frequency might actually form a more complex pattern in the awake than in the anesthetized state, with fewer neurons showing the classical V-shaped tuning to best frequency (Gaese and Ostwald, 2003). Perhaps such rather more complex maps measured during the awake, behaving state should be used to provide a more appropriate reference for interpreting human tonotopy. Third, human results have not quite fully sampled the entire human audible frequency range, leaving the highest-frequency portion of the frequency-gradient unmeasured. The audible range for human listeners extends from 60 Hz to up to 20 kHz, yet the widest range used for fMRI frequency mapping has been from 125 Hz to 8 kHz (Schönwiesner *et al.*, 2002; Talavage *et al.*, 2000, 2004). In contrast, best frequency maps in the primate auditory field have been measured for tone frequencies from below 100 Hz to about 32 kHz, which is nearly the full frequency range over which that species is known to hear (Merzenich and Brugge, 1973). Future research in this field should attempt to clearly ascertain the spatial pattern of human frequency-sensitivity across the audible range of frequencies, presented close to the listener's hearing threshold.

V. Wide-Spectrum Signals

In the macaque brain, the belt and parabelt regions extend out from the core to the exposed surface of the STG. Although all regions receive some thalamic input, belt regions receive their major projections from the core, while parabelt regions receive inputs from adjacent belt regions. The anatomical connectivity of the auditory cortex suggests that each of the three regions represents a distinct level of auditory processing. This organization is apparent in the response characteristics of the three regions, especially in their response to the frequency spectrum of the acoustic signal. Whereas neurons in the core regions are highly tuned to single tone frequencies at low SPLs, neurons in lateral belt areas prefer broadband stimuli (Rauschecker *et al.*, 1995). The difference in response properties is most likely to be explained by the convergence of afferents from the core region to adjacent belt regions. The lateral belt regions do preserve some tonotopic structure, revealed by mapping isofrequency bands using narrow-spectrum noise at different center frequencies, but the isofrequency bands are much less tightly defined than in the core regions. The broad frequency tuning of non-primary neurons has been demonstrated many times. For example, Kosaki *et al.* (1997) reported that many belt neurons responded at threshold to a wide range of tone frequencies that often extended over a range of 300 to 1000 Hz. In the surrounding parabelt regions, neurons responded very poorly to pure tones but could often be driven by wide-spectrum noise (see also Merzenich and Brugge, 1973).

Neuroimaging studies have sought to confirm the presence of the same hierarchical pattern of frequency tuning across the human auditory cortex. A neuroimaging measure of population synaptic activity would predict more activation within the primary cortex in response to wide-spectrum noise than to single frequency tones (due to the greater inhibitory and excitatory activity within frequency channels) and would also predict greater activation in surrounding regions of the non-primary cortex. PET and fMRI have been used to probe the response sensitivity across primary and non-primary auditory cortical regions. Generally, results are in good agreement with our *a priori* expectations.

An early study by Mirz *et al.* (1999) used PET to measure activations by single frequency tones and wide-spectrum (50 to 5000 Hz) noise. The conclusions from this study are limited because the conditions of interest were each contrasted with a common silent baseline rather than directly with one another, and the tone-evoked activation was illustrated, but the noise-evoked activation was not. Nevertheless, the statistical value of the peak of activation was reported to be greater in HG for the noise than for the tone, a result that is at least consistent with the model predictions. More recently, Wessinger *et al.* (2001) presented the spatial distribution of responses to single frequency tones and one octave, bandpass noises with the same center frequency, but again each condition was contrasted only with a silent baseline condition, rather than directly with one another. The results showed that single frequency tones generated restricted foci of activation that generally included the medial portion of HG, but was somewhat variable across subjects. In contrast, bandpass noise produced larger regions of activation that incorporated the single frequency activation and extended around it onto surrounding parts of the auditory cortex. A subsequent study confirmed that the difference between single frequency tones and wider-spectrum signals does reach statistical significance. By directly contrasting single frequency tones and center-frequency matched, 2.6-octave wide, harmonic complex tones, Hall *et al.* (2002b) were able to show that spectral complexity was associated with an increase in activation in the primary auditory cortex on HG and in portions of the adjacent posterior and lateral non-primary fields. Both harmonic complex tones and wide-spectrum noise are equally effective at generating greater activity than a single-frequency tone baseline; the latter comparison is illustrated in Fig. 7 (unpublished data from the author, 2004).

Wessinger *et al.* (2001) claimed that comparisons between single frequency and wide-spectrum based activation might provide a marker for differentiating primary (core) from non-primary (belt) areas. However, this ambition is unlikely to be realized on several counts: (1) the borders are determined by simple statistical criteria, (2) the analysis is applied to highly smoothed image data, and (3) relative to single frequency tones, wide-spectrum signals significantly increase the amount of activation in both primary and non-primary regions (Hall *et al.*, 2003).

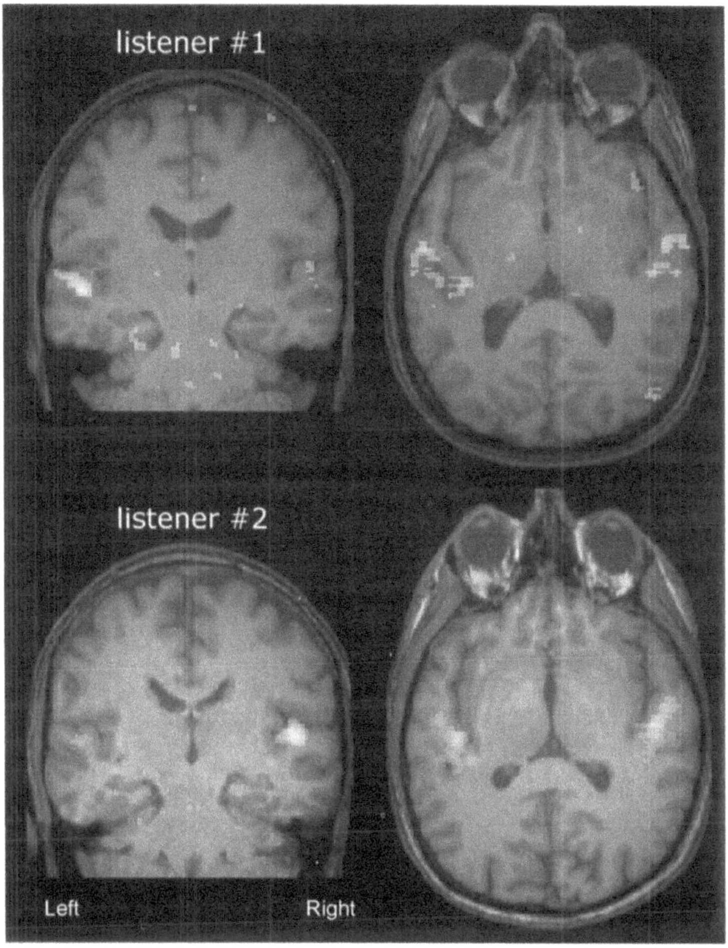

FIG. 7. A comparison between single-frequency tones and wide-spectrum noise that had a 0.75 octave bandwidth centered at the same frequency ($p < 0.001$). Coronal and axial views are shown for two listeners. Listener #1 is the same as that presented in Fig. 5. Noise evokes greater activation in medial and lateral subdivisions close to HG, although its precise distribution is not identical across listeners (based on unpublished data from the author). (See Color Insert.)

VI. Pitch

Unlike frequency which is a physical attribute of sound, pitch is a perceptual attribute that is related to the temporal regularity of the sound. It is likely that at least two complementary neural mechanisms are responsible for the perception of pitch, based on the location and timing of neural excitation, respectively. The location (place) account proposes that a pitch is related to the position along the basilar membrane activated by the spectral components present in the signal. This peripheral mechanism can explain pitch perception for pure tones and harmonic-complex sounds where the individual spectral components are resolved within separate frequency channels. For these sounds, pitch arises from a detection of the harmonically related spectral peaks. For complex sounds with unresolved harmonics which violate this rule, pitch can be determined centrally by the temporal pattern of neuronal spiking generated by the periodic envelope (amplitude modulation) of the unresolved spectral components. For example, amplitude modulations of a carrier signal elicit the perception of a periodicity pitch that corresponds to the pitch of a sine wave whose frequency matches the modulation envelope. Pitch processing mechanisms most probably exploit both spectral and fine-timing information (Shamma and Klein, 2000). What biologically plausible mechanisms might perform these computations are unknown, but the auditory midbrain is a candidate for the convergence of information to form an abstract representation of pitch (Ehret, 1997). In the cortex, the response to pitch is primarily driven by perceptual attributes of the sound, such as the pitch strength, rather than by physical factors such as whether the harmonics are peripherally resolved or unresolved (Penagos *et al.*, 2004). Although the biological substrate of pitch is not yet known, the pattern of pitch-related activation in the auditory cortex suggests that the cortex might deal with pitch in a single manner, irrespective of whether the acoustic cues are spectral or temporal, because activity is strongly driven by the perceptual attributes of sound (Hall *et al.*, 2004). Neuroimaging studies first followed the animal model in searching for a cortical map of pitch sensitivity, while more recent investigations have more modestly identified which region might be sensitive to pitch information.

The animal model is based on a neural theory that neurons which are tuned to best modulation frequencies contribute to maps of periodicity in the auditory cortex (Langner, 1992). Specifically, this theory addresses the mechanism for coding the pitch of unresolved harmonics because the animal experiments have exclusively used amplitude-modulated tones containing periodicity pitch information. Claims have been made that the cortical response to periodicity is laid out in a spatially systematic manner, just as it is to frequency (Schulze and Langner, 1997). Using this technique, in the primary auditory cortex of the gerbil, Schulze *et al.* (2002) proposed a circular periodicity gradient that overlaps onto the linear tonotopic gradient, especially in the low-frequency region. The claim for a central systematic representation of periodicity, based on the evidence from these experiments, is not wholly accepted in the field. First, the periodicity map is highly inconsistent. Unlike tonotopic gradients, a periodotopic gradient is not always reproducible in every brain and even when it has been demonstrated, the precise geometric relationship with the tonotopic gradient varies across animals (Schulze and Langner, 1997; Schulze *et al.*, 2002). Second, different stimuli with the same periodicity produce different patterns of cortical activation in the same animal (Bizley *et al.*, 2003). Third, the common technique used to measure neural sensitivity to periodicity has been criticized because the amplitude-modulated, high-frequency stimuli also happen to generate cochlear distortion in the form of low-frequency spectral components that correspond precisely to the rate of the amplitude modulation (McAlpine, 2004). This spectral phenomenon can provide a simple peripheral account for the apparent differential sensitivity to periodicity.

What is the evidence for a systematic representation of periodicity in the human auditory cortex? First reports indicated that both pure tones and harmonic-complex tones with the identical pitch had a common source of the M100 in the auditory cortex, and this result led to the conclusion that the apparent tonotopic organization of the M100 signal represented the pitch rather than the spectral components of the sound (Pantev *et al.*, 1989). This claim was later challenged by Langner *et al.* (1997), who proposed that linear tonotopic and periodotopic gradients were present across human auditory cortex and were oriented at approximately 90° to one another (Fig. 8A). However, both these proposals have received general criticism of the modeling methods used to analyze the scalp magnetic field distributions. First, frequency sensitivity was

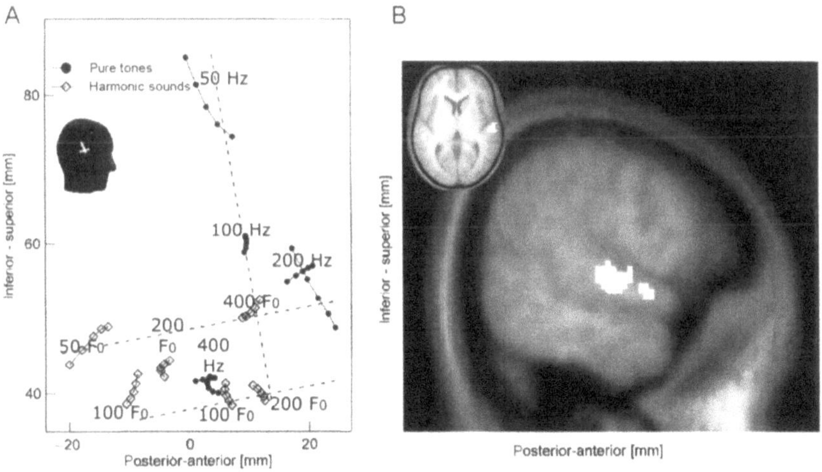

FIG. 8. Pitch representations displayed in a coronal view of the human auditory cortex, right hemisphere. (A) illustrates the putative relationship between frequency (pure tones) and periodicity (harmonic-complex tones) in a single listener. Reproduced with permission from Langner *et al.* (1997). (B) illustrates pitch-related activation obtained by contrasting a RIN stimulus with a noise matched in spectral energy in a group of 14 listeners ($p < 0.001$). The axial insert illustrates the chosen plane for the coronal slice and demonstrates that the activity is centered on lateral HG, especially in the right hemisphere (based on unpublished data from the author).

mapped using single sinusoids which do not separate frequency from periodicity representations, and second, the conclusion rests on the assumption that the mean orientation of a single dipole is the most representative summary of the individual results (Lütkenhöner, 2003). Subsequent electromagnetic studies have addressed the first criticism by independently manipulating frequency and periodicity using harmonic-complex tones, but have failed to replicate this orthogonality. Instead, the results reveal that the cortical representation of spectral content dominates over periodicity pitch, but that pitch seems to have a greater effect on the latency of the neuronal activity than on its spatial location (Cansino *et al.*, 2003; Crottaz-Herbette and Ragot, 2000). Recent psychophysical evidence suggests that timing information cannot be represented without regard to the tonotopic organization of the auditory system, since pitch discrimination for harmonic-complex tones is poor when the low-frequency temporal periodicity is carried on a high-frequency signal (Oxenham *et al.*, 2004).

Neuroimaging has been most successful in defining the region of the human auditory cortex that is sensitive to pitch processing, rather than seeking to define any periodotopic map per se. A series of studies have measured pitch sensitivity by creating a temporal pitch from a filtered noise signal by delaying and adding the noise back onto itself to generate temporal regularity (Yost *et al.*, 1996). This stimulus is referred to as a regular-interval noise (RIN). High-pass filtered RIN produces reasonably uniform excitation along the BM so that there are no harmonically related spectral peaks in the excitation pattern. When contrasted with a random noise, the filtered RIN provides a well-controlled stimulus for isolating the effects of temporal pitch. Data support a role for lateral HG in pitch processing. Lateral HG responds strongly to a RIN relative to its spectrally matched noise control (e.g., Patterson *et al.*, 2002), a finding that has been replicated several times in our laboratory (Fig. 8B). It also shows a linear response to pitch salience in the absence of changes in spectral structure, when pitch strength is manipulated by increasing the number of delay-and-add iterations of the RIN (Griffiths *et al.*, 1998). Recent MEG data also support a role for lateral HG in pitch processing. Notably, the strength of both transient and sustained fields at this generator site depend on the occurrence of temporal regularity in the click train (Gutschalk *et al.*, 2002, 2004) or in noise (Krumbholz *et al.*, 2003).

In summary, the evidence points to a distinct region of the auditory cortex, immediately anterolateral to the medial portion of HG, that represents pitch information. We hesitate to label this region a 'pitch center'—a term used by Patterson *et al.* (2002)—because this term implies that the region is specialized for pitch processing, to the exclusion of other types of auditory analysis. Response sensitivity to low-frequency signals, harmonic-complex tones, amplitude and frequency modulated tones, and wide-spectrum signals have all been associated with the same lateral region (Hall *et al.*, 2003; Scott and Johnsrude, 2003), suggesting that pitch sensitivity is just one of this region's many response properties.

VII. Spectral Cues for Sound Localization

Most commonly, headphone studies of sound localization manipulate the time of arrival and/or the level of the sound at the two ears to generate the percept of a sound source lateralized to a specific position on the azimuthal plane. In the real world, spectral cues provide an additional important contribution to spatial hearing. Specifically, the folds of the pinnae provide a complex direction-dependent filter that causes spectral changes which can be used as a cue to the location of a sound source, especially to its elevation and to front-back distinctions (Blauert, 1969; also see chapter by Carlile *et al.*). The effect of the filter is often measured by comparing the spectrum of the original signal and the spectrum of the signal reaching the eardrum. The ratio of these two spectra gives the head-related transfer function (HRTF) which shows a complex pattern of sharp peaks and dips that varies systematically with the position of the sound source relative to the head (Wightman and Kistler, 1989). The spectral cues are also important for the percept of a free-field sound source, even when that sound is presented over headphones, because the cues recreate the filtering effect of the ears. Externalized sounds have been created for experimental purposes by convolving the sound signal with a generic HRTF based on a manikin head. However, the percept is only partially realistic because the generic filter fails to capture each individual's unique HRTF that is based on the size of the head and the shape of their own pinnae. A body of human neuroimaging research has used the generalized virtual acoustic space to determine the neural substrate of sound motion processing (Griffiths and Green, 1999; Pavani *et al.*, 2002; Warren *et al.*, 2002). Essentially, these studies generally agree that, relative to stationary sounds, horizontal and vertical sound movements engage a bilateral region of medial PT, plus inferior parietal areas, premotor areas, and the superior posterior parietal cortex. Pavani *et al.* (2002) failed to find any significant differences in activation between sound sources moving in azimuth or in elevation suggesting that PT has a nonspecific spatial processing role. However, the PT region could be particularly sensitive to the presence of spectral cues for sound localization and motion processing, since left PT was more activated by speech lateralized in virtual acoustic space than to speech lateralized by interaural level cues alone, irrespective of which hemispace the sound occurred in (Hunter *et al.*, 2003). The left-sided dominance might have little bearing on spatial coding asymmetries, but more on the category of stimulus being used (i.e., speech). Nevertheless, the results of Hunter *et al.* are consistent with the view that PT is critically involved in the neural representation of complex spectrotemporal sound structure, including those that are associated with spatial percepts (Griffiths and Warren, 2002).

VIII. Speech and Music

Speech and music are both complex multi-dimensional acoustic signals that carry information in the time-varying patterns of frequency and level. For example, speech contains quasi-periodic and aperiodic segments (due to the presence or absence of voicing), amplitude and frequency modulations, and considerable spectral formant structure (due to the movements of the articulators). The acoustic pattern can be represented by a spectrogram of frequency and level variations over time (Fig. 9). The resulting waveforms for speech and music can be described as codes which must be unraveled by the listener to reveal their meaning. Complexity is built of rule-based permutations of a relatively small number of perceptual elements (phonemes or notes) to yield meaningful structure (words, sentences, phrases, and melodies). Perceiving speech and music, therefore, commonly requires a transformation of the continuous time-varying input into representations of discrete perceptual elements. Moreover, these elements are not necessarily clearly present in the acoustic waveform. Despite these similar

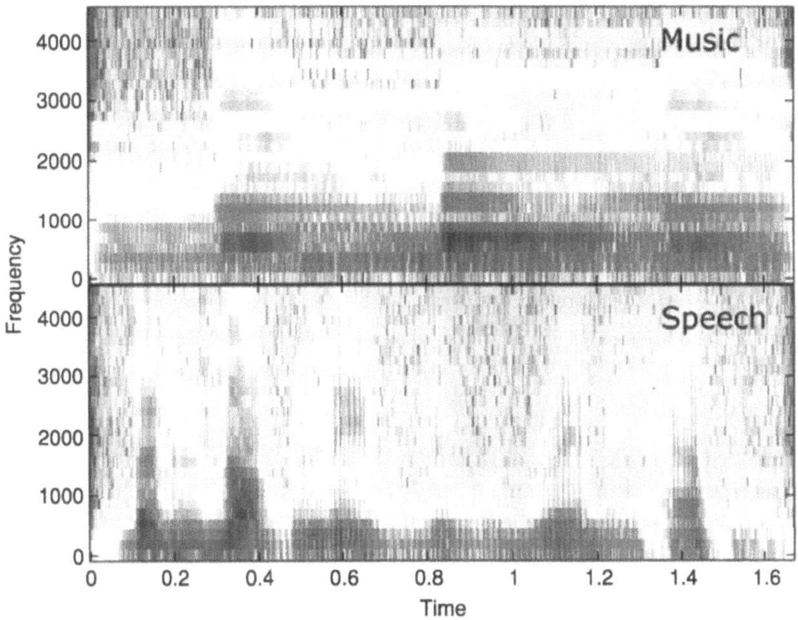

FIG. 9. Spectrograms of speech ("... but had not believed that it could really happen") and music (excerpt from Rachmaninoff's 2nd piano concerto). Time is represented on the abscissa (0 to 1.65 s) and frequency on the ordinate (0.0 to 4.4 kHz). The darkness of the trace in each time/frequency region represents the amount of sound energy in the signal.

computational requirements, some critical acoustic features differ for the perception of speech and music. For example, in Chapter 4, Shannon describes how fine spectral detail is much more important for music recognition than it is for speech decoding.

Ever since the observations of selective language deficits following left hemisphere damage made by Paul Broca and Carl Wernicke in the late 19th century, the prevailing view has been that the brain handles speech differently from other classes of sound such as music. Hemispheric specialization provides one solution to the computational problem posed by the need for precise coding of both spectral and temporal information. Agnosic symptoms after brain damage have generally favored a relative hemispheric asymmetry in that unilateral lesions to the left side are associated with word problems whereas lesions to the right are more likely to lead to non-verbal auditory deficits (Zatorre *et al.*, 2002). In the 1960s, dichotic listening studies in normally hearing listeners revived research in this field. Listeners' performance revealed that when competing speech stimuli were simultaneously presented to the two ears the right ear signal was generally better identified than the left ear signal (Broadbent and Gregory, 1964), while the reverse was true for melodies (Kimura, 1964). Given the dominant excitatory auditory inputs to the contralateral hemisphere, these findings support the view that speech is preferably coded in the left hemisphere and melody in the right. The hemispheric dissociation is not an absolute one, since left-sided damage can also impair performance on some musical tasks, particularly those involving discriminating temporal asynchrony at rapid presentation rates (Ehrlé *et al.*, 2001). However, many neuroimaging studies support the notion that discrimination of short pitch sequences and melody processing are primarily dealt with by the right hemisphere. Many of these studies are reviewed by Zatorre *et al.* (2002) and two examples are described here. For example, imagining familiar melodies generates greater activation in the right auditory cortex than listening to a random permutation of the sequence of notes in those same melodies (Halpern and Zatorre, 1999) and listening to a mixed sequence of frequent and infrequent chords, while performing a distracting visual task, generates greater right- than left-sided activation, but not when the sequence contains all identical chords (Tervaniemi *et al.*, 2000).

Melody processing involves determining both the pitch of each note and how the pitch values change from note to note. Patterson *et al.* (2002) recently attempted to distinguish the neural substrates for pitch coding and for melody processing using fMRI data from nine listeners. Periodicity pitch was generated using RIN stimuli. When sequences of same-pitch RIN were contrasted with a no-pitch noise, pitch processing was shown to differentially activate lateral HG, equally in both hemispheres. This result concurs with previous findings that lateral HG is sensitive to the presence of pitch in a sound and to its salience. To reveal regions associated with melody processing, Patterson *et al.* (2002)

conducted two statistical comparisons; one comparing novel melodies with the same-pitch sequence and one comparing random-note melodies with the same-pitch sequence. Generally, both melody comparisons generated more activity in the right than the left hemisphere and both revealed significant activation in the anterior STG, including the superior temporal sulcus and the planum polare, with no significant difference between melody types. This result suggests that the anterior auditory regions beyond lateral HG are probably involved in tracking pitch changes on a longer time scale across a sequence of notes. This result also indicates that the right hemisphere dominance in music processing, observed in neuropsychological and functional neuroimaging studies (Zatorre *et al.*, 2002), is most likely to be due to the melodic pitch changes in the sound, rather than the pitch per se.

Speech perception was one of the first topic studies using functional neuroimaging (e.g., Wise *et al.*, 1991). In order to separate the pattern of speech-specific responses from the more generalized responses to the complex structure of the acoustic signal, it has proved insufficient to simply contrast speech perception with a resting baseline or even with listening to tones or noise (Scott *et al.*, 2000). Speech is such a spectrotemporally rich acoustic signal, that designing stimuli that are equally matched for acoustic complexity but that lack the potential for intelligibility, has proved to be difficult. Even signal-correlated noise which controls many of the temporal properties of speech (see Belin *et al.*, 2000; Mummery *et al.*, 1999), is less acoustically complex because it has a flat spectrum and does not match the distinctive pitch and harmonic structure found in normal speech. Pseudowords and backward words would seem to differ from normal speech only along the dimension of speech comprehension, yet significant activation differences have not been clearly revealed using passive listening tasks (Binder, 1995).

A spectral inversion of the speech preserves most of the spectral and temporal complexity of the speech, including some phonetic features and intonation, but destroys its intelligibility, at least to the untrained listener. This acoustic manipulation has been used successfully to investigate the neural substrate of speech acoustic intelligibility (Davis and Johnsrude, 2003; Narain *et al.*, 2003; Scott *et al.*, 2000). Scott *et al.* performed the spectral inversion not only on normal speech, but also on speech signals that had previously been vocoded (described by Shannon *et al.*, 1995). Noise-vocoded speech reduces the spectral complexity and pitch salience of speech, while preserving sufficient temporal fluctuations for it to remain intelligible. By contrasting speech and vocoded speech, the effect of the dynamic pitch variations in the signal was shown to be associated with activity in the lateral STG of the right hemisphere. This right hemisphere pitch-related effect probably identifies why preceding studies, contrasting speech with a signal-correlated noise baseline, find 'speech-related' activation in both the left and right hemispheres. Activation in the lateral STG is unlikely to be speech specific, since it is also engaged by non-speech sounds such as harmonic tones and frequency-modulated tones that share some of the fundamental acoustic properties of speech (Hall *et al.*, 2002b). In contrast, the speech-specific effect of intelligibility, identified by contrasting speech and rotated speech, revealed activity in distinct anterior and posterior regions of the left temporal lobe, including the superior temporal sulcus. This result has also been confirmed by Davis and Johnsrude (2003) who systematically reduced speech intelligibility using not just one, but three different types of acoustic alteration: replacing segments of the speech with signal-correlated noise, noise-vocoded speech, and speech in a noise background. Again, speech intelligibility revealed a systematic left-hemisphere bias and the anterior and posterior aspects of the temporal lobe activity were shown to be somewhat insensitive to the different types of acoustic distortion.

The left-sided temporal lobe activation is understood to represent a speech-specific processing level beyond that which purely extracts the physical acoustic elements of the signal. Anterior and posterior regions of the superior temporal sulcus are well-placed to perform a higher level function because they are heavily interconnected with the prefrontal cortex and receive multimodal inputs. Speech is processed both as a sound and as an action and the precise nature of the task will largely determine those higher-level neural systems engaged by speech perception. A large body of electrophysiological, neuropsychological, and neuroimaging evidence supports the view that speech perception and production are intimately associated, so that the articulatory mechanisms can be engaged, even if the task does not involve speaking aloud. The cortical regions supporting speech perception appear to vary according to two different task demands; one that favors access to a mental lexicon—coding speech for meaning—and one that favors access to sub-lexical segmental information—coding speech in an articulatory context (Hickok and Poeppel, 2000). Within this model, the left posterior temporal region is responsible for interfacing between the semantic or the articulatory systems. To acknowledge the important contribution of the left anterior region to speech perception, Scott and Johnsrude (2003) propose a speech-related functional dichotomy for the left superior temporal lobe. According to this view, the anterior system is important for mapping acoustic-phonetic cues onto lexical representations and the posterior system is important for processing articulatory representations of speech acts.

Can a unifying account of the functional neuroanatomy of speech and music be put forward? Griffiths *et al.* (1999) claimed that the difficulties in perceiving music and speech after brain injury are more likely to reflect a single disorder in perceiving spectrotemporal sound patterns than they are to reflect separate deficits. They argued that published cases have rarely performed the appropriate tests for revealing a core common problem in analyzing basic sound patterns. In contrast, three frameworks emphasize the importance of different low-level acoustic cues for speech and music perception. First, laterality has been proposed to follow from the asymmetric sampling operations performed on the waveform in the time domain. Dichotic listening performance suggests that the left hemisphere supports fine temporal discriminations necessary for identifying stop consonants; hence, the left hemisphere probably has a short temporal window of integration. The right hemisphere supports the processing of intonation contour; hence, it is claimed to hold information over a longer period of time (Allard and Scott, 1975). Second, the sound frequency account of laterality is derived from a popular account of laterality in visual perception in which computational biases exist across hemispheres. In the visual domain, the left hemisphere is held to favor high spatial frequency information while the right favors low spatial frequencies (Sergent, 1982). Likewise, in the auditory domain, it is claimed that the left hemisphere is biased for processing high sound frequencies and the right for low frequencies (Ivry and Lebby, 1993). The authors use this model to explain hemispheric asymmetries in the analysis of phonetic and prosodic speech information, involving the left and right hemispheres, respectively. The sound frequency hypothesis does not clearly define the mechanism for music perception, but the third model does account for both speech and music asymmetries. According to the third view, the left hemisphere is selectively better at coding temporal differences of the order of tens of milliseconds—necessary for perceiving phonetic categories—while the right hemisphere has much better spectral resolving power, necessary for detecting changes in timbre and phrasing (Zatorre *et al.*, 2002). Hence, the asymmetry reflects hemispheric specialization for rapid temporal and fine spectral processing. Patterson *et al.* (2002) assert that it is difficult to reconcile the processing of the temporal pitch in RIN stimuli with this version of the hemispheric specialization hypothesis. According to Zatorre *et al.*'s hypothesis, the right hemisphere advantage for fine pitch judgments is a result of fine spectral processing, yet the right hemisphere RIN melody effects demonstrated by Patterson *et al.* (2002) are not a consequence of spectral processing because the RIN pitch arises from the temporal regularity in the sound pattern. A modified version of the hypothesis in which the specialization involves fine pitch tracking, rather than fine spectral processing, would provide an alternative interpretation of the data.

IX. Conclusions

With respect to the question posed in the introduction of this chapter, subdivisions of the human auditory cortex *can* be defined on the basis of their differential sensitivity to specific acoustical features. Arguably one of the most successful examples of this work, in terms of its high spatial precision, has been the topographic mapping of frequency responses around HG in individual listeners. Recent neuroimaging data also provide clear progress in drawing up the boundaries to classify which auditory regions deal with certain classes of spectral information. This chapter highlights the importance of lateral HG and PT in representing many of these sound properties. For example, lateral HG responds preferentially to low-frequency tones, wide-spectrum signals, and pitch information. Likewise, PT has been associated with at least two tonotopic gradients, wide-spectrum signals, cues for sound localization and motion, and speech intelligibility. Clearly, the evidence does not support a simple view of auditory cortical modularity in which subregions are highly specialized for processing one acoustic property. There are two alternative views. First, organizational modularity still exists but it is represented at the neuronal level (for example, by distributed or inter-digitated subpopulations of specialized neurons within a subregion). Electrophysiological studies in the primary auditory cortex reveal this type of organization with tuning to sound level, binaural characteristics, and frequency-modulation rate embedded within an isofrequency band (Read *et al.*, 2002). Second, auditory cortical regions support neural computations that are based on aspects of the sound signal other than its acoustic properties per se. Again, electrophysiological data suggest that the behavioral context and significance of the spectral cues in a sound can affect neuronal activity in the primary auditory cortex (Read *et al.*, 2002). Based on human neuroimaging data, a recent proposal claims that PT plays a generalized role in processing complex sound information by segregating the sound input into individual auditory objects, segregating those objects across time and space and matching sound patterns with stored knowledge (Griffiths and Warren, 2002). This view predicts that PT might perform similar operations on sounds that have different spectral properties if they shared the same context dependence. Hence, one way in which neuroimaging studies could significantly contribute to our understanding of the functional role of PT and other auditory regions in spectral coding would be to systematically manipulate not only the type of spectral sound information, but also the context in which those sound cues are presented.

References

Allard, F., and Scott, B. L. (1975). Burst cues, transition cues, and hemispheric specialization with real speech sounds. *Q. J. Exp. Psychol.* **27,** 487–497.

Amaro, E., Williams, S. C. R., Shergill, S. S., Fu, C. H. Y., MacSweeney, M., Piccioni, M. M., Brammer, M. J., and McGuire, P. K. (2002). Acoustic noise and functional magnetic resonance imaging: Current strategies and future prospects. *J. Magn. Reson. Imaging* **16,** 497–510.

Arthurs, O. J., and Boniface, S. (2002). How well do we understand the neural origins of the fMRI BOLD signal? *Trends Neurosci.* **25,** 27–31.

Baumgart, F., Kaulisch, T., Tempelmann, C., Gaschler-Markefski, B., Tegeler, C., Schindler, F., Stiller, D., and Scheich, H. (1998). Electrodynamic headphones and woofers for application in magnetic resonance imaging scanners. *Med. Phys.* **25,** 2068–2070.

Belin, P., Zatorre, R. J., Lafaille, P., Ahad, P., and Pike, B. (2000). Voice-selective areas in human auditory cortex. *Nature* **403,** 309–312.

Bilecen, D., Scheffler, K., Schmid, N., Tschopp, K., and Seelig, J. (1998). Tonotopic organization of the human auditory cortex as detected by BOLD-FMRI. *Hear. Res.* **126,** 19–27.

Binder, J. R. (1995). Functional magnetic resonance imaging of language cortex. *Int. J. Imag. Syst. Technol.* **6,** 280–288.

Bizley, J. K., Nelken, I., Nodal, F. R., Ahmed, B., Schnupp, J. W. H., and King, A. J. (2003). Optical imaging of tonotopicity and periodotopicity in primary and nonprimary areas of the ferret auditory cortex. *J. Physiol.* **547P,** C86.

Blauert, J. (1969). Sound localization in the median plane. *Acustica* **22,** 205–213.

Broadbent, D. E., and Gregory, M. (1964). Accuracy of recognition for speech presented to the right and left ears. *Q. J. Exp. Psychol.* **16,** 359–360.

Brodmann, K. (1909). "Vergleichende Lokalisationslehre der Grosshirnrind." Barth, Leipzig. Translated version. Brodmann's 'Localisation in the cerebral cortex.' Garey, L. J. (1909). Imperial College Press, London.

Budd, T. W., Hall, D. A., Gonçalves, M. S., Akeroyd, M. A., Foster, J. R., Palmer, A. R., Head, K., and Summerfield, A. Q. (2003). Binaural specialisation in human auditory cortex: An fMRI investigation of interaural correlation sensitivity. *NeuroImage* **20,** 1783–1794.

Campain, R., and Minckler, J. (1976). A note on the gross configurations of the human auditory cortex. *Brain Lang.* **3,** 318–323.

Cansino, S., DuCorps, A., and Ragot, R. (2003). Tonotopic representation of periodic complex sounds. *Hum. Brain Mapp.* **20,** 71–81.

Celesia, G. G. (1976). Organization of auditory cortical areas in man. *Brain* **99,** 403–414.

Crottaz-Herbette, S., and Ragot, R. (2000). Perception of complex sounds: N1 latency codes pitch and topography codes spectra. *Clin. Neurophysiol.* **111,** 1759–1766.

Davis, M. H., and Johnsrude, I. S. (2003). Hierarchical processing in spoken language comprehension. *J. Neurosci.* **23,** 3423–3431.

de Zwart, J. A., van Gelderen, P., Kellman, P., and Duyn, J. H. (2002). Reduction of gradient acoustic noise in MRI using SENSE-EPI. *NeuroImage* **16,** 1151–1155.

Ehret, G. (1997). The auditory midbrain, a "shunting yard" of acoustical information processing. *In* "The Central Auditory System" (G. Ehret and R. Romand, Eds.), pp. 259–316. Oxford University Press, Oxford.

Ehrlé, N., Samson, S., and Baulac, M. (2001). Processing of rapid auditory information in epileptic patients with left temporal lobe damage. *Neuropsychologia* **39,** 525–531.

Formisano, E., Kim, D-S., Di Salle, F., van de Moortele, P-F., Urgibil, K., and Goebel, R. (2003). Mirror-symmetric tonotopic maps in human primary auditory cortex. *Neuron* **40,** 859–869.

Foster, J. R., Hall, D. A., Summerfield, A. Q., Palmer, A. R., and Bowtell, R. W. (2000). Sound-level measurements and calculations of safe noise dosage during EPI at 3T. *J. Magn. Reson. Imaging* **12,** 157–163.

Gaese, B. H., and Ostwald, J. (2003). Complexity and temporal dynamics of frequency coding in the awake rat auditory cortex. *Europ. J. Neurosci.* **18,** 2638–2652.

Galaburda, A. M., LeMay, M., Kemper, T. L., and Geschwind, N. (1978). Right-left asymmetries in the brain: Structural differences between the hemispheres may underlie cerebral dominance. *Science* **199,** 852–856.

Galaburda, A., and Sanides, F. (1980). Cytoarchitectonic organization of the human auditory cortex. *J. Comp. Neurol.* **190,** 597–610.

Godey, B., Schwartz, D., de Graaf, J. B., and Liegeois-Chauvel, C. (2001). Neuromagnetic source localization of auditory evoked fields and intracerebral evoked potentials: A comparison of data in the same patients. *Clin. Neurophysiol.* **112,** 1850–1859.

Griffiths, T. D., Büchel, C., Frackowiak, R. S. J., and Patterson, R. D. (1998). Analysis of temporal structure in sound by the brain. *Nat. Neurosci.* **1,** 422–427.

Griffiths, T. D., and Green, G. G. R. (1999). Cortical activation during perception of a rotating wide-field acoustic stimulus. *Neuroimage* **10,** 84–90.

Griffiths, T. D., Rees, A., and Green, G. G. R. (1999). Disorders of human complex sound processing. *Neurocase* **5,** 365–378.

Griffiths, T. D., and Warren, J. D. (2002). The planum temporale as a computational hub. *Trends Neurosci.* **25,** 348–353.

Gutschalk, A., Patterson, R. D., Rupp, A., Uppenkamp, S., and Scherg, M. (2002). Sustained magnetic fields reveal separate sites for sound level and temporal regularity in human auditory cortex. *NeuroImage* **15,** 207–216.

Gutschalk, A., Patterson, R. D., Scherg, M., Uppenkamp, S., and Rupp, A. (2004). Temporal dynamics of pitch in human auditory cortex. *NeuroImage* **22,** 755–766.

Hackett, T. A. (2003). The comparative anatomy of the primate auditory cortex. *In* "Primate Audition: Ethology and Neurobiology" (A. A. Ghazanfar, Ed.), pp. 199–225. CRC Press, Boca Raton, FL.

Hall, D. A. (2004). Sensitivity to the spectral and temporal properties of sound stimuli across the human non-primary auditory cortex measured using functional magnetic resonance imaging. *In* "Auditory Cortex—A Zynthesis of Human and Animal Research" (R. König, P. Heil, E. Budinger, and H. Scheich, Eds.), pp. 51–76. Lawrence Erlbaum Associates, Hillsdale, NJ.

Hall, D. A., Gonçalves, M. S., Smith, S., Jezzard, P., Haggard, M. P., and Kornak, J. (2002a). A method for determining venous contribution to BOLD contrast sensory activation. *Magn. Reson. Imaging* **20,** 695–706.

Hall, D. A., Johnsrude, I. S., Haggard, M. P., Palmer, A. R., Akeroyd, M. A., and Summerfield, A. Q. (2002b). Spectral and temporal processing in human auditory cortex. *Cereb. Cortex* **12,** 140–149.

Hall, D. A., Hart, H. C., and Johnsrude, I. S. (2003). Relationships between human auditory cortical structure and function. *Audiol. Neurootol.* **8,** 1–18.

Halpern, A. R., and Zatorre, R. J. (1999). When that tune runs through your head: A PET investigation of auditory imagery of familiar melodies. *Cereb. Cortex* **9,** 697–704.

Heeger, D. J., and Ress, D. (2002). What does fMRI tell us about neuronal activity? *Nat. Rev. Neurosci.* **3,** 142–151.

Hennel, F., Girard, F., and Loenneker, T. (1999). "Silent" MRI with soft gradient pulses. *Magn. Reson. Med.* **42,** 6–10.

Hickok, G., and Poeppel, D. (2000). Towards a functional neuroanatomy of speech perception. *Trends Cog. Sci.* **4,** 131–138.

Howard, M. A., Volkov, I. O., Abbas, P. J., Damasio, H., Ollendieck, M. C., and Granner, M. A. (1996). A chronic microelectrode investigation of the tonotopic organisation of human auditory cortex. *Brain Res.* **724,** 260–264.

Hunter, M. D., Griffiths, T. D., Farrow, T. F. D., Zheng, Y., Wilkinson, I. D., Hegde, N., Woods, W., Spence, S. A., and Woodruff, P. W. R. (2003). A neural basis for the perception of voices in external auditory space. *Brain* **126,** 161–169.

Huotilainen, M., Tiitinen, H., Lavikainen, J., Ilmoniemi, R., Pekkonen, E., Sinkkonen, J., Laine, P., and Näätänen, R. (1995). Sustained fields of tones and glides reflect tonotopy of the auditory cortex. *Neuroreport* **6,** 841–844.

Huotilainen, M., Winkler, I., Alho, K., Escera, C., Virtanen, J., Ilmoniemi, R. J., Jääskeläinen, I. P., Pekkonen, E., and Näätänen, R. (1998). Combined mapping of human auditory EEG and MEG responses. *Electroenceph. Clin. Neurophysiol.* **108,** 370–379.

Ivry, R., and Lebby, P. C. (1993). Hemispheric differences in auditory perception are similar to those found in visual perception. *Psychol. Sci.* **4,** 41–45.

Johnsrude, I. S., Giraud, A. L., and Frackowiak, R. S. J. (2002). Functional imaging of the auditory system: The use of positron emission tomography. *Audiol. Neurootol.* **7,** 251–276.

Jones, S. J., Sprague, L., and Vaz Pato, M. (2002). Electrophysiological evidence for a defect in the processing of temporal sound patterns in multiple sclerosis. *J. Neurol. Neurosurg. Psychiatry* **73,** 561–567.

Kaas, J. H., and Hackett, T. A. (1998). Subdivisions and levels of processing in primate auditory cortex. *Audiol. Neurootol.* **3,** 73–85.

Kaas, J. H., Hackett, T. A., and Tramo, M. J. (1999). Auditory processing in primate cerebral cortex. *Curr. Opin. Neurobiol.* **9,** 164–170.

Kimura, D. (1964). Left-right differences in the perception of melodies. *Q. J. Exp. Psychol.* **16,** 355–358.

Kosaki, H., Hashikawa, T., He, J., and Jones, E. G. (1997). Tonotopic organisation of auditory cortical fields delineated by parvalbumin immunoreactivity in macaque monkeys. *J. Comp. Neurol.* **386,** 304–316.

Krumbholz, K., Patterson, R. D., Seither-Preisler, A., Lammertmann, C., and Lütkenhöner, B. (2003). Neuromagnetic evidence for a pitch processing center in Heschl's gyrus. *Cereb. Cortex* **13,** 765–772.

Langner, G. (1992). Periodicity coding in the auditory system. *Hear. Res.* **60,** 115–142.

Langner, G., Sams, M., Heil, P., and Schulze, H. (1997). Frequency and periodicity are represented in orthogonal maps in the human auditory cortex: Evidence from magnetoencephalography. *J. Comp. Physiol. A* **181,** 665–676.

Lauter, J. L., Herscovitch, P., Formby, C., and Raichle, M. E. (1985). Tonotopic organisation in human auditory cortex revealed by positron emission tomography. *Hear. Res.* **20,** 199–205.

Leonard, C. M., Puranik, C., Kuldau, J. M., and Lombardino, L. J. (1998). Normal variation in the frequency and location of human auditory cortex landmarks. Heschl's gyrus: Where is it? *Cereb. Cortex* **8,** 397–406.

Liegeois-Chauvel, C., Musolino, A., and Chauvel, P. (1991). Localization of the primary auditory area in man. *Brain* **114,** 139–151.

Liegeois-Chauvel, C., Musolino, A., Badier, J. M., Marquis, P., and Chauvel, P. (1994). Evoked potentials recorded from the auditory cortex in man: Evaluation and topography of the middle latency components. *Electroenceph. Clin. Neurophysiol.* **92,** 204–214.

Lockwood, A. H., Salvi, R. J., Coad, M. L., Arnold, S. A., Wack, D. S., Murphy, B. W., and Burkard, R. F. (1999). The functional anatomy of the normal human auditory system: Responses to 0.5 and 4.0 kHz tones at varied intensities. *Cereb. Cortex* **9,** 65–76.

Logothetis, N. K., and Wandell, B. A. (2004). Interpreting the BOLD signal. *Annu. Rev. Physiol.* **66,** 735–769.

Lütkenhöner, B. (2003). Single-dipole analyses of the N100m are not suitable for characterizing the cortical representation of pitch. *Audiol. Neurootol.* **8,** 222–233.

Lütkenhöner, B., Krumbholz, K., and Seither-Preisler, A. (2003). Studies of tonotopy based on wave N100 of the auditory evoked field are problematic. *NeuroImage* **19,** 935–949.

Lütkenhöner, B., and Steinsträter, O. (1998). High-precision neuromagnetic study of the functional organization of the human auditory cortex. *Audiol. Neurootol.* **3,** 191–213.

Mansfield, P., Chapman, B. L. W., Bowtell, R., Glover, P., Coxon, R., and Harvey, P. R. (1995). Active acoustic screening: Reduction of noise in gradient coils by Lorentz force balancing. *Mag. Reson. Med.* **33,** 276–281.

McAlpine, D. (2004). Neural sensitivity to periodicity in the inferior colliculus: Evidence for the role of cochlear distortions. *J. Neurophysiol.* **92,** 1295–1311.

Merzenich, M. M., and Brugge, J. F. (1973). Representation of the cochlear partition on the superior temporal plane of the macaque monkey. *Brain Res.* **50,** 275–296.

Mirz, F., Ovesen, T., Ishizu, K., Johannsen, P., Madsen, S., Gjedde, A., and Pedersen, C. B. (1999). Stimulus-dependent central processing of auditory stimuli. A PET study. *Scand. Audiol.* **28,** 161–169.

Morel, A., Garraghty, P. E., and Kaas, J. H. (1993). Tonotopic organisation, architectonic fields, and connections of auditory cortex in macaque monkeys. *J. Comp. Neurol.* **335,** 437–459.

Mummery, C. J., Ashburner, J., Scott, S. K., and Wise, R. J. S. (1999). Functional neuroimaging of speech perception in six normal and two aphasic patients. *J. Acoust. Soc. Am.* **106,** 449–457.

Narain, C., Scott, S. K., Wise, R. J. S., Rosen, S., Leff, A., Iversen, S. D., and Matthews, P. M. (2003). Defining a left-lateralised response specific to intelligible speech using fMRI. *Cereb. Cortex* **13,** 1362–1368.

Nunez, P. L. (1981). Electric fields of the brain: The neurophysics of EEG. Oxford University Press, London.

Oxenham, A. J., Bernstein, J. G. W., and Penagos, H. (2004). Correct tonotopic representation is necessary for complex pitch perception. *Proc. Natl. Acad. Sci.* **101,** 1421–1425.

Pantev, C., Bertrand, O., Eulitz, C., Verkindt, C., Hampson, S., Schuierer, G., and Elbert, T. (1995). Specific tonotopic organizations of different areas of the human auditory cortex revealed by simultaneous magnetic and electric recordings. *Electroencephalogr. Clin. Neurophysiol.* **94,** 26–40.

Pantev, C., Eulitz, C., Elbert, T., and Hoke, M. (1994). The auditory evoked sustained field: Origin and frequency dependence. *Electroencephalogr. Clin. Neurophysiol.* **90,** 82–90.

Pantev, C., Hoke, M., Lehnertz, K., Lütkenhöner, B., Anogianakis, G., and Wittowski, W. (1988). Tonotopic organization of the human auditory cortex revealed by transient auditory evoked magnetic fields. *Electroencephalogr. Clin. Neurophysiol.* **69,** 160–170.

Pantev, C., Hoke, M., Lutkenhoner, B., and Lehnertz, K. (1989). Tonotopic organization of the auditory cortex: Pitch versus frequency representation. *Science* **246,** 486–488.

Pantev, C., Roberts, L., Elbert, T., Ross, B., and Wienbruch, C. (1996). Tonotopic organization of the sources of human auditory steady-state responses. *Hear. Res.* **101,** 62–74.

Patterson, R. D., Uppenkamp, S., Johnsrude, I. S., and Griffiths, T. D. (2002). The processing of temporal pitch and melody information in auditory cortex. *Neuron* **36,** 767–776.

Pavani, F., Macalusco, E., Warren, J. D., Driver, J., and Griffiths, T. D. (2002). A common cortical substrate activated by horizontal and vertical sound movement in the human brain. *Curr. Biol.* **12,** 1584–1590.

Penagos, H., Melcher, J. R., and Oxenham, A. J. (2004). A neural representation of pitch salience in nonprimary human auditory cortex revealed with functional magnetic resonance imaging. *J. Neurosci.* **24,** 6810–6815.

Phillips, D. P., and Cynader, M. S. (1985). Some neural mechanisms in the cat's auditory cortex underlying sensitivity to combined tone and wide-spectrum noise stimuli. *Hear. Res.* **18,** 87–102.

Phillips, D. P., Semple, M. N., Calford, M. B., and Kitzes, L. M. (1994). Level-dependent representation of stimulus frequency in cat primary auditory cortex. *Exp. Brain Res.* **102,** 210–226.

Poremba, A., Saunders, R. C., Crane, A. M., Cook, M., Sokoloff, L., and Mishkin, M. (2003). Functional mapping of the primate auditory system. *Science* **299,** 568–572.

Quine, D. B., Regan, D., Beverly, K. I., and Murray, T. J. (1984). Patients with multiple sclerosis experience hearing loss for shifts of tone frequency. *Arch. Neurol.* **41,** 506–507.

Rademacher, J., Morosan, P., Schormann, T., Schleicher, A., Werner, C., Freund, H-J., and Zilles, K. (2001). Probabilistic mapping and volume measurement of human primary auditory cortex. *NeuroImage* **13,** 669–683.

Rauschecker, J. P., Tian, B., and Hauser, M. (1995). Processing of complex sounds in the macaque nonprimary auditory cortex. *Science* **268,** 111–114.

Read, H. L., Winer, J. A., and Schreiner, C. E. (2002). Functional architecture of auditory cortex. *Curr. Opin. Neurobiol.* **12,** 433–440.

Rivier, F., and Clarke, S. (1997). Cytochrome oxidase, acetylcholinesterase, and NADPH-diaphorase staining in human supratemporal and insular cortex: Evidence for multiple auditory areas. *NeuroImage* **6,** 288–304.

Roberts, T. P. L., and Poeppel, D. (1996). Latency of auditory evoked M100 as a function of tone frequency. *NeuroReport* **7,** 1138–1140.

Rogers, R. L., Papanicolaou, A. C., Baumann, S. B., Saydjari, C., and Eisenberg, H. M. (1990). Neuromagnetic evidence of a dynamic excitation pattern generating the N100 auditory response. *Electroencephalogr. Clin. Neurophysiol.* **77,** 237–240.

Romani, G. L., Williamson, S. J., and Kaufman, L. (1982). Tonotopic organization of the human auditory-cortex. *Science* **216,** 1339–1340.

Scherg, M. (1990). Fundamentals of dipole source potential analysis. *In* "Auditory evoked magnetic fields and electric potentials. Advances in audiology. Volume 6" (F. Grandori, M. Hoke, and G. L. Romani, Eds.), pp. 40–69. Karger, Basel.

Schönwiesner, M., von Cramon, Y. D., and Rübsamen, R. (2002). Is it tonotopy after all? *NeuroImage* **17,** 1144–1161.

Schulze, H., Hess, A., Ohl, F. W., and Scheich, H. (2002). Superposition of horseshoe-like periodicity and linear tonotopic maps in auditory cortex of the Mongolian gerbil. *Eur. J. Neurosci.* **15,** 1077–1084.

Schulze, H., and Langner, G. (1997). Periodicity coding in the primary auditory cortex of the Mongolian gerbil (*Meriones unguiculatus*): Two different coding strategies for pitch and rhythm? *J. Comp. Physiol. A* **181,** 651–663.

Scott, S. K., Blank, C. C., Rosen, S., and Wise, R. J. S. (2000). Identification of a pathway for intelligible speech in the left temporal lobe. *Brain* **123,** 2400–2406.

Scott, S. K., and Johnsrude, I. S. (2003). The neuroanatomical and functional organization of speech perception. *Trends Neurosci.* **26,** 100–107.

Sergent, J. (1982). The cerebral balance of power: Confrontation or cooperation. *J. Exp. Psychol. Hum. Percept. Perform.* **8,** 253–272.

Shamma, S, and Klein, D. (2000). The case of the missing pitch templates: How harmonic templates emerge in the early auditory system. *J. Acoust. Am.* **107,** 2631–2644.

Shannon, R. V., Zeng, F. G., Hamath, V., Wygonski, J., and Ekelid, M. (1995). Speech recognition with primarily temporal cues. *Science* **270,** 303–304.

Steinmetz, H., Rademacher, J., Huang, Y., Hefter, H., Zilles, K., Thron, A., and Freund, H-J. (1989). Cerebral asymmetry: MR planimetry of the human planum temporale. *J. Comput. Assist. Tomogr.* **13,** 996–1005.

Stufflebeam, S. M., Poeppel, D., Rowley, H. A., and Roberts, T. P. L. (1998). Peri-threshold encoding of stimulus frequency and intensity in the M100 latency. *NeuroReport* **9,** 91–94.

Talavage, T. M., Ledden, P. J., Benson, R. R., Rosen, B. R., and Melcher, J. R. (2000). Frequency-dependent responses exhibited by multiple regions in human auditory cortex. *Hear. Res.* **150,** 225–244.

Talavage, T. M., Sereno, M. I., Melcher, J. R., Ledden, P. J., Rosen, B. R., and Dale, A. M. (2004). Tonotopic organization in human auditory cortex revealed by progressions of frequency sensitivity. *J. Neurophysiol.* **91,** 1282–1296.

Tervaniemi, M., Medvedev, S. V., Alho, K., Pakhomov, S. V., Roudas, M. S., van Zuijen, T. L., and Näätänen, R. (2000). Lateralized automatic auditory processing of phonetic versus musical information: A PET study. *Hum. Brain Mapp.* **10,** 74–79.

Van der Poel, J. C., Jones, S. J., and Miller, D. H. (1988). Sound lateralisation, brainstem auditory evoked potentials and magnetic resonance imaging in patients with multiple sclerosis. *Brain* **11,** 1453–1474.

Verkindt, C., Bertrand, O., Perrin, F., Echallier, J-F., and Pernier, J. (1995). Tonotopic organization of the human auditory cortex: N100 topography and multiple dipole model analysis. *Electroencephalogr. Clin. Neurophysiol.* **96,** 143–156.

Von Békésy, G. (1947). The variations of phase along the basilar membrane with sinusoidal vibrations. *J. Acoust. Soc. Am.* **19,** 452–460.

Wallace, M. N., Johnston, P. W., and Palmer, A. R. (2002). Histochemical identification of cortical areas in the auditory region of the human brain. *Exp. Brain Res.* **143,** 499–508.

Warren, J. D., Zielinski, B. A., Green, G. G. R., Rauschecker, J. P., and Griffiths, T. D. (2002). Perception of sound source motion by the human brain. *Neuron* **34,** 139–148.

Wessinger, C. M., VanMeter, J., Tian, B., Van Lare, J., Pekar, J., and Rauschecker, J. P. (2001). Hierarchical organisation of the human auditory cortex revealed by functional magnetic resonance imaging. *J. Cog. Neurosci.* **13,** 1–7.

Westbury, C. F., Zatorre, R. J., and Evans, A. C. (1999). Quantifying variability in the planum temporale: A probability map. *Cereb. Cortex* **9,** 392–405.

Wightman, F. J., and Kistler, D. J. (1989). Headphone simulation of free-field listening. I: Stimulus synthesis. *J. Acoust. Soc. Am.* **85,** 858–867.

Wise, R., Chollet, F., Hadar, U., Friston, K., Hoffner, E., and Frackowiak, R. (1991). Distribution of cortical neural networks involved in word comprehension and word retrieval. *Brain* **114,** 1803–1817.

Yost, W. A., Patterson, R. D., and Sheft, S. (1996). A time-domain description for the pitch strength of iterated rippled noise. *J. Acoust. Soc. Am.* **99,** 1066–1078.

Zatorre, R. J., Belin, P., and Penhune, V. B. (2002). Structure and function of auditory cortex: Music and speech. *Trends Cog. Sci.* **6,** 37–46.

Zilles, K., Palomero-Gallagher, N., Grefkes, C., Scheperjans, F., Boy, C., Amunts, K., and Schleicher, A. (2002). Architectonics of the human cerebral cortex and transmitter receptor fingerprints: Reconciling functional neuroanatomy and neurochemistry. *Eur. Neuropsychopharmacol.* **12,** 587–599.

SPECTRAL PROCESSING AND SOUND SOURCE DETERMINATION

Donal G. Sinex

Department of Psychology, Utah State University, Logan, Utah 84322, USA

Humans normally listen in mixed environments, in which sounds originating from more than one source overlap in time and in frequency. The auditory system is able to extract information specific to the individual sources that contribute to the composite signal and process the information for each source separately; this is called "auditory scene analysis" or "sound-source determination." Sounds that are simultaneously present but generated independently tend to differ along relatively simple acoustic dimensions. These dimensions may be temporal, as when sounds begin or end asynchronously, or spectral, as when the sounds have different fundamental frequencies. Psychophysical experiments have identified some of the ways in which human listeners use these dimensions to isolate sources of sound. A simple but useful stimulus, a harmonic complex tone with or without a mistuned component, can be used for parametric investigation of the processing of spectral structure. This "mistuned tone" stimulus has been used in several psychophysical experiments, and more recently in studies that specifically address the neural mechanisms that underlie segregation based on harmonicity. Studies of the responses of single neurons in the chinchilla auditory system to mistuned tones are reviewed here in detail. The results of those

INTERNATIONAL REVIEW OF
NEUROBIOLOGY, VOL. 70
DOI: 10.1016/S0074-7742(05)70011-8

0074-7742/05 $35.00

experiments support the view that neurons in the inferior colliculus (IC) exhibit responses to mistuned tones that are larger and temporally more complex than the same neurons' responses to harmonic tones. Mistuning does not produce comparable changes in the discharge patterns of auditory nerve (AN) fibers, indicating that a major transformation in the neural representation of harmonic structure occurs in the auditory brainstem. The brainstem processing that accomplishes this transformation may contribute to the segregation of competing sounds and ultimately to the identification of sound sources.

I. Introduction

We usually listen to sound in situations in which multiple sound-generating objects are present. For example, a person may be talking in the presence of other talkers or music, or two persons may be holding a conversation in the presence of unrelated environmental sounds. In these situations, the signals generated by multiple simultaneous sound sources add linearly and arrive at each of the listener's ears as a single waveform whose spectrum includes all the components associated with all the sources. The signal that arises from the addition of multiple sounds will be referred to here as a "composite" signal. In the composite signal, the waveforms produced by the individual sources often overlap in time and in frequency. However, in what is a remarkable feat of signal processing, we tend to perceive the individual sound-generating objects accurately. We do not hear a single sound that has the characteristics of the entire mixture, or sounds that have the characteristics of inappropriate subsets of the mixture.

Different terminology has been used to describe the process of sorting out the spectrum and identifying the individual sound-generating objects that contribute to the composite signal. Bregman (1990) referred to the process as auditory scene analysis, a term that is widely used. The outcome of the process was called the perception of auditory entities by Hartmann (1988), auditory object perception by Handel (1995), and auditory image perception by Yost (1991). Yost later suggested the term sound-source determination (Yost, 1993; Yost and Sheft, 1993); this phrase will be used in this chapter. Setting aside differences in terminology, these authors agree that sound source determination is an essential function of the auditory system.

Although the psychophysical and perceptual literature emphasizes sound-source determination by humans, it is highly likely that individuals of any species that use hearing must have the ability to determine what sound sources exist in a composite signal. Without that ability, it would be difficult or impossible to identify the spectral profile (i.e., the pattern of amplitude as a function of frequency). The spectral profile, in turn, provides information about who or what made the sound: predator, prey, a conspecific, or something that may safely be ignored. The adaptive value of distinguishing enemies from friends even when listening conditions are less than perfect is obvious. The spectral profile also conveys meaning beyond simply identifying the source of the sound. For example, human speech would be less understandable if its spectral profile as represented by the auditory system were not stable in the presence of competing sounds.

Excellent reviews of perceptual and psychophysical phenomena relevant for sound source determination are already available (Bregman, 1990; Darwin and Carlyon, 1995; Hartmann, 1988; Yost, 1991, 1993; Yost and Sheft, 1993). This chapter will extend the scope of those reviews by emphasizing neural processing that may be relevant for understanding sound-source determination. The focus will be on microelectrode studies of the representation of complex tones conducted in the author's laboratory. Neural mechanisms that underlie sound-source determination are only beginning to be understood, but it is likely that activity in this area of research will increase.

II. Processes That Underlie Sound-Source Determination

Sound-source determination surely involves more than one auditory mechanism. Bregman (1990) made a distinction between innate mechanisms, which he called "primitive scene analysis," and those that are influenced by factors such as learning or attention, which he called "schema-based scene analysis." In addition to being innate, primitive scene analysis includes processes that are tightly linked to properties of the acoustic signal. In contrast, schema-based scene analysis incorporates cognitive factors, such as the use of stored knowledge and hypothesis testing. The focus of this chapter will be on primitive mechanisms that can be investigated in non-human animal models. Since the general plan of the auditory pathway is relatively similar across species, especially across mammals, it can be argued that the neural circuitry and processing algorithms that underlie primitive auditory scene analysis are also highly conserved, making it likely that mechanisms identified in non-human animals can be generalized to humans. This is not always the case in observations of the neural processing of complex sounds. The processing of animal vocalizations, for example, may involve species-specific mechanisms even at relatively early stages of the auditory pathway (Portfors and Sinex, 2005).

As the first step toward the determination of sound sources, the parts of the composite signal originating from the individual sound generators must be isolated and extracted; this step is called "spectral segregation" or simply segregation. The spectral components that are segregated are often non-contiguous in the spectrum of the composite signal. This can be illustrated with examples from vowel perception. A typical voiced vowel is spoken with periodic vocal fold activity, and its spectrum will consist of components at integer multiples of the fundamental frequency (f0) of voicing, whose amplitudes are shaped by the resonances of the talker's vocal tract—the spectral profile described in the previous text. In the case of two simultaneous vowels produced with slightly different f0, harmonics of one f0 will be interspersed with harmonics of the other f0. However, listeners are able to identify simultaneous vowels at levels well above chance (Assmann and Summerfield, 1989, 1990; Chalikia and Bregman, 1989, 1993; Culling and Darwin, 1994; de Cheveigné *et al.*, 1997). To identify any single vowel presented in isolation, the listener makes use of the vowel's spectral profile. A listener who can successfully identify two vowels presented simultaneously must have had access to information about the spectral profiles of each vowel separately. It follows that the listener must have been able to identify and associate the non-adjacent spectral components that belong to each vowel.

Segregation will usually be followed by "grouping," a separate step in which the signal components originating from each sound source are combined, or fused, to produce a unitary percept. In the previous example of two simultaneous vowels, the identity of an individual vowel would not actually have been determined until the appropriately segregated components had been grouped. Grouping, although obviously important, is less easily studied in neurophysiological experiments than primitive segregation, and so little more will be said about it in this chapter.

A. Physical Cues That Promote Segregation

When two sounds are generated independently in real-world situations, they are likely to exhibit differences along relatively simple acoustic dimensions such as duration or periodicity that can be exploited for segregation and eventual sound-source determination. Table I lists six acoustic variables that could be used to segregate sound sources, as summarized by Yost and Sheft (1993). Some of these variables are temporal; for example, independently generated sounds usually do not begin and end simultaneously. A difference of as little as 30 msec in the relative onset times of two sounds promotes segregation, even when listeners are unaware of the onset disparity itself (Rasch, 1978). Ongoing amplitude modulation, another temporal cue, also contributes to segregation and grouping. Incoherently modulated components tend to be perceived as arising from separate sources. For example, it is difficult or impossible to hear the individual components in a harmonic signal, such as a vowel or complex tone. However, if amplitude modulation is applied to a single component in a harmonic complex

TABLE I
SUMMARY OF SIX ACOUSTIC DIMENSIONS DESCRIBED BY YOST AND SHEFT (1993) THAT MAY BE USED FOR SEGREGATION AND GROUPING. BOTH SPECTRAL AND TEMPORAL CUES ARE AVAILABLE

	Physical variable	Type of cue
1	Harmonicity	Spectral
2	Spectral separation	Spectral
3	Spectral profile	Spectral
4	Amplitude modulation	Temporal
5	Temporal separation	Temporal
6	Spatial separation	Spectral and temporal

tone, the modulated component can more easily be heard as separate from the remaining, unmodulated components (Duifhuis, 1970). Coherent modulation applied to more than one component in a complex sound promotes fusion or grouping of those components (Cohen, 1980; Hall and Grose, 1990; Hall *et al.*, 1984). The intelligibility of sounds with components that are not easily grouped, such as sinewave speech, is increased by coherent modulation of the sinewave components (Carrell and Opie, 1992).

Spatial separation also facilitates segregation. Cherry (1953) found that competing sounds presented to the same ear as the target sound produced more interference than competing sounds presented to the other ear. Cherry's paper introduced the phrase "cocktail party effect" as a general term to describe the ability to process one voice in the presence of others. Because the cocktail party effect is well known, the significance of binaural cues for sound-source determination may be occasionally overstated. Spatial separation improves the recognition of sounds in noise, in part because the listener's head can be oriented to improve the ratio of the signal intensity to the noise intensity (Plomp and Mimpen, 1981). However, individual sources can be segregated from composite sounds that lack any interaural cues for location; for example, different musical instruments or voices can be identified in monophonic recordings or radio broadcasts. Listeners presented with simultaneous vowels can identify one or both vowels when they are presented to the same ear (Assmann and Summerfield, 1989, 1990) or diotically (Chalikia and Bregman, 1989; de Cheveigné, 1999).

Harmonic structure provides important spectral cues that can be used for segregation. The spectra of sounds produced by periodic vibration consist of equally spaced lines falling at integer multiples of the f0 of the vibration. These sounds are said to be harmonic, or to exhibit harmonicity (Kidd *et al.*, 2003). Many natural sounds are periodic, or nearly so, and when heard in isolation, they have a low pitch that is determined by f0. Listeners typically are not able to hear the individual components in a harmonic complex tone; the harmonics are fused or grouped to produce a single integrated percept. Two objects that independently generate periodic signals typically do not vibrate with exactly the same f0, and the resulting sounds will have different pitches. Simultaneous periodic sounds that differ in f0, that is, that differ in harmonicity, are more easily segregated than sounds that have the same f0. The introduction of small differences in f0, on the order of 1% to 3%, leads to rapid improvement in identification accuracy, and f0 differences greater than 1 semitone usually produce only modest additional improvement (Assmann and Summerfield, 1989, 1990; Chalikia and Bregman, 1989, 1993; Culling and Darwin, 1994; de Cheveigné *et al.*, 1997, 1999). Ongoing speech is more intelligible in the presence of a competing voice if the pitch of the second voice is different (Brokx and Nooteboom, 1982). Listeners can also use small differences in f0 to segregate pairs of non-speech harmonic complex tones presented simultaneously (Chalikia and Bregman, 1989; Carlyon and Shackleton, 1994).

The segregation of simultaneous harmonic sounds is usually thought to require the estimation of f0 for one or both sounds. Models of segregation usually make explicit use of an estimate of f0 (Meddis and Hewitt, 1992); if f0 is known, it can be used to enhance one harmonic sound or cancel the other, thereby improving recognition (de Cheveigné, 1993). One hypothetical mechanism that could enhance the representation of one harmonic sound within a composite sound is a "harmonic sieve" (Duifhuis *et al.*, 1982; Goldstein, 1973). The harmonic sieve, usually invoked in the context of pitch perception, identifies spectral components that fall at integer multiples of a fundamental frequency, within a tolerance usually estimated to be about ±3%. The components passed by the sieve would normally belong to a single source and should be grouped, while those falling outside the sieve presumably arise from another sound generator and should be processed separately (Lin and Hartmann, 1998; Moore *et al.*, 1986). However, not all observations of the processing of simultaneous vowels with differences in f0 are consistent with the expectations of a harmonic sieve. For example, Culling and Darwin (1994) presented listeners with "interleaved vowels," in which the spectral profile of each individual vowel was defined by some harmonics of one f0 and some harmonics of the other f0. The interleaved vowels were presented in pairs, such that the total number and frequencies of harmonics in a pair was the same as in a control condition consisting of two simultaneous unmodified vowels. With these stimuli, a harmonic sieve would identify an unrelated set of spectral components, from which the intended vowel could not be identified. In fact, for small f0 separations, listeners were able to identify interleaved vowels as accurately as they identified the unmodified vowels. A computational model that used the pattern of beats produced by unresolved stimulus components as the basis for identification successfully predicted the listeners' performance for interleaved vowels, and for normal vowels at small f0 separations. However, the model underestimated listeners' performance at f0 separations of 1 semitone or more for the unmodified vowels, suggesting that a process more like a harmonic sieve may be usable for larger differences in f0.

B. The Perception and Processing of Mistuned Harmonics

Segregation based on harmonicity can be studied with a relatively simple stimulus, a "mistuned tone" (Grube *et al.*, 2003; Hartmann, 1988, 2004; Hartmann *et al.*, 1990; Moore *et al.*, 1985, 1986; Roberts and Brunstrom, 1998). A mistuned tone is a harmonic complex tone in which the frequency of one component has been shifted so that it no longer falls at an integer multiple of f0. As noted in the previous section, the individual spectral components in a strictly harmonic tone are not perceived separately. However, under certain conditions, the mistuned component will be heard as a separate tone with its own pitch and timbre, against the background of the original tone, which has a lower pitch (Moore *et al.*, 1986). When that happens, the auditory system can be said to have segregated the mistuned component from the harmonic tone. Presumably, this is accomplished by the same mechanisms that are engaged by naturally occurring composite sounds. Parameters of the mistuned tone can be varied systematically, to determine the conditions under which segregation does or does not occur. Results summarized in Fig. 1A indicate that mistuning one harmonic by 1% to 2% is sufficient to produce the percept of a second tone (Hartmann, 2004; Moore *et al.*, 1986). Hartmann *et al.* (1990) reported that when one component of a harmonic tone was mistuned by 4%, subjects could consistently and accurately adjust a pure tone to match the pitch of the mistuned harmonic. The pattern of matches provided convincing evidence that the listeners based their responses on a perceived second sound, rather than some change in the overall quality of the complex tone. The threshold at which the mistuned harmonic emerges as a separate sound is relatively independent of the absolute frequency of the original harmonic component, for frequencies up to about 2 kHz (Hartmann *et al.*, 1990). It is also relatively independent of f0, over the range from 0.1 to 0.4 kHz (Moore *et al.*, 1986). The threshold is relatively independent of harmonic number, at least for harmonics above the second (Hartmann *et al.*, 1990; Moore *et al.*, 1986). It varies non-monotonically with the stimulus level, and is lowest over a range of levels between about 10 to 50 dB SPL per component (Hartmann, 2004).

The interpretation of experiments involving mistuned tones is complicated by the fact that it is possible to detect mistuning as a change in timbre, even if a second sound source is not perceived (Lee and Green, 1994; Moore *et al.*, 1985). When listeners are allowed to report the detection of any change in the complex tone, their thresholds may be much smaller than 1% or 2% (Fig. 1B). Also, rather than being independent of frequency, thresholds for the detection of mistuning (expressed in percent) decrease with frequency, to values that can be an order of

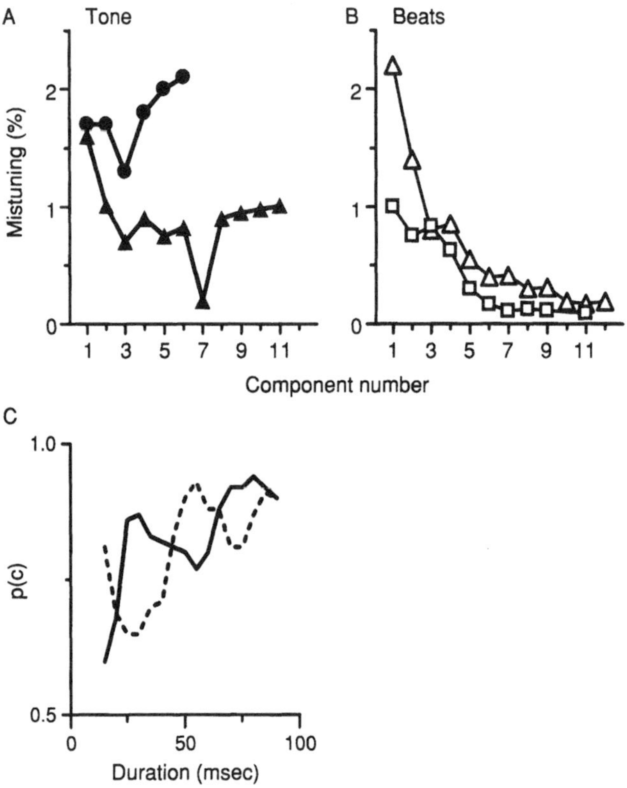

FIG. 1. Psychophysics of mistuning. In each study summarized here, f0 for the harmonic tone was 0.2 kHz. (A) Thresholds at which a mistuned component is first heard as a separate tone. Circles = Data from Moore *et al.* (1986), Table I. Triangles = data from Hartmann *et al.* (1990), Fig. 5. Thresholds judged by Hartmann *et al.*, to be based on the detection of roughness or beats were omitted. (B) Thresholds at which mistuning first produces any audible change. Triangles = data from Moore *et al.* (1985), Fig. 1, duration = 410 msec. Squares = data from Lee and Green (1994), "simultaneous-monotic" condition. (C) Effect of tone duration and phase on the detectability of a mistuned component. Data were redrawn from Hartmann (1988), Fig. 3. Solid line = all stimulus components added in sine phase. Dashed line = phase of the mistuned component was shifted by 180°.

magnitude lower than the threshold at which a second tone was first heard (Grube *et al.*, 2003; Lee and Green, 1994; Moore *et al.*, 1985). Thresholds for detecting mistuning decrease as stimulus duration increases (Moore *et al.*, 1985), but thresholds for perceiving a second sound source are less strongly dependent on duration (Moore *et al.*, 1986). Moore *et al.* (1985, 1986) suggested that very low thresholds for detection may occur when listeners can hear beats or roughness produced by adjacent unresolved components of the complex sound. They suggested that a mistuned component will emerge as a second tone only if it is resolved by auditory filtering and if the mistuning exceeds the width of the slots in a hypothetical harmonic sieve. The dependence of detection thresholds on frequency and duration is consistent with that interpretation. As the frequency of the mistuned component increases, a smaller fractional change is required to produce a constant beat frequency. The decrease in threshold with duration can be accounted for because, other factors being equal, a longer stimulus includes more cycles of a low-frequency beat, providing the listener with more opportunities to detect the beat.

Hartmann (1988) reported the results of a preliminary experiment showing that the detectability of a mistuned component was affected by a phase shift applied to the mistuned component (but not to other components). Phase interacted with duration. For tones with durations between approximately 20 to 40 or 65 to 85 msec, the mistuned component was easier to detect if it was in-phase with the remaining harmonics, but for tones 40 to 65 msec long, the mistuned component was easier to detect if it was 180° out-of-phase with the other components (Fig. 1C). If the mistuned component were resolved, its phase relative to the adjacent components should have been irrelevant, so the result strongly suggests that the mistuned component was not resolved. Hartmann (2004) has also found that the detectability of mistuning is degraded if the tuned harmonics immediately adjacent to the mistuned component are deleted from the stimulus. Deleting components increases the frequency distance between the mistuned component and the adjacent components, which should make the mistuned component more resolvable. The design of these two experiments did not require listeners to indicate that they perceived the mistuned component as a second tone. However, the same 2.5% change in the fourth harmonic of 0.2 kHz exceeded the threshold for hearing a second tone in other experiments in which listeners were explicitly required to report what they heard (Hartman *et al.,* 1990; Moore *et al.,* 1986; Roberts and Brunstrom, 1998; also see Fig. 1A). These results are difficult to explain if the mistuned component must be resolved in order to be heard as a separate tone. Hartmann (2004) concluded that the findings were best accounted for by assuming that mistuning produces a tonotopically local perturbation of a neural response synchronized to f0, caused by interactions between adjacent unresolved components.

Up to this point, mistuning has been defined as a change in the frequency of one component in an otherwise harmonic tone. A slightly different experiment was conducted by Roberts and Brunstrom (1998), who measured listeners' abilities to segregate components from inharmonic tones. In one condition they presented "frequency-shifted" tones, in which the spacing between components was a constant 0.2 kHz although the components were not multiples of 0.2 kHz (0.23, 0.43, 0.63 kHz, and so on). In another condition, the tones were "stretched," so that the frequency interval between adjacent components increased by 3 Hz with each successive component number (0.200, 0.403, 0.609 kHz, and so on). For the two inharmonic conditions, the component that was shifted from its expected frequency was considered to be mistuned. Standard harmonic and mistuned tones were also presented as control conditions. Listeners were able to segregate mistuned components from the inharmonic tones, with sensitivity similar to that observed in the control conditions. The result suggested that harmonic structure was evaluated within restricted regions of the spectrum, and that interactions between adjacent components were more important than deviations from a global spectral pattern tied to f0, the pattern that would be detected by a harmonic sieve.

C. Evoked Potentials in Response to Mistuned Tones

Neural activity evoked by complex tones and recorded non-invasively from human listeners may vary with the harmonic structure of the evoking stimulus. Jones (2003) measured cortical evoked potentials during presentation of sequences of harmonic and inharmonic tones. The latencies of evoked potentials N1 and P1 measured from electrode Fz at the onset of a steady-state tone within the sequence were significantly longer when the evoking stimulus was an inharmonic tone. Harmonicity did not affect evoked potentials recorded at other electrode sites and at other times during the stimulus sequence. The stimulus configuration used in that study does not allow simple comparison to the psychophysical studies reviewed in the previous section.

Dyson and Alain (2004) collected middle-latency auditory-evoked responses (MLAER) in response to complex tones with and without mistuned components, stimuli similar to those used in previous psychophysical studies (Hartmann *et al.*, 1990; Moore *et al.*, 1986). They reported that the amplitude of component Pa of the MLAER increased with mistuning. Significant changes in Pa amplitude occurred at latencies as short as 32 msec, extending the authors' previous studies in which mistuning had produced changes in evoked and event-related potentials with longer latencies (Alain *et al.*, 2001, 2002). Perceptual segregation of the mistuned component was measured in a separate series of trials, from a subset of the listeners who provided the evoked-potential data. The probability that the listeners reported hearing two tones increased with the amount of mistuning, reaching 0.5 for approximately 6% mistuning, and the increase in Pa amplitude with mistuning roughly followed the increase in probability. The response magnitude increased, even though mistuning did not increase the intensity of the stimulus. The increase in MLAER magnitude with mistuning was not dependent on attention, suggesting that the amplitude increase was a correlate of primitive, rather than schema-based, segregation (Bregman, 1990). The shorter response latency, along with independent evidence that the MLAER originates in the primary auditory cortex (Eggermont and Ponton, 2002), was interpreted as indicating that mistuning is registered at a relatively early, preattentive stage of auditory processing.

III. Neural Representation of Harmonic and Mistuned Complex Tones

The effects on evoked potentials described in the previous section suggest that mistuning produces changes in the responses of neurons in the auditory pathway that are larger than expected given the actual physical change in the stimulus. However, only limited conclusions about the contributions of particular auditory nuclei or neurons within those nuclei to the encoding of mistuned tones can be drawn from evoked potential measurements. To provide information about cellular-level processing that may contribute to spectral segregation, the responses of individual neurons to harmonic and mistuned complex tones have been measured in my laboratory. As Hartmann and others have suggested, the mistuned tone is a relatively simple stimulus that appears to engage the same neural mechanisms that underlie the segregation of mixtures of natural sounds. The observations made so far indicate that mistuning produces disproportionate changes in the discharge patterns of neurons in the IC (Sinex *et al.*, 2002, 2005). By disproportionate changes, it is meant that the discharge rates of individual IC neurons increase even though stimulus intensity does not, that mistuned tones are more likely to elicit temporally patterned responses than harmonic tones, that mistuning changes the periodicity of the response in a way that does not directly follow the change in the frequency of the mistuned component, and that mistuning affects the responses of neurons with characteristic frequencies (CF) remote from the frequency at which the mistuning was applied. The same stimulus manipulations that produce these changes in the IC have more modest effects in the peripheral auditory system; the responses of AN fibers change in proportion to the stimulus, and the changes that do occur are tonotopically local (Sinex *et al.*, 2003). Thus, the representation of harmonic structure undergoes a major transformation at or just below the level of the IC, which appears to result from broadband integration of temporally patterned excitation and inhibition (Sinex *et al.*, 2005). Consequently the IC appears to play a critical role in spectral segregation and, ultimately, sound-source determination.

A. Responses of Individual Neurons to Tones with Mistuned Harmonics

In general, mistuned tones elicit responses from IC neurons that are larger and exhibit unexpected temporal patterning, compared to the responses of the same neurons to harmonic tones. An example is shown in Fig. 2. The top panel

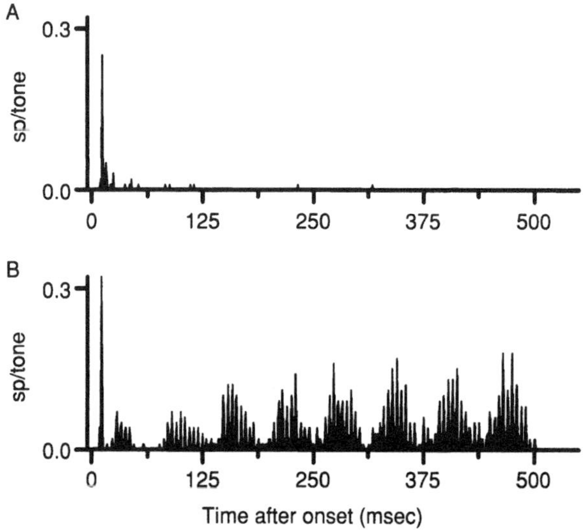

FIG. 2. Discharge patterns elicited by complex tones from one IC neuron, shown as peristimulus time histograms. Neuron 0410-11, CF = 2 kHz. (A) Responses to a 500-msec harmonic tone. The tone included the first eight harmonics of 0.4 kHz, with all components having sine phase and equal amplitude. (B) Responses to the same tone with 12% mistuning of the 4th harmonic, from 1.600 to 1.792 kHz.

(Fig. 2A) is the discharge pattern elicited by a complex tone consisting of the first eight harmonics of 0.4 kHz. The duration of the harmonic tone was 500 msec, but it elicited only a modest transient response from this neuron, similar to the responses elicited from the same neuron by pure tones. The lower panel illustrates the discharge pattern obtained when the fourth harmonic of the tone was mistuned by 12%, from 1.6 kHz to 1.792 kHz. Each presentation of the mistuned tone elicited approximately 10 spikes from the neuron, compared to a rate of <1 spike per presentation of the harmonic tone. The rate increase occurred in the absence of any change in stimulus intensity.

The mistuned tone also elicited a distinctive temporally patterned response characterized by peaks separated by about 5 msec and a slowly modulated envelope with a period of 62.5 msec. A Fourier transform (FFT) of the cycle histogram indicated that the discharge pattern included large components at 0.208, 0.192, and 0.016 kHz, frequencies that are generated directly or indirectly by beating between adjacent stimulus components. One of these frequencies, 0.208 kHz, matches the difference in frequency between the mistuned fourth harmonic and the tuned fifth harmonic. The stimulus does not have a component or potential beat at 0.192 kHz, but a response at that frequency could be generated by beating between response components synchronized to 0.208 kHz and another synchronized to f0, 0.4 kHz. The 0.016 kHz component could arise from beating between response components at 0.208 and 0.192 kHz. Although the particular frequencies in this example are different, all of these features observed in the response to this tone with f0 = 0.4 kHz have also been observed in the responses to mistuned tones with f0 = 0.25 kHz and with various amounts of mistuning by Sinex *et al.* (2002, 2005).

Although mistuning can produce dramatic changes in the responses of IC neurons, it has a much less effect on the responses of AN fibers. Figure 3 shows isointensity response areas for one AN fiber (Fig. 3A) and one IC neuron (Fig. 3B) that had the same CF, 1.2 kHz, and similar overall frequency selectivity. The responses of these two units to complex tones are directly compared in Fig. 4. The AN fiber responded with a rate of about 35 spikes per presentation of the harmonic tone (Fig. 4A). The response was highly synchronized to the stimulus, and FFT analysis of the cycle histogram elicited by the harmonic tone revealed

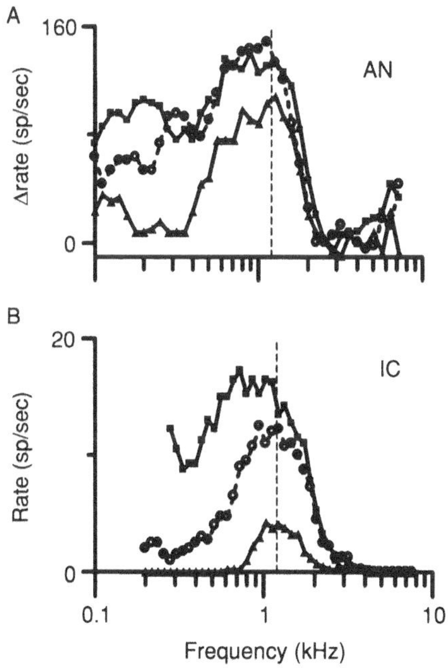

FIG. 3. Isointensity response areas recorded at two levels of the chinchilla auditory pathway. (A) Responses from an auditory nerve fiber. Each line represents discharge rate elicited by single tones of variable frequency but constant level. Each rate in spikes/sec has been normalized by subtracting the fiber's spontaneous rate. Triangles = 30 dB SPL; circles = 50 dB SPL; squares = 70 dB SPL. Fiber 0017-11, CF = 1.2 kHz (vertical dashed line). (B) Similar to A for a neuron in the IC. Discharge rates were not normalized, since the neuron was not spontaneously active. Neuron 0302-20, CF = 1.2 kHz (dashed line).

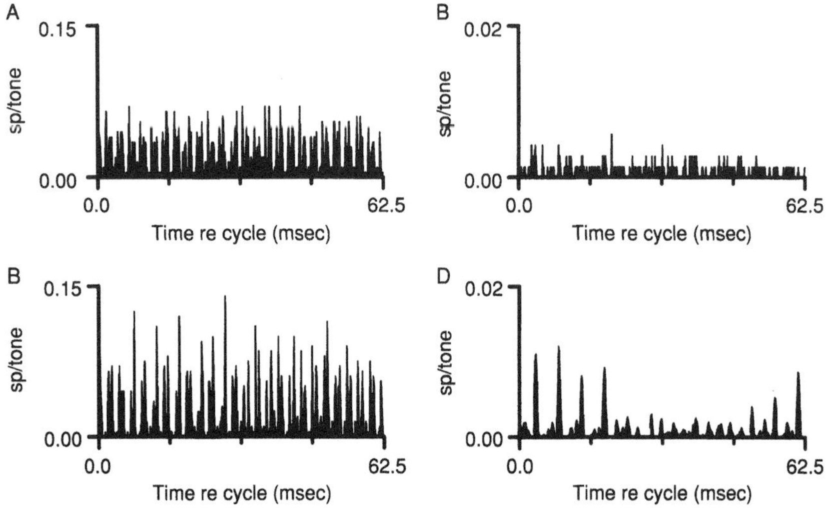

FIG. 4. Discharge patterns elicited by complex tones, shown as cycle histograms. The cycle length for the histograms, 62.5 msec, includes 25 cycles of 0.4 kHz, the fundamental frequency of the harmonic tone, and one cycle of 0.016 kHz, the actual fundamental frequency for the mistuned tone. (A) Responses to the harmonic tone, from the chinchilla auditory nerve fiber whose response area was shown in Fig. 3A. (B) Responses of the same fiber to the mistuned tone. (C) Responses to the harmonic tone, from the chinchilla IC neuron whose response area was shown in Fig. 3B. (D) Responses of the same IC neuron to the mistuned tone.

large response components at the frequencies of components near CF, 1.2 and 1.6 kHz. As a result of beating between these two adjacent components, the envelope of the response was modulated at 0.4 kHz, the difference frequency and also f0 for the harmonic tone. Response components were also observed at 0.4 and 0.8 kHz that were more likely to be rectifier distortion products that are usually present in the FFT of neural response histograms (Young and Sachs, 1979).

The discharge pattern elicited by the mistuned tone (Fig. 4B) was qualitatively and quantitatively similar to the response to the harmonic tone. Although the peaks in the cycle histogram appear to be slightly larger than those in Fig. 4A, the average rate of 37 spikes per mistuned tone was not reliably higher than the rate in response to the harmonic tone. The major change was that the periodicity of the synchronized response changed as expected from the increase in the frequency of the fourth harmonic; a new response component appeared at the frequency of the mistuned harmonic, 1.792 kHz. Additional components most likely to be rectifier distortion products were also observed. Overall, these responses could be described as resulting from a relatively simple narrowband analysis, in which the complex stimulus was bandpass-filtered and half-wave rectified, comparable to responses observed in previous studies of the representation of harmonic complex tones (Deng and Geisler, 1987; Horst *et al.*, 1986; Wang and Sachs, 1993).

For the IC neuron, the harmonic tone elicited a transient response at the onset of the harmonic tone, similar to the one in Fig. 2A, and a small sustained response (Fig. 4C). Although the pure-tone frequency selectivity of the IC neuron closely matched that of the AN fiber, its discharge rate was much lower, 8 spikes per tone. The cycle histogram (which excluded the onset response) exhibited no obvious temporal pattern that could be attributed to the stimulus. The absence of synchrony to stimulus components with frequencies >1 kHz or to the 0.4-kHz waveform envelope is consistent with previous measurements showing that synchrony to stimulus components and envelopes decreases at successive levels of the auditory pathway including the IC (Kuwada *et al.*, 1984; Langner and Schreiner, 1988; Rees and Møller, 1987). As in the previous example, mistuning had a pronounced effect on the response of the IC neuron. The neuron's discharge rate increased from 8 to 13 spikes per tone. Mistuning changed the fine structure and the envelope of the response, resulting in a temporal pattern consisting of low-frequency components similar to the example shown in Fig. 2B. These changes were not as easily predictable as they were for the AN fiber, and neither the fine structure nor the envelope appeared to have any counterpart in the responses of the AN fiber with the same CF. It is clear that in the case of the IC neuron, mistuning led to changes in the IC discharge pattern that could not be directly explained by the change in the frequency of one component, or attributed to simple features of IC neurons that distinguish them from AN fibers such as reduced temporal resolution (Joris and Yin, 1992; Langner and Schreiner, 1988). Discharge patterns analogous to the one illustrated in Fig. 4B have been observed in IC neurons that differ in CF and in other pure-tone response properties (Sinex *et al.*, 2005). As a result, it appears that the responses elicited across the population of IC neurons by a mistuned tone are more homogeneous than the responses elicited from the same neurons by pure tones or harmonic complex tones. It is apparent from Fig. 4 that the effect of mistuning is greater at the level of the IC than at the level of the AN.

B. What Accounts for Differences Between AN Fiber and IC Discharge Patterns?

Many characteristics of the discharge patterns of AN fibers can be accounted for, at least to a first approximation. Their frequency selectivity, including the contribution of suppressive sidebands, is determined by the non-linear mechanics of the basilar membrane (BM) and organ of Corti (Ruggero *et al.*, 2000). The discharges of AN fibers will synchronize to the waveforms of simple (Johnson,

1980) and complex sounds (Sinex and Geisler, 1983; Young and Sachs, 1979), within limits imposed by cochlear filtering and the properties of inner hair cells. Because these properties are relatively well understood, at least at the phenomenological level, it has been possible to develop computational models that provide accurate approximations of AN fiber discharge patterns (Meddis *et al.*, 1990; Sumner *et al.*, 2003; Tan and Carney, 2003; Zhang *et al.*, 2001).

Although the frequency selectivity of IC neurons superficially resembles that of AN fibers (Fig. 3), individual IC neurons receive ascending inputs from several different brainstem nuclei (Oliver and Shneiderman, 1991; Roth *et al.*, 1978), some of which are excitatory and some of which are inhibitory (Burger and Pollak, 2001; Li and Kelly, 1992; Saint Marie, 1996; Saint Marie and Baker, 1990). Small spiral ganglion lesions produce CF shifts and other unexpected changes in the shapes of IC frequency tuning curves that could not occur in peripheral neurons, demonstrating that IC neurons integrate frequency information in a complex way (Snyder and Sinex, 2002). In addition, IC neurons receive descending inputs whose effects on the discharge pattern are less well understood but may be as complex as those provided by the ascending innervation (Winer, 2005). Thus, for IC neurons, frequency selectivity measured with single tones usually does not provide a complete picture of the bandwidth or complexity of the inputs that the neuron receives.

A better appreciation of the integrative bandwidth of an IC neuron may be obtained if the frequency pattern of excitation and inhibition is estimated (Burger and Pollak, 2001; LeBeau *et al.*, 2001; Portfors and Wenstrup, 2002; Ramachandran *et al.*, 1999). Inhibitory effects of single tones can be measured if neurons are spontaneously active (Spirou and Young, 1991), but they are difficult or impossible to see in neurons with little or no spontaneous activity. However, inhibition can be seen if the frequency response area is remeasured in the presence of a second, excitatory tone that elicits a modest background discharge rate (Li *et al.*, 2002). An example of the effects of inhibition on the frequency response area of a neuron from the chinchilla IC is shown in Fig. 5. The top panel (Fig. 5A) shows an isointensity contour obtained with single tones presented at 60 dB SPL. The neuron's CF (estimated from similar contours obtained at lower SPL but not shown in the figure) was 2.3 kHz. The lower panel (Fig. 5B) shows a similar contour, obtained by presenting the same variable frequency, 60 dB SPL tones simultaneously with a fixed tone at 2.3 kHz and 30 dB SPL. Variable tones at frequencies between 0.2 to 0.7 kHz and 3 to 6 kHz strongly inhibited the response to the CF tone. The shaded areas shown in the figure indicate the width of the chinchilla's psychophysical critical band, measured at 2 kHz with a variable bandwidth masker by Niemiec *et al.* (1992); the critical bandwidth was 0.92 kHz, or approximately 0.64 octaves (oct). Niemiec *et al.* (1992) also measured the width of the auditory filter using a notched-noise masker; that bandwidth was even smaller, 0.304 kHz or approximately 0.21 oct. In contrast to these measurements,

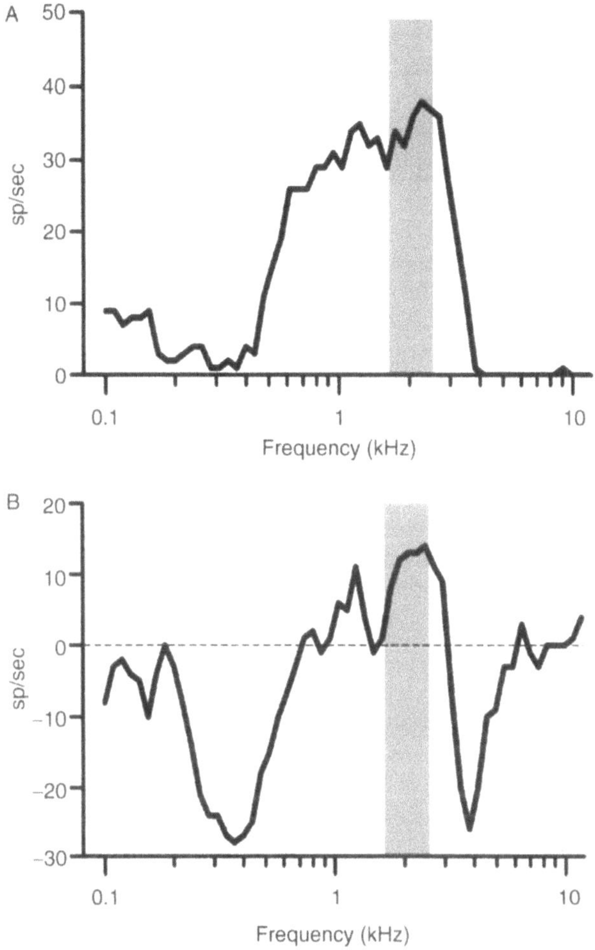

FIG. 5. Neural and psychophysical frequency selectivity. (A) Isointensity response areas obtained at 60 dB SPL from one IC neuron, 0410-12, CF = 2.3 kHz. The shaded area marks the extent of the chinchilla's critical band centered near the neuron's CF (as reported by Niemiec *et al.*, 1992). (B) Responses to the same tones presented with a simultaneous fixed tone at 2.3 kHz, 30 dB SPL. The average discharge rate elicited by the fixed tone was 29 sp/sec. The ordinate is the increase or decrease from that background discharge rate produced by the second, variable tone. Shaded area: chinchilla critical bandwidth, as in Panel A.

the bandwidth over which inhibition was observed was nearly 5 oct, a value that far exceeded the width of the psychophysical estimates but was representative of the integrative bandwidth of neurons in our sample. Broadband inhibition like that illustrated in the figure is hypothesized to be essential for producing discharge patterns like the ones shown in Figs. 3B and 4B, as described in the following text.

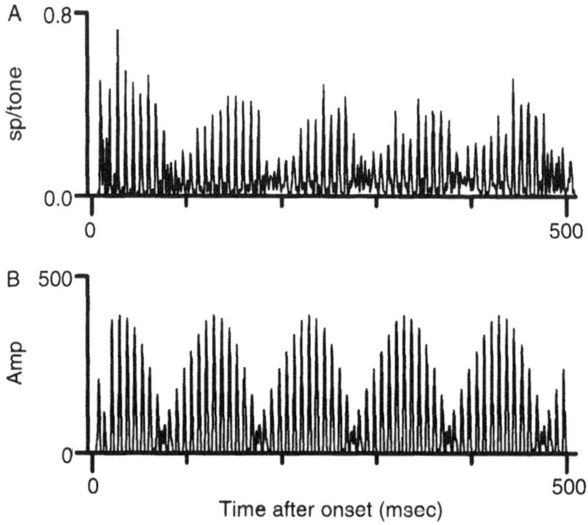

FIG. 6. Responses of the computational model compared to neural discharge patterns. (A) Responses of the neuron from the previous figure to a mistuned tone with f0 = 0.25 kHz and component 4 mistuned by 12% (from 1.000 to 1.120 kHz). (B) Response of the computational model to the same mistuned tone.

C. A Model That Duplicates the Discharge Patterns of IC Neurons to Mistuned Tones

A computer model that could duplicate the complex temporal discharge patterns of IC neurons was developed and reported by Sinex *et al.* (2002), and a simplified version of the model was described in greater detail by Sinex *et al.* (2005). The key features of the model are envelope extraction across a series of narrow bands with different center frequencies, integration over two or more of those bands, and the inclusion of temporally patterned excitatory and inhibitory input. The plots in Fig. 6 illustrate that the output of the model accurately reproduces the temporal discharge pattern elicited by a mistuned tone from a typical IC neuron. The significance of the integration bandwidth is considered further in Section IV.A.

IV. The Possible Relation Between the Responses of IC Neurons and Psychoacoustic Data

Psychoacoustic experiments have established that mistuning creates a spectral cue that promotes perceptual segregation. Mistuning has now also been shown to produce disproportionate changes in the temporal discharge patterns of IC neurons. Manipulations of the complex tone that do not affect the perception of a second tone have more modest effects on IC discharge patterns. Changes in component phases or stimulus level over a moderate range, for example, do not change the frequency content of IC responses, although they can produce shifts in the phase of the response envelope (Sinex *et al.*, 2005). The data support the following conclusions. First, the processing of complex composite signals undergoes a major transformation in the brainstem. Second, stimuli that are heard as more than one sound source generate highly modulated, partially coherent, spatially distributed discharge patterns at the level of the IC, and stimuli such as harmonic tones that do not lead to segregation generate weakly patterned or unpatterned responses in the IC.

A. Interactions Between Stimulus Components in Psychoacoustic and Neurophysiological Observations

There is disagreement about the role played by interactions between adjacent stimulus components in the processing of mistuned tones. Some psychophysical results have been interpreted as suggesting that a mistuned harmonic must be resolved in order to be heard as a separate tone (Moore *et al.*, 1986), while other evidence argues for an important contribution from unresolved components (Hartmann, 1988, 2004). In the microelectrode studies summarized in the previous text, the stimuli were such that neural interactions between adjacent stimulus components nearly always occurred, and so it is fair to ask whether that limits the generality of the neurophysiological observations. A "resolved" component is one that conforms to an operational definition based on other psychophysical measurements (Carlyon and Shackleton, 1994; Plomp and Mimpen, 1968). Shackleton and Carlyon (1994), for example, considered components in harmonic tones to be completely resolved if fewer than two harmonics fell within the 10-dB bandwidths of the auditory filter originally reported by Glasberg and Moore (1990). The distinction between resolved and unresolved components is valid and useful, but it would be an error to assume that two tones that are resolved in the psychoacoustic sense cannot simultaneously influence the responses of a single auditory neuron. As Fig. 5 illustrated, the responses of an IC neuron to one tone or spectral component are routinely affected by other tones separated by much more than the width of the auditory filter. It seems plausible that discharge patterns like those reported by Sinex *et al.* and summarized here could also be generated in the human IC by harmonic and mistuned tones, including tones consisting only of psychophysically resolved components.

B. Simulated Neural Responses to Complex Tones Used in Psychophysical Experiments

The computational model described in Section III.C was used to estimate the responses of IC neurons to the mistuned-tone stimulus most often used in psychophysical studies. Figure 7A shows the response of the model to a harmonic tone with f0 = 0.2 kHz (Hartmann *et al.,* 1990; Moore *et al.,* 1985, 1986). In response to the harmonic tone, this simulated neuron exhibited a simple modulated envelope whose frequency, 0.2 kHz, was determined by the spacing of components but also matched f0. For the simulation shown in Fig. 7B, the model parameters remained the same but the fourth component of the tone was mistuned by 2.5%. The result was a deeply modulated response similar to those shown previously but with temporal details determined by the particular parameters of this mistuned tone. Peaks in the envelope occurred at approximately 25, 75, 125, and 175 msec.

The simulation was repeated with frequency-shifted inharmonic tones, as used by Roberts and Brunstrom (1998). Figure 7C shows the response of the model to a tone in which each component was shifted by 0.03 kHz before the

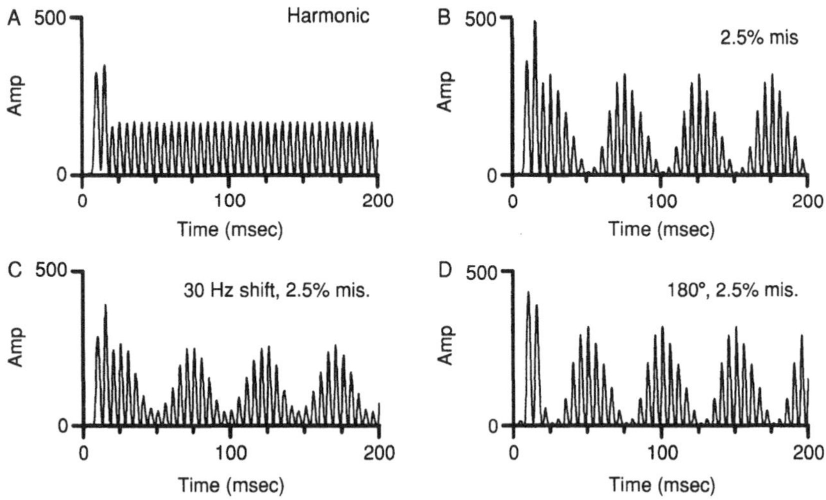

FIG. 7. Responses of the computational model to mistuned-tone stimuli used in psychophysical experiments including those by Moore *et al.* (1986), Hartmann *et al.* (1990), and Roberts and Brunstrom (1998). Identical model parameters were used for each calculation. (A) Response to a harmonic tone with f0 = 0.2 kHz. (B) Response to a tone with f0 = 0.2 kHz and 2.5% mistuning of component 4. (C) Response to an inharmonic, frequency-shifted tone with component 1 at 0.230 kHz, frequency interval = 0.2 kHz, and 2.5% mistuning of component 4, as in Roberts and Brunstrom (1998). (D) Same as Panel B, except the starting phase of the mistuned component was shifted by 180°, as in Hartmann (1988).

fourth component was mistuned by 2.5%. The frequency shift had little effect on the simulated response; slow modulation was again apparent in the response envelope, and envelope peaks occurred with latencies that were nearly identical to those elicited by the original mistuned tone and shown in panel B. As noted in Section II.B, Roberts and Brunstrom (1998) found that the same frequency shift had little effect on the psychophysical processing of the mistuned component, and the results of the simulation appear to be consistent with that.

An additional set of simulations was based on Hartmann's (1988) study of the effect of the phase of the mistuned component, results of which were summarized in Fig. 1C. Figure 7D shows the response to a mistuned tone in which a 180° phase shift was applied to the mistuned component. This pattern can also be compared to the pattern in Fig. 7B, the response to the in-phase condition in Hartmann's experiment. Shifting the phase of one component by 180° produced a 180° shift in the phase of response envelope; envelope peaks occurred at about 50, 100, 150, and 200 msec, a shift of approximately 25 msec, or one-half of an envelope cycle, compared to the envelope of the response in Fig. 7B.

This difference in envelope phase may be related to Hartmann's (1988) finding that the detectability of the mistuned component varied with duration. Meddis and Hewitt (1991) had previously simulated responses to the stimuli used in that experiment; their analysis based on autocorrelation of simulated AN fiber responses generally reproduced the phase effect, suggesting that psychophysical performance was constrained by the temporal information in the response. Here, simulations of IC responses like those shown in Fig. 7A, B, and D were used to estimate the detectability of mistuning as a function of phase and duration, to determine whether the effect of phase replotted in Fig. 8A might be produced by temporally based processes other than autocorrelation. The detectability of mistuning was assumed to be proportional to the difference between the neural representation of the harmonic tone and the mistuned tone. That was quantified by calculating the RMS difference in the amplitude of the response to the harmonic tone and the response to each mistuned tone, repeating the calculation for tones of varying duration. Each panel of Fig. 8B displays the results of the calculations for two simulated neurons with slightly different CFs. For each phase condition, the amplitude difference measure varied non-monotonically with duration. Across the two phase conditions, large RMS differences for one condition tended to occur for the durations that produced smaller RMS differences for the other condition. Although the patterns did not precisely duplicate the pattern of psychophysical sensitivity, there was a trend for the simulated responses to the in-phase mistuned tone to be larger for the tone durations at which the psychophysical detectability of the in-phase tone was higher, and vice-versa. This preliminary analysis suggests a possible link between the neural representation observed in the IC and psychophysical segregation based on harmonicity.

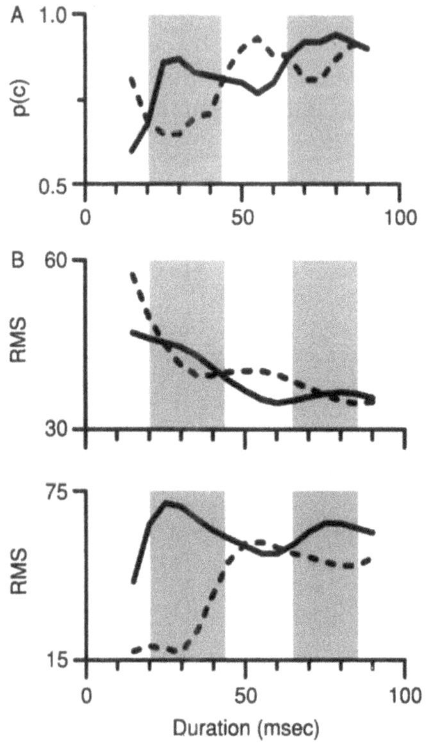

FIG. 8. The effect of the phase of the mistuned component on the responses of two simulated IC neurons. (A) Hartmann's (1988) psychophysical data showing an effect of the phase of the mistuned component on the detection of mistuning (also shown in Fig. 1C). Shaded areas highlight the tone durations at which a mistuned component with phase = 0° was more detectable than the same mistuned component with phase = 180°. (B) Estimates of the detectability of mistuning for the same two phase conditions, derived from the responses of two simulated IC neurons. See Section IV.B. for details. Solid line = mistuned component phase = 0°. Dashed line = mistuned component phase = 180°. Shaded areas highlight the same tone durations as in Panel A.

C. Strategies for Representing Spectral Structure

The fine structure of AN fiber responses encodes the frequencies of individual components of the spectrum. The precision of that representation as measured with pure tones begins to decline at about 1 to 2 kHz (Johnson, 1980; Palmer and Russell, 1986). The frequency range over which synchronized discharges are available is more or less in agreement with the range over which mistuned components can reliably be segregated (Hartmann *et al.*, 1990), suggesting that the initial representation of those components is temporal. At the level of the AN, the temporal representation of components in complex sounds is affected by cochlear filtering but is not influenced by whether the components are part of a harmonic series or not (Sinex *et al.,* 2003).

In contrast, the temporal representation of spectral structure and harmonicity in the IC is quite different from the representation at the level of the AN. IC neurons, because of their lower temporal resolution, do not represent individual components by discharge synchrony. Because their integrative bandwidths are large, they represent envelopes produced by interactions among stimulus components, rather than the components themselves. Harmonic structure determines how many unique response envelopes can be created. For harmonic tones, the frequency of the response envelope can only match f0. Sinex *et al.* (2005) found that when $f0 = 0.25$ kHz, a response component with the periodicity of f0 was common throughout the IC, although its magnitude was very low. Fewer observations of responses to the harmonic tone with $f0 = 0.4$ kHz are available, but one would expect even less synchrony in that case, since the ability of IC neurons to follow modulated envelopes falls rapidly in that frequency range (Langer and Schreiner, 1988; Rees and Møller, 1987). Stimulus-related periodicities other than the one determined by f0 have not been observed in IC responses to harmonic tones.

Unlike harmonic tones, mistuned tones generate response envelopes with multiple frequencies. Adjacent harmonic components beat with one periodicity, as described in the previous paragraph, but the mistuned component can beat with the adjacent harmonic to produce an envelope with a different frequency. The modeling results of Sinex *et al.* (2005) suggest that those first-order response envelopes also beat with one another, producing the stereotypical low-frequency envelope that is prominent in discharge patterns like the examples shown in Figs. 2B, 4D, and 6A.

According to this view of the determinants of IC discharge patterns, spectral structure is represented by a pattern of low-frequency beats that strongly reflects local spectral patterns. Beating responses also appear to play an important role in the perception of simultaneous vowels (Culling and Darwin, 1994) and in the neural representation of a similar composite stimulus consisting of two simultaneous harmonic tones (Sinex and Li, 2005). In the latter study, adjacent components in the composite signal produced low-frequency modulation in the responses of IC neurons that was quantitatively similar to the modulation produced by mistuned tones. When two harmonic sounds are summed, adjacent components in the composite signal nearly always arise from separate sources, and in most cases the frequency differences are unresolvable by the psychophysical definition, even if the original tones consisted of resolved harmonics. Since the frequency difference between adjacent components in the composite spectrum varies with component number, the range of possible envelope frequencies is even greater than it is for mistuned tones. However, despite this additional complexity in the stimulus, the periodicity of the responses of IC neurons followed the same rules that successfully accounted for the temporal patterns elicited by mistuned tones. As with mistuned tones, the pattern of low-frequency modulation in the responses of IC neurons encoded the relation between adjacent components in composite signals; the discharge patterns did not provide information about individual components.

V. Conclusions

Information about harmonic structure is conveyed in the IC by a code created by interactions between adjacent spectral components. This suggests that the auditory system must work backwards from the pattern of frequency differences to determine the spectra of the original sound sources; that is, an inverse problem must be solved. In the IC, information about individual spectral components is reduced or lost, and the representation of the mixture of envelopes produced by the components in a composite sound is widely distributed across the tonotopic axis. These observations seem to argue against mechanisms that resemble the harmonic sieve. As noted, some psychophysical data also argue against that kind of mechanism. It is not known how the information conveyed by the discharge patterns described here might be extracted at levels central to the IC, or to what extent the decoded information might underlie the determination of sound sources. Those are important issues that could be addressed by future research.

Acknowledgments

Supported by DC00341 from the NIDCD. Drs. Susan Nittrouer and Russell Snyder provided helpful comments on a draft of the manuscript.

References

Alain, C., McDonald, K. L., Ostroff, J. M., and Schneider, B. (2001). Age-related changes in detecting a mistuned harmonic. *J. Acoust. Soc. Am.* **109,** 2211–2216.

Alain, C., Schuler, B. M., and McDonald, K. L. (2002). Neural activity associated with distinguishing concurrent auditory objects. *J. Acoust. Soc. Am.* **111,** 990–995.

Assmann, P. F., and Summerfield, Q. (1990). Modeling the perception of concurrent vowels: Vowels with different fundamental frequencies. *J. Acoust. Soc. Am.* **88,** 680–697.

Assmann, P. F., and Summerfield, Q. (1989). Modeling the perception of concurrent vowels: Vowels with the same fundamental frequency. *J. Acoust. Soc. Am.* **85,** 327–338.

Bregman, A. S. (1990). Auditory Scene Analysis MIT Press, Cambridge, MA.

Brokx, J., and Nooteboom, S. (1982). Intonation and the perceptual separation of simultaneous voices. *J. Phonet.* **10,** 23–26.

Burger, R. M., and Pollak, G. D. (2001). Reversible inactivation of the dorsal nucleus of the lateral lemniscus reveals its role in the processing of multiple sound sources in the inferior colliculus of bats. *J. Neurosci.* **21,** 4830–4843.

Carlyon, R. P., and Shackleton, T. M. (1994). Comparing the fundamental frequencies of resolved and unresolved harmonics: Evidence for two pitch mechanisms? *J. Acoust. Soc. Am.* **95,** 3541–3554.

Carrell, T. D., and Opie, J. M. (1992). The effect of amplitude comodulation on auditory object formation in sentence perception. *Percept. Psychophys.* **52,** 437–445.

Chalikia, M. H., and Bregman, A. S. (1989). The perceptual segregation of simultaneous auditory signals: Pulse train segregation and vowel segregation. *Percept. Psychophys.* **46,** 487–496.

Chalikia, M. H., and Bregman, A. S. (1993). The perceptual segregation of simultaneous vowels with harmonic, shifted, or random components. *Percept. Psychophys.* **53,** 125–133.

Cherry, E. C. (1953). Some experiments on the recognition of speech, with one and two ears. *J. Acoust. Soc. Am.* **25,** 975–979.

Cohen, E. (1980). The influence of nonharmonic partials on tone perception. Dissertation, Stanford University.

Culling, J. F., and Darwin, C. J. (1994). Perceptual and computational separation of simultaneous vowels: Cues arising from low-frequency beating. *J. Acoust. Soc. Am.* **95,** 1559–1569.

Darwin, C. J., and Carlyon, R. P. (1995). Auditory grouping. *In* "Hearing" (B. C. Moore, Ed.), pp. 387–424. Academic Press, San Diego.

de Cheveigné, A. (1993). Separation of concurrent harmonic sounds: Fundamental frequency estimation and a time-domain cancellation model of auditory processing. *J. Acoust. Soc. Am.* **93,** 3271–3290.

de Cheveigné, A. (1999). Vowel-specific effects in concurrent vowel identification. *J. Acoust. Soc. Am.* **106,** 327–340.

de Cheveigné, A., Kawahara, H., Tsuzaki, M., and Aikawa, K. (1997). Concurrent vowel identification: I. Effects of relative amplitude and f0 difference. *J. Acoust. Soc. Am.* **101,** 2839–2847.

Deng, L., and Geisler, C. D. (1987). Responses of auditory-nerve fibers to multiple-tone complexes. *J. Acoust. Soc. Am.* **82,** 1989–2000.

Duifhuis, H. (1970). Audibility of high harmonics in a periodic pulse. *J. Acoust. Soc. Am.* **48,** 888–893.

Duifhuis, H., Willems, L. F., and Sluyter, R. J. (1982). Measurement of pitch in speech: An implementation of Goldstein's theory of pitch perception. *J. Acoust. Soc. Am.* **71,** 1568–1580.

Dyson, B. J., and Alain, C. (2004). Representation of concurrent acoustic objects in primary auditory cortex. *J. Acoust. Soc. Am.* **115,** 280–288.

Eggermont, J. J., and Ponton, C. W. (2002). The neurophysiology of auditory perception: From single units to evoked potentials. *Audiol. Neurootol.* **7,** 71–99.

Glasberg, B. R., and Moore, B. C. (1990). Derivation of auditory filter shapes from notched-noise data. *Hear. Res.* **47,** 103–138.

Goldstein, J. L. (1973). An optimum processor theory for the central formation of the pitch of complex tones. *J. Acoust. Soc. Am.* **54,** 1496–1516.

Grube, M., von Cramon, D. Y., and Rubsamen, R. (2003). Inharmonicity detection. Effects of age and contralateral distractor sounds. *Exp. Brain. Res.* **153,** 637–642.

Hall, J. W., and Grose, J. H. (1990). Comodulation masking release and auditory grouping. *J. Acoust. Soc. Am.* **88,** 119–125.

Hall, J. W., Haggard, M. P., and Fernandes, M. A. (1984). Detection in noise by spectro-temporal pattern analysis. *J. Acoust. Soc. Am.* **76,** 50–56.

Handel, S. (1995). Timbre perception and auditory object identification. *In* "Hearing" (B. Moore, Ed.), pp. 425–461. Academic Press, San Diego.

Hartmann, W. M., McAdams, S., and Smith, B. K. (1990). Hearing a mistuned harmonic in an otherwise periodic complex tone. *J. Acoust. Soc. Am.* **88,** 1712–1724.

Hartmann, W. M. (2004). Detecting a Mistuned Harmonic. *Michigan State Psychoacoustics. Report* **129.**

Hartmann, W. M. (1988). Pitch perception and the segregation and integration of auditory entities. *In* "Auditory Function: Neurobiological Bases of Hearing" (G. Edelman, W. Gall, and W. Cowan, Eds.), pp. 623–645. John Wiley & Sons, New York.

Horst, J. W., Javel, E., and Farley, G. R. (1986). Coding of spectral fine structure in the auditory nerve. I. Fourier analysis of period and interspike interval histograms. *J. Acoust. Soc. Am.* **79,** 398–416.

Johnson, D. H. (1980). The relationship between spike rate and synchrony in responses of auditory-nerve fibers to single tones. *J. Acoust. Soc. Am.* **68,** 1115–1122.

Jones, S. J. (2003). Sensitivity of human auditory evoked potentials to the harmonicity of complex tones: Evidence for dissociated cortical processes of spectral and periodicity analysis. *Exp. Brain Res.* **150,** 506–514.

Joris, P. X., and Yin, T. C. (1992). Responses to amplitude-modulated tones in the auditory nerve of the cat. *J. Acoust. Soc. Am.* **91,** 215–232.

Kidd, G., Mason, C. R., Brughera, A., and Chiu, C. Y. (2003). Discriminating harmonicity. *J. Acoust. Soc. Am.* **114,** 967–977.

Kuwada, S., Yin, T. C., Syka, J., Buunen, T. J., and Wickesberg, R. E. (1984). Binaural interaction in low-frequency neurons in inferior colliculus of the cat. IV. Comparison of monaural and binaural response properties. *J. Neurophysiol.* **51,** 1306–1325.

Langner, G., and Schreiner, C. E. (1988). Periodicity coding in the inferior colliculus of the cat. I. Neuronal mechanisms. *J. Neurophysiol.* **60,** 1799–1822.

LeBeau, F. E., Malmierca, M. S., and Rees, A. (2001). Iontophoresis in vivo demonstrates a key role for GABA(A) and glycinergic inhibition in shaping frequency response areas in the inferior colliculus of guinea pig. *J. Neurosci.* **21,** 7303–7312.

Lee, J., and Green, D. M. (1994). Detection of a mistuned component in a harmonic complex. *J. Acoust. Soc. Am.* **96,** 716–725.

Li, H., Henderson, J., and Sinex, D. G. (2002). Responses of Inferior Colliculus Neurons to SAM Tones Located in Inhibitory Response Area. *Abstr. Assoc. Research. Otolaryngol.* **25,** 38.

Li, L., and Kelly, J. B. (1992). Inhibitory influence of the dorsal nucleus of the lateral lemniscus on binaural responses in the rat's inferior colliculus. *J. Neurosci.* **12,** 4530–4539.

Lin, J. Y., and Hartmann, W. M. (1998). The pitch of a mistuned harmonic: Evidence for a template model. *J. Acoust. Soc. Am.* **103,** 2608–2617.

Meddis, R., and Hewitt, M. J. (1992). Modeling the identification of concurrent vowels with different fundamental frequencies. *J. Acoust. Soc. Am.* **91,** 233–245.

Meddis, R., and Hewitt, M. J. (1991). Virtual pitch and phase sensitivity of a computer model of the auditory periphery. II: Phase sensitivity. *J. Acoust. Soc. Am.* **89,** 2883–2894.

Meddis, R., Hewitt, M. J., and Schackleton, T. M. (1990). Implementation details of a computational model of the inner hair cell/auditory nerve synapse. *J. Acoust. Soc. Am.* **87,** 1813–1816.

Moore, B. C., Glasberg, B. R., and Peters, R. W. (1986). Thresholds for hearing mistuned partials as separate tones in harmonic complexes. *J. Acoust. Soc. Am.* **80,** 479–483.

Moore, B. C., Peters, R. W., and Glasberg, B. R. (1985). Thresholds for the detection of inharmonicity in complex tones. *J. Acoust. Soc. Am.* **77,** 1861–1867.

Niemiec, A. J., Yost, W. A., and Shofner, W. P. (1992). Behavioral measures of frequency selectivity in the chinchilla. *J. Acoust. Soc. Am.* **92,** 2636–2649.

Oliver, D. L., and Shneiderman, A. (1991). The anatomy of the inferior colliculus: A cellular basis for integration of monaural and binaural information. *In* "Neurobiology of Hearing: The Central Auditory System" (R. A. Altschuler, R. P. Bobbin, B. M. Clopton, and D. W. Hoffman, Eds.), pp. 195–222. Raven Press, New York.

Palmer, A. R., and Russell, I. J. (1986). Phase-locking in the cochlear nerve of the guinea-pig and its relation to the receptor potential of inner hair-cells. *Hear. Res.* **24,** 1–15.

Plomp, R., and Mimpen, A. M. (1968). The ear as a frequency analyzer. II. *J. Acoust. Soc. Am.* **43,** 764–767.

Plomp, R., and Mimpen, A. M. (1981). Effect of the orientation of the speaker's head and the azimuth of a sound source on the speech reception threshold for sentences. *Acustica* **48,** 325–328.

Portfors, C. V., and Sinex, D. G. (2005). Coding of communication sounds in the inferior colliculus. *In* "The Inferior Colliculus" (J. A. Winer and C. E. Schreiner, Eds.). Springer-Verlag, New York.

Portfors, C. V., and Wenstrup, J. J. (2002). Excitatory and facilitatory frequency response areas in the inferior colliculus of the mustached bat. *Hear. Res.* **168,** 131–138.

Ramachandran, R., Davis, K. A., and May, B. J. (1999). Single-unit responses in the inferior colliculus of decerebrate cats. I. Classification based on frequency response maps. *J. Neurophysiol.* **82,** 152–163.

Rasch, R. A. (1978). The perception of simultaneous notes such as in polyphonic music. *Acustica* **40,** 21–33.

Rees, A., and Møller, A. R. (1987). Stimulus properties influencing the responses of inferior colliculus neurons to amplitude-modulated sounds. *Hear. Res.* **27,** 129–143.

Roberts, B., and Brunstrom, J. M. (1998). Perceptual segregation and pitch shifts of mistuned components in harmonic complexes and in regular inharmonic complexes. *J. Acoust. Soc. Am.* **104,** 2326–2338.

Roth, G. L., Aitkin, L. M., Andersen, R. A., and Merzenich, M. M. (1978). Some features of the spatial organization of the central nucleus of the inferior colliculus of the cat. *J. Comp. Neurol.* **182,** 661–680.

Ruggero, M. A., Narayan, S. S., Temchin, A. N., and Recio, A. (2000). Mechanical bases of frequency tuning and neural excitation at the base of the cochlea: Comparison of basilar-membrane vibrations and auditory- nerve-fiber responses in chinchilla. *Proc. Natl. Acad. Sci. USA* **97,** 11744–11750.

Saint Marie, R. L. (1996). Glutamatergic connections of the auditory midbrain: Selective uptake and axonal transport of D-[3H]aspartate. *J. Comp. Neurol.* **373,** 255–270.

Saint Marie, R. L., and Baker, R. A. (1990). Neurotransmitter-specific uptake and retrograde transport of [3H] glycine from the inferior colliculus by ipsilateral projections of the superior olivary complex and nuclei of the lateral lemniscus. *Brain. Res.* **524,** 244–253.

Shackleton, T. M., and Carlyon, R. P. (1994). The role of resolved and unresolved harmonics in pitch perception and frequency modulation discrimination. *J. Acoust. Soc. Am.* **95,** 3529–3540.

Sinex, D. G., and Geisler, C. D. (1983). Responses of auditory nerve fibers to consonant-vowel syllables. *J. Acoust. Soc. Am.* **73,** 602–615.

Sinex, D. G., Guzik, H., Li, H., and Henderson Sabes, J. (2003). Responses of auditory nerve fibers to harmonic and mistuned complex tones. *Hear. Res.* **182,** 130–139.

Sinex, D. G., and Li, H. (2005). Responses of inferior colliculus neurons to simultaneous complex tones with different fundamental frequencies. *Abstr. Assoc. Research. Otolaryngol.* **28,** 340.

Sinex, D. G., Li, H., and Velenovsky, D. S. (2005). Prevalence of stereotypical responses to mistuned tones in the inferior colliculus. *J. Neurophysiol.*, in press.

Sinex, D. G., Sabes, J. H., and Li, H. (2002). Responses of inferior colliculus neurons to harmonic and mistuned complex tones. *Hear. Res.* **168,** 150–162.

Snyder, R. L., and Sinex, D. G. (2002). Immediate changes in tuning of inferior colliculus neurons following acute lesions of cat spiral ganglion. *J. Neurophysiol.* **87,** 434–452.

Spirou, G. A., and Young, E. D. (1991). Organization of dorsal cochlear nucleus type IV unit response maps and their relationship to activation by bandlimited noise. *J. Neurophysiol.* **66,** 1750–1768.

Sumner, C. J., O'Mard, L. P., Lopez-Poveda, E. A., and Meddis, R. (2003). A nonlinear filter-bank model of the guinea-pig cochlear nerve: Rate responses. *J. Acoust. Soc. Am.* **113,** 3264–3274.

Tan, Q., and Carney, L. H. (2003). A phenomenological model for the responses of auditory nerve fibers. II. Nonlinear tuning with a frequency glide. *J. Acoust. Soc. Am.* **114,** 2007–2020.

Wang, X., and Sachs, M. B. (1993). Neural encoding of single-formant stimuli in the cat. I. Responses of auditory nerve fibers. *J. Neurophysiol.* **70,** 1054–1075.

Winer, J. A. (2005). Descending connections of the cortex and thalamus. *In* "The Inferior Colliculus" (J. A. Winer and C. E. Schreiner, Eds.). Springer-Verlag, New York.

Yost, W. A. (1991). Auditory image perception and analysis: The basis for hearing. *Hear. Res.* **56,** 8–18.

Yost, W. A. (1993). Overview: Psychoacoustics. *In* "Human Psychophysics" (W. Yost, A. Popper, and R. Fay, Eds.), pp. 1–12. Springer-Verlag, New York.

Yost, W. A., and Sheft, S. (1993). Auditory Perception. *In* "Human Psychophysics" (W. A. Yost, A. N. Popper, and R. R. Fay, Eds.), pp. 193–236. Springer-Verlag, New York.

Young, E. D., and Sachs, M. B. (1979). Representation of steady-state vowels in the temporal aspects of the discharge patterns of populations of auditory nerve fibers. *J. Acoust. Soc. Am.* **66,** 1381–1403.

Zhang, X., Heinz, M. G., Bruce, I. C., and Carney, L. H. (2001). A phenomenological model for the responses of auditory-nerve fibers: I. Nonlinear tuning with compression and suppression. *J. Acoust. Soc. Am.* **109,** 648–670.

SPECTRAL INFORMATION IN SOUND LOCALIZATION

Simon Carlile,* Russell Martin,† and Ken McAnally†

*Auditory Neuroscience Laboratory, School of Medical Sciences
University of Sydney, Sydney NSW 2006, Australia
†Defense Science and Technology Organisation, Fisherman's Bend
Victoria 3207, Australia

The perception and neural representations of auditory space are based upon an analysis of the acoustical cues that arise at each ear. The spectral cues to sound location arise as a function of the filtering of the incoming sound by the auditory periphery. In this chapter, we review the psychophysical evidence for the role of these cues in sound localization and the manner in which they are combined with the other binaural cues to sound location. In addition, we review some of the bioacoustic, psychophysical, and neurophysiological studies that have examined the mechanisms of encoding and processing of these cues.

I. Introduction

Neural coding of auditory space is quite different from the encoding of space by the other spatial senses, notably vision and somatosensation. For the latter modalities, space is coded directly at the sensory epithelia, the retina, and the skin

INTERNATIONAL REVIEW OF
NEUROBIOLOGY, VOL. 70
DOI: 10.1016/S0074-7742(05)70012-X

0074-7742/05 $35.00

surface, respectively. An ordered topographic projection from the receptor epithelia into the central nervous system gives rise to topographic representations of two-dimensional space. For example, stimuli in the adjacent portions of the visual field will excite adjacent photoreceptors on the retina which in turn excite adjacent retinal ganglion cells and as a consequence of the ordered projections of the optic nerve, adjacent neurons in the visual thalamus and visual cortex. By contrast, the auditory epithelia encodes the amplitude of energy entering the ears as a function of frequency. The location of a sound source in space is computed from the characteristics of the inputs to the two ears. In the case of a single source in space, the levels and time of arrival of the sound at each ear provide cues to location. So, for a sound to the right of the frontal midline, the sound will occur earlier in the right ear than the left ear, giving rise to an interaural time difference (ITD), and will have a greater amplitude in the right than the left ear, giving rise to an interaural level difference (ILD).

ITDs and ILDs provide information about the so-called *cone of confusion* on which a sound source lies (Mills, 1972; Woodworth and Schlosberg, 1954). A cone of confusion is a set of points at a common lateral angle, where lateral angle is the angle subtended at the center of the head between a point and the median sagittal plane. Within a cone of confusion the binaural cues are approximately constant. It is clear that the auditory localization system is capable of resolving locations within cones of confusion because the elevations and front-back hemifields of several types of sound can be accurately determined (e.g., Makous and Middlebrooks, 1990; Oldfield and Parker, 1984a).

The information used by listeners to resolve the ambiguity inherent in ITD and ILD cues (i.e., to distinguish locations within the cone of confusion) must extend across a broad range of frequencies, because judgments of sound-source elevation and front-back hemifield become less accurate as sound bandwidth is reduced (Butler, 1986; Butler and Planert, 1976; Hebrank and Wright, 1974a; King and Oldfield, 1997; Roffler and Butler, 1968a). That the information is spectral in nature is indicated by several lines of evidence. First, the perceived elevation of a source of narrow-band sound can be influenced more by the sound's center frequency than by the true elevation of its source (Blauert, 1969/1970; Middlebrooks, 1992; Pratt, 1930; Roffler and Butler, 1968b). Second, the accuracy with which the elevation of a broadband stimulus can be discerned is disrupted when a narrow band of frequencies is removed (Burlinghame and Butler, 1998; Butler and Musicant, 1993; Hebrank and Wright, 1974a) or its spectrum is scrambled (Macpherson, 1996; Macpherson and Middlebrooks, 2003; McAnally *et al.,* 2000; Wightman and Kistler, 1997).Third, filling the cavities of the pinnae, which interact with incident sound to alter its spectrum, almost completely abolishes the ability to determine sound-source elevation and front-back hemifield (Gardner, 1973; Gardner and Gardner, 1973; Hofman and van Opstal, 2003; Humanski and Butler, 1988; Morimoto, 2001; Musicant and Butler, 1984a; Oldfield and Parker, 1984b).

Studies in which the spatial dependence of human free-field to eardrum transfer functions have been measured have shown that the pinnae, head, and torso act as a directional acoustic filter, imposing location-dependent spectral patterns on incident sound (e.g., Blauert, 1969/1970; Hebrank and Wright, 1974a; Mehrgardt and Mellert, 1977; Shaw and Teranishi, 1968; Carlile and Pralong, 1994; Duda, 1997; Fig. 1a). These are often referred to as the head-related transfer functions (HRTF). Prominent features of these filter patterns are likely represented at the level of the auditory nerve (AN) despite the spectral smoothing imposed by the encoding in the inner ear (Carlile and Pralong, 1994; Fig. 1b). These features include significant peaks and troughs in the excitation patterns that vary in frequency as a function of spatial location. An important feature illustrated in Fig. 1 is that each location in space is associated with a unique transfer function. Importantly, for locations reflected around the interaural axis (e.g., 30° and 150° azimuth) there are significant variations in the spectral profiles that have the potential to disambiguate ITD and ILD cues.

II. Spectral Cues to Sound Location

As noted in the previous section, there is strong evidence that the information used by listeners to resolve the ambiguity in ITD and ILD cues is spectral in nature. In this section, several additional issues concerning the nature of that information are considered.

A. Are Spectral Cues Monaural or Interaural?

The spatially dependent acoustical filtering properties of the body give rise to an independent set of monaural spectral cues at each of the two ears as well as a set of interaural spectral difference (ISD) cues (Fig. 1). A serious difficulty associated with the successful exploitation of monaural spectral cues is that the spectrum of a sound at its source is rarely known. In the absence of this information it is impossible to determine the spectral filtering imposed on a sound by the body on the basis of the signal at a single ear. ISD cues, in contrast, are robust to spectral variation in at-source spectra and, therefore, have potential to support the localization of unfamiliar sounds (e.g., Searle *et al.,* 1975). The extent to which the auditory system makes use of monaural spectral and ISD cues is an important issue that has received considerable attention in the literature.

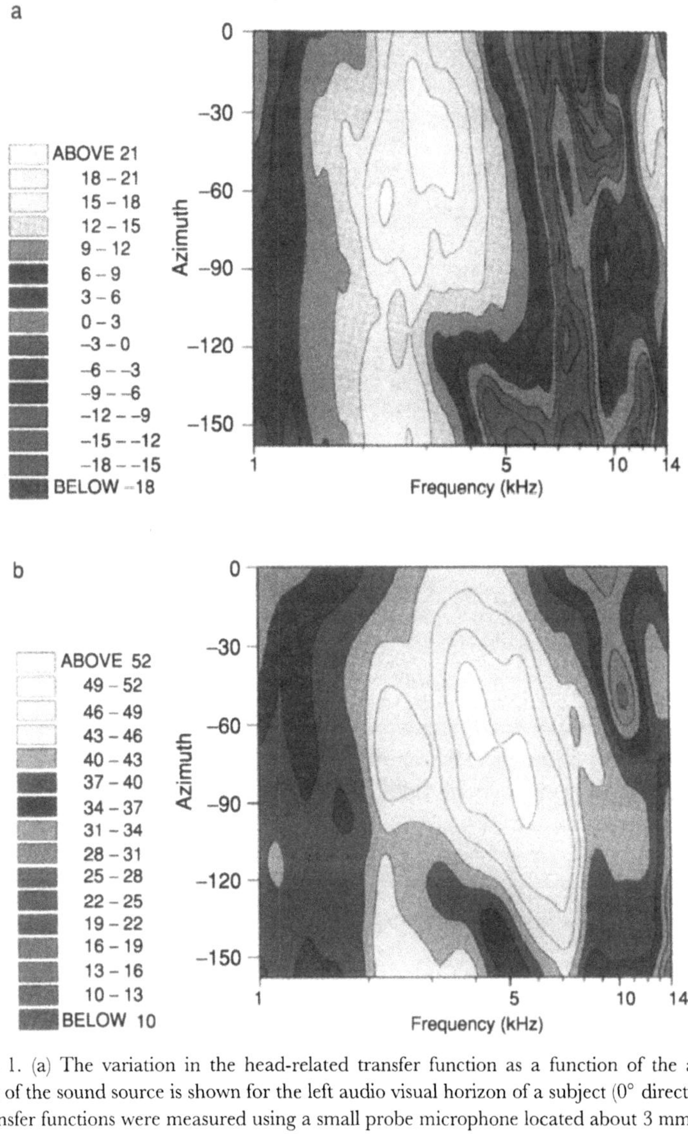

FIG. 1. (a) The variation in the head-related transfer function as a function of the azimuthal location of the sound source is shown for the left audio visual horizon of a subject (0° directly ahead). The transfer functions were measured using a small probe microphone located about 3 mm from the subject's ear drum. Color scale is in dB. (b) To illustrate the features of the transfer functions that are likely to be encoded by the auditory system the data in (a) has been passed through a cochlear filter model (see Carlile and Pralong, 1994, for more details). (See Color Insert.)

1. *Monaural Localization*

Studies that have examined the abilities of listeners to localize sound using the cues available at a single ear provide information about both the sufficiency of monaural and the necessity of binaural cues for localization. In a number of such studies unilaterally deaf individuals were tested. Unilaterally deaf individuals, of course, will have had extensive experience localizing sounds monaurally and may make more effective use of monaural cues than do normal-hearing individuals. In other studies, individuals with normal hearing were tested under conditions in which one of their ears was temporarily blocked. Many of the studies involving normal-hearing individuals were compromised by the use of sound levels too high to ensure that only one ear was stimulated. A few, however, were carefully designed to avoid this shortcoming and it is the results of those that are referred to here.

Both unilaterally deaf and normal-hearing individuals with one ear blocked have generally been found to: (1) show greater disruption of the ability to determine the azimuth than the elevation of sounds (Butler *et al.*, 1990; Gatehouse, 1976; Oldfield and Parker, 1986; Slattery and Middlebrooks, 1994), and (2) localize sounds more accurately when sources are ipsilateral to the intact, or unoccluded, ear (Angell and Fite, 1901; Slattery and Middlebrooks, 1994). It appears, however, that there is considerable variation in sound-localization ability across unilaterally deaf individuals, and several individuals have been described who defy these trends (Slattery and Middlebrooks, 1994). It is clear that both unilaterally deaf and normal-hearing individuals with one ear blocked can determine the azimuths and elevations of sounds with considerably greater than chance accuracy (Angell and Fite, 1901; Butler *et al.*, 1990; Gatehouse, 1976; Oldfield and Parker, 1986; Slattery and Middlebrooks, 1994; van Wanrooij and van Opstal, 2004). Unilaterally deaf individuals have been described who have the ability to determine a sound's elevation as accurately as can binaural listeners (Van Wanrooij and van Opstal, 2004) and there is some evidence (Hebrank and Wright, 1974b) that normal-hearing individuals with one ear blocked can be trained to discern elevation (at least in the median sagittal plane) as accurately as can untrained binaural listeners. Nevertheless, most studies of localization by unilaterally deaf or normal-hearing individuals with one ear blocked have concluded that localization using only one ear is inferior to that using two, with respect to both azimuth and elevation judgments (Butler *et al.*, 1990; Oldfield and Parker, 1986; Slattery and Middlebrooks, 1994; van Wanrooij and van Opstal, 2004).

Monaural localization has also been examined using virtual auditory space techniques (e.g., Wightman and Kistler, 1999), which provide greater monauralization than that readily achieved by ear blocking. Virtual auditory space stimulation exploits the recorded HRTFs of the listener. The appropriate filters for the left and right ear are convolved with the stimulus which is then presented over headphones. In contrast to the normal listening experience using headphones, this filtering generates the perception of a sound outside the head and away from the body in virtual auditory space. Wightman and Kistler (1997) found that normally-binaural listeners were *almost completely unable* to localize virtual sources of sound when sound was presented to only one of their ears. They noted that listeners localizing under free-field conditions have access to spectral cues associated with small head movements and argued that such cues may have supported the accurate elevation judgments observed in previous (i.e., free-field) monaural localization studies. A more recent study by Martin *et al.* (2004), however, indicates that monaural localization under virtual conditions can be substantially more accurate than would be expected on the basis of chance.

2. *Contribution of Near- and Far-Ear Spectral Information*

That monaural localization has been found to be more accurate for sound sources ipsilateral to the intact, or unoccluded, ear (Angell and Fite, 1901; Slattery and Middlebrooks, 1994) raises the issue of the relative contribution of near- and far-ear spectral information in binaural localization. This issue has been the subject of a number of studies, most of which have examined the effects of degrading the spectral information at one or both ears by filling the cavities of the pinna with a mould that leaves the ear canal open. An important feature of this technique is that it allows assessment of the utility of a single set of spectral cues in the presence of veridical interaural cues to sound-source lateral angle. (In the monaural localization studies referred to in the previous section, interaural cues would always have indicated that the sound source was located at an extreme lateral angle.)

Filling the cavities of the pinna nearer to a sound source has been found to dramatically reduce the accuracy with which the elevation and front-back hemifield of the source can be discerned (Gardner and Gardner, 1973; Hofman and van Opstal, 2003; Humanski and Butler, 1988; Morimoto, 2001; Musicant and Butler, 1984a). Filling the cavities of both pinnae has been found to have an even greater effect and almost completely abolishes the ability to discern sound-source elevation and front-back hemifield (Hofman and van Opstal, 2003; Humanski and Butler, 1988; Morimoto, 2001; Musicant and Butler, 1984a). The effect on localization of filling the cavities of the pinna farther from a sound source, in contrast, appears to be slight and has been reported to decrease as a source is moved progressively more laterally (Hofman and van Opstal, 2003; Morimoto, 2001). For sources more than about 60° from the median sagittal plane, the spectral information at the far ear has been found to make no measurable contribution to localization (Morimoto, 2001).

On observing that the contribution of far-ear spectral information to localization in elevation decreases with increasing source laterality, Hofman and van Opstal (2003) proposed that judgments of source elevation are based on weighted averages of near- and far-ear spectral cues, with weights being dependent on perceived source laterality. Data presented by Hofman and van Opstal suggest that weights change with perceived source laterality in a listener-dependent way. Macpherson *et al.* (2004) have demonstrated that listeners place greater weight on the spectral cue at the ear they perceive to be nearer to the source when the cues at the two ears indicate source elevations that differ by as much as 40°.

3. *Evidence for and against the Exploitation of ISD Cues*

That localization performance is generally superior when veridical spectral cues are available at both ears suggests that either two sets of monaural spectral cues are more useful than one or use is made of ISD cues.

The various experimental results that have been cited as evidence of the use of ISD cues, when considered carefully, do not provide strong support for that contention. Searle *et al.* (1975) presented in-ear recordings of spectrally scrambled and unscrambled broadband stimuli that had been generated by sources on the median sagittal plane to listeners via headphones under each of two conditions. In a *dichotic* condition, normal binaural listening conditions were simulated, whereas in a *diotic* condition, the signal recorded at the left or the right ear was presented to both ears. Searle *et al.* (1975) found that performance was always better for the dichotic condition and argued that this means that ISD cues are important. However, there are at least two other plausible explanations of this finding. First, performance in the diotic condition may have been degraded because the monaural spectral cue presented to one of the ears was inappropriate. Second, and as noted by Hebrank (1976), performance in the dichotic condition may have been assisted by the presence of small ITD and/or ILD cues. Indeed, even when midline alignment is very carefully controlled median sagittal plane ILDs can be as large as 10 dB presumably as a result of the morphological asymmetry of the ears (Carlile and Pralong, 1994).

A number of studies have demonstrated that the amount of spectral scrambling (i.e., the extent of variation in spectrum level across frequency) required to seriously disrupt binaural localization is quite large (Macpherson, 1996; Macpherson and Middlebrooks, 2003; McAnally *et al.*, 2000; Wightman and Kistler, 1997). This robustness of binaural localization to spectral scrambling is suggestive of the exploitation of ISD cues. However, a surprising robustness to spectral scrambling has also been demonstrated for monaural localization (Hebrank and Wright, 1974b; Paterson, 2000), where ISD cues are not available. That spectral scrambling has been found to disrupt binaural and monaural localization to similar extents (Hebrank and Wright, 1974b; Searle *et al.*, 1975) has been cited as evidence that ISD cues are *not* exploited (Hebrank and Wright, 1974b).

A recent study by Jin *et al.* (2004) indicates that ISD cues are insufficient to maintain accurate elevation and front-back hemifield judgments in the presence of a non-veridical spectral cue at one ear. Jin *et al.* examined the abilities of listeners to localize virtual sources of broadband sound under each of three conditions. Those conditions were: (1) veridical ITDs, ILDs, and left- and right-ear spectral cues (i.e., a baseline condition in which free-field binaural listening was simulated), (2) veridical ITDs, ILDs, and ISDs (which involved presenting a flat spectrum to the left ear and non-veridical spectral cues to the right ear), and (3) veridical ITDs, ILDs, and right-ear spectral cues (which involved presenting a flat spectrum to the left ear and non-veridical ISDs). Listeners were found to localize with reasonable accuracy under condition 1 but with little accuracy under condition 2, where ISD cues were preserved. It is important to note, however, that poor localization performance was also observed under condition 3, where right-ear spectral cues were preserved. However, localization performance under condition 3 was substantially better for sources located on the right compared with the left hemifield. Other studies in which it has been shown that the spectral cues at the far ear can, under certain circumstances, be degraded without disrupting localization in elevation (Humanski and Butler, 1988; Morimoto, 2001) indicate that ISD cues are not always necessary for accurate localization.

Jin *et al.* (2004) presented an analysis, based on information theory, of the information inherent in ISD and near- and far-ear spectral cues as a function of sound-source lateral angle (i.e., an analysis of the information available to support discrimination of locations on the cone of confusion). The analysis indicates that monaural spectral cues provide more information than ISD cues within 20° of the median sagittal plane but the information in ISD cues tends to increase with laterality and can exceed that in monaural spectral cues at lateral angles more extreme than about ±30°. It is interesting that ISD cues are close to maximally informative at lateral angles more extreme than ±60°, where far-ear spectral cues do not appear to be important for accurate localization (Morimoto, 2001).

B. Are Spectral Cues to Sound Location Lateral-Angle Specific?

The possibility that spectral cues to sound location are specific to the lateral angle of the source has been examined in two recent studies.

Martin *et al.* (2004) asked whether cues to elevation and front-back hemifield are lateral-angle specific by comparing the accuracy with which five listeners localized virtual sources of broadband sound in the absence and presence of interaural cues to lateral angle. They found that elevations and front-back hemifields were discerned inaccurately under monaural conditions, where listeners also had considerable difficulty discerning lateral angles. The accuracy with which lateral angles, elevations, and front-back hemifields could be determined improved dramatically when low-frequency (i.e., 2.5 kHz low pass filtered) interaural cues to lateral angle were made available. Low-frequency interaural cues in isolation were found to be insufficient to support accurate elevation or front-back hemifield judgments. Martin *et al.* concluded that information regarding sound-source lateral angle is required for monaural spectral cues to elevation and front-back hemifield to be correctly interpreted.

Morimoto *et al.* (2003) asked whether spectral cues to polar angle are lateral-angle specific. They examined the accuracy with which three listeners localized virtual sources of broadband sound generated using filters constructed from measurements of spectral cues on the median sagittal plane and frequency-independent ITDs and ILDs on the interaural horizontal plane. (This is a more direct approach than that taken by Martin *et al.* [2004].) For each listener, Morimoto *et al.* found that the patterns of polar errors for sources at 30° and 60° lateral angles were similar to that for sources on the median sagittal plane. They also noted that average polar errors for sources at 0°, 30°, and 60° lateral angles were similar to those described by Morimoto and Ando (1980) for localization of free-field sources. Morimoto *et al.* concluded that spectral cues to sound location are common to all lateral angles.

The results of the Martin *et al.* (2004) and Morimoto *et al.* (2003) studies are difficult to reconcile. However, in one important respect they are not as incompatible as might at first appear to be the case. The Morimoto *et al.* data suggest that spectral cues provide information about polar angle rather than elevation and that those cues are common to all lateral angles. On the other hand, Martin *et al.* found that listeners could not determine the elevation of a sound source in the absence of information regarding its lateral angle. Importantly, the mapping of polar angle to elevation varies with lateral angle, and the two studies used different coordinate systems to indicate the deviation of the location from the audio-visual horizon (polar angle and elevation respectively). A re-analysis of the Martin *et al.* data focusing on polar rather than elevation errors (not shown here) are consistent with this interpretation and provide further evidence of spectral-cue commonality across lateral angle. With respect to spectral cues to the front-back hemifield, however, the Martin *et al.* and Morimoto *et al.* studies remain at odds.

C. How Much Spectral Detail Is Required for Accurate Sound Localization?

The question of the amount of spectral detail necessary to support accurate localization of sound has arisen in the context of the development of efficient virtual-audio techniques, but is clearly of broader relevance. As previously discussed, the upper limit of this detail is set by the spectral resolution of the cochlea. Making use of the auditory filter model developed by Glasberg and Moore (1990), Carlile and Pralong (1994) have shown that much of the measurable detail in the spectral patterns imposed on sound by the body does not survive cochlear filtering (see Fig. 1b). The extent to which the surviving detail is utilized by listeners when localizing sound has been the subject of several studies.

Macpherson and Middlebrooks (2003) explored the effect of varying the density of logarithmic ripples imposed on the spectra of broadband sounds on the accuracy with which the sounds could be localized. They found that polar errors were substantially larger for ripple- compared with flat-spectrum sound when ripple density was in the range from 0.5 to 2 ripples/octave, with a ripple density of 1 ripple/octave being most disruptive. The effect on polar errors of imposing spectral ripples decreased at both lower and higher ripple densities, despite the fact that the sensitivity of listeners to spectral ripples appears to extend beyond the 0.5 to 2 ripples/octave range (e.g., Supin *et al.*, 1999). Macpherson and Middlebrooks concluded that spectral details finer than 2 ripples/octave or coarser than 0.5 ripples/octave are not particularly important for localization.

Other studies have examined the effect on various auditory tasks of removing spectral detail from the filters used to generate virtual sound sources. Kulkarni and Colburn (1998) found that listeners could not tell which of two sources was free-field and which was virtual until the spectral resolution of the filters was reduced to 1562 Hz (i.e., 16 Fourier coefficients at a sample rate of 50 kHz). Zahorik *et al.* (1995), however, found that free-field and virtual sources could be discriminated when filter resolution was reduced to 195 Hz (i.e., 5.12 ms duration at a sample rate of 50 kHz) and Senova *et al.* (2002) found that the accuracy of localization was significantly lower for virtual compared with free-field sources (for one of the three participants in their study) when filter resolution was reduced to 195 Hz (i.e., 5.12 ms duration at a sample rate of 50 kHz).

It is difficult to compare the results of these three studies in which the spectral resolution of a given filter was constant on a linear frequency scale, with the results of the Macpherson and Middlebrooks (2003) study, in which the ripple density for a given sound was constant on a logarithmic frequency scale. Results more readily compared with those described by Macpherson and Middlebrooks have been reported by Senova (2002), who examined the accuracy with which virtual sound sources generated using frequency-warped filters of various spectral resolutions could be localized. Filters were frequency warped, using techniques described by Huopaniemi *et al.* (1999) and Harma *et al.* (2000), in such a way that information was distributed across frequency in a manner resembling the distribution of critical bands. Senova (2002) found that the accuracy with which virtual and free-field sources could be localized remained more or less the same provided filter resolution was at least as fine as about 4.6 points/octave (i.e., 32 coefficients across approximately 7 octaves from 0.2 to 25 kHz). This is generally consistent with the Macpherson and Middlebrooks conclusion that spectral details finer than 2 ripples/octave (i.e., 4 points/octave) are not particularly important for localization.

Asano *et al.* (1990) modeled median-sagittal-plane HRTFs with autoregressive moving-average (ARMA) filters of various orders. They found that spectral regions above 3 kHz could be replaced with 7.3 point/octave (i.e., 10/10 order, or about 20 points, across approximately 2.75 octaves from 3 to 20 kHz) ARMA approximations without disrupting localization to any great extent. The ARMA modeling technique distributes points across frequency in a manner that minimizes error in approximations. As Asano *et al.* applied the technique to approximate HRTFs that appear to have been measured on a linear frequency scale, the approximations are likely to have contained excessive detail at high frequencies. (On a linear frequency scale most of the measurable detail in HRTFs is at high frequencies but, as noted previously, much of this detail does not survive cochlear filtering.) This could explain the slight discrepancy between the Asano *et al.* and Senova (2002) studies with respect to the average filter resolution required to maintain accurate localization (i.e., about 7.3 versus 4.6 points/octave). In addition, the slopes of the high and low pass filters used in the Asano *et al.* study were fairly shallow indicating the need for caution in the interpretation of these data (see Jin, 2001, Chapter 3).

D. Are Locations Cued by Specific Spectral Features?

A number of models of human sound localization have incorporated processes that recognize broadband spectral profiles of at-ear signals (e.g., Hofman and van Opstal, 1998; Jin *et al.*, 2000; Jin, 2001; Langendijk *et al.*, 2001; Middlebrooks, 1992, 1999; van Schaik *et al.*, 1999; Watkins, 1978; Zakarauskas and Cynader, 1993). Despite the success of these models at predicting localization behavior, it remains unclear whether source locations are cued by entire spectral patterns or discrete spectral features. Candidate spectral features for localization cues include overt peaks, notches, steep gradients, and covert peaks. These features are defined and discussed in more detail in the following paragraphs.

One of the earliest studies to explore the possibility that specific *overt* spectral features, which are defined as features evident in individual HRTFs, act as cues to particular source locations was that described by Hebrank and Wright (1974a). On the basis of observations of listeners localizing sources of various highpass-, lowpass-, bandpass-, and bandstop-filtered noises in the median sagittal plane and inspection of HRTFs for several artificial ears, Hebrank and Wright proposed that: (1) the cue to front in the median sagittal plane is a 1-octave notch with a low-frequency cutoff between 4 and 10 kHz and increased energy above 13 kHz, (2) the cue to above is a 1/4-octave peak between 7 and 9 kHz and a high-frequency cutoff at 10 kHz, (3) the cue to behind is a small peak between 10 and 12 kHz, and (4) the elevation of a source in front is signaled by the location in frequency of the low-frequency cutoff of the 1-octave notch.

The presence in HRTFs of a prominent spectral notch that could provide a cue to sound-source elevation has been noted by several authors (e.g., Bloom, 1977a; Butler and Belendiuk, 1977; Hebrank and Wright, 1974a). The center frequency of this notch increases systematically from about 5 to 14 kHz with increases in source elevation. An illusion of varying source elevation can be created by varying the center frequency of a notch inserted in a broadband noise and presented to a listener from a free-field source at a constant elevation (e.g., Bloom, 1977b).

The proposition that localization in elevation is based entirely on the center frequency of the primary spectral notch, however, is not well supported. Macpherson (1994) argued that if the proposition is true, the relationship between notch center frequency and source elevation in an individual listener should be predictive of the listener's localization behavior. Specifically, he suggested that: (1) the rate at which notch center frequency changes with source elevation should be predictive of the scatter of elevation judgments for individual source locations, and (2) where the relationship between source elevation and notch center frequency differs substantially across front and back hemifields, front-back confusions should be associated with predictable relative biases in perceived elevation. Macpherson found only a weak correlation between the scatter of elevation judgments and the rate of change of notch center frequency with source elevation, and no evidence of relative biases in perceived elevation for front-back confusions in two listeners for whom the relationship between source elevation and notch center frequency was markedly different across front and back hemifields. It is also clear that sounds must include frequencies outside the range in which the primary spectral notch is embedded for elevation judgments to be optimally accurate (King and Oldfield, 1997), and that sounds that include only frequencies outside that range can be localized in elevation with reasonable accuracy (Algazi *et al.,* 2001; King and Oldfield, 1997). More generally, Macpherson (1997) found that the performance of listeners localizing sounds having a variety of irregular spectra was predicted more accurately by a model that matched the derivatives of entire at-ear spectra with stored templates than by one that matched at-ear signals and stored templates on the basis of prominent local features such as peaks and notches.

Covert spectral features are defined as features that become evident when HRTFs are compared across locations. The covert peak indicates the spatial locations where the at-ear level for a given frequency is greatest (Humanski and Butler, 1988) or alternatively, the location at which a particular frequency is maximally transmitted by the ear to the ear drum. A number of studies have shown that listeners operating under monaural conditions tend to localize band-limited sound according to the location of the covert peak (Flannery and Butler, 1981; Humanski and Butler, 1988; Musicant and Butler, 1984b; Rogers and Butler, 1992). That a covert-peak analysis underlies the ability of listeners operating under binaural conditions to resolve locations on the cone of confusion has been suggested by Musicant and Butler (1984b) and Humanski and Butler (1988).

Models that incorporate covert-peak analyses have been found to be more predictive of the performance of binaural listeners localizing band-limited sounds than are models that incorporate processes that match entire spectra (Jin, 2001; Middlebrooks, 1992). Nevertheless, the value of covert-peak models has been questioned on the grounds that it is difficult to see how a covert-peak analysis could be applied to a broadband sound (Middlebrooks, 1992). Although the covert-peak model has been widely attributed to Butler and his colleagues, a model that is essentially the same was described by Blauert (1969/1970) several years prior to Butler's first description of the covert-peak model (as acknowledged by Musicant and Butler in their 1984b paper). Blauert observed that listeners tend to perceive narrowband sounds presented from sources on the median sagittal plane as emanating from locations that are influenced more by the center frequency of the sound than by the true location of its source. He referred to frequency bands that are reliably localized to particular locations as *directional bands* and noted that the location associated with a given directional band tended to be that, from a limited set, for which the at-ear levels of the frequencies within the band were maximal. Of particular interest, Blauert suggested that the median-sagittal-plane locations of broadband sounds are determined by calculating the energy levels in the directional bands associated with each plausible location (a given location may be associated with more than one directional band) and assigning the sound to the location for which the total energy level is greatest.

III. Processing of Spectral Cues to Sound Location

A. Human Psychophysical Studies

1. *Temporal Aspects of Spectral Processing for Sound Localization*

The studies considered so far, and the majority of all previous studies of human sound localization, were concerned with sounds that are stationary with respect to both their spectra and spatial locations. Almost all naturally occurring sounds, however, have time-varying spectra and many sources of sound can move relative to a listener. This raises the issue of the duration of the time window associated with the localization system's spectral analysis. Spectrally based estimates of location for spectrally non-stationary sounds are likely to be more accurate for longer time windows because at-source spectra tend to become flatter as sounds are sampled over longer periods. On the other hand, estimates of location for moving sound sources are likely to be more accurate for shorter time windows because sound-source movement results in at-ear spectral smearing.

An analysis by Jin (2001) suggests that the lower limit for the duration of the localization system's time window for spectral analysis is of the order of 5 ms. Jin noted that the response and decay times of cochlear filters are related to their bandwidths and that a filter's output provides a reliable measure of spectrum level only if a signal is integrated over a period at least as long as the filter's decay time (which is about three times as long as its response time). A 5-ms integration time would result in a spectral analysis that is reliable for frequencies down to about 2.6 kHz. For each halving of this integration time, the frequency below which the spectral analysis would be unreliable would double. Clearly, integration times much shorter than 5 ms would result in spectral analyses that are unreliable for frequency ranges within which important spectral cues to sound-source location are thought to reside.

In one of the few studies concerned with the duration of the localization system's time window for spectral analysis, Hofman and van Opstal (1998) examined the accuracy with which three types of stimuli can be localized. These stimulus types were: (1) frequency-modulated (16-0 kHz downward sweeping) tones of 500-ms duration and sweep periods ranging from 1.28 to 40.96 ms, (2) single bursts of broadband noise of durations ranging from 3 to 500 ms, and (3) 500-ms (approximately) trains of 3-ms broadband-noise bursts separated by intervals ranging from 0 to 77 ms. All stimuli were presented at 70 dB SPL (A-weighted) from locations restricted to the oculomotor range (i.e., from −26 to +26° in azimuth and elevation). Localization performance was analyzed primarily in terms of a measure, which Hofman and van Opstal referred to as *response gain*, that describes the slope of the line of best fit relating true and perceived sound-source locations (with respect to either their azimuth or elevation components). Hofman and van Opstal found that tone sweeps with 1.28- or 2.56-ms periods were localized in elevation almost as accurately as were 500-ms noise bursts and that tone sweeps with longer periods (from 5.12 to 20.48 ms) were localized in elevation increasingly less accurately. This suggests that the duration of the localization system's time window for spectral analysis is <5.12 ms.

Hofman and van Opstal's (1998) observations in relation to the effect of tone-sweep period on sound-localization accuracy have been confirmed and extended by Martin and McAnally (2004), who examined the accuracy with which sinusoidally frequency-modulated (0.5 to 19.5 kHz) tones of 400 ms duration and sweep half-periods ranging from 0.25 to 100 ms can be localized. Sounds were presented in virtual auditory space from sources extending from 0° to 359° in azimuth and −40° to +70° in elevation. The accuracy with which the elevation and front-back hemifield of sound sources could be discerned (as indicated by absolute elevation errors and proportions of front-back confusions, respectively) was found to be greatest for tone sweeps with 2.5-ms half periods, for which localization accuracy was equivalent to that for broadband-noise bursts. This suggests that the duration of the localization system's time window for spectral analysis is between 1 and 5 ms.

Hofman and van Opstal (1998) also found that the accuracy with which the elevation of a single noise burst could be discerned increased steadily as the duration of the burst was increased from 3 to 80 ms and that 80-ms noise bursts were localized almost as accurately as were 500-ms noise bursts. In addition, they observed that the accuracy with which the elevation of trains of 3-ms noise bursts could be discerned steadily decreased as the inter-burst interval was increased from 0 to 77 ms. On the basis of their findings Hofman and van Opstal proposed that the localization system estimates sound-source elevation from short-term (i.e., <5.12 ms) spectra then averages short-term estimates over a period of about 80 ms to arrive at a final estimate. They suggested that this process incorporates a power-dependent gating mechanism (because silent periods did not dramatically affect the accuracy of final elevation estimates) and a leaky procedure for averaging short-term estimates (because silent periods did have some effect on the accuracy of final elevation estimates).

An alternative explanation of some of the Hofman and van Opstal (1998) results has been proposed by Macpherson and Middlebrooks (2000). They suggested that the relatively inaccurate elevation judgments for short-duration noise bursts reported by Hofman and van Opstal may be related to the *negative level effect* described by Hartmann and Rakerd (1993). The negative level effect refers to the decrease in accuracy of sagittal-plane localization judgments that Hartmann and Rakerd observed to be associated with increases in level for short-duration sounds such as clicks. (No equivalent effect was observed for long-duration sounds.) The effect has been attributed to a failure of the peripheral auditory system to resolve the spectral details of high-level brief sounds, possibly as a consequence of the saturation of AN fibers. Longer duration sounds may initiate adaptive processes that enhance the resolution of the peripheral auditory system's spectral analysis (Hartmann and Rakerd, 1993; Macpherson and Middlebrooks, 2000).

In concordance with the results of Hofman and van Opstal (1998), Macpherson and Middlebrooks (2000) found that the accuracy with which the polar angle of 3- and 100-ms broadband noise bursts could be discerned (as indicated by polar-angle response gains) differed significantly when bursts were presented at a relatively high level (i.e., 55 dB SL). They also found, however, that the accuracy with which the polar angle of 3- and 100-ms noise bursts could be discerned did *not* differ significantly when bursts were presented at a low level (i.e., 30 dB SL). In addition, the introduction of low-level (i.e., 40 dB below masking level) spatially diffuse flanking noise was observed to result in an increase in the accuracy with which the polar angle of high-level 3-ms noise bursts could be discerned for five of the six participants in the Macpherson and Middlebrooks study. These latter findings are consistent with the explanation of the Hofman and van Opstal results in terms of the negative level effect, which we will call the *adaptation account*, but are inconsistent with the explanation proposed by Hofman and van Opstal, which we will call the *estimate-integration account.*

More recently, Vliegen and van Opstal (2004) extended the range of noise-burst levels across which the accuracy of judgments of sound-source elevation have been examined. In contrast to Macpherson and Middlebrooks (2000), they found that elevation judgments became less accurate as burst duration was decreased, irrespective of level (at least across the range from 26 to 73 dB SPL). The Vliegen and van Opstal results are convincing and strongly favor the estimate-integration over the adaptation account. The cause of the discrepancy between the Vliegen and van Opstal and Macpherson and Middlebrooks findings is unclear.

Interestingly, Vliegen and van Opstal (2004) observed a negative level effect for long as well as short duration noise bursts at levels above about 60 dB SPL and a *positive level effect* for both long and short duration noise bursts at levels below about 45 dB SPL. They noted that both effects could result from processes occurring in the peripheral auditory system (e.g., saturation of AN fibers at high levels leading to a negative level effect or nonlinear cochlear amplification at low levels leading to a positive level effect), but also offered an explanation of the positive level effect that involves more central processes. To be specific, they suggested that elevation judgments are a weighted sum of elevation estimates based on acoustic and non-acoustic (e.g., knowledge of likely source locations) information and that greater weight is placed on acoustically based estimates as signal-to-noise ratio increases. Accordingly, Vliegen and van Opstal (2004) proposed an extension to the Hofman and van Opstal (1998) localization model allowing for level-dependent variation in the accuracy of short-term spectral analyses and weighting of final elevation estimates (i.e., estimates arrived at by averaging short-term estimates over periods of about 80 ms) against estimates based on nonacoustic factors.

It should be noted that increases in signal-to-noise ratio could lead to increases in the accuracy of elevation judgments through a process more direct than that just described. Low signal-to-noise ratios would be expected to be associated with noisy short-term spectral analyses and elevation estimates. Vliegen and van Opstal (2004) considered but rejected this possibility on the basis of a lack of similarity between the ways in which the accuracy and variability of elevation judgments varied with noise-burst duration and level. Inspection of the relevant data presented by Vliegen and van Opstal (i.e., their Figs. 5 and 7), however, suggests that the accuracy and variability of elevation judgments were moderately correlated.

2. *Recovering Source Spectrum from At-Ear Signals*

When listening to a broadband sound source that moves in space around us we do not usually perceive dramatic changes in the sound's timbre. This indicates that changes in the spectra of at-ear signals resulting from changes in source location are interpreted correctly as artifacts of sound-source movement. In principle, listeners could deconvolve the location-dependent spectral changes imposed on a sound (i.e., the HRTFs) from the at-ear signals to recover the spectrum of the sound at its source. At least two studies have explicitly examined the ability of listeners to do this.

Rakerd *et al.* (1999) trained listeners to recognize particular broadband sounds that had distinct spectral structures. They then measured the accuracy with which listeners identified those sounds when they were presented to them from a free-field source at either a constant location or varying locations on the median sagittal plane. They found that sounds were generally identified correctly when the location of the source was fixed. When the location of the source was varied, sounds were identified correctly only if their distinguishing spectral features were at low frequencies, where free-field to eardrum transfer functions show little variation across locations within the median sagittal plane. For sounds with their distinguishing spectral features at high frequencies, identification performance under roved-location conditions was little better than that expected on the basis of chance. Rakerd *et al.* also demonstrated a clear dissociation between the accuracy with which sounds could be identified and localized, ruling out the possibility that the reason some sounds could not be identified under roved-location conditions was that they could not be localized.

The Rakerd *et al.* (1999) results suggest that the spectrum of a sound at its source cannot be recovered from at-ear signals. A similar conclusion can be drawn from an earlier study by Macpherson (1995), in which discrimination of spectrally complex sounds was found to be poorer under roved-, in comparison with fixed-, location conditions for sounds presented from virtual sources.

B. Non-Human Studies of the Processing of Spectral Cues

In this section, we will consider several aspects of spatial cue processing examined neurophysiologically that are particularly relevant to the processing of spectral information. This review is necessarily limited and will focus on results obtained using the mammalian auditory system. However, it is important to note that there is a significant body of work examining sound localization in a diverse array of other classes such as frogs and toads, bats, birds, and many insects (in particular, the grasshopper) that are not considered here for reasons of space. Over the past decade there has also been increasing interest in the vital imaging of sound processing in the human brain using PET and fMRI techniques (for example, see Griffiths *et al.*, 1994, 1996; Griffiths and Warren, 2002; Pavani *et al.*, 2002). Much of the research has focused on stationary and moving stimuli and the processing of spectral cues is addressed by implication and will not be considered further here (but see Pavani *et al.*, 2002).

1. *Ethological Factors and Species Specialization*

In general, when localization acuity is compared across a range of species, it appears that the carnivorous predators have a greater spatial acuity than the prey species (Heffner and Heffner, 1987). The presence of a well-defined superior olivary complex also appears to co-vary across similar lines. Thus, the ability to accurately localize prey so as to better manage approach and strike is likely to provide a stronger evolutionary driver on localization ability than the ability to simply localize a predator to direct flight. This provides an important context for understanding the results of studies examining the bio-acoustic and neuronal basis of auditory localization using different animal models. Obviously, this also cautions against conclusions regarding the generality of findings from any particular animal model. However, such a context can also provide a powerful heuristic to explain the structural and functional variations that are evident in even a cursory comparative review of the morphology of the outer ears of species and of auditory function in general.

Regardless of the species, the mechanisms that give rise to the capacity to localize a sound source depend on the nature of the acoustic cues available at the ears. The auditory periphery is generally characterized by an asymmetrical or complex three-dimensional structure which results in a spatially non-uniform receptivity. This results in modifications of the sound over some biologically relevant frequency range before it reaches the eardrum. The complexity of the interaction between evolutionary pressure in a particular niche and evolutionary heritage goes some way to explaining the range of variation that we see in the shapes and dispositions of the outer ears of different species. The, at times, beautifully baroque structures that comprise the auditory periphery of different species serve the same fundamental functional requirements. Namely, to provide the location dependent spectral cues that, when integrated with other acoustic cues, result in an appropriate acuity of spatial perception. Indeed, this is some distance from Darwin's original assertion that the pinna was vestigial!

The sorts of acoustic mechanisms that either dominate and/or interact to produce the spectral cues are highly dependent on the precise morphology of the outer ear. How we understand these interactions also depends on the sorts of approximations we apply to the modeling of these complex acoustic structures (Fletcher and Thwaites, 1979). For instance, the cat outer ear can be modeled in terms of a number of features: (1) as a truncated conical horn which provides gain over frequency ranges where the wavelengths are related to the depth and flare of the horn (e.g., Fletcher and Thwaites, 1979; see also for wallaby, Coles and Guppy, 1986), and (2) as a roughly circular aperture where the acoustic behavior, at least at the mid- to high frequency, can be well described by the diffraction at the circular aperture of the pinna rim (Calford and Pettigrew, 1984; see also for the guinea pig at high frequencies, Carlile and Pettigrew, 1987a). Quite different mechanisms relying on the depth and transverse resonances of the concavities of the concha and pinna are thought to underlie the dominant acoustic mechanisms in the human ear (e.g., Shaw, 1982; reviewed in Carlile, 1996). Similar mechanisms may also dominate the acoustic responses of other species which have flattened, fixed pinna such as the ferret (see Carlile, 1990a). Some of these mechanisms are considered in more detail below.

An important issue in respect of the cues to location is that, in addition to the ambiguity in the binaural cues (the so-called cones of confusion) the spectral features in the HRTF which vary with location can also be spatially ambiguous. Therefore, to uniquely specify a location in space it is necessary to integrate this information across frequency and across ears. Given the non-stationary and band-limited nature of the sounds in the real biosphere it is likely that auditory systems will use whatever combination of cue processing that is likely to provide the most stable and reliable indication of location.

2. *Acoustics of the Cat Pinna*

Over the last decade or so, significant work has been focused on examining the role of spectral cues in the sound localization behavior of cats and the manner in which this cue is processed at various levels in the auditory system. Much of this work has been motivated by the work of Young and colleagues on the processing of spectral information in the dorsal cochlear nucleus (DCN) (Young *et al.*, 1992). These animal physiological studies emphasize the encoding of spectral notches as opposed to other spectral characteristics identified in the human psychophysical studies (see the following text and in the chapter by Young and Davis). This reflects the quite different acoustics of the outer ears of the cat compared to the human. Notches do not appear to play an especially dominant role in human localization because the spatially dependent variation in spectral notches in the human HRTF is far less prominent (see Carlile and Pralong, 1994) whereas other spectral features are more strongly linked to spatial location.

The auditory periphery of the cat is dominated by a relatively large and mobile pinna. First, the HRTFs demonstrate a strong on-axis gain with the frequency and magnitude of the gain related to the dimensions of the horn. Second, diffraction at the (roughly) circular aperture results in spectral nulls at off-axis locations whose frequencies vary with location. This acoustic behavior has been confirmed in a number of studies where the directionality of the pinna has been examined (Calford and Pettigrew, 1984; Musicant *et al.*, 1990; Phillips *et al.*, 1982; Rice *et al.*, 1992; Xu and Middlebrooks, 2000; and see also Wiener, 1965). A critical feature is that the center frequency of a mid-frequency notch varies systematically with spatial location from 5 kHz to around 18 kHz (Musicant *et al.*, 1990; Rice *et al.*, 1992). The width and depth of this spectral notch is such that despite the spectral smoothing of the encoding in the inner ear, it is very likely preserved in the AN fiber activity (Poon and Brugge, 1993). Moreover, using signal detection theory, May and Huang (1997) demonstrated that the pairs of HRTFs that corresponded to the behavioral spatial acuity of the cat (the just noticeable differences in locations) result in statistically significant variations in the spectral patterns encoded by the AN activity.

Plots of notch frequency as a function of spatial location demonstrate iso-frequency notch contours that lie diagonally across the frontal hemisphere. Therefore, any single notch frequency is ambiguous with respect to space. Rice *et al.* (1992) proposed a model whereby spatial location could be uniquely specified by comparing the notch frequencies in the left and right ears. It is unclear from the current data what spectral features are available to the cat to resolve the front-back ambiguity in the binaural cues. However, on inspection of the published HRTFs it would appear likely that the spectral notches for sounds behind the pinna would be very different from those on the same ILD or ITD contour (cone of confusion).

A further complication with the cat auditory periphery is that the pinna is highly mobile and the cat is able to direct each ear independently to spatial locations of interest. Pinna movements produce changes in the orientation of the principal gain features (Musicant *et al.*, 1990; Young *et al.*, 1996) of the ear and some changes in the frequency dispersion of the acoustic axis (Calford and Pettigrew, 1984). Also, the patterns of the spectral notches are varied by pulling the pinna forward or backward, resulting in a shift in the notch frequency contours towards the anterior mid-line and interaural axis, respectively (Young *et al.*, 1996). Young *et al.* (1996) conclude that spectral variations resulting from changes in pinna position would result in ambiguity in the spectral and ILD location information by as much as 60° in azimuth and 30° in elevation. These changes substantially increase the range of spectral cues associated with each specific location in space. Presumably, information regarding pinna position would provide a means of resolving this ambiguity.

Young *et al.* (1996) have suggested that there are a number of advantages of pinna mobility that may relate to different listening tasks. First, changing the direction of a conical horn will increase the received amplitude of the sources located on the acoustic axis of the horn, which will aid in detection and scanning localization of very quiet sounds. This can also reduce the masking produced by off-axis sources. Second, localization can be optimized by placing the pinna in a position that will optimize the spectral features available given the variable spectra of the source. Third, the pinna could also be placed to minimize the spectral distortions produced by the pinna, presumably to aid in the recognition of a sound. Fourth, varying the pinna position will provide a means of disambiguating the source spectrum from the spectrum received at the eardrum by allowing two or more samples of the sound where the spectral distortion at each sample is known (the HRTF for each different pinna position).

3. *Localization Behavior of the Cat*

The cat, as a predator, is an efficient hunter and enjoys quite accurate localization abilities. Behavioral studies of auditory localization of cats in anechoic environments indicate absolute localization errors of around 5° in azimuth and 15° in elevation for short noise bursts presented in the frontal hemisphere (May and Huang, 1996). Furthermore, manipulation of the stimulus spectrum to either remove or degrade the spectral notch cues in the 5 kHz to 18 kHz range resulted in significant mis-localization of the stimulus. In contrast, band-passing the stimulus to preserve the spectral notch information had little effect on localization accuracy (Huang and May, 1996). However, elevation errors were found to increase when the stimuli were high passed at 18 kHz. In contrast, in a minimal audible angle task, cats were able to detect a change in the relative position of short (40 ms) concurrent sounds regardless of the presence or absence of mid-frequency notch information.

Localization behavior has also been studied in cats that have had lesions to the brainstem that interrupt information flow from the DCN (May, 2000). The localization accuracy for short duration noise bursts was studied before and after unilateral and bilateral lesion of the dorsal and intermediate acoustic stria (DAS). Following unilateral lesion, localization accuracy was more variable with large elevations errors, particularly for locations in the ipsilateral hemifield. Bilateral lesions produced an increase in errors for all locations in the frontal hemifield. Interestingly, the cats showed no deficit in a minimal audible angle discrimination test for either the horizontal or median plane. Clearly, binaural cues are available for the horizontal MAA but the near normal MAA for the median plane is harder to explain. The spectral information provided by the HRTF will still be encoded in other parallel ascending auditory pathways and the failure to show an MAA discrimination deficit in cats with bilateral DAS lesions suggests that the difference detection processes normally involves other structures, or can be compensated for by structures other than the DCN.

4. *Spectral Processing in the DCN of the Cat*

There has been considerable work examining the processing of spectral information in the DCN and this has been recently reviewed in detail (see Young and Davis, 2002; Oertel and Young, 2004; also see the chapter by Young and Davis). Therefore, we will only briefly review spectral notch processing in DCN and the evidence for the modulation of the output cells by somatosensory input. The neural structure of the DCN is not unlike the neighboring cerebellum and the inputs to the DCN are composed of the AN which contacts the smooth dendrites of the fusiform and giant cells in the deep layers of the DCN. These are also the principal output cells of the DCN. Orthogonally arranged parallel fibers arising from granule cells make dendritic contact with the fusiform and giant cells as well as the intermingled cartwheel interneurons. The granule cells receive input from a wide range of structures, including the dorsal column nucleus, vestibular afferents, octopus cells in the VCN, and unmyelinated auditory nerve fibers, as well as descending inputs from the IC and auditory cortex. The synaptic coupling between the parallel fibers and their targets is modulated by activity with both long-term potentiation (LTP) and long-term depression (LTD) evident in different target populations. By contrast, the ascending inputs from the AN to the principal cells are not synaptically plastic (Fujino and Oertel, 2003; reviewed in Oertel and Young, 2004).

Not surprisingly, the response of the principal output cells reflect a complex interplay of excitatory and inhibitory inputs and exhibit a so-called type IV response pattern (Fig. 2). Characteristically, when stimulated with pure tones they exhibit excitation only for a narrow band of frequencies—the best frequency (BF)—within 5 dB to 10 dB of threshold (see Young *et al.,* 1992a, b). At greater sound levels the response is generally inhibitory with the firing rate of the unit falling below the spontaneous rate. At higher levels, type IV units can also demonstrate a narrow band of excitation above BF and at very high levels (>60 dB above threshold) can also demonstrate excitation at low frequencies. In contrast to the mainly inhibitory response to pure tones, type IV units are excited by broadband stimuli over a wide dynamic range. Critically, this excitatory behavior is also modulated by the frequency of a spectral notch in the broadband noise. When the notch frequency equals the BF of the unit, the response is strongly inhibited so that the response falls well below the spontaneous rate. Varying the notch frequency by as little as 5% from the BF results in the response again becoming excitatory (reviewed in Young *et al.,* 1997; Young and Davis, 2002). This response behavior is in sharp contrast to the responses of other neurons in the DCN such as type II (vertical cells) and type III (probably principal cells) response types which exhibit more classical 'V' shaped tuning curves with inhibitory sidebands.

The spectral notch selectivity has been well modeled from the interplay between the level non-montonicities of the type II units, the variations in thresholds of the different unit types, the different spectral sensitivities, and the combination of inhibitory and excitatory interconnections (Nelken and Young, 1994; also see Young and Davis, 2002, for an excellent and extensive discussion). This has also been modeled using a system identification method which indicates that the type IV units sum acoustic information in a highly nonlinear way that signals the presence of the spectral notch at BF (Yu and Young, 2000).

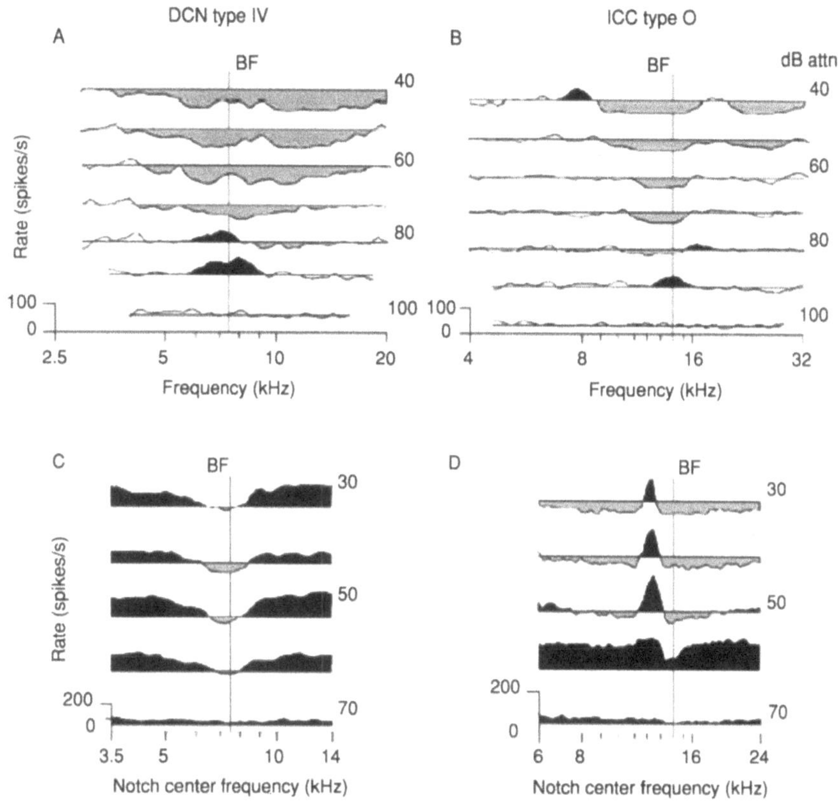

FIG. 2. The sensitivity to pure tone stimulation (top) and notched noise (bottom) is shown for a DCN type IV (left) and an ICC type O (right) neuron. The characteristic frequency of the DCN neuron was 7 kHz and 14 kHz for the ICC neuron. In the bottom panels the bandwidth of the notch was 3.2 kHz, The excitatory (black) and inhibitory (grey) variation around the average spontaneous rate (horizontal lines) is plotted for each tone frequency (top) or the center frequency of the notch in the broadband noise (bottom) stimulus. The response rate is indicated on the bottom left of each plot. Stimulus levels are to the right of each plot in dB attenuation relative to 0 dB. Reproduced with permission from Davis *et al.*, 2003.

Using a free-field broadband stimuli, Imig *et al.* (2000) have shown that the responses of DCN principal cells (notably pauser type units) were modulated by varying the azimuth location of the source. This was not the case for units recorded from the VCN, which demonstrated a very low spatial dependence of their responses. The spatial response nulls evident in the DCN principal cells corresponded to the spatial locations of the spectral notches corresponding to the unit's BF. This is consistent with the overall hypothesis that these units are selective for the notch spectral cue. Notably, in this study some units also showed response nulls that were not associated with spectral notch locations indicating that spectral information off-BF may also be playing a role.

Interestingly, the range of notch frequencies to which DCN type IV units have been shown to be sensitive extends to at least an octave above the upper limit of the pinna spectral notch (Young and Davis, 2002). This might suggest a more general spectral processing role for the DCN rather than one specialized only for the processing of pinna based spectral cues.

As the cat's pinna is highly mobile, the pinna position has to be accounted for if the HRTF is to provide a useful cue to the location of a sound source. It has been shown that stretching muscles attached to the pinna result in significant modulation of the type IV activity producing changes that are comparable to the variations in firing produced by auditory stimulation (Kanold and Young, 2001). The DCN receives somatosensory afferents from both the ipsilateral dorsal column nuclei and the ipsilateral spinal trigeminal nuclei (or medullary somatosensory nuclei [MSN]). This modulation is probably mediated by the inhibitory cartwheel cells which are excited by electrical stimulation of sites in the MSN which correspond to the representation of the pinna (Davis and Young, 1997). Cutaneous stimulation of the region in and around the pinna produced evoked potential responses, but not variations in the spiking of principal cells, in the DCN. In addition, the DCN also receives vestibular and other inputs, but the role of these is not yet known. The range of auditory and somatosensory inputs to the DCN and the laminated cerebellar-like structure strongly suggest that this structure is involved in sensorimotor coordination with the modulation of the DCN processing according to pinna position being a strong candidate function (see Oertel and Young, 2004, for discussion).

5. *Spectral Notch Processing at Other Levels of the Auditory System*

In the cat, the principal output cells of the DCN, the fusiform and giant cells, project to the contralateral central nucleus of the inferior colliculus (ICC) via the dorsal acoustic stria (DAS) (Oliver, 1984). There is a range of response types in the ICC and a number of different nomenclature are used in describing them. Using the three categories of May and colleagues (Ramachandran *et al.*, 1999; see also the chapter by Young and Davis), the type O responses appear to have very similar characteristics to the type IV responses of the DCN based on their tone frequency response maps (Davis *et al.*, 2003; Fig. 2). Unlike the type IV units, type O units when tested with broadband notched noise are excited by a notch at frequencies just below the BF of the unit, rather than inhibited. For stimulus levels 20 dB or so above threshold, notches that are above or below the excitatory notch frequency inhibit the response of the type O unit well below the spontaneous rate. However, at levels close to threshold the type O units were excited at all notch frequencies with a dip in excitation at BF. This latter response pattern is common with type IV units. May and colleagues conclude that type IV and type

O units in DCN and ICC respectively contribute to an auditory pathway specialized for the analysis of the notch like cues present in the HRTF of the cat.

The cells in ICC project to the auditory thalamus, the medial geniculate body (MGB), which then projects to the auditory cortex. Prior to the identification of the type O responses in the ICC, Imig *et al.* (1997) reported a subclass of units recorded from the ventral nucleus (VN) and lateral part of the posterior group of thalamic nuclei (PO) of the MGB that demonstrated restricted excitatory receptive fields for free-field broadband stimuli. The frequency response maps of these units, determined using pure tones, demonstrated multiple excitatory and inhibitory frequency response areas, some of which were narrowly tuned in frequency. The manner in which these units were modulated by variation in the spatial location of the stimulus is consistent with the spatial variation of the spectral peaks and notches that occur in cat's HRTF. To test the hypothesis that the spectral selectivity was the result of contrasting levels in two or more frequency regions, the strength of two-tone inhibition was measured. Consistent with the hypothesis, two-tone inhibition for frequencies at an inhibitory and excitatory frequency region was found to be most effective for locations which demonstrated a minimal response to the free-field broadband stimuli but was relatively ineffective at other locations. That is, the location dependent decrease in the responses to broadband noise could be explained by a relative increase in the sound energy in the inhibitory frequency ranges when compared to the energy in the excitatory frequency regions.

6. *The Representation of Auditory Space in the Mammalian Superior Colliculus*

There have been a number of studies examining the neural representation of auditory space and we will consider in some detail the so-called auditory space map in the superior colliculus (SC) as this provides some insights into how the various cues to spatial location are integrated in this midbrain nucleus.

The SC is a multisensory nucleus which integrates auditory, somatosensory, and visual information and whose motor output plays a role in driving pinna, eye, and head movements towards novel stimuli (see Stein and Meredith, 1991). The sensory representations share a common motor output pool and this presents an interesting problem of coordinate alignment across the different modalities to achieve an efficient sensory motor integration. Likely as a consequence of this requirement, the representation of auditory space in the deep layers of the SC (SCd) takes the form of a two-dimensional representation of space that aligns well with the visual representation of space. That is, the units recorded in the rostral portion of the SCd have restricted spatial receptive fields that are located in the anterior region of space. For locations progressively more caudal in the nucleus, the spatial receptive fields are located in progressively more posterior locations in space. Importantly, the auditory representation extends beyond that of the visual representation to locations at the side and behind the head so that the rostro-caudal axis in the nucleus codes for frontal to posterior azimuthal location. In a similar manner, the medio-lateral axis of the nucleus codes for a superior to inferior variation in the elevation of the sound source. Such an arrangement has been found in the superior colliculi in a range of mammals including the guinea pig (King and Palmer, 1983; Palmer and King, 1982), cat (Middlebrooks and Knudsen, 1984), ferret (King and Hutchings, 1987), and rat (Gaese and Johnen, 2000). An important feature of this representation is that the map of space is unambiguous. That is, the ambiguities in the binaural cues that give rise to the cones of confusion and front-back confusions have been resolved. This suggests the integration of spectral cues provided by the auditory periphery.

7. *The Auditory Periphery of the Guinea Pig and the Ferret*

Many of the neurophysiological investigations of SC function have used the guinea pig and the ferret. The auditory periphery of these animals is quite different to that of the cat and is briefly considered here. The HRTFs of the guinea pig have been measured both acoustically (Dickson *et al.*, 2002, 2003; Sinyor and Laszlo, 1973) and estimated using the directional changes in the cochlear microphonic (Carlile and Pettigrew, 1987; Palmer and King, 1985). The acoustics of the auditory periphery of the ferret have also been studied extensively (Carlile, 1990a,b, 1991; Schnupp *et al.*, 2003). In terms of the overall spectral patterns and their location dependence, the HRTFs of both species are not unlike the human HRTF, albeit at a higher frequency. This presumably reflects the similarities in structures, but on a smaller scale when compared to the human auditory periphery. Critically, the HRTFs of both species demonstrate significant differences in the middle frequency range (5 to 15 kHz) when compared along the cones of confusion, particularly for locations on the audio-visual horizon (Carlile, 1990b). Additionally, similar to the human HRTF, the spatial rate of change of the HRTFs is also greatest for locations around the midline and on the audio-visual horizon.

Behavioral studies of ferret localization performance have indicated minimal audible angles around the midline of around 10° to 12° (Kavanagh, 1987; King *et al.*, 2000) which indicates a reasonable facility for sound localization (see Fig. 9 in Heffner, 2004). This is corroborated by absolute localization data obtained with a speaker array separated by 30° between speakers where ferrets have on average a 75% accuracy rate for broadband stimuli (King *et al.*, 2000). The localization ability of the adult guinea pig does not appear to have been examined. A study of approach behavior in neonatal guinea pigs (1 to 4 days old) indicated above chance approach towards the sounds of guinea pig chirps presented by loudspeaker. An approach was scored if the animal moved into a 120° sector containing the loudspeaker (Clements and Kelly, 1978).

8. *Effects of Manipulating the Spectral Cues on the Map of Space in the SC*

The auditory units in the SCd are generally broadly tuned for frequency and in many cases exhibit binaural sensitivity to ILD that is arranged along the rostro-caural axis of the nucleus (see for the cat, Wise and Irvine, 1983, 1984, 1985; Hirsch *et al.*, 1985). In the guinea pig, the frequency sensitivity of the neuronal population at different rostro-caudal locations along the rostro-caudal axis of the SCd was found to vary in a manner that matched the estimated spectral shape functions of the HRTF from anterior to posterior locations on the audio-visual horizon (Carlile and Pettigrew, 1987a,b).

Narrow-band stimuli cannot provide the spectral information necessary to resolve the ambiguities in the binaural cues. Consistent with this observation, the auditory spatial response of multi-unit clusters to narrow band stimuli demonstrated a number of regions of space where maximal or near maximal activity could be elicited by sources in the contralateral hemifield (King and Carlile, 1994). Also, when the pinnae were surgically removed in anesthetized ferrets, the spatial receptive fields of single units responding to broadband noise demonstrated two locations at which maximal or near-maximal activity could be elicited for locations in the contralateral field. In general, these locations were symmetrically arranged around the interaural axis and on the cone of confusion for a particular interaural disparity to which the unit was presumably tuned (Carlile and King, 1994). An examination of the HRTFs, following removal of the pinna, indicated that spectral cues involved in resolving the ambiguities in the binaural cues had been eliminated by this procedure.

These data provide strong evidence that the spectral information derived from the HRTF is integrated with the ILD cues to support the emergent unambiguous representation of auditory space in the SCd. Consistent with this conclusion is the finding that a near-threshold map of auditory space can be demonstrated in the SCd of guinea pigs following either monaural ear plugging or destruction of the cochlea (Palmer and King, 1985). This near-threshold representation was also demonstrated to be dependent on an intact contralateral pinna. What is less clear is whether this spectral processing is emergent in the SCd or results from processing at lower levels in the auditory brainstem.

Studies in the brachium of the inferior colliculus (BIC) of the ferret (Schnupp and King, 1997) and the external nucleus of the inferior colliculus (ICX) of guinea pigs (Binns *et al.*, 1992) have demonstrated a topographic representation of the azimuthal dimensions of space. In the case of the ferret, selectivity for the elevation of the sound source was also demonstrated in single units, although no evidence of topography could be demonstrated. This representation of space appears to be coarser than that demonstrated in the SCd and, curiously, there is an absence of the near-threshold topography evident in the SCd. The BIC receives inputs from the ICX and projects to SC as well as to the medial geniculate and auditory cortex (see King *et al.*, 1998). However, given the differences in the precision of the topographies observed in these nuclei it appears that the unambiguous, two-dimensional auditory space map in the SCd results from the convergence and integration of information from a number of sources. The lack of an elevation code together with the lack of a threshold map of azimuth space suggests that the BIC may not be processing spectral cues. By contrast, units in the ICX are broadly tuned for frequency (Aitkin *et al.*, 1975) and integration of information across frequency could provide the basis of the threshold map observed in that nucleus. Importantly, at suprathreshold sound levels, the representation of auditory azimuth was shown to be dependent on binaural inputs to preserve the neural spatial tuning underlying the topographic representation (Binns *et al.*, 1992). These data suggest that spectral cues to location are processed to some extent in the ICX but not the BIC prior to projecting to the SCd. However, given that the ICX projects to the BIC it remains somewhat anomalous that the spectral processing evident in the ICX is not evident in the near threshold response of the BIC units (although this may reflect species differences in these two studies). Furthermore, it is unclear as to whether there is additional spectral processing in the SCd which refines the sensitivity of these units to spectral shape or whether the spectral topography evident in the SCd (Carlile and Pettigrew, 1987) simply reflects the projections from the ICX.

IV. Functional Plasticity and Learning New Spectral Cues

A. Re-Learning to Use Spectral Cues to Sound Location

Recent human psychophysical work has shown that under some conditions the auditory system has a remarkable ability to relearn to use modified spectral cues to a sound location. Shinn-Cunningham *et al.* (1998) exploited virtual auditory space (VAS) and training with visual feedback to "reassign" spectral cues to new spatial locations. Although training did not result in complete adaptation, there were persistent shifts in the perceived locations of test stimuli and response biases could be observed in subsequent testing using normal HRTF/location conjunctions.

In an ingenious experiment, van Opstal and colleagues modified the shape of the outer ears of human subjects using a chronically wearable mold. Acoustic measurements confirmed that the modified HRTFs of the subjects still contained significant spectral cues to location but that these were significantly different to the subjects normal HRTFs, particularly in the mid- to high frequency range.

Unsurprisingly, subjects were initially unable to accurately localize the elevation of targets presented in the frontal hemisphere. However, over a period of between 20 to 40 days of chronically wearing the ear molds, subjects recovered the ability to accurately localize the sound sources. Thus, the adult auditory system was able to adapt to the new set of spectral cues in much the same way as the visual system is able to adapt to significant changes in the visual input produced by wearing inverting prisms (e.g., Linden *et al.*, 1999). However, in contrast to the persistence of compensatory changes in the visual system, immediately following the reinstatement of normal HRTFs by removal of the ear molds, subjects were able to accurately localize the source of the sound. This implies that multiple spectral cues can be associated with specific positions and poses interesting questions about how the auditory system "selects" between the different libraries of spectral cues when mapping the cues to spatial location. Furthermore, as van Opstal and colleagues only tested locations in a relatively restricted region of space (the area to which subjects were able to make a fixed head visual saccade), it is unclear as to whether these new spectral/location associations can also be obtained for locations outside the visual field.

B. Neurophysiological Evidence of Auditory Functional Plasticity

Studies of developmental plasticity of the space map in the SCd have shown that modification of the visual and/or auditory input within a limited range results in adaptive changes in the map so as to maintain the normal topographic register. Adaptive changes to such manipulation are also seen in behavioral performance (reviewed by King *et al.*, 2000). Such plasticity would allow the auditory system to adapt to the progressive changes in the acoustic cues to location resulting from the developmental changes in the size, shape, and placement of the ears. Certainly, this is consistent with the neurophysiological evidence of auditory experience-induced plasticity in the cortex of mature animals. Following auditory discrimination training over several weeks, an increase in both the frequency discrimination capability and the size of the related auditory cortical areas has been reported for primates (Recanzone *et al.*, 1993). Interestingly, for cats it has been recently demonstrated that substantial improvements in discrimination ability following training is not necessarily associated with significant neuronal change in auditory cortex (Brown *et al.*, 2004). Nonetheless, given that there are also age-related changes in auditory function, such as the decrease in sensitivity to high frequencies and an increase in the size of the pinna in the aged, such re-calibration of the spectral cues over time at some level in the auditory system would be essential for the system to maintain a high level of localization performance.

Acknowledgments

The authors would like to thank Craig Jin and Dexter Irvine for their most constructive and helpful comments on an earlier version of this chapter.

References

Aitkin, L. M., Webster, W. R., Veale, J. L., and Crosby, D. C. (1975). Inferior colliculus. I: Comparison of response properties in central, pericentral and external nuclei of the adult cat. *J. Neurophysiol.* **38,** 1196–1207.

Algazi, V. R., Avendano, C., and Duda, R. O. (2001). Elevation localization and head-related transfer function analysis at low frequencies. *J. Acoust. Soc. Am.* **109,** 1110–1122.

Angell, J. R., and Fite, W. (1901). The monaural localization of sound. *Psychol. Rev.* **8,** 225–246.

Asano, F., Suzuki, Y., and Sone, T. (1990). Role of spectral cues in median plane localization. *J. Acoust. Soc. Am.* **88,** 159–168.

Binns, K. E., Grant, S., Withington, D. J., and Keating, M. J. (1992). A topographic representation of auditory space in the external nucleus of the inferior colliculus of the guinea-pig. *Brain Res.* **589**(2), 231–242.

Blauert, J. (1969/70). Sound localization in the median plane. *Acustica* **22,** 205–213.

Bloom, P. J. (1977a). Determination of monaural sensitivity changes due to the pinna by use of minimum-audible-field measurements in the lateral vertical plane. *J. Acoust. Soc. Am.* **61,** 820–828.

Bloom, P. J. (1977b). Creating source elevation illusions by spectral manipulation. *J. Audio Eng. Soc.* **25,** 560–565.

Brown, M., Irvine, D. R. F., and Park, V. N. (2004). Perceptual learning on an auditory frequency discrimination task by cats: Association with changes in primary auditory cortex. *Cerebral Cortex* **14,** 952–965.

Burlinghame, J. A., and Butler, R. A. (1998). The effects of attenuation of frequency segments on binaural localization of sound. *Percept. Psychophys.* **60,** 1374–1383.

Butler, R. A. (1986). The bandwidth effect on monaural and binaural localization. *Hear. Res.* **21,** 67–73.

Butler, R. A., and Belendiuk, K. (1977). Spectral cues utilized in the localization of sound in the median sagittal plane. *J. Acoust. Soc. Am.* **61,** 1264–1269.

Butler, R. A., Humanski, R. A., and Musicant, A. D. (1990). Binaural and monaural localization of sound in two-dimensional space. *Perception* **19,** 241–256.

Butler, R. A., and Musicant, A. D. (1993). Binaural localization: Influence of stimulus frequency and the linkage to covert peak areas. *Hear. Res.* **67,** 220–229.

Butler, R. A., and Planert, N. (1976). The influence of stimulus bandwidth on localization of sound in space. *Percept. Psychophys.* **19,** 103–108.

Calford, M. B., and Pettigrew, J. D. (1984). Frequency dependence of directional amplification at the cat's pinna. *Hearing Res.* **14,** 13–19.

Carlile, S. (1990a). The auditory periphery of the ferret I: Directional response properties and the pattern of interaural level differences. *J. Acoust. Soc. Am.* **88**(5), 2180–2195.

Carlile, S. (1990b). The auditory periphery of the ferret II: The spectral transformations of the external ear and their implications for sound localization. *J. Acoust. Soc. Am.* **88**(5), 2196–2204.

Carlile, S. (1991). The auditory periphery of the ferret: Postnatal development of acoustic properties. *Hearing Res.* **51,** 265–278.

Carlile, S. (1996). The physical and psychophysical basis of sound localization. Ch 2. Virtual auditory space: Generation and applications. S. Carlile, Austin, Landes.

Carlile, S., and King, A. J. (1994). Monaural and binaural spectrum level cues in the ferret: Acoustics and the neural representation of auditory space. *J. Neurophysiol.* **71,** 785–801.

Carlile, S., and Pettigrew, A. G. (1987a). Directional properties of the auditory periphery in the guinea pig. *Hearing Res.* **31,** 111–122.

Carlile, S., and Pettigrew, A. G. (1987b). Distribution of frequency sensitivity in the superior colliculus of the guinea pig. *Hear Res.* **31,** 123–136.

Carlile, S., and Pralong, D. (1994). The location-dependent nature of perceptually salient features of the human head-related transfer function. *J. Acoust. Soc. Am.* **95**(6), 3445–3459.

Clements, M., and Kelly, J. B. (1978). Auditory spatial responses of young guinea pigs (*Cavia porcellus*) during and after ear blocking. *J. Comp. Physiol. Psychol.* **92**(1), 34–44.

Coles, R. B., and Guppy, A. (1986). Biophysical aspects of directional hearing in the Tammar wallaby, Macropus eugenii. *J. Exp. Biol.* **121,** 371–394.

Davis, K. A., Ramachandran, R., and May, B. J. (2003). Auditory processing of spectral cues for sound localization in the inferior colliculus. *JARO* **4**(2), 148–163.

Davis, K. A., and Young, E. D. (1997). Granule cell activation of complex-spiking neurones in dorsal cochlear nucleus. *J. Neurosci.* **17**(17), 6796–6806.

Dickson, B., Behrend, O., Anandan, A., and Carlile, S. (2002). Changes in the spectral filtering due to variation in the auditory periphery of the guinea pig. *Proc. Aust. Neurosci. Soc.* **13,** 209.

Dickson, B., Behrend, O., and Carlile, S. (2003). Acoustic properties of the outer of the guinea pig. *Proc. Aust. Neurosci. Soc.* **14,** 267.

Duda, R. O. (1997). Elevation dependence of the interaural transfer function. *In* "Binaural and Spatial Hearing in Real and Virtual Environments" (R. H. Gilkey and T. R. Anderson, Eds.), pp. 49–75. Lawrence Erlbaum Assoc., Mahwah, NJ.

Flannery, R., and Butler, R. A. (1981). Spectral cues provided by the pinna for monaural localization in the horizontal plane. *Percept. Psychophys.* **29,** 438–444.

Fletcher, N. H., and Thwaites, S. (1979). Physical models for the analysis of acoustical systems in biology. *Quarterly Reviews of Biophysics* **12**(1), 25–65.

Fujino, K., and Oertel, D. (2003). Bidirectional synaptic plasticity in the cerebellum-like mammalian dorsal cochlear nucleus. *Proc. Natl. Acad. Sci. USA* **100,** 265–270.

Gaese, B. H., and Johnen, A. (2000). Coding for auditory space in the superior colliculus of the rat. *European J. Neurosci.* **12**(5), 1739–1752.

Gardner, M. B. (1973). Some monaural and binaural facets of median plane localization. *J. Acoust. Soc. Am.* **54,** 1489–1495.

Gardner, M. B., and Gardner, R. S. (1973). Problem of localization in the median plane: Effect of pinnae cavity occlusion. *J. Acoust. Soc. Am.* **53,** 400–408.

Gatehouse, R. W. (1976). Further research in localization of sound by completely monaural subjects. *J. Aud. Res.* **16,** 263–273.

Glasberg, B. R., and Moore, B. C. J. (1990). Derivation of auditory filter shapes from notched-noise data. *Hear. Res.* **47,** 103–138.

Griffiths, T. D., Bench, C. J., and Frackowiak, R. S. J. (1994). Human cortical areas selectively activated by apparent sound movement. *Curr. Biol.* **4,** 892–895.

Griffiths, T. D., Rees, A., Witton, C., Shakir, R. A., Henning, G. B., and Green, G. G. R. (1996). Evidence for a sound movement area in the human cerebral cortex. *Nature* **383,** 425–427.

Griffiths, T. D., and Warren, J. D. (2002). The planum temporale as a computational hub. *Trends Neurosci.* **25,** 348–353.

Harma, A., Karjalainen, M., Savioja, L., Valimaki, V., Laine, U. K., and Huopaniemi, J. (2000). Frequency-warped signal processing for audio applications. *J. Audio Eng. Soc.* **48,** 1011–1029.

Hartmann, W. M., and Rakerd, B. (1993). Auditory spectral discrimination and the localization of clicks in the sagittal plane. *J. Acoust. Soc. Am.* **94,** 2083–2092.

Hebrank, J., and Wright, D. (1974a). Spectral cues used in the localization of sound sources on the median plane. *J. Acoust. Soc. Am.* **56,** 1829–1834.

Hebrank, J., and Wright, D. (1974b). Are two ears necessary for localization of sound sources on the median plane? *J. Acoust. Soc. Am.* **56,** 935–938.

Hebrank, J. H. (1976). Pinna disparity processing: A case of mistaken identity? *J. Acoust. Soc. Am.* **59,** 220–221.

Heffner, R. S. (2004). Primate hearing from a mammalian perspective. *Anat. Rec. Part A.* **281A,** 1111–1122.

Heffner, R. S., and Heffner, H. E. (1987). Localization of noise, use of binaural cues, and a description of the superior olive complex in the smallest carnivore, the least weasel (Mustella Nivalis). *Behav. Neurosci.* **101**(5), 701–708.

Hirsch, J. A., Chan, J. C. K., and Yin, T. C. T. (1985). Responses of neurons in the cat's superior colliculus to acoustic stimuli. I. Monaural and binaural response properties. *J. Neurophysiol.* **53**(3), 726–745.

Hofman, P. M., and van Opstal, J. (1998). Spectro-temporal factors in two-dimensional human sound localization. *J. Acoust. Soc. Am.* **103,** 2634–2648.

Hofman, P. M., and van Opstal, J. (2003). Binaural weighting of pinna cues in human sound localization. *Exp. Brain Res.* **148,** 458–470.

Huang, A. Y., and May, B. J. (1996). Sound orientation behavior in cats. II. Mid frequency spectral cues for sound localization. *J. Acoust. Soc. Am.* **100**(2), 1070–1080.

Humanski, R. A., and Butler, R. A. (1988). The contribution of the near and far ear toward localization of sound in the sagittal plane. *J. Acoust. Soc. Am.* **83,** 2300–2310.

Huopaniemi, J., Zacharov, N., and Karjalainen, M. (1999). Objective and subjective evaluation of head-related transfer function filter design. *J. Audio Eng. Soc.* **47,** 218–239.

Imig, T. J., Poirier, P., Irons, W. A., and K, S. F. (1997). Monaural spectral contrast mechanisms for neural sensitivity to sound direction in the medial geniculate body of the cat. *J. Neurophys.* **78,** 2754–2771.

Imig, T. J., Bibikov, N., Poirier, P., and K, S. F. (2000). Directionality derived from pinna-cue spectral notches in cat dorsal cochlear nucleus. *J. Neurophysiol.* **83,** 907–925.

Jin, C. T. (2001). "Spectral Analysis and Resolving Spatial Ambiguities in Human Sound Localization." Doctoral dissertation, University of Sydney, Sydney, Australia.

Jin, C. T., Corderoy, A., Carlile, S., and van Schaik, A. (2000). Spectral cues in human sound localization. *In* "Advances in Neural Information Processing Systems 12" (S. A. Solla, T. K. Leen, and K.-R. Muller, Eds.), pp. 768–774. MIT Press, Cambridge.

Jin, C., Corderoy, A., Carlile, S., and van Schaik, A. (2004). Contrasting monaural and interaural spectral cues for human sound localization. *J. Acoust. Soc. Am.* **115,** 3124–3141.

Kanold, P. O., and Young, E. D. (2001). Proprioceptive information from the pinna provides somatosensory input to the cat dorsal cochlear nucleus. *J. Neurosi.* **21**(19), 7848–7858.

Kavanagh, G. L., and Kelly, J. B. (1987). Contribution of auditory cortex to sound localization by the ferret (*Mustela putorius*). *J. Neurophysiol.* **57,** 1746–1766.

King, A. J., and Carlile, S. (1994). Responses of neurones in the ferret superior colliculus to the spatial location of tonal stimuli. *Hear Res.* **81,** 137–149.

King, A. J., and Hutchings, M. E. (1987). Spatial response properties of acoustically responsive neurons in the superior colliculus of the ferret: A map of auditory space. *J. Neurophysiol.* **57**(2), 596–624.

King, A. J., Jiang, Z. D., and Moore, D. R. (1998). Auditory brainstem projections to the ferret superior colliculus: Anatomical contributions to the neural coding of sound azimuth. *J. Comp. Neurol.* **390,** 342–365.

King, A. J., and Palmer, A. R. (1983). Cells responsive to free-field auditory stimuli in guinea pig superior colliculus: Distribution and response properties. *J. Physiol.* **342,** 361–381.

King, A. J., Parsons, C. H., and Moore, D. R. (2000). Plasticity in the neural coding of auditory space in the mammalian brain. *Proc. Natl. Acad. Sci. USA* **97**(22), 11821–11828.

King, R. B., and Oldfield, S. R. (1997). The impact of signal bandwidth on auditory localization: Implications for the design of three-dimensional audio displays. *Human Factors* **39,** 287–295.

Kulkarni, A., and Colburn, H. S. (1998). Role of spectral detail on sound-source localization. *Nature* **396,** 747–749.

Langendijk, E. H. A., Kistler, D. J., and Wightman, F. L. (2001). Sound localization in the presence of one or two distracters. *J. Acoust. Soc. Am.* **109,** 2123–2134.

Linden, D. E., Kallenbach, U., Heinecke, A., Singer, W., and Goebel, R. (1999). The myth of upright vision. A psychological and functional imaging study of adaptation to inverting spectacles. *Perception* **28,** 469–481.

Macpherson, E. A. (1994). On the role of head-related transfer function spectral notches in the judgment of sound source elevation. *In* "Proceedings of the Second International Conference on Auditory Display, ICAD '94" (G. Kramer and S. Smith, Eds.), pp. 187–194. Santa Fe Institute, Santa Fe.

Macpherson, E. A. (1995). Source spectrum recovery at different spatial locations. *J. Acoust. Soc. Am.* **98,** 2946.

Macpherson, E. A. (1996). Effects of source spectrum irregularity and uncertainty on sound localization. *J. Acoust. Soc. Am.* **99,** 2515.

Macpherson, E. A. (1997). A comparison of spectral correlation and local feature-matching models of pinna cue processing. *J. Acoust. Soc. Am.* **101,** 3104.

Macpherson, E. A., and Middlebrooks, J. C. (2000). Localization of brief sounds: Effects of level and background noise. *J. Acoust. Soc. Am.* **108,** 1834–1849.

Macpherson, E. A., and Middlebrooks, J. C. (2003). Vertical-plane sound localization probed with ripple-spectrum noise. *J. Acoust. Soc. Am.* **114,** 430–445.

Macpherson, E. A., Sabine, A. T., and Middlebrooks, J. C. (2004). Binaural weighting of monaural spectral cues for sound localization. *Assoc. Res. Otolaryngol. Abs.* **1081**.

Makous, J. C., and Middlebrooks, J. C. (1990). Two-dimensional sound localization by human listeners. *J. Acoust. Soc. Am.* **87,** 2188–2200.

Martin, R. L., and McAnally, K. I. (2004). Localization of frequency-modulated sweeps as a function of sweep period. Unpublished data.

Martin, R. L., Paterson, M., and McAnally, K. I. (2004). Utility of monaural spectral cues is enhanced in the presence of cues to sound-source lateral angle. *J. Assoc. Res. Otolaryngol.* **5,** 80–89.

McAnally, K. I., Martin, R. L., Watt, T., and Flanagan, P. (2000). The effect of spectral roughness on sound localization. *Proc. Aust. Neurosci. Soc.* **11,** 145.

May, B. J. (2000). Role of the dorsal cochlear nucleus in the sound localization behavior of cats. *Hear Res.* **148,** 74–87.

May, B. J., and Huang, A. Y. (1996). Sound orientation behavior in cats. I. Localization of broadband noise. *J. Acoust. Soc. Am.* **100**(2), 1059–1069.

May, B. J., and Huang, A. Y. (1997). Spectral cues for sound localization in cats: A model for discharge rate representation in the auditory nerve. *J. Acoust. Soc. Am.* **101**(5), 2705–2719.

Mehrgardt, S., and Mellert, V. (1977). Transformation characteristics of the external human ear. *J. Acoust. Soc. Am.* **86,** 89–108.

Middlebrooks, J. C. (1992). Narrow-band sound localization related to external ear acoustics. *J. Acoust. Soc. Am.* **92,** 2607–2624.

Middlebrooks, J. C. (1999). Virtual localization improved by scaling non-individualized external-ear transfer functions in frequency. *J. Acoust. Soc. Am.* **106,** 1493–1510.

Middlebrooks, J. C., and Knudsen, E. I. (1984). A neural code for auditory space in the cat's superior colliculus. *J. Neurosci.* **4,** 2621–2634.

Mills, A. W. (1972). Auditory localization. *In* "Foundations of Modern Auditory Theory" (J. V. Tobias, Ed.), Vol. 2, pp. 303–348. Academic Press, New York.

Morimoto, M. (2001). The contribution of two ears to the perception of vertical angle in sagittal planes. *J. Acoust. Soc. Am.* **109,** 1596–1603.

Morimoto, M., and Ando, Y. (1980). On the simulation of sound localization. *J. Acoust. Soc. Jpn. (E)* **1,** 167–174.

Morimoto, M., Iida, K., and Itoh, M. (2003). Upper hemisphere sound localization using head-related transfer functions in the median plane and interaural differences. *Acoust. Sci. & Tech.* **24,** 267–275.

Musicant, A. D., and Butler, R. A. (1984a). The influence of pinnae-based spectral cues on sound localization. *J. Acoust. Soc. Am.* **75,** 1195–1200.

Musicant, A. D., and Butler, R. A. (1984b). The psychophysical basis of monaural localization. *Hear. Res.* **14,** 185–190.

Musicant, A. D., Chan, J. C. K., and Hind, J. E. (1990). Direction-dependent spectral properties of cat external ear: New data and cross-species comparisons. *J. Acoust. Soc. Am.* **87** (2), 757–781.

Nelken, I., and Young, E. D. (1994). Two separate inhibitory mechanisms shape the responses of dorsal cochlear nucleus type IV units to narrowband and wideband noise. *J. Neurophys.* **71,** 2446–2462.

Oertel, D., and Young, E. D. (2004). What's a cerebellar circuit doing in the auditory system. *Trends Neurosci.* **27**(2), 104–110.

Oldfield, S. R., and Parker, S. P. A. (1984a). Acuity of sound localization: A topography of auditory space. I. Normal hearing conditions. *Perception* **13,** 581–600.

Oldfield, S. R., and Parker, S. P. A. (1984b). Acuity of sound localization: A topography of auditory space. II. Pinna cues absent. *Perception* **13,** 601–617.

Oldfield, S. R., and Parker, S. P. A. (1986). Acuity of sound localization: A topography of auditory space. III. Monaural hearing conditions. *Perception* **15,** 67–81.

Oliver, D. L. (1984). Dorsal cochlear nucleus projections to the inferior colliculus in the cat: A light and electron microscope study. *J. Comp. Neurol.* **224,** 155–172.

Palmer, A. R., and King, A. J. (1982). The representation of auditory space in the mammalian superior colliculus. *Nature* **299,** 248–249.

Palmer, A. R., and King, A. J. (1985). A monaural space map in the guinea-pig superior colliculus. *Hear Res.* **17,** 267–280.

Pavani, F., Macaluso, E., Warren, J. D., Driver, J., and Griffiths, T. D. (2002). A common cortical substrate activated by horizontal and vertical sound movement in the human brain. *Curr. Biol.* **12** (18), 1584–1590.

Phillips, D. P., Calford, M. B., Pettigrew, J. D., Aitkin, L. M., and Semple, M. N. (1982). Directionality of sound pressure transformation at the cat's pinna. *Hearing Res.* **8,** 13–28.

Poon, P. W., and Brugge, J. F. (1993). Sensitivity of auditory nerve fibers to spectral notches. *J. Neurophysiol.* **20**(2), 655–665.

Pratt, C. C. (1930). The spatial character of high and low tones. *J. Exp. Psychol.* **13,** 278–285.

Paterson, M. A. (2000). "Cues to Sound-Source Direction: The Role of Spectral Information." Unpublished B.A. (Honours) dissertation, Deakin University, Geelong, Australia.

Rakerd, B., Hartmann, W. M., and McCaskey, T. L. (1999). Identification and localization of sound sources in the median sagittal plane. *Acoust. Soc. Am.* **106,** 2812–2820.

Ramachandran, R., Davis, K. A., and May, B. J. (1999). Single-unit response in the inferior colliculus of decerebrate cats I. Classification based on frequency response maps. *J. Neurophys.* **82,** 125–163.

Recanzone, G. H., Schreiner, C. E., and Merzenich, M. M. (1993). Plasticity in the frequency representation of primary auditory cortex following discrimination training in adult owl monkeys. *J. Neurosci.* **13**(1), 87–103.

Rice, J. J., May, B. J., Spirou, G. A., and Young, E. D. (1992). Pinna-based spectral cues for sound localization in cat. *Hearing Res.* **58**(2), 132–152.

Roffler, S. K., and Butler, R. A. (1968a). Factors that influence the localization of sound in the vertical plane. *J. Acoust. Soc. Am.* **43,** 1255–1259.

Roffler, S. K., and Butler, R. A. (1968b). Localization of tonal stimuli in the vertical plane. *J. Acoust. Soc. Am.* **43,** 1260–1266.

Rogers, M. E., and Butler, R. A. (1992). The linkage between stimulus frequency and covert peak areas as it relates to monaural localization. *Percept. Psychophys.* **52,** 536–546.

Schnupp, J. W. H., Booth, J., and King, A. J. (2003). Modeling individual difference in ferret external ear transfer functions. *J. Acoust. Soc. Am.* **113**(4), 2021–2030.

Schnupp, J. W. H., and King, A. J. (1997). Coding for auditory space in the nucleus of the brachium of the inferior colliculus in the ferret. *J. Neurophysiol.* **78,** 2717–2731.

Searle, C. L., Braida, L. D., Cuddy, D. R., and Davis, M. F. (1975). Binaural pinna disparity: Another auditory localization cue. *J. Acoust. Soc. Am.* **57,** 448–455.

Senova, M. (2002). "The Effect of Smoothing Head-Related Transfer Functions on the Spatial Fidelity of Virtual Sounds." Unpublished doctoral dissertation, Swinburne University of Technology, Melbourne, Australia.

Senova, M. A., McAnally, K. I., and Martin, R. L. (2002). Localization of virtual sound as a function of head-related impulse response duration. *J. Audio Eng. Soc.* **50,** 57–66.

Shaw, E. (1982). "External Ear Response and Sound Localization. Localization of Sound: Theory and Applications," pp. 30–41. R. Gatehouse. Amphora, New York.

Shaw, E. A. G., and Teranishi, R. (1968). Sound pressure generated in an external-ear replica and real human ears by a nearby point source. *J. Acoust Soc. Am.* **44,** 240–249.

Shinn-Cunningham, B. G., Durlach, N., and Held, R. M. (1998). Adapting to supernormal auditory localization cues: Bias and resolution. *J. Acoust. Soc. Am.* **103,** 3656–3666.

Sinyor, A., and Laszlo, C. A. (1973). Acoustic behavior of the outer ear of the guinea pig and the influences of the middle ear. *J. Acoust. Soc. Am.* **54,** 916–921.

Slattery, W. H., III, and Middlebrooks, J. C. (1994). Monaural sound localization: Acute versus chronic unilateral impairment. *Hear. Res.* **75,** 38–46.

Stein, B., and Meredith, M. (1991). "Functional Organisation of the Superior Colliculus. The Neural Basis of Visual Function," pp. 85–110. J. Cronly-Dillon. Macmillan Press, London.

Supin, A. Y., Popov, V. V., Milekhina, O. N., and Tarakanov, M. B. (1999). Ripple depth and density resolution of rippled noise. *J. Acoust. Soc. Am.* **106,** 2800–2804.

van Schaik, A., Jin, C., and Carlile, S. (1999). Human localization of band-pass filtered noise. *Int. J. Neural Sys.* **9,** 441–446.

Van Wanrooij, M. M., and van Opstal, A. J. (2004). Contribution of head shadow and pinna cues to chronic monaural sound localization. *J. Neurosci.* **24,** 4163–4171.

Vliegen, J., and van Opstal, A. J. (2004). The influence of duration and level on human sound localization. *J. Acoust. Soc. Am.* **115,** 1705–1713.

Watkins, A. J. (1978). Psychoacoustical aspects of synthesized vertical locale cues. *J. Acoust. Soc. Am.* **63,** 1152–1165.

Wiener, F. M., Pfeiffer, R. R., and Backus, A. S. N. (1965). On the sound pressure transformation by the head and auditory meatus of the cat. *Acta Oto-laryng.* **61,** 255–269.

Wightman, F. L., and Kistler, D. J. (1989). Headphone simulation of free-field listening. I: Stimulus synthesis. *J. Acoust Soc. Am.* **85,** 858–867.
Wightman, F. L., and Kistler, D. J. (1997). Monaural sound localization revisited. *J. Acoust. Soc. Am.* **101,** 1050–1063.
Wightman, F. L., and Kistler, D. J. (1999). Sound localization with unilaterally degraded spectral cues. *J. Acoust. Soc. Am.* **105,** 1162.
Wise, L. Z., and Irvine, D. R. F. (1983). Auditory response properties of neurones in the deep layers of cat superior colliculus. *J. Neurophysiol.* **49,** 674–685.
Wise, L. Z., and Irvine, D. R. F. (1984). Interaural intensity difference sensitivity based on facilitatory binaural interaction in cat superior colliculus. *Hear Res.* **16,** 181–188.
Wise, L. Z., and Irvine, D. R. F. (1985). Topographic organization of interaural intensity difference sensitivity in deep layers of cat superior colliculus: Implications for auditory spatial representation. *J. Neurophysiol.* **54,** 185–211.
Woodworth, R. S., and Schlosberg, H. (1954). "Experimental Psychology, Third Ed.," p. 352, Methuen and Co. Ltd., London.
Xu, L., and Middlebrooks, J. C. (2000). Individual difference in external-ear transfer function of cats. *J. Acoust. Soc. Am.* **107**(3), 1451–1459.
Young, E. D., and Davis, K. A. (2002). Circuitry and function of the Dorsal cochlear nucleus. *In* "Integrative Functions in the Mammalian Auditory Pathway" (D. Oertel, R. R. Fay, and A. N. Popper, Eds.), pp. 160–206. Springer-Verlag, New York.
Young, E. D., Rice, J. J., Spirou, G. A., Nelken, I., and Conley, R. A. (1997). Head-related transfer functions in cat: Neural representation and the effects of pinna movement. *In* "Binaural and Spatial Hearing in Real and Virtual Environments" (R. H. Gilkey and T. R. Anderson, Eds.), pp. 475–498. Lawrence Erlbaum Assoc., Mahwah, NJ.
Young, E. D., Rice, J. J., and Tong, S. C. (1996). Effects of pinna position on head-related transfer functions in the cat. *J. Acoust. Soc. Am.* **99**(5), 3064–3076.
Young, E. D., Spirou, G. A., and Rice, J. J. (1992a). Processing of spectral sound localization cues in the dorsal cochlear nucleus. *Proc. Physiol. Soc.* **8**.
Young, E. D., Spirou, G. A., Rice, J. J., and Voigt, H. F. (1992b). Neural organization and responses to complex stimuli in the dorsal cochlear nucleus. *Philos. Trans. R. Soc. London B* **336**(1278), 407–413.
Yu, J. J., and Young, E. D. (2000). Linear and nonlinear pathways of spectral information transmission in the cochlear nucleus. *PNAS* **97**(22), 11780–11786.
Zakarauskas, P., and Cynader, M. S. (1993). A computational theory of spectral cue localization. *J. Acoust. Soc. Am.* **94,** 1323–1331.
Zahorik, P., Wightman, F., and Kistler, D. (1995). On the discriminability of virtual and real sound sources. *In* "Proceedings of the ASSP (IEEE) Workshop on Applications of Signal Processing to Audio and Acoustics." IEEE Press, New York.

PLASTICITY OF SPECTRAL PROCESSING

Dexter R. F. Irvine* and Beverly A. Wright[†]

*Department of Psychology, Faculty of Medicine, Nursing, and Health Sciences
Monash University, Victoria 3800, Australia

[†]Department of Communication Sciences and Disorders, Northwestern University
Evanston, Illinois 60208, USA

In the last two decades, a substantial body of evidence has demonstrated that the spectral processing ability of human and animal listeners and the response characteristics of neurons in circuits at higher brain levels that are involved in auditory spectral processing can be modified by experience. This plasticity can occur as a consequence of particular forms of training or perceptual experience, or of damage to the cochlea that results in changes in input to the central nervous system and in partial hearing loss. In this chapter, physiological and psychophysical evidence for the occurrence of such spectral processing plasticity is reviewed, the possible underlying mechanisms are examined, and therapeutic applications are considered.

INTERNATIONAL REVIEW OF
NEUROBIOLOGY, VOL. 70
DOI: 10.1016/S0074-7742(05)70013-1

0074-7742/05 $35.00

I. Introduction

A. Background

One of the major discoveries in sensory neuroscience in the last two decades has been the extent to which the stimulus selectivity of sensory cortical neurons and the functional organization of sensory cortices can be modified in adult animals as a consequence of altered patterns of input or of procedures that alter the significance of particular sensory inputs. Prior to recognition of this plasticity in adult sensory systems, it was generally assumed that these characteristics, while modifiable during development, were stable features of the adult brain. In parallel with the physiological studies of adult sensory plasticity, psychophysical studies of improvements in perceptual discriminative ability with practice also indicated that this "perceptual learning" might reflect plastic changes in sensory cortical processing mechanisms. In the case of the auditory system, physiological studies of adult plasticity have been focused on the frequency selectivity of individual neurons and the frequency organization of the primary auditory cortex (AI), changes which would be expected to have consequences for an organism's spectral processing ability. Such changes in performance have been demonstrated directly in psychophysical studies of various forms of spectral processing in humans and in non-human animals. In this chapter, we will review the evidence from physiological and psychophysical studies for plasticity in spectral processing, the mechanisms underlying this plasticity, and its therapeutic applications. Before examining this evidence, the nature of the changes that constitute evidence of plasticity requires consideration.

B. Definition of Plasticity

The term "plasticity" is used in this chapter to refer to a range of experience- and injury-induced changes in auditory processing at the psychophysical and neurophysiological levels. The most ubiquitous manifestation of behavioral plasticity is learning, and the bulk of the evidence to be reviewed in both domains is derived from studies of various forms of learning. Just as learning must be distinguished from changes in behavior attributable to other factors, neural plasticity (a phrase which is widely used but seldom clearly defined) must be distinguished from other forms of neural change.

In an early review of synaptic plasticity, Tsukahara (1981) defined plasticity as "any persistent change in the functional properties of single neurons or neuronal aggregates" (p. 351). The use of the term "functional properties" serves to indicate that not all changes in neuronal responsiveness constitute cases of plasticity. For example, the response of a cortical or subcortical neuron in any sensory system might change (increase or decrease in magnitude) as a function of an arousal state or of other variables having a general influence on brain excitability. Generalized changes in discharge rate of this sort without any change in the neuron's stimulus selectivity are not evidence of plasticity.

This definition is still rather too broad, however, as not all changes in stimulus selectivity constitute evidence for plasticity. For example, the frequency selectivity of auditory nerve (AN) fibers and of neurons at higher levels of the pathway is immediately changed by damage to outer hair cells in the region of the cochlea from which their input is derived (e.g., Dallos and Harris, 1978). These changes are a direct (or passive) reflection of the altered input to the AN fibers from the inner hair cells, and should not be regarded as instances of plasticity. In the same way, the reduced sensitivity and spectral processing ability of a listener with this form of hearing loss would not be regarded as an instance of plasticity. For a change in a neuron's response selectivity consequent on altered patterns of input to be considered a manifestation of plasticity, the change must not be explicable as a direct reflection of the altered input, but must involve some kind of *active* process that is initiated or triggered by the change in input. Such active processes might involve changes in intrinsic neuronal characteristics or in the number or efficacy of synapses, either at the level of the neuron from which recordings are made or in neuron(s) in the pathways over which its input is derived. Unfortunately, the distinction between active and passive processes, the extent to which they can be separated, and the determination of their involvement in particular changes, are less simple matters than might at first appear (see Calford, 2002). Nevertheless, the definition of physiological plasticity as changes in neuronal stimulus selectivity and/or in the associated functional organization of populations of neurons that are not explicable in terms of general changes in excitability or as passive reflections of changes in input provides an adequate basis for examination of the evidence on plastic changes in spectral processing mechanisms.

In the case of behavioral plasticity, it is also important, as previously noted, to distinguish plastic changes in sensory performance from those associated with changes in state or that are explicable as direct consequences of altered input. A more complex issue arises from the fact that perceptual discriminative ability is always measured experimentally using some sort of task on which the participants (human or animal) must be trained. Learning to perform that task—what is commonly termed "procedural learning"—is itself a form of plasticity and will lead to improved performance, but improved performance attributable to such learning is not a manifestation of the plasticity in sensory processing that is of interest here.

II. Neurophysiological Studies of Plasticity in Spectral Processing Mechanisms

A. Overview

Neurophysiological studies of plasticity in spectral processing have examined the effects of a variety of manipulations of auditory input on the frequency selectivity of single neurons or multi-neuron "clusters" at various levels of the auditory pathway. Although there is a large body of evidence on the occurrence of these effects at subcortical auditory centers and in different auditory cortical fields, this account will concentrate on data relating to primary auditory cortex (AI). Most of these studies have been performed using anesthetized or unanesthetized non-human animals, but there are also limited data from imaging studies in humans. In some cases the emphasis has been on the frequency selectivity of individual neurons or multi-neuron "clusters," while in other cases it has been on the distribution of frequency selectivity across a cortical area or areas, i.e., on the frequency organization (or "map") in the area(s) under investigation. The major techniques that have been used to manipulate auditory input have been: (1) training the animal on behavioral conditioning or detection tasks which alter the significance of a particular stimulus frequency or frequencies for the organism; (2) eliminating or reducing input to the central nervous system over a restricted range of frequencies by damaging a restricted region of the cochlea (which produces a partial hearing loss); (3) training the animal on a perceptual discrimination task; and (4) pairing acoustic stimulation at a particular frequency with activation of a neuromodulatory system. In the following sections, a brief and highly selective review of the evidence provided by each of these procedures will be provided.

B. Neural Plasticity Associated with Behavioral Conditioning or Detection Tasks

There is a long history of research, dating at least from the report of Galambos *et al.* (1955), describing changes in auditory neural responses as a function of conditioning procedures which alter the significance of acoustic stimuli. These studies, and the control procedures that are necessary to establish that the observed changes are associative in nature and are selective (i.e., are specific to the training stimulus or to some limited range of stimuli), have been the subject of a number of reviews (e.g., Weinberger, 1995, 2004a,b; Edeline, 1999, 2003). Therefore, these issues will not be dealt with here, but the major results on the occurrence of plastic changes in the frequency selectivity of AI neurons will be briefly summarized.

Although a great variety of behavioral conditioning tasks have been utilized in these studies, the most common procedure has been to use a classical aversive conditioning paradigm with an acoustic stimulus as the conditioned stimulus (CS), foot-shock as the unconditioned stimulus (UCS), and either a flexion response or heart-rate decrease (bradycardia) as the conditioned response (CR). Diamond and Weinberger (1986) introduced the procedure of determining the effect of conditioning on what they called the "frequency receptive field" (FRF) of a single neuron or cluster of neurons. This involves determination of a frequency response function at a fixed sound pressure level (SPL) prior to and following conditioning with a CS at a frequency within the neuron's pre-conditioning FRF but other than the best frequency (BF) (that eliciting the maximum response). In a classical conditioning discrimination-learning task, a tone of one frequency is used as the CS^+ (i.e., paired with the UCS) and a tone of another frequency is unpaired (CS^-). In a variant on this procedure designed to control for differential exposure to different frequencies, Ohl and Scheich (1996, 1997) used all of the frequencies (other than that used as CS^+) that were presented in determining the pre-conditioning frequency selectivity as multiple CS^-s.

In a number of early studies of the effects of behavioral conditioning on the responses of neurons and neuron clusters in AI and surrounding fields of a range of species (and using a range of conditioning tasks), individual neurons (or clusters) were reported to exhibit increases, decreases, or no change in the response to the frequency of a tonal CS (e.g., Kraus and Disterhoft, 1982; Weinberger *et al.*, 1984; Diamond and Weinberger, 1984, 1986). In more recent studies restricted to AI and using the heart rate conditioning paradigm previously described, the most commonly reported result has been an increase in the response to the CS^+, in most cases combined with a decrease in response at the pre-training BF and at other frequencies (e.g., Bakin and Weinberger, 1990; Edeline and Weinberger, 1993; Weinberger *et al.*, 1993). Representative data illustrating this pattern of effects are presented in Fig. 1. In Fig. 1A, the pre- and post-training FRFs for a single AI neuron reveal an increase in the response at the CS frequency and a decrease in response at the pre-training BF (also see insert in panel A) such that the CS became the post-training BF. Normalized functions for a sample of AI neurons in Fig. 1B reveal a general pattern of an increase in the response to the CS and a decrease at surrounding frequencies. Non-selective changes in responsiveness as a consequence of sensitization, and a selective decrease in response as a consequence of habituation, are illustrated in the other panels in Fig. 1B. The changes in frequency selectivity associated with behavioral conditioning can occur extremely rapidly, occurring after as few as five training trials (over a period of approximately 10 min) (e.g., Edeline *et al.*, 1993).

Suga and colleagues have reported similar shifts in BF towards the CS frequency in auditory cortical neurons in the big brown bat as a consequence of conditioning procedures (Gao and Suga, 2000; Ji *et al.*, 2001). Changes in

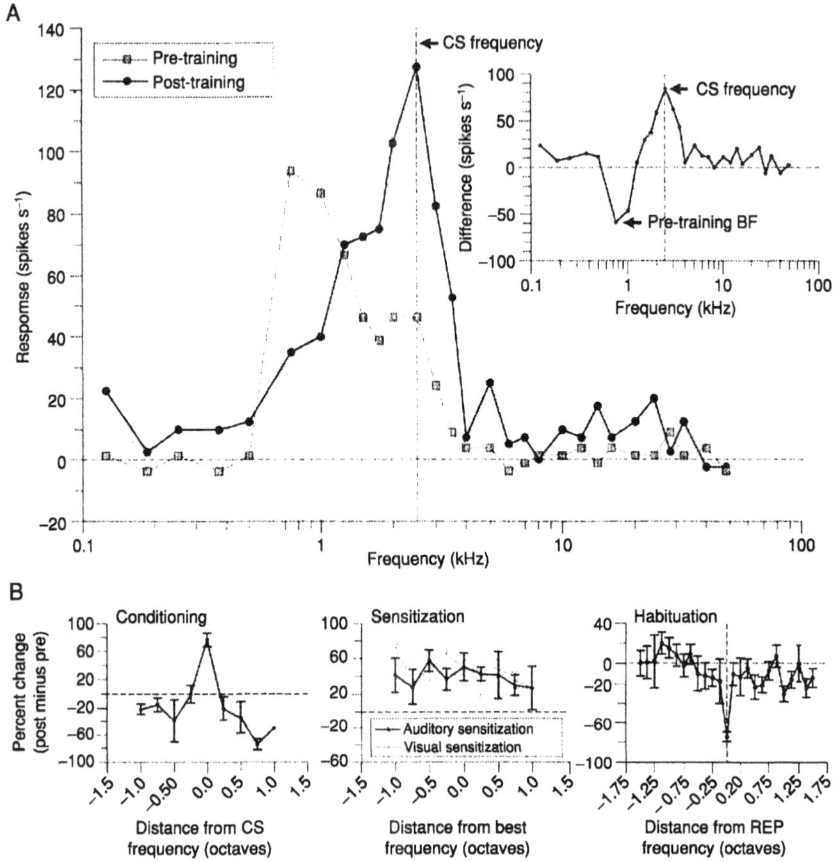

FIG. 1. The effects of behavioral conditioning on the frequency selectivity of neurons in AI of the guinea pig. (A) Receptive field plasticity of a single AI neuron is manifested in iso-level functions obtained (at 70 dB SPL) before and after tone-shock conditioning: the BF shifts from below 1 kHz to the CS frequency. The inset shows the difference in frequency selectivity (post minus pre), with the maximum increase in response at the frequency of the CS. (B) Normalized group difference functions (post-pre) showing change in response as a function of the distance of the probe frequency from the frequency used in behavioral training in conditioning (left), sensitization (middle), and habituation (right) conditions. Conditioning (left) produces a selective increase in response at the CS frequency, with reduced responses at most other frequencies. Sensitization training produces a non-selective increase in response at all frequencies both for auditory sensitization (tone-shock unpaired) and visual sensitization (light-shock unpaired), showing that this non-associative effect is transmodal. Habituation (repeated presentation of the same tone alone) produces a selective decrease in response at that frequency. Reproduced with permission from Weinberger (2004b).

frequency tuning (and consequently in the cortical frequency map) are also produced by focal electrical stimulation of the cortex in big brown bats (Chowdhury and Suga, 2000; Ma and Suga, 2001) and in gerbils (Sakai and Suga, 2002). In most areas of the cortex, the BF of surrounding neurons shifts towards that of the neurons at the stimulation site, resulting in an enlarged representation of that frequency (see Suga and Ma, 2003, for review).

A major exception to the reports of selective increases in response to the CS frequency as a consequence of conditioning is the report by Ohl and Scheich (1997), who studied the effects of discrimination training (using the multiple CS^- procedure previously described) on single units in AI of unanesthetized gerbils. They also examined changes in neural responses throughout their temporal time course, using what they termed "response planes," which are effectively spectrotemporal receptive fields (STRFs) (see Chapters 5, 7, and 8) obtained with a restricted set of tonal stimuli rather than from the responses to broadband stimuli by reverse correlation. Although individual neurons showed either increases or decreases in spike rate at CS^+ at different times during the response, the pooled data from the entire sample of neurons showed a decrease in the early (onset) part of the response at the CS^+ frequency, whereas increases in response at this time were greatest at frequencies above and below the CS frequency, such that the CS^+ frequency came to lie at a local minimum. Whether this different pattern of results is attributable to differences in task requirements or in other experimental procedures is unclear (see Weinberger, 2004a,b, for discussion).

Ohl and Scheich (1997) discuss the changes they observed in terms of dynamic inhibition and disinhibition. Direct evidence of plastic changes in inhibition resulting in changes in neuronal frequency selectivity has recently been presented by Fritz *et al.* (2003, 2005). They trained ferrets to detect a target tone of a particular frequency embedded in a sequence of broadband noise-like stimuli ("temporally orthogonal ripple combinations" or TORCs), using a conditioned-suppression-of-drinking task. The STRF of an AI neuron was determined from the responses to the TORC stimuli recorded during performance of the detection task and while the animal was in a "passive" resting state. The target tone frequency in a given testing session was generally placed within or in close proximity to the major excitatory region of the STRF obtained in the resting state or within an inhibitory sideband. Approximately 70% of neurons showed significant changes in STRF shape during the detection task, and for the majority of these units the change involved either a local (i.e., restricted to the region of the target tone; see Fig. 2A to C) enhancement or expansion of the excitatory field or a reduction in an inhibitory sideband. In the remaining cases, the changes were more global, in some cases involving the reduction or elimination of all inhibitory regions in the STRF (e.g., Fig. 2D). The average spectral change in the STRF relative to the detection frequency (Fig. 2F) shows a pattern of changes (facilitation at frequencies close to the target frequency and suppression at

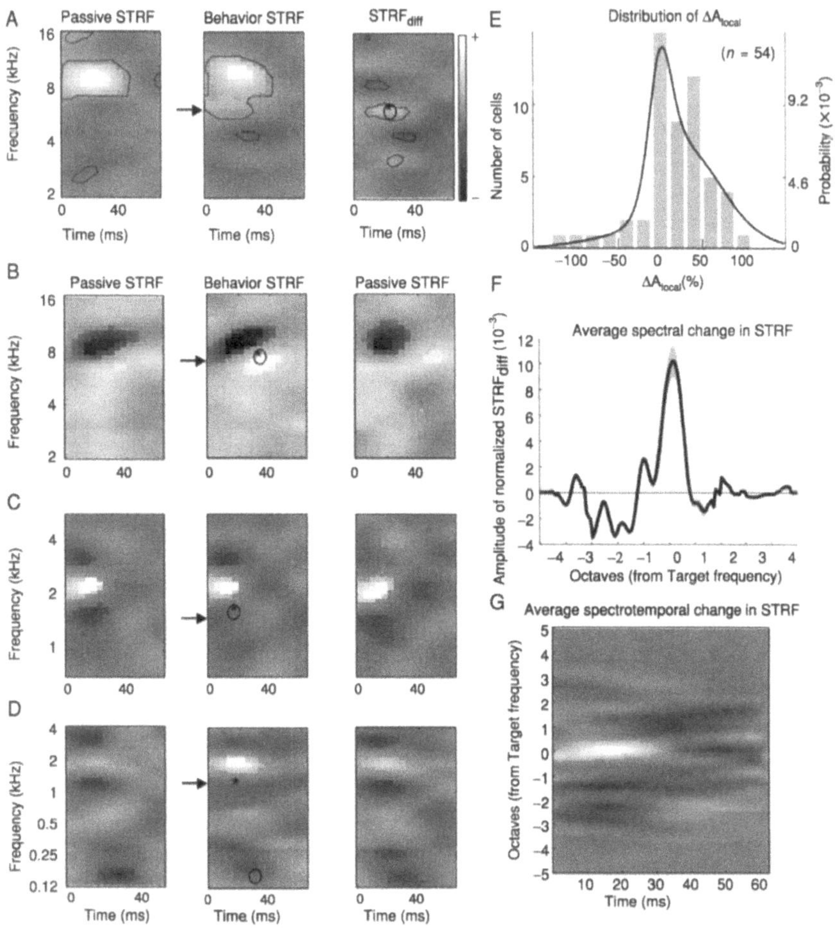

FIG. 2. Plasticity of spectrotemporal receptive fields (STRFs) of neurons in AI of the ferret. (A to D) STRFs of four single neurons showing changes typical of those observed during performance of the detection task. For each neuron a pre-behavior "passive STRF" is shown in the left column, and a "behavior STRF" recorded during performance of the detection task is shown in the middle column. The STRF in each panel was normalized, and all STRFs were then depicted using the same grey scale. The contours in (A) demarcate the excitatory (white) and inhibitory (black) regions with statistically significant fluctuations (± 3 SD) from the mean (see Methods in Fritz *et al.*, 2003, for details). These contours are not drawn on the lower panels to avoid cluttering the figures. The black arrows in the middle column indicate the frequency of the target tone during the detection task. (A) In the right column, the difference between the normalized passive and behavior STRFS ($STRF_{diff}$) is shown. An asterisk marks the location of maximum local change (A_{local}; 45%), and a circle marks the global change (A_{global}; 45%), defined in Fritz *et al.* (2003). In this case the local and global maximal changes were both at the target frequency, as in about one-half of all neurons. (B) Localized enhancement of an excitatory region in the STRF during behavior (left and middle). The post-behavior passive STRF (right column) reverted immediately to its original shape (A_{local}, 37%; A_{global}, 42%). The

adjacent frequencies) that is similar to that reported by Weinberger and his colleagues for frequencies relative to CS^+ in the conditioning paradigm (Fig. 1B).

One of the advantages of the behavioral task and measurement procedures used by Fritz *et al.* (2003, 2005) is that the effects of changing the target tone frequency can be assessed within a single testing session. In experiments of this sort, changes in the STRF were found to occur very rapidly, within 10 min or less of initiating training with one target frequency or of changing to another target frequency. In some neurons, changes induced during training persisted after training for a period of some hours, while in others the STRF returned to its pre-training form within a relatively short period (e.g., Fig. 2B and D). These experiments, therefore, confirm not only the occurrence of plasticity in the spectral sensitivity of AI neurons but the rapidity with which it can occur.

The different patterns of results reported in various studies of the effects of behavioral conditioning procedures on the frequency selectivity of auditory cortical neurons undoubtedly reflect methodological and contextual factors that remain incompletely understood (see Weinberger, 2004a, for a detailed discussion). In the present context, however, the important point is that the conditioning and target detection data constitute one of the lines of evidence that the frequency selectivity of individual neurons, and hence the way in which spectral information is represented by populations of neurons, exhibits a high degree of plasticity in adult animals.

Although it is possible that the increased response at or around the CS^+ or target frequency in these experiments reflects increased synaptic excitation,

---

two maxima were nearly coincident at the target frequency. (C) Local decrease or elimination of inhibitory sidebands in the behavior STRF. The inhibition recovered quickly afterwards, but the overall STRF shape was different (A_{local}, 40%; A_{global}, 40%; coincident maxima at target frequency). (D) A global weakening of inhibitory fields during behavior. Immediately following behavior the STRF recovered its pre-behavior shape (A_{local}, 64%; A_{global}, 86%). In this example the local maximum difference occurred at the target tone frequency, whereas the global maximum was located over a low-frequency inhibitory field (which was also knocked out during behavior). (E) Summary histogram and smoothed distribution of local STRF changes from all STRFs in the study. The histogram (left ordinate) and distribution (right ordinate) are significantly skewed toward positive changes (overall mean, +20.2%). (F) Average spectral change in the STRF at all frequencies relative to the target frequency. This curve shows the average plastic change in STRF amplitude as a function of log-frequency distance from the target frequency (centered at 0 and marked by the dot). The shaded region around the curve represents the variance around the mean. There was facilitation for about one octave around the target (half-width of ± 0.56 octave) and asymmetric suppressive sidebands outside of this range. (G) Average spectrotemporal changes in the STRF derived from all neurons. The image was constructed by computing the $STRF_{diff}$ for each neuron, centering it around its target frequency, and then averaging over all neurons. The facilitative and suppressive changes near the target frequency, as well as the relatively rapid onset of these STRF changes, can be seen here. Reproduced with permission from Fritz *et al.* (2003). (See Color Insert.)

it could also reflect a release from inhibition. Indeed, the rapidity of this facilitatory effect suggests that release from inhibition is the most likely cause. The reduction in inhibitory sidebands or the global reduction in inhibition described by Fritz *et al.* (2003, 2005) also clearly involve a decrease in inhibition. Thus, as suggested by Ohl and Scheich (1997), most of the effects of conditioning and related procedures appear to be explicable in terms of dynamic inhibition and disinhibition.

C. Lesion-Induced Neural Plasticity

Recognition of the fact that plasticity in the sensory cortex is not restricted to critical periods during development was originally driven by evidence for reorganization in primary sensory cortices as a consequence of elimination of input from a restricted region of the receptor epithelium. Such reorganization was first demonstrated in somatosensory cortex after digit amputation or peripheral nerve section, but was subsequently demonstrated in both auditory and visual cortices after analogous lesions (see Buonomano and Merzenich, 1998; Gilbert, 1998 and Kaas and Florence, 2001, for reviews). In the auditory system, unilateral mechanical lesions to a restricted region of the cochlea in adult guinea pigs (Robertson and Irvine, 1989) and cats (Rajan *et al.*, 1993) result in a reorganization of the frequency map in AI contralateral to the lesioned cochlea. The general form of this reorganization is that the neurons in the region of cortex in which the lesioned section of the cochlea would normally be represented, which has been termed the lesion projection zone (LPZ) (Schmid *et al.*, 1996), have new CFs at frequencies represented at the edge(s) of the cochlear lesion. As a consequence of this change in tuning, the LPZ is wholly or partly occupied by an expanded representation of the lesion edge frequency or frequencies. This pattern of results is illustrated in Fig. 3. The upper and lower left panels show compound action potential (CAP) audiograms for normal and lesioned cochleas, respectively: the latter exhibit the losses in high-frequency sensitivity produced by mechanical lesions of the basal cochlea. The losses in these two cases start in the region of 18 kHz and increase steeply at higher frequencies. The upper right panel shows CF distance plots for AI produced by stimulation of the contralateral ear in normal cats. CF increases progressively with caudo-rostral distance across AI, and reverses at the border with the anterior auditory field. In the lower right panel, CF distance plots for AI contralateral to the lesioned cochleas for which the cap audiograms are presented in the lower left panel are shown, and are plotted against spline functions fitted to the normal data. It is apparent that the region in which frequencies in the range 18 to 38 kHz would normally be represented is occupied by a massively enlarged representation of the lesion-edge frequencies (i.e., frequencies in the range 18 to 20 kHz).

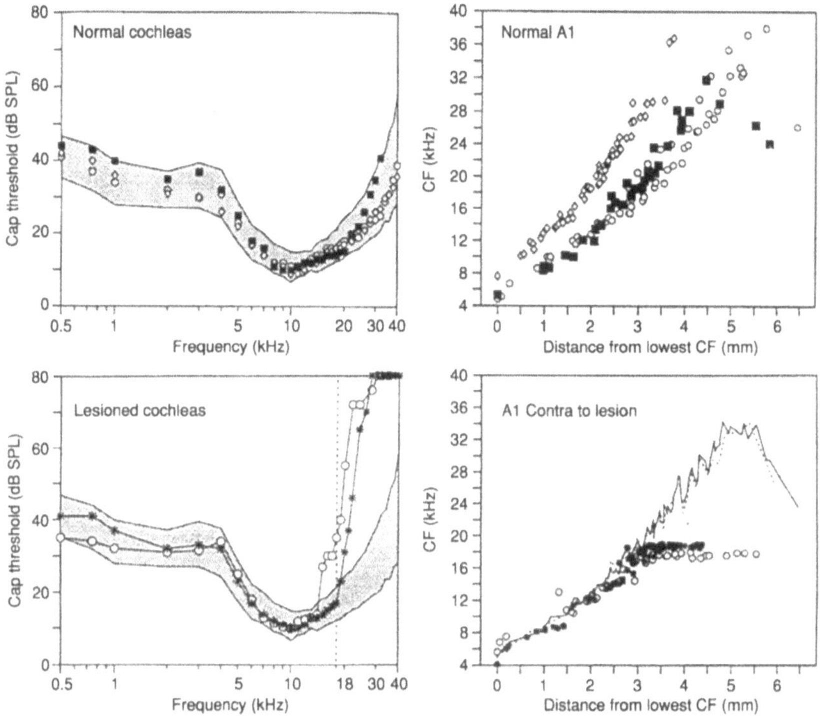

FIG. 3. Cochlear sensitivity (left column) and AI frequency organization (right column) in normal (upper panels) and chronically lesioned (lower panels) cats. In the left panels, compound action potential (CAP) thresholds are shown as a function of frequency (CAP "audiograms") for three normal (top panel) and two chronically lesioned (lower panel) cats. The shaded region in the left panels represent the mean ± 1.96 standard deviations (SD) range of CAP thresholds from a large sample of normal cats. In the lesioned cats, thresholds fall within the normal range up to a frequency in the region of 18 kHz, beyond which they rise steeply, reflecting the high-frequency hearing losses produced by basal cochlear lesions. In the right panels, CFs recorded at AI loci in the these cats (corresponding symbols) are plotted as a function of distance of the recording locus across AI. Distance was measured from the isofrequency contour for the lowest CF recorded in each animal along an axis orthogonal to the predominant orientation of the isofrequency contours. In the lower right panel, the lines are those fitted to the normal CF-distance plots shown in the lower-left panel, using either a three-point smoothing function (dotted line) or a Beiter-spline fit (full line). In the normal animals, CF increases monotonically with distance up to a reversal point marking the border of AI and the anterior auditory field. In the lesioned animals, CF increases with distance in a normal fashion up to the region of the lesion edge frequency (17 to 19 kHz) but then remains near constant across the cortex. The region in which higher frequencies would normally be represented is therefore occupied by an expanded representation of lesion-edge frequencies. Reproduced with permission from Rajan and Irvine (1996).

Changes in frequency organization of the type shown here could also occur as a passive consequence of cochlear lesions, because the responses of cortical neurons at suprathreshold SPLs reflect convergent input derived from regions of the cochlea other than that from which their CF input is derived. For neurons with CF at and above the lesion-edge frequencies, elimination of input over the frequency range affected by the cochlear lesion, therefore, would be expected to leave intact input derived from lower-frequency channels. If the changed frequency organization observed in these studies reflected a passive process of this sort it would be associated with substantially increased thresholds in the LPZ (see Rajan *et al.*,1993, for detailed discussion). In the Robertson and Irvine (1989) and Rajan *et al.* (1993) studies, however, the thresholds of neurons in the LPZ at their new CFs (and their other response characteristics) were normal or near-normal, indicating that the changes were not simply passive consequences of the cochlear lesion but reflected a dynamic process of reorganization.

Analogous reorganization has been described in primary auditory cortex (field AI) of adult animals of a variety of species and as a consequence of different forms of cochlear insult. Schwaber *et al.* (1993) mapped AI in adult macaques before and after inducing a broad high-frequency hearing loss by injection of the ototoxic agents kanamycin and furosemide. When the cortex was remapped 80 to 90 days after the ototoxic treatment, the high-frequency region of AI was found to be occupied by an enlarged representation of lesion-edge frequencies. Reorganization has also been observed in cat AI after neonatal ototoxic injections producing high frequency losses (Harrison *et al.*, 1991, 1992). Although evidence on the thresholds of neurons in the enlarged representations was not presented in these studies, the extent of the reorganization makes it unlikely that it simply reflected passive consequences of the lesions. Lesions produced by noise trauma in juvenile cats (Eggermont and Komiya, 2000) have also been shown to result in such reorganization, with normal or near-normal neuronal thresholds and sharpness of tuning in the reorganized region.

Further evidence for auditory cortical reorganization in adults is provided by data from animals with naturally occurring peripheral hearing losses. Willott *et al.* (1993) described auditory cortical plasticity in C57BL/6J (C57) mice, which show normal auditory system development up to an age of 1 to 2 months, followed by a progressive sensorineural pathology that begins in the basal turn. These animals, therefore, suffer a progressive loss of high-frequency hearing, beginning in young adulthood. The frequency representation in AI of 3- to 12-month-old C57 mice was characterized by an enlarged representation of a middle-frequency range (10 to 13 kHz), which occupied the cortical regions that were devoted to higher frequencies in younger C57 mice and in older mice of other strains that do not exhibit this sensorineural pathology. Thresholds in these regions of expanded representation were reported to be "relatively low," indicating that the responses could not be interpreted as passive consequences of the hearing loss.

In all of these studies, the cortex was mapped some time (minimally 3 to 4 weeks) after the cochlear insult, so the time course of the changes in the frequency selectivity of neurons in the LPZ is not known. However, the time course of plastic changes in neuronal RFs in primary somatosensory cortex after peripheral lesions has been described in detail in experiments utilizing chronic indwelling electrodes. Calford and Tweedale (1988, 1991; see Calford, 2002, for review) reported that digit amputation resulted in an immediate expansion onto peri-lesion skin areas of the RFs of neurons in the LPZ. The rapidity of this expansion could be explained only in terms of the unmasking (presumably, the release from inhibition) of pre-existing excitatory input. Over a period of weeks after the lesion, the expanded RFs gradually reduced in size, a process which Calford and Tweedale (1988) attributed to a re-establishment of excitatory-inhibitory balance.

Unmasking of previously inhibited excitatory inputs to auditory cortical neurons as a consequence of loud-tone induced hearing losses has been demonstrated in the immediate expansion of the frequency-SPL response areas of some AI neurons and associated changes in CF (Calford *et al.*, 1993; Noreña *et al.*, 2003). Changes in frequency selectivity of this sort must contribute to the reorganization of cortical frequency maps seen after longer survival times, but do not account for the full extent of such reorganization. Snyder and colleagues (Snyder and Sinex, 2002; Snyder *et al.*, 2000) have also reported that restricted spiral ganglion lesions result in the immediate post-lesion appearance of low-threshold excitation at frequencies below (or, in fewer cases, above) the borders of the pre-lesion frequency tuning curve in neurons in the LPZ in the central nucleus of the inferior colliculus (ICC). The relationship of these changes to long-term reorganization is less clear, as evidence of reorganization was seen in only a small proportion of electrode penetrations through ICC in adult cats with mechanical cochlear lesions (Irvine *et al.*, 2003).

There is little information concerning the synaptic mechanisms underlying long-term lesion-induced reorganization. In the visual system, in which restricted retinal lesions produce minimal changes in the thalamus (Gilbert and Wiesel, 1992), the extensive reorganization seen in the cortex reflects changes in the efficacy of long-range horizontal intracortical connections (Calford *et al.*, 2003; Darian-Smith and Gilbert, 1994, 1995; Obata *et al.*, 1999). These changes are associated with increased levels of growth factors and their receptors, transcription factors, and growth- and vesicle-related proteins, in the LPZ (Obata *et al.*, 1999). The possible involvement of intracortical connections in auditory reorganization is indicated by the Kaur *et al.* (2004) intracellular data, which show that the RFs of AI neurons defined in terms of subthreshold excitatory postsynaptic potentials (EPSPs) are much broader than those defined in terms of spike responses, and that this breadth of tuning reflects input over intracortical pathways. In the auditory system, however, cochlear lesions in adult cats result in reorganization in the medial geniculate body (MGB) that is indistinguishable from that in AI (Kamke *et al.*, 2003). If the reorganization in cortex partially or wholly reflects that in MGB, it would indicate that reorganization in AI involves mechanisms other than changes in intracortical circuitry. However, it could be that the cortical changes are primary and are transferred to the thalamus via corticofugal projections. The possible importance of corticofugal projections in subcortical plasticity is indicated by the demonstration of Suga and colleagues that conditioning-produced changes in the frequency organization of ICC in the big brown bat depend on corticofugal projections from AI (e.g., Gao and Suga, 1998; see Suga *et al.*, 2000, and Suga and Ma, 2003, for reviews).

Whereas the changes in the frequency selectivity of auditory cortical neurons that are seen in behavioral conditioning and attention experiments are clearly adaptive, in that the changes result in a stronger response to, or better discrimination of, the frequency of behavioral significance (the CS^+ or target frequency), it is not clear that the injury-induced reorganization seen after cochlear lesions is in any sense adaptive. Although it is tempting to think of the reorganization as a central compensation for the peripheral loss, the fact remains that sensitivity to the frequencies that would normally produce activation in the part of the cochlea damaged by the lesion is in no way restored by central reorganization.

Although they do not appear to be compensatory, the marked changes in the pattern of cortical activation produced by peri-lesion frequencies would be expected to have perceptual consequences. There has been no study of such perceptual changes in animals in which injury-induced reorganization has been shown to have occurred, but there have been a number of studies in humans with hearing losses of the type that would be expected to result in auditory cortical reorganization. These studies have focused on frequency discrimination, on the assumption that if perceptual discrimination of two frequencies depends on their producing different spatial distributions of activity in the cortex, then such differences would be likely to be enhanced when the representation of one of the frequencies was enlarged. In accordance with this expectation, McDermott *et al.* (1998) reported that listeners with steep-sloping hearing losses exhibited enhanced frequency difference limens (DLFs) in the region of the cut-off (edge) frequency of their hearing loss, a result confirmed by Thai-Van *et al.* (2002, 2003). Thai-Van *et al.* (2003) also confirmed that the improved DLFs were not attributable to loudness cues, and found that they did not reflect spontaneous otoacoustic emissions at or near the edge frequency. These observations strengthen the possibility that the enhanced DLFs might reflect cortical reorganization, but it must be emphasized that there is no *direct* evidence of such reorganization having occurred in the participants in these studies. As discussed in more detail in the following text, however, Dietrich *et al.* (2001) have presented evidence of an expanded representation of edge frequencies in listeners with similar hearing losses. In another study directed at the possible perceptual consequences of putative reorganization in humans with steep-sloping hearing losses, Buss *et al.* (1998) found no differences in frequency sweep detection, intensity discrimination, or gap detection and discrimination, at the edge frequency in observers with steep-sloping losses, as compared with normal observers in which the effects of such losses were simulated with masking noise.

D. Neural Plasticity Associated with Training on Perceptual Discriminations

Although improvement of perceptual discriminative ability with training or practice is a ubiquitous feature both of everyday life and of psychophysical experiments, the neural mechanisms involved have only recently become a focus of experimental interest. The specificity of many forms of visual perceptual learning to the region of the receptor surface to which the training stimuli are presented, or to particular parameters of the stimuli, has prompted the proposal that the learning might reflect plastic changes at early stages of cortical processing (see Karni and Bertini, 1997, for review). However, direct evidence from studies of the response characteristics of neurons in primary visual cortex of monkeys exhibiting perceptual learning on various discrimination tasks have shown either no changes or relatively subtle changes in RFs and no changes in cortical topography (Crist *et al.*, 2001; Ghose *et al.*, 2002; Schoups *et al.*, 2001). In the auditory system, however, there is conflicting evidence. Recanzone *et al.* (1993) reported that in owl monkeys that were trained, and showed perceptual learning, on a frequency discrimination task the area of representation in AI of the frequencies used in training was enlarged (by a factor of $\sim$7 to $\sim$9, depending on frequency) relative to untrained (control) monkeys, and that larger areas of representation were associated with superior discrimination performance. The frequency tuning of multi-neuron clusters in these enlarged areas of representation was sharper, and response latency was longer, relative to those of neurons in untrained animals in the areas in which these frequencies were represented. In contrast to these results, Brown *et al.* (2004) found no change in the frequency organization of AI in cats that were trained on a somewhat different frequency discrimination task. The cats all showed perceptual learning and reached asymptotic performance levels, but the only change in neural response characteristics was a tendency for neurons with CF immediately above the training frequency to have slightly broader tuning in the trained cats than in controls. This pattern of results is similar to those reported in the studies of visual perceptual learning referred to in the previous text, but the reasons for the differences between the Recanzone *et al.* (1993) and Brown *et al.* (2004) results remain unclear (see Brown *et al.*, 2004, for discussion).

In both of the studies previously described, auditory cortex was mapped in detail in acute experiments at the completion of training, and the results from trained animals were compared with those from untrained controls. Blake *et al.* (2002) presented data from a study of frequency discrimination training in two owl monkeys in which chronic recordings were made from neurons at the same cortical loci throughout the course of training. Although these experiments were directed at elucidating the neural correlates of reinforcement rather than those of perceptual learning *per se*, they shed light on the latter issue. In both animals, the summed responses of all neurons sampled by the electrode array decreased during training, but the decrease was non-monotonic: at the time at which the monkey's performance began to improve, responses to both the target frequency and control frequencies transiently increased, with a greater increase in the response to the target frequency. The method of data summation does not allow it to be determined whether this increase reflected a change in the frequency selectivity of some or all of the neurons, or simply a greater increase in responsiveness of those neurons tuned to the target frequency. Response strength declined over further training trials, while behavioral discrimination continued to improve, suggesting that the period of increased responsiveness was not the substrate of the improved discrimination but was a correlate of the behavior coming under the control of the reinforcement contingency.

A similar decrease in AI responses as a consequence of discrimination training was reported by Beitel *et al.* (2003) in a study in which owl monkeys were trained to detect an increase in the envelope frequency of a sinusoidally amplitude modulated (SAM) 1-kHz tone. When the responses of AI neurons with CFs within a 1-octave band centered at 1 kHz were recorded in acute experiments at the end of training, the responses to the SAM stimuli were substantially weaker than responses in AI of untrained monkeys. However, the reduction was smallest for the highest envelope frequencies, so that the relative amplitude in the trained animals was greatest at the high SAM frequencies that were most easy to discriminate and that, therefore, were most often associated with reward.

The studies reviewed here have used a range of different tasks and procedures. The two studies in which spectral processing mechanisms were directly examined have produced conflicting results with respect to changes that are associated with perceptual learning on frequency discrimination tasks. Recanzone *et al.* (1993) reported an increased representation of the training frequency and increased sharpness of tuning in that region, whereas Brown *et al.* (2004) reported no change in the area of representation and data suggestive of decreased sharpness of tuning at frequencies above the training frequency. The studies of Blake *et al.* (2002) and Beitel *et al.* (2003) have reported changes in response strength when behavior on spectral discrimination tasks came under the control of the reinforcer, but in neither case is it clear if the changes in responses strength reflected changes in spectral sensitivity. Evidence for plasticity of spectral processing mechanisms from perceptual learning studies in animals, therefore, remains equivocal.

E. Neural Plasticity Associated with Activation of Neuromodulatory Systems

Both developmental and adult plasticity in sensory cortices have been related to the influence of a number of neuromodulatory systems (e.g., Dykes, 1990; Rasmusson, 2000; Gu, 2002). The neocortex receives diffuse extrathalamic projections from four different subcortical cell groups: cholinergic neurons in the basal forebrain (BF), noradrenergic neurons in the locus coeruleus (LC), serotonergic neurons in the raphe nuclei, and dopaminergic neurons in the ventral tegmental area (VTA). The role of the BF cholinergic system (BFCS) in the modulation of auditory cortical responses and frequency selectivity has been studied in much greater detail than that of the other systems, and these data will be briefly reviewed here with emphasis on the effects relevant to spectral processing. More detailed reviews can be found in Edeline (1999, 2003) and Weinberger (2003).

Pairing a tonal stimulus of a particular frequency with direct application of either cholinergic agents (e.g., McKenna *et al.*, 1989; Metherate and Weinberger, 1990) or of acetylcholinesterase inhibitors (Ashe *et al.*, 1989) alters the FRFs of a large proportion of auditory cortical neurons by selectively modifying the response to tones at or near the paired frequency. These observations imply that endogenous acetylcholine (ACh) has a role in modulating the frequency selectivity of auditory cortical neurons. This interpretation is supported by data obtained under more physiological conditions by pairing electrical stimulation of the BF with tonal stimulation. Such pairing has been shown to result in frequency-selective changes in FRFs in a number of species (e.g., rat: Edeline *et al.*, 1994; Metherate and Ashe, 1991; guinea pig: Bakin and Weinberger; 1996; Dimyan and Weinberger, 1999; bat: Ma and Suga, 2003) and in an enlarged representation of the paired frequency in AI of the rat (Kilgard and Merzenich, 1998). Although such stimulation also activates non-cholinergic BF neurons (notably the large population of GABAergic neurons that also project to the cortex), the fact that these effects were in a number of cases shown to be blocked by application of ACh antagonists to the cortex indicates that they were mediated by the BFCS.

Although they have been studied far less extensively, the available evidence indicates that other neuromodulators can have similar effects on auditory cortical frequency selectivity and organization. Thus, Bao *et al.* (2001) reported that pairing pulsed tones at a particular frequency with electrical microstimulation of the VTA in rats resulted in an enlargement of the representation of that frequency in AI. It also resulted in the emergence of an area characterized by strong responses to that frequency in a region adjacent to AI in which neurons in naive animals responded poorly and in a non-frequency selective fashion to tones. Manunta and Edeline (2004) found that pairing tonal stimulation with brief iontophoretic pulses of noradrenaline to auditory cortical neurons in the rat resulted in frequency-specific changes in their FRFs.

The effects of activation of these neuromodulatory systems, or of the direct application of their transmitters, on the frequency selectivity of auditory cortical neurons and the frequency organization of auditory cortical fields suggest that they are likely to be involved in the range of injury- and use-related plasticity in auditory cortex described in previous sections. Although there is evidence for the role of the cholinergic system in various forms of auditory learning in both animals (e.g., Kudoh *et al.*, 2004) and humans (e.g., Thiel *et al.*, 2002), there is only limited *direct* evidence for the involvement of this or the other neuromodulatory systems in injury- and use-related changes in frequency selectivity and organization. Ji *et al.* (2001) reported that the changes in the frequency selectivity of auditory cortical neurons produced by behavioral conditioning in the big brown bat were eliminated by application of atropine (which blocks muscarinic ACh receptors) to the cortex. However, Kamke *et al.* (2004) found that immunotoxic lesions of the BFCS did not prevent cochlear lesion-induced reorganization in AI of adult cats. The latter finding is apparently at variance with evidence from the somatosensory system that injury-induced reorganization is blocked by lesions of the BF (Webster *et al.*, 1991; Juliano *et al.*, 1991). That effect cannot unequivocally be attributed to destruction of the cholinergic system, however, as the neurotoxic lesions used in those studies would also have destroyed non-cholinergic neurons in the BF.

F. Imaging Evidence of Injury- and Use-Related Plasticity in Human Spectral Processing

All of the evidence for physiological plasticity of spectral processing mechanisms reviewed in the preceding sections has been derived from studies in non-human animals. The advent of a range of techniques for imaging cortical activity in awake humans has generated a small but rapidly growing body of data on plasticity of spectral processing mechanisms in human auditory cortex. In what appears to be the only study of frequency-specific changes in human auditory cortex associated with behavioral conditioning, Morris *et al.* (1998) used positron emission topography (PET) to investigate the effects of discriminatory classical conditioning using a noxious intense noise as the unconditioned stimulus (UCS) and tones of different frequencies as CS^+ and CS^-. They reported that behavioral conditioning of the skin conductance response was associated with a decrease in the response to the CS^+ frequency. This result would appear to be in accord with the Ohl and Scheich (1997) report of a decreased response to the CS^+, although it might also reflect the limited spatial resolution of PET or the use of an unusual conditioning paradigm using an acoustic UCS (see Morris *et al.*, 1998, for discussion).

There is a larger body of imaging data indicating differences in auditory cortical maps, which might be associated with forms of perceptual learning, in musicians and non-musicians. Pantev *et al.* (1998) presented magnetoencephalographic (MEG) data indicating that the late component (peaking at about 100 ms) of the auditory cortical response to piano tones, but not to pure tones, was larger in musicians than in non-musicians. In a more recent study, Schneider *et al.* (2002) have reported a larger early (19 to 30-ms) component of the MEG response to pure tones in musicians, while confirming the Pantev *et al.* (1998) finding of no difference between the two groups in the late component. They also reported a massive increase in the grey matter volume of Heschl's gyrus in musicians, and strong correlations between response amplitude, grey matter volume, and musical aptitude. Shahin *et al.* (2003) have similarly reported larger late components of the auditory-evoked potential (AEP) to piano tones, violin tones, and pure tones in musicians, although these potentials were localized to secondary auditory cortex.

Larger MEG and AEP responses could reflect either an increase in the number of responsive neurons, which the correlation with grey matter volume in the Schneider *et al.* (2002) study suggests is the case, or increased synchrony in the discharge of the responding population. Whichever is the case, these data indicate differences in the representations of spectral information in the auditory cortex of musicians and non-musicians. However, the correlations reported in these studies could indicate either that musical training had produced plastic changes in auditory cortical responses and morphology or that individuals with these traits are more likely to pursue musical careers. In an attempt to resolve this issue, Pantev *et al.* (2001) subsequently investigated the timbre-specificity of the enhancement of the late component of MEG responses to musical tones. They found that responses to tones of different timbre (violin and trumpet) were preferentially enhanced in musicians trained on the particular instrument, an observation strongly suggesting that the changes in these responses are the result of training.

This conclusion is supported by other studies of the effects of within-experiment training using non-musically trained participants. Pantev *et al.* (1999) reported a decreased 100-ms MEG response to 1 kHz tones in participants who had listened for 3 hours per day for 3 days to music in which a narrow frequency band centered at that frequency had been removed, whereas Menning *et al.* (2000) reported an increased response at 1 kHz in participants trained on a frequency discrimination task at that frequency. Finally, Bosnyak *et al.* (2004) reported that in participants trained on a frequency discrimination task using

2-kHz tones amplitude modulated at 40 Hz, later components of the AEP localized to secondary auditory cortex were enhanced.

Although these studies undoubtedly demonstrate short-term training effects on auditory cortical responses and thus on various features of spectral representation, it is likely that the effects seen in musicians with much longer periods of training, often starting in early childhood, reflect various combinations of genetic predispositions and training-induced plasticity. If the early-component and associated grey-matter changes reported by Schneider *et al.* (2002) were predominantly the effect of training rather than of a genetic predisposition it would indicate a remarkable degree of *structural* plasticity in auditory cortex.

Although a number of perceptual phenomena consequent on restricted receptor damage in adult humans have been attributed to cortical reorganization, the only direct evidence of such reorganization in humans is that provided by imaging studies. Most of this evidence has been derived from studies of the somatosensory cortex of amputees (e.g., Elbert *et al.*, 1994; Weiss *et al.*, 1998; Yang *et al.*, 1994). There appears to have been only one study directed to the effects on the frequency organization of the auditory cortex in humans of peripheral hearing losses of the type studied in animal experiments. Dietrich *et al.* (2001) used MEG to examine the organization of auditory cortex in adult subjects with a steeply-sloping high-frequency hearing loss (analogous to those produced by mechanical cochlear lesions in animals) of at least 6-months standing. They interpreted their results in terms of an expansion of the cortical representation of the lesion edge frequency of the type seen in the animal studies. Further evidence for injury-related reorganization in human auditory cortex is provided by the Mühlnickel *et al.* (1998) study of a group of subjects with tonal tinnitus. They reported a change in the representation of the "tinnitus frequency," described as either a shift or an expansion of the representation into adjacent cortical areas, in a subgroup of their subjects. However, this reorganization differed in a number of ways from that described in the animal studies. The maximum hearing loss at the tinnitus frequency in the Mühlnickel *et al.* subjects was 25 dB (mean 13.8 dB), losses much smaller than those studied in animals and of a magnitude that does not result in reorganization of the frequency map in cats (Rajan and Irvine, 1996; Rajan, 1998). Furthermore, the reorganization described in the tinnitus subjects was a shift or expansion of the representation of the loss-frequency itself into adjacent cortical areas rather than an expansion of the representation of adjacent peri-lesion frequencies into the area normally occupied by the representation of the loss frequency or frequencies.

For obvious reasons, the imaging evidence for plasticity in human auditory cortex is more limited than that from animal studies, and numerous issues remain to be resolved. Nevertheless this evidence strongly suggests that the function of spectral processing mechanisms in the human cortex is modified by experience and by altered input in a manner analogous to the changes produced by such interventions in animals.

III. Psychophysical Studies of Plasticity in Spectral Processing

A. Overview

Psychophysical studies of plasticity in spectral processing have examined the effects of practice on a variety of spectral discrimination and detection tasks. The following brief review of these reports will focus on data obtained from humans trained over multiple daily sessions (for shorter training periods, see e.g., Hawkey *et al.*, 2004) in purely behavioral investigations. The evidence for spectral-processing plasticity derives from demonstrations of training-induced improvements on: (1) frequency discrimination with tonal and more complex stimuli, and (2) the detection of tones in the quiet and in the presence of various background sounds.

B. Frequency Discrimination

1. *Pure Tones*

The largest body of purely psychophysical evidence of plasticity in human spectral processing comes from reports of training-induced improvements on pure-tone frequency discrimination. In a typical frequency-discrimination task, the listener is presented with two tones sequentially on a trial. One tone, called the standard, has the same frequency on every trial. The other tone, called the signal, has a different frequency than the standard. The standard and signal tones are presented in random order, and the listener is required to indicate which of the two tones was the signal. The frequency of the signal tone is adjusted over trials to establish the frequency difference between the standard and signal tones that the listener can detect on a given percentage of trials. This frequency difference is referred to as the frequency-discrimination threshold. Reductions in this threshold with practice indicate that some aspect of the representation of the spectra of the tones has changed. Such improvements have been noted in a number of reports. In some cases, these improvements are merely mentioned, with statements that naive listeners either improved with experience or had higher thresholds than listeners with extensive training (Harris, 1952; Moore, 1973). In a few others, however, learning on frequency discrimination has been the topic of interest. From these experiments arises a preliminary picture of the malleability of this form of spectral processing in humans.

The majority of investigations of frequency-discrimination learning have focused on the generalization of learning from trained to untrained standard stimuli. Generalization refers to the improvement in performance with an untrained standard that results from learning with a trained standard. When generalization does not occur, learning is said to be specific to the particular characteristic that differed between the trained and untrained standards. In the present context, the interest in patterns of generalization comes from the realization that they provide a window into the neural processing that is affected by training (Karni and Bertini, 1997; Wright, 2001). The idea is that the neural circuitry that is modified by training with one particular standard stimulus likely also responds to some, but not all, other stimuli. Therefore, the ability to perform the trained task, such as frequency discrimination, with an untrained standard should be improved only if that standard also engages the modified neuronal circuitry. Thus, the untrained standards to which learning generalizes indicate the response selectivity of the neuronal circuitry modified by training. Three tentative conclusions about the plasticity affecting frequency-discrimination performance can be drawn from the results of these experiments.

First, there seems to be systematically less generalization to untrained standard frequencies as the amount of practice increases, suggesting that, through the course of training, improvements in frequency discrimination may be mediated by different neural circuits with progressively greater frequency selectivity. Demany and Semal (2002) first noted the inverse relationship between the breadth of frequency generalization and training amount in a comparison of their data to those of Demany (1985). The pattern holds across other frequency-discrimination experiments as well. The clearest generalization to untrained frequencies has occurred with fewer than ~3,000 trials of training: learning generalized from 0.36 and 2.5 kHz (but not 6 kHz) to 0.2 kHz with 700 training trials (Demany, 1985), and from 1 kHz to 1.1 and 2.0 kHz with ~2,800 training trials (Roth *et al.*, 2003). Following ~4,000 to 5,000 training trials, however, learning generalized only partially from 5 to 8 kHz, or the reverse (Irvine *et al.*, 2000). Finally, with ~11,000 trials of training, learning at 3 kHz did not generalize to either 1.2 or 6.5 kHz (Demany and Semal, 2002). It should be noted that much of this work was inspired by the proposal of Demany (1985) that learning on frequency discrimination may be specific to the type of processing—place of excitation along the basilar membrane for high frequencies and temporal patterns of neural firing for low frequencies—rather than to the stimulus frequency itself.

Second, learning on frequency discrimination appears to generalize only partially to some untrained stimulus durations, providing an indication that improvements on this task may result from modification of neural circuits that are specialized for encoding stimuli of particular durations. Delhommeau *et al.* (2002) trained listeners for ~4,500 trials on frequency discrimination with a

200-ms, 1-kHz standard and then tested for generalization to standards with the trained frequency (1 kHz), but untrained durations (100 and 40 ms). The magnitude of learning was similar for the trained 200-ms and the untrained 100-ms standards, and the final thresholds with those standards matched thresholds previously reported for highly trained listeners. In contrast, for the 40-ms standard, the amount of improvement was smaller than with the other stimulus durations, and the final thresholds were considerably higher than those of highly trained listeners. These results led the authors to conclude that learning generalized well from the 200-ms to the 100-ms stimulus, but only partially to the 40-ms stimulus. They offered two possible explanations for this specificity to stimulus duration. One is that different mechanisms could encode stimulus frequency for long tones (viz., the temporal pattern of peripheral neural responses) and short tones (viz., the place of excitation along the basilar membrane) and learning could affect only the mechanism that processes long tones. The other is that a single mechanism could encode tones of all durations, but could do so through neurons that are tuned to specific durations and that are affected separately by learning.

Third, there appears to be cross-ear generalization of improvements on frequency discrimination, suggesting that this learning arises from changes in neural circuits that receive input from both ears. Frequency-discrimination learning in the untrained ear has been observed at trained standard frequencies of 1 and 3 kHz, and following from ~2,800 to ~11,000 training trials (Delhommeau *et al.*, 2002; Demany and Semal, 2002; Roth *et al.*, 2003). Improvements at untrained frequencies to which learning generalized in the trained ear also have been claimed to generalize to the untrained ear (Roth *et al.*, 2003). There has been some question as to whether a small ear-specific component associated with the trained stimulus duration may exist (Delhommeau *et al.*, 2002); if it does, it is clearly less than the specificity seen for stimulus frequency with the same number of training trials (Demany and Semal, 2002). A curious result that seems worthy of further investigation is the observation by Delhommeau *et al.* (2002) that the frequency-discrimination thresholds measured in each ear of naive listeners depended on which ear was tested first. In an initial testing session designed to establish baseline performance with three stimulus durations in each ear, performance differences between the ears were larger in listeners who were tested in the right ear before the left than in listeners tested with the opposite order.

Two additional preliminary conclusions about the plasticity affecting frequency-discrimination performance come from examinations of learning on the trained standard itself. One is that this plasticity occurs over a longer time course at low (<4 kHz) than at high standard frequencies. This pattern was first observed by Olsho *et al.* (1988) who trained every listener at 0.5, 1, and 4 kHz, varying the training order randomly among the three frequencies within each session. They reported a systematic decrease in the training time needed for asymptotic performance as the standard frequency increased. The same relationship between training time and standard frequency also is apparent across other experiments. The number of training trials needed for asymptotic performance has been estimated at ~3,000 to 5,000 for 1, 2, and 3 kHz tones (Campbell and Small, 1963; Delhommeau *et al.*, 2002; Demany and Semal, 2002), but at only ~2,000 for 5 and 8 kHz tones (Irvine *et al.*, 2000).

The final tentative conclusion is that learning on frequency discrimination may not reflect an improvement in the ultimate ability to detect changes in frequency, but rather an increase in the ability to perform well consistently. Olsho *et al.* (1988) evaluated frequency-discrimination learning with an unusual analysis. Instead of examining only the standard threshold estimates obtained from the adaptive track within each block of trials, as is customary, they also examined the average of the five lowest reversal values from each track. Early in training at 0.5 and 1 kHz (the only frequencies at which there was learning), the standard threshold estimates were clearly above those of highly trained listeners, but the average of the lowest reversal values were not. Thus, the lowest reversal values at the beginning of training could be used to predict performance at the end of training. This result suggests that the best performance on frequency discrimination does not improve. Rather, the ability to consistently achieve this best performance does.

2. *Complex Stimuli*

In addition to the demonstrations of learning on frequency discrimination with pure tones presented in isolation are similar examples obtained with more complex stimuli. Among these are several reports that the ability to discriminate the fundamental frequency (F_0) of harmonic complex tones improves with practice. Differences in the F_0s of these complexes come from changes in the frequencies of the individual harmonics. From the currently available data it appears that there is substantially more learning on F_0 discrimination when the individual harmonics in a complex pass through separate peripheral channels (resolved harmonics) than when several harmonics pass through the same channel (unresolved harmonics) (Grimault *et al.*, 2002). It also appears that the improvements on F_0 discrimination with resolved harmonics occur over a longer time course than those for pure-tone frequency discrimination (Demany and Semal, 2002; Grimault *et al.*, 2002). However, the changes underlying the learning on F_0 discrimination are a matter of debate, owing to contradictory conclusions as to the patterns of generalization among resolved complexes, unresolved complexes, and pure tones. Grimault and colleagues concluded that learning with harmonic complexes generalized more readily to untrained complexes with the same resolvability status as those used in training regardless of F_0 (Grimault *et al.*, 2002), and that learning on pure tones generalized more readily to resolved than unresolved harmonic complexes (Grimault *et al.*, 2003).

Therefore, they proposed that learning with harmonic complexes represents an improvement in the encoding of the individual components of the complex, a process that can occur only if the individual components can be resolved. In contrast, Demany and Semal (2002) concluded that learning did not generalize from resolved complexes to pure tones. Given this result, they suggested that learning with both harmonic complexes and pure tones results from refined processing of the temporal fine structure of the output of the peripheral filters. They attributed the longer time course of learning for resolved harmonics to the greater variation in temporal responses across filters for that waveform type than for unresolved harmonics and pure tones.

Listeners also show substantial training-induced improvements in their ability to discriminate the frequencies of individual tones presented in tonal sequences. The longer time course of this learning compared to that for tones in isolation suggests that additional factors affect this form of spectral-processing plasticity. In particular, the increasing familiarity with the tonal sequence used in training, and accompanying improvement in the ability to focus attention selectively on the target tone, appear to contribute substantially to learning in these circumstances. Watson and colleagues (Spiegel and Watson, 1981; Watson, 1976, 1980; Watson *et al.*, 1976) reported dramatic training-induced improvements in pure-tone frequency discrimination thresholds for brief tones presented in 10-tone sequences. This learning occurred over a longer time course than that observed for tones in isolation, especially when the sequence varied from trial to trial. For example, when the same sequence was used throughout training, the average frequency-discrimination threshold for a 40-ms, 920-Hz component in that sequence decreased from 165 Hz to 9 Hz (Watson *et al.*, 1976). That final value is near to the thresholds of listeners trained on brief tones in isolation (Watson, 1980). However, the listeners trained with the tonal sequence reached asymptotic performance only after 10 to 14 hours of training (Watson *et al.*, 1976), at least twice the training required for asymptotic performance with isolated tones (Watson, 1980). Learning occurred even more slowly when the tonal sequence varied across trials. For instance, when the standard sequence on each trial was randomly selected from eight possibilities, but the temporal position of the target tone was fixed for each particular sequence, improvements continued over ~25 testing sessions (Spiegel and Watson, 1981). In that case, performance at the end of training approached that obtained with training on only a single sequence. Adding uncertainty about the temporal position of the target tone within each of those eight sequences resulted in frequency-discrimination thresholds that were still substantially higher than in the single-sequence condition, even after 50 practice sessions (Spiegel and Watson, 1981). Thus, frequency-discrimination performance for tones in sequences improved markedly with practice, but that learning occurred much more slowly than for isolated tones.

There is some hint that the improvements in frequency discrimination observed for tones in sequences, similar to those for tones in isolation, are frequency specific. Although not a focus of study in this line of experiments, Watson (1976) mentioned in passing that listeners trained on a mid-frequency tone, apparently in a single sequence, were subsequently better at frequency discrimination with mid-frequency tones in conditions in which several sequences and several temporal positions of the target were presented. At the same time, performance was somewhat degraded for low-frequency tones and unaffected for high-frequency ones. Thus, the learning primarily influenced the perception of tones within the restricted frequency range emphasized during training.

It appears that the improvements in performance with these tonal sequences result in large part from the increasing ability to recognize each sequence, and the presumably accompanying growth in the capacity to focus attention selectively on the target tone. Indicating that familiarity with the sequences contributes to these improvements, listeners who were sequentially trained on frequency discrimination with eight different sequences, and allowed to reach asymptotic performance on each before proceeding to the next, were later able to perform as well when these trained sequences were presented randomly in another set of tests (Spiegel and Watson, 1981). Correspondingly, Watson (1976) briefly mentioned that listeners trained separately on five sequences could subsequently select them from 50 other sequences (see Leek and Watson, 1988, for an investigation of learning on a sequence-identification task). Additional evidence that stimulus familiarity affects performance comes from reports that listeners are better able to discriminate frequency differences in stimuli that are encountered frequently compared to those rarely heard in daily life. For example, frequency-discrimination thresholds tend to be lower for individual components of consonant than dissonant chords or melodies (Acker *et al.*, 1995; Divenyi and Hirsh, 1974; McFadden and Callaway, 1999; Schellenberg and Trehub, 1994).

The idea that improvements in the ability to attend selectively to the target tone within a sequence contribute to learning on frequency discrimination in these conditions initially arose from listeners' descriptions of how their perception of the sequence changed during the course of training. Listeners trained with a single target component in a single sequence reported that the sequence first divided into several separate frequency streams, that the salience of the target stream then improved, and finally that the loudness of the target increased (Spiegel and Watson, 1981). Consistent with this description, frequency discrimination thresholds were unaffected by a 40-dB decrease in the level of the target tone in relation to the other tones within a trained sequence (Spiegel and Watson, 1981). To account for these results, Spiegel and Watson (1981) proposed that listeners first identify a stimulus by its overall structure and only subsequently analyze its details.

3. *Signal Detection*

Psychophysical evidence for plasticity in human spectral processing can also be found in reports of learning on signal-detection tasks. Adding or removing spectral components of a stimulus changes the stimulus spectrum. Thus, demonstrations that the ability to detect such changes improves with practice provide another indication that experience can alter spectral processing. In a typical detection task, the listener is presented with a background, masking stimulus in each of two observation intervals in a trial. In one of the two intervals, at random, an additional sound, the signal, is presented. The listener is required to indicate which of the two intervals contained the signal. The signal level is adjusted across trials to establish the listener's detection threshold, the level at which the listener can detect the signal on a given percentage of trials. Examples of learning on signal-detection tasks range from improvements in the detection of tones in the quiet (absolute threshold) to their detection in complex maskers. The descriptions of each provide insights into different facets of this aspect of spectral-processing plasticity.

Improvements in absolute threshold with practice have been reported for tonal signals at 0.1, 0.5, and 1 kHz (Ward, 1957; Zwislocki, 1955; Zwislocki *et al.*, 1958). This learning typically occurred over three to four testing sessions, had a magnitude of 2.5 to 10 dB, and was most marked in listeners with the highest starting thresholds. Reinforcement reduced initial thresholds and increased learning with a subjective measure of performance (Békésy tracking) (Zwislocki *et al.*, 1958). However, learning also occurred with an objective measure (forced-choice), indicating that the improvements were not the result of a simple shift in response criterion (Zwislocki *et al.*, 1958). The slope of the psychometric function did not change over the course of training (Zwislocki *et al.*, 1958). As with frequency discrimination, there was more learning on absolute threshold at low ($\sim<1$ kHz) than at higher signal frequencies: the improvement was greater at 0.1 than 1 kHz (Zwislocki *et al.*, 1958) and at 0.5 than 4 kHz (Ward, 1957). Further, this learning appeared to be frequency specific. In the cleanest test, learning at 1 kHz did not generalize to 0.1 kHz (Zwislocki *et al.*, 1958). Interpretation of a report of partial generalization in the other direction, from 0.1 kHz to other signal frequencies <1 kHz (Zwislocki, 1955), is complicated both because listeners were exposed to a wide range of stimulus frequencies on each testing session, albeit with the most exposures at 0.1 kHz, and because less learning is expected at higher than at lower stimulus frequencies even with direct training on each frequency. Zwislocki *et al.* (1958) interpreted learning in measures of absolute threshold as reflecting the acquisition of better cues to the detection of tones in physiological noise.

The ability to detect tonal signals presented during a noise also improves with practice, but the apparently substantially shorter learning course for signal detection in continuous than in gated noise indicates that different processes may be involved in these two cases. Improvements in the detection of tonal signals in continuous noise appear to occur only early in training. Gundy (1961) reported improvements in the detection of a 1-kHz tone presented in continuous noise during a single 400-trial training session. The learning rate was accelerated by providing clearly audible examples of the tone to be detected at the beginning of every 100-trial block, but was relatively unaffected by the presence or absence of feedback (Gundy, 1961). Examinations of performance over longer periods of training with continuous noise, sometimes extending over several months, show little to no additional learning (Egan *et al.*, 1961; Neff and Dethlefs, 1995). In contrast, the ability to detect a tonal signal presented at the beginning of a gated noise improves over weeks or even months of practice (Carlyon, 1989; McFadden and Wright, 1992; Tucker *et al.*, 1968; Wright, 1995; Zwicker, 1965). This learning has not been specifically investigated, but large individual differences in the magnitude of improvement have been noted (Carlyon, 1989; McFadden and Wright, 1992). Interestingly, there is also some indication that this learning may be specific to the trained signal, and thus gated-noise, duration. Tucker *et al.* (1968) observed that listeners who were given extensive practice detecting 0.5-kHz tones of short, but unspecified, duration in gated noise were then better at detecting shorter than longer tones in the tested duration range of 5 to 50 ms. Although not mentioned by the authors, the performance of the trained listeners with the longer stimulus durations was similar to that of listeners with less extensive experience, suggesting that the learning with the short stimuli did not generalize to longer untrained stimulus durations.

In addition to the training-induced improvements reported for signal detection in noise, learning has also been observed in the detection of tonal signals presented within complex spectra. Suggesting that this learning arises primarily from strengthening the representation of the standard spectrum in long-term memory (Kidd *et al.*, 1986), performance improvements have been observed more consistently when the standard spectrum was held constant, than when it varied, across presentations. The clearest examples of learning on signal detection in complex spectra come from investigations of the ability to detect changes in the spectral profile of a stimulus, a task referred to as profile analysis (Green, 1983). In these investigations, the change was an increment in the intensity of a single, signal tone gated synchronously with a set of other tones. Because the aim was to measure sensitivity to changes in the spectral shape of the entire stimulus, rather than to only the level of the signal tone itself, the overall stimulus level was selected randomly on every presentation. However, the spectral shape of the standard and the frequency of the signal remained constant. Both Kidd *et al.* (1986) and Drennan and Watson (2001) observed learning on profile-analysis tasks that followed similar time courses. For the majority of listeners, the greatest improvements occurred during the first 500 to 700 trials, but learning continued through 2,000 to 3,000 trials of training. Listeners with the highest starting thresholds learned more slowly, with improvements continuing through 9,000 training trials (Drennan and Watson, 2001). However, in the Drennan and Watson report, the standard spectrum differed from the target one, on which learning was assessed, on trials 2,000 to 8,000 of the 9,000 trials. Therefore, it is not known whether the additional learning resulted from generalization of learning from the other trained standard spectra to the target standard spectrum, or simply from the additional trials on the target spectrum, with or without the intervening rest from training on that spectrum. Finally, while performance in profile-analysis tasks improved with practice, the learning did not reduce the threshold differences observed across different standard spectral profiles (Kidd *et al.*, 1986), nor the large individual differences in performance between listeners (Drennan and Watson, 2001).

In contrast to the reports that most listeners improve on signal-detection tasks when the standard spectrum is held constant, few listeners learn when that spectrum differs on every presentation. Neff and colleagues (Neff and Callaghan, 1988; Neff and Dethlefs, 1995) examined learning on a condition in which the listener's task was to detect the presence of a fixed-frequency tone presented simultaneously with 10 other tones, the frequencies of which were selected randomly on every stimulus presentation. Only one of four listeners improved during the first 600 trials of testing (Neff and Callaghan, 1988), a period when learning is typically most rapid. Further, only about one-quarter of a total of 49 listeners showed any improvement on that condition during the next ~800 trials of testing on it, with those trials distributed over several months during which the listeners also received extensive experience on similar conditions (records of performance on the first ~600 testing trials were unavailable for these listeners; Neff and Dethlefs, 1995). The few listeners who did learn could not be distinguished from those who did not based on starting thresholds (Neff and Dethlefs, 1995). Suggesting that learning on this condition may require extensive and focused training, four of five listeners tested over a period of eight months did improve on it or closely related conditions, with the greatest improvements occurring after a period in which only this condition was tested. Taken together, the more limited learning with unstable than stable standard spectra supports the idea that listeners improve on signal detection in complex spectra by forming a better representation of the standard spectrum in long-term memory. Thus, this conclusion is consistent with the idea that learning on frequency discrimination in tonal sequences is aided by familiarity with the sequence.

Finally, an examination of improvements in the ability to detect individual brief target tones in a 10-tone sequence provides insights into individual differences in learning strategy. Leek and Watson (1984) presented the same sequence throughout training, but randomized the temporal position of the tone to be detected. Listeners' ability to detect each target tone clearly improved over the

~8,000 training trials (~800 trials per target), and that learning generalized to the detection of the same tones presented in a different order. However, none of the listeners improved on all of the temporal positions at the same time, and the pattern of learning differed across listeners. For example, one listener learned slowly over time on a few temporal positions, while another had very high thresholds for individual positions that then suddenly dropped to near asymptotic values, apparently as a consequence of focusing attention on that particular position at the temporary expense of performance on the other positions. Interestingly, two listeners who had particular trouble detecting the tone at one particular position benefited from training with a longer version of the tone at that position as the target on every trial in the otherwise unaltered sequence. Those listeners subsequently were able to detect the tone at that position, at its normal duration, even when it was randomly presented with target tones at other positions. Echoing the description of changes in tone salience with training on frequency discrimination (Spiegel and Watson, 1981), the listeners in this detection experiment reported that the sequence changed from one sound to a series of individually perceived components over the course of training.

IV. Therapeutic Applications of Spectral Processing Plasticity

The plasticity of spectral processing is receiving increasing attention as a basis for the treatment of auditory disorders. There are several reports that various populations known to have poor frequency-discrimination skills improve those abilities with training. For example, both children (Gengel, 1969) and adults (Turner and Nelson, 1982) with hearing impairment showed training-induced learning on pure-tone frequency discrimination. However, even after 25 to 40 hours of practice in some cases, frequency discrimination performance was poorer in regions of hearing loss than in regions of normal hearing (Turner and Nelson, 1982). It is worth noting that the learning of the adults with hearing impairment was frequency specific, because their practice effects looked similar at each of three sequentially trained frequencies. Performance on tasks involving spectral discrimination also improved with practice in individuals with language-based learning disorders, such as dyslexia (Merzenich *et al.*, 1996; Schaffler *et al.*, 2004). Training on these spectral tasks, in combination with other tasks, was associated with improvements on a variety of language skills (Merzenich *et al.*, 1996; Schaffler *et al.*, 2004), but it is not known whether training on spectral tasks alone would yield this benefit.

The cases described so far have arisen from the recognition that populations that have particular difficulty with frequency discrimination may profit from training on that task. The final case is based on the idea that the disorder itself, tinnitus, results from a maladaptive cortical reorganization associated with plasticity. Based on their observation from MEG recordings that the representation of the tinnitus frequency was shifted from its expected location in auditory cortex, and that the degree of shift was correlated with the subjective tinnitus strength, Mühlnickel *et al.* (1998) proposed that tinnitus results from an abnormal alteration of the cortical tonotopic map. The evidence that training on frequency-discrimination in owl monkeys also changed cortical organization by increasing the representation of the trained frequency (Recanzone *et al.*, 1993; but see Brown *et al.*, 2004) then led to the idea that such training might be an effective treatment for tinnitus. The reasoning was that training on frequency discrimination at a frequency other than the tinnitus frequency would increase the representation of the trained frequency by recruiting neurons that had been representing the tinnitus frequency (Flor *et al.*, 2004; Mühlnickel *et al.*, 1998). The prediction that training at a frequency near to the tinnitus frequency should be more effective at reducing tinnitus than training at more remote frequencies was not supported (Flor *et al.*, 2004). However, there was a significant decrease in a daily estimate of tinnitus severity, and in the global score of a tinnitus questionnaire, in individuals with tinnitus who had practiced for more than 26 2-hour sessions in a 4-week period (Flor *et al.*, 2004). Unfortunately, individuals who practiced for fewer sessions actually showed an increase in tinnitus severity with the training. Therefore, while these preliminary results are promising, further investigation is needed to unravel the relationship between frequency discrimination training and tinnitus.

V. Summary and Conclusions

In this chapter we have reviewed what is now a considerable body of evidence for plasticity in various perceptual aspects of human spectral processing and in the characteristics of the neural circuits in which such spectral processing occurs. It must be acknowledged, however, that there remains a substantial division between the human psychophysical and animal electrophysiological evidence for such plasticity. The bulk of the human psychophysical evidence has been derived from studies of perceptual learning on basic frequency discrimination or more complex spectral processing tasks. In contrast, the overwhelming bulk of animal electrophysiological evidence has been derived from studies of behavioral conditioning or of injury-induced plasticity. The number of studies that bridge the gap between these two research domains remains small, and it cannot be assumed with any certainty that the same mechanisms are involved in both. A major need for future research, therefore, is to bring about a convergence between these two areas. The prospects of such convergence are strong, given the increasing sophistication of techniques both for recording neural activity from animals (particularly non-human primates) while they are performing perceptual tasks and for recording precisely localizable activity in the cortex of humans performing such tasks. Progress in understanding the synaptic, cellular, and molecular mechanisms of plasticity will continue to depend on animal models in the foreseeable future. Despite these qualifications on our understanding of spectral processing plasticity, it is clear that many forms of such plasticity are highly adaptive and beneficial, and that some have potential therapeutic applications.

References

Acker, B. E., Pastore, R. E., and Hall, M. D. (1995). Within-category discrimination of musical chords: Perceptual magnet or anchor? *Percept. Psychophys.* **57,** 863–874.

Ashe, J. H., McKenna, T. M., and Weinberger, N. M. (1989). Cholinergic modulation of frequency receptive fields in auditory cortex: II. Frequency-specific effects of anticholinesterases provide evidence for a modulatory action of endogenous ACh. *Synapse* **4,** 44–54.

Bakin, J. S., and Weinberger, N. M. (1990). Classical conditioning induces CS-specific receptive field plasticity in the auditory cortex of the guinea pig. *Brain Res. (Amsterdam)* **536,** 271–286.

Bakin, J. S., and Weinberger, N. M. (1996). Induction of a physiological memory in the cerebral cortex by stimulation of the nucleus basalis. *Proc. Natl. Acad. Sci. USA* **93,** 11219–11224.

Bao, S., Chan, V. T., and Merzenich, M. M. (2001). Cortical remodeling induced by activity of ventral tegmental dopamine neurons. *Nature* **412,** 79–83.

Beitel, R. E., Schreiner, C. E., Cheung, S. W., Wang, X., and Merzenich, M. M. (2003). Reward-dependent plasticity in the primary auditory cortex of adult monkeys trained to discriminate temporally modulated signals. *Proc. Natl. Acad. Sci. USA* **100,** 11070–11075.

Blake, D. T., Strata, F., Churchland, A. K., and Merzenich, M. M. (2002). Neural correlates of instrumental learning in primary auditory cortex. *Proc. Natl. Acad. Sci. USA* **99,** 10114–10119.

Bosnyak, D. J., Eaton, R. A., and Roberts, L. E. (2004). Distributed auditory cortical representations are modified when non-musicians are trained at pitch discrimination with 40 Hz amplitude modulated tones. *Cerebral Cortex* **14,** 1088–1099.

Brown, M., Irvine, D. R. F., and Park, V. N. (2004). Perceptual learning on an auditory frequency discrimination task by cats: Association with changes in primary auditory cortex. *Cerebral Cortex* **14,** 952–965.

Buonomano, D. V., and Merzenich, M. M. (1998). Cortical plasticity: From synapses to maps. *Annl. Rev. Neurosci.* **21,** 149–186.

Buss, E., Hall, J. W., III, Grose, J. H., and Hatch, D. R. (1998). Perceptual consequences of peripheral hearing loss: Do edge effects exist for abrupt cochlear lesions? *Hear. Res.* **125,** 98–108.

Calford, M. B. (2002). Dynamic representational plasticity in sensory cortex. *Neuroscience* **111,** 709–738.

Calford, M. B., Rajan, R., and Irvine, D. R. F. (1993). Rapid changes in the frequency tuning of neurons in cat auditory cortex resulting from pure-tone-induced temporary threshold shift. *Neuroscience* **55,** 953–964.

Calford, M. B., and Tweedale, R. (1988). Immediate and chronic changes in responses of somatosensory cortex in adult flying-fox after digit amputation. *Nature* **332,** 446–448.

Calford, M. B., and Tweedale, R. (1991). Acute changes in cutaneous receptive fields in primary somatosensory cortex after digit denervation in adult flying fox. *J. Neurophysiol.* **65,** 178–187.

Calford, M. B., Wright, L. L., Metha, A. B., and Taglianetti, V. (2003). Topographic plasticity in primary visual cortex is mediated by local corticocortical connections. *J. Neurosci.* **23,** 6434–6442.

Campbell, R. A., and Small, A. M., Jr. (1963). Effect of practice and feedback on frequency discrimination. *J. Acoust. Soc. Am.* **35,** 1511–1514.

Carlyon, R. P. (1989). Changes in the masked thresholds of brief tones produced by prior bursts of noise. *Hear. Res.* **41,** 223–235.

Chowdhury, S. A., and Suga, N. (2000). Reorganization of the frequency map of the auditory cortex evoked by cortical electrical stimulation in the big brown bat. *J. Neurophysiol.* **83,** 1856–1863.

Crist, R. E., Li, W., and Gilbert, C. D. (2001). Learning to see: Experience and attention in primary visual cortex. *Nat. Neurosci.* **4,** 519–525.

Dallos, P., and Harris, D. (1978). Properties of auditory nerve responses in absence of outer hair cells. *J. Neurophysiol.* **41,** 365–383.

Darian-Smith, C., and Gilbert, C. D. (1994). Axonal sprouting accompanies functional reorganization in adult cat striate cortex. *Nature* **368,** 737–740.

Darian-Smith, C., and Gilbert, C. D. (1995). Topographic reorganization in the striate cortex of the adult cat and monkey is cortically mediated. *J. Neurosci.* **15,** 1631–1647.

Delhommeau, K., Micheyl, C., Jouvent, R., and Collet, L. (2002). Transfer of learning across durations and ears in auditory frequency discrimination. *Percept. Psychophys.* **64,** 426–436.

Demany, L. (1985). Perceptual learning in frequency discrimination. *J. Acoust. Soc. Am.* **78,** 1118–1120.

Demany, L., and Semal, C. (2002). Learning to perceive pitch differences. *J. Acoust. Soc. Am.* **111,** 1377–1388.

Divenyi, P. L., and Hirsh, I. J. (1974). Identification of temporal order in three-tone sequences. *J. Acoust. Soc. Am.* **56,** 144–151.

Diamond, D. M., and Weinberger, N. M. (1984). Physiological plasticity of single neurons in auditory cortex of the cat during acquisition of the pupillary conditioned response: II. Secondary field (AII). *Behav. Neurosci.* **98,** 189–210.

Diamond, D. M., and Weinberger, N. M. (1986). Classical conditioning rapidly induces specific changes in frequency receptive fields of single neurons in secondary and ventral ectosylvian auditory cortical fields. *Brain Res. (Amsterdam)* **372,** 357–360.

Dietrich, V., Nieschalk, M., Stoll, W., Rajan, R., and Pantev, C. (2001). Cortical reorganization in patients with high frequency cochlear hearing loss. *Hear. Res.* **158,** 1–7.

Dimyan, M. A., and Weinberger, N. M. (1999). Basal forebrain stimulation induces discriminative receptive field plasticity in the auditory cortex. *Behav. Neurosci.* **113,** 691–702.

Drennan, W. R., and Watson, C. S. (2001). Sources of variation in profile analysis. I. Individual differences and extended training. *J. Acoust. Soc. Am.* **110,** 2491–2497.

Dykes, R. W. (1990). Acetylcholine and neuronal plasticity in somatosensory cortex. *In* "Brain Cholinergic Systems" (M. Steriade and D. Biesold, Eds.), pp. 294–313. Oxford University, Oxford.

Edeline, J.-M. (1999). Learning-induced physiological plasticity in the thalamo-cortical sensory systems: A critical evaluation of receptive field plasticity, map changes and their potential mechanisms. *Prog. Neurobiol.* **57,** 165–224.

Edeline, J.-M. (2003). The thalamo-cortical auditory receptive fields: Regulation by the states of vigilance, learning and the neuromodulatory systems. *Exp. Brain. Res.* **153,** 554–572.

Edeline, J.-M., Hars, B., Maho, C., and Hennevin, E. (1994). Transient and prolonged facilitation of tone-evoked responses induced by basal forebrain stimulations in the rat auditory cortex. *Exp. Brain Res.* **97,** 373–386.

Edeline, J.-M., Pham, P., and Weinberger, N. M. (1993). Rapid development of learning-induced receptive field plasticity in the auditory cortex. *Behav. Neurosci.* **107,** 539–551.

Edeline, J.-M., and Weinberger, N. M. (1993). Receptive field plasticity in the auditory cortex during frequency discrimination training: Selective retuning independent of task difficulty. *Behav. Neurosci.* **107,** 82–103.

Egan, J. P., Greenberg, G. Z., and Schulman, A. I. (1961). Interval of time uncertainty in auditory detection. *J. Acoust. Soc. Am.* **33,** 771–778.

Eggermont, J. J., and Komiya, H. (2000). Moderate noise trauma in juvenile cats results in profound cortical topographic map changes in adulthood. *Hear. Res.* **142,** 89–101.

Elbert, T., Flor, H., Birbaumer, N., Knecht, S., Hampson, S., Larbig, W., and Taub, E. (1994). Extensive reorganization of the somatosensory cortex in adult humans after nervous system injury. *NeuroReport* **5,** 2593–2597.

Flor, H., Hoffmann, D., Struve, M., and Diesch, E. (2004). Auditory discrimination training for the treatment of tinnitus. *Appl. Psychophysiol. Biofeedback* **29,** 113–120.

Fritz, J., Elhilali, M., and Shamma, S. (2005). Task-dependent adaptive plasticity of receptive fields in primary auditory cortex of the ferret. *In* "Auditory Cortex: Towards a Synthesis of Human and Animal Research" (R. Koenig, P. Heil, E. Budinger, and H. Scheich, Eds.), pp. 445–466. Lawrence Erlbaum, Mahwah, New Jersey.

Fritz, J., Shamma, S., Elhilali, M., and Klein, D. (2003). Rapid task-related plasticity of spectrotemporal receptive fields in primary auditory cortex. *Nat. Neurosci.* **6,** 1216–1223.

Galambos, R., Sheatz, G., and Vernier, V. G. (1955). Electrophysiological correlates of a conditioned response in cats. *Science* **123,** 376–377.

Gao, E. Q., and Suga, N. (1998). Experience-dependent corticofugal adjustment of midbrain frequency map in bat auditory system. *Proc. Natl. Acad. Sci. USA* **95,** 12663–12670.

Gao, E. Q., and Suga, N. (2000). Experience-dependent plasticity in the auditory cortex and the inferior colliculus of bats: Role of the corticofugal system. *Proc. Natl. Acad. Sci. USA* **97,** 8081–8086.

Gengel, R. W. (1969). Practice effects in frequency discrimination by hearing impaired children. *J. Speech. Hear. Res.* **12,** 847–856.

Ghose, G. M., Yang, T., and Maunsell, J. H. R. (2002). Physiological correlates of perceptual learning in monkey V1 and V2. *J. Physiol.* **87,** 1867–1888.

Gilbert, C. D. (1998). Adult cortical dynamics. *Physiol. Rev.* **78,** 467–485.

Gilbert, C. D., and Wiesel, T. N. (1992). Receptive field dynamics in adult primary visual cortex. *Nature* **356,** 150–152.

Green, D. M. (1983). Profile analysis. A different view of auditory intensity discrimination. *Am. Psychol.* **38,** 133–142.

Grimault, N., Micheyl, C., Carlyon, R. P., Bacon, S. P., and Collet, L. (2003). Learning in discrimination of frequency or modulation rate: Generalization to fundamental frequency discrimination. *Hear. Res.* **184,** 41–50.

Grimault, N., Micheyl, C., Carlyon, R. P., and Collet, L. (2002). Evidence for two pitch encoding mechanisms using a selective auditory training paradigm. *Percept. Psychophys.* **64,** 189–197.

Gu, Q. (2002). Neuromodulatory transmitter systems in the cortex and their role in cortical plasticity. *Neuroscience* **111,** 815–835.

Gundy, R. F. (1961). Auditory detection of an unspecified signal. *J. Acoust. Soc. Am.* **33,** 1008–1012.

Harris, J. (1952). Pitch discrimination. *J. Acoust. Soc. Am.* **24,** 750–755.

Harrison, R. V., Nagasawa, A., Smith, D. W., Stanton, S., and Mount, R. J. (1991). Reorganization of auditory cortex after neonatal high frequency cochlear hearing loss. *Hear. Res.* **54,** 11–19.

Harrison, R. V., Smith, D. W., Nagasawa, A., Stanton, S., and Mount, R. J. (1992). Developmental plasticity of auditory cortex in cochlear hearing loss: Physiological and psychophysical findings.

In "Auditory Physiology and Perception" (Y. Cazals, K. Horner, and L. Demany, Eds.), pp. 625–633. Pergamon, Oxford.

Hawkey, D. J., Amitay, S., and Moore, D. R. (2004). Early and rapid perceptual learning. *Nat. Neurosci.* **7,** 1055–1056.

Irvine, D. R., Martin, R. L., Klimkeit, E., and Smith, R. (2000). Specificity of perceptual learning in a frequency discrimination task. *J. Acoust. Soc. Am.* **108,** 2964–2968.

Irvine, D. R. F., Rajan, R., and Smith, S. (2003). Effects of restricted cochlear lesions in adult cats on the frequency organization of the inferior colliculus. *J. Comp. Neurol.* **467,** 354–374.

Ji, W. Q., Gao, E. Q., and Suga, N. B. (2001). Effects of acetylcholine and atropine on plasticity of central auditory neurons caused by conditioning in bats. *J. Neurophysiol.* **86,** 211–225.

Juliano, S. L., Ma, W., and Eslin, D. (1991). Cholinergic depletion prevents expansion of topographic maps in somatosensory cortex. *Proc. Natl. Acad. Sci. USA* **88,** 780–784.

Kaas, J. H., and Florence, S. L. (2001). Reorganization of sensory and motor systems in adult mammals after injury. *In* "The Mutable Brain" (J. H. Kaas, Ed.), pp. 165–242. Harwood Academic Publishers.

Kamke, M. R., Brown, M., and Irvine, D. R. F. (2003). Plasticity in the tonotopic organization of the medial geniculate body in adult cats following restricted unilateral cochlear lesions. *J. Comp. Neurol.* **459,** 355–367.

Kamke, M. R., Brown, M., and Irvine, D. R. F. (2004). Effects of cholinergic basal forebrain lesions on cochlear-lesion-induced reorganisation in the primary auditory cortex of adult cats. *Proc. Aust. Neurosci. Soc.* **15,** 92.

Karni, A., and Bertini, G. (1997). Learning perceptual skills: Behavioral probes into adult cortical plasticity. *Curr. Opin. Neurobiol.* **7,** 530–535.

Kaur, S., Lazar, R., and Metherate, R. (2004). Intracortical pathways determine breadth of subthreshold frequency receptive fields in primary auditory cortex. *J. Neurophysiol.* **91,** 2551–2567.

Kidd, G., Jr., Mason, C. R., and Green, D. M. (1986). Auditory profile analysis of irregular sound spectra. *J. Acoust. Soc. Am.* **79,** 1045–1053.

Kilgard, M. P., and Merzenich, M. M. (1998). Cortical map reorganization enabled by nucleus basalis activity. *Science* **279,** 1714–1718.

Kraus, N., and Disterhoft, J. F. (1982). Response plasticity of single neurons in rabbit auditory association cortex during tone-signaled learning. *Brain Res. (Amsterdam)* **246,** 205–215.

Kudoh, M., Seki, K., and Shibuki, K. (2004). Sound-sequence discrimination learning is dependent on cholinergic inputs to the rat auditory cortex. *Neurosci. Res.* **50,** 113–123.

Leek, M. R., and Watson, C. S. (1984). Learning to detect auditory pattern components. *J. Acoust. Soc. Am.* **76,** 1037–1044.

Leek, M. R., and Watson, C. S. (1988). Auditory perceptual learning of tonal patterns. *Percept. Psychophys.* **43,** 389–394.

Ma, X., and Suga, N. (2003). Augmentation of plasticity of the central auditory system by the basal forebrain and/or somatosensory cortex. *J. Neurophysiol.* **89,** 90–103.

Ma, X. F., and Suga, N. (2001). Plasticity of bat's central auditory system evoked by focal electric stimulation of auditory and/or somatosensory cortices. *J. Neurophysiol.* **85,** 1078–1087.

Manunta, Y., and Edeline, J.-M. (2004). Noradrenergic induction of selective plasticity in the frequency tuning of auditory cortex neurons. *J. Neurophysiol.* **92,** 1445–1463.

McDermott, H. J., Lech, M., Kornblum, M. S., and Irvine, D. R. F. (1998). Loudness perception and frequency discrimination in subjects with steeply sloping hearing loss: Possible correlates of neural plasticity. *J. Acoust. Soc. Am.* **104,** 2314–2325.

McFadden, D., and Callaway, N. L. (1999). Better discrimination of small changes in commonly encountered than in less commonly encountered auditory stimuli. *J. Exp. Psychol. Hum. Percept. Perform.* **25,** 543–560.

McFadden, D., and Wright, B. A. (1992). Temporal decline of masking and comodulation masking release. *J. Acoust. Soc. Am.* **92,** 144–156.

McKenna, T. M., Ashe, J. H., and Weinberger, N. M. (1989). Cholinergic modulation of frequency receptive fields in auditory cortex: I. Frequency-specific effects of muscarinic agonists. *Synapse* **4,** 30–43.

Menning, H., Roberts, L. E., and Pantev, C. (2000). Plastic changes in the auditory cortex induced by intensive frequency discrimination training. *NeuroReport* **11,** 817–822.

Merzenich, M. M., Jenkins, W. M., Johnston, P., Schreiner, C., Miller, S. L., and Tallal, P. (1996). Temporal processing deficits of language-learning impaired children ameliorated by training. *Science* **271,** 77–81.

Metherate, R., and Ashe, J. H. (1991). Basal forebrain stimulation modifies auditory cortex responsiveness by an action at muscarinic receptors. *Brain Res.* **559,** 163–167.

Metherate, R., and Weinberger, N. M. (1990). Cholinergic modulation of responses to single tones produces tone-specific receptive field alterations in cat auditory cortex. *Synapse* **6,** 133–145.

Moore, B. C. (1973). Frequency difference limens for short-duration tones. *J. Acoust. Soc. Am.* **54,** 610–619.

Morris, J. S., Friston, K. J., and Dolan, R. J. (1998). Experience-dependent modulation of tonotopic neural responses in human auditory cortex. *Proc. R. Soc. London, Series B* **265,** 649–657.

Mühlnickel, W., Elbert, T., Taub, E., and Flor, H. (1998). Reorganization of auditory cortex in tinnitus. *Proc. Natl. Acad. Sci. USA* **95,** 10340–10343.

Neff, D. L., and Callaghan, B. P. (1988). Effective properties of multicomponent simultaneous maskers under conditions of uncertainty. *J. Acoust. Soc. Am.* **83,** 1833–1838.

Neff, D. L., and Dethlefs, T. M. (1995). Individual differences in simultaneous masking with random-frequency, multicomponent maskers. *J. Acoust. Soc. Am.* **98,** 125–134.

Noreña, A., Tomita, M., and Eggermont, J. J. (2003). Neural changes in cat auditory cortex after a transient pure-tone trauma. *J. Neurophysiol.* **90,** 2387–2401.

Obata, S., Obata, J., Das, A., and Gilbert, C. D. (1999). Molecular correlates of topographic reorganization in primary visual cortex following retinal lesions. *Cerebral Cortex* **9,** 238–248.

Ohl, F. W., and Scheich, H. (1996). Differential frequency conditioning enhances spectral contrast sensitivity of units in auditory cortex (field AI) of the alert Mongolian gerbil. *Eur. J. Neurosci.* **8,** 1001–1017.

Ohl, F. W., and Scheich, H. (1997). Learning-induced dynamic receptive field changes in primary auditory cortex of the unanaesthetized Mongolian gerbil. *J. Comp. Physiol. A Sensory. Neural. Behav. Physiol.* **181,** 685–696.

Olsho, L. W., Koch, E. G., and Carter, E. A. (1988). Nonsensory factors in infant frequency discrimination. *Infant. Behav. Develop.* **11,** 205–222.

Pantev, C., Oostenveld, R., Engelien, A., Ross, B., Roberts, L. E., and Hoke, M. (1998). Increased auditory cortical representation in musicians. *Nature* **392,** 811–814.

Pantev, C., Roberts, L. E., Schulz, M., Engelien, A., and Ross, B. (2001). Timbre-specific enhancement of auditory cortical representations in musicians. *NeuroReport* **12,** 169–174.

Pantev, C., Wollbrink, A., Roberts, L. E., Engelien, A., and Luetkenhoener, B. (1999). Short-term plasticity of the human auditory cortex. *Brain Res.* **842,** 192–199.

Rajan, R. (1998). Receptor organ damage causes loss of cortical surround inhibition without topographic map plasticity. *Nat. Neurosci.* **1,** 138–143.

Rajan, R., and Irvine, D. R. F. (1996). Features of and boundary conditions for lesion-induced reorganization of adult auditory cortical maps. *In* "Auditory System Plasticity and Regeneration" (R. J. Salvi, D. Henderson, F. Fiorino, and V. Colletti, Eds.), pp. 224–237. Thieme Medical, New York.

Rajan, R., Irvine, D. R. F., Wise, L. Z., and Heil, P. (1993). Effect of unilateral partial cochlear lesions in adult cats on the representation of lesioned and unlesioned cochleas in primary auditory cortex. *J. Comp. Neurol.* **338,** 17–49.

Rasmusson, D. D. (2000). The role of acetylcholine in cortical synaptic plasticity. *Behav. Brain Res.* **115,** 205–218.

Recanzone, G. H., Schreiner, C. E., and Merzenich, M. M. (1993). Plasticity in the frequency representation of primary auditory cortex following discrimination training in adult owl monkeys. *J. Neurosci.* **13,** 87–103.

Robertson, D., and Irvine, D. R. F. (1989). Plasticity of frequency organization in auditory cortex of guinea pigs with partial unilateral deafness. *J. Comp. Neurol.* **282,** 456–471.

Roth, D. A.-E., Amir, O., Alaluf, L., Buchsenspanner, S., and Kishon-Rabin, L. (2003). The effect of training on frequency discrimination: Generalization to untrained frequencies and to the untrained ear. *J. Basic Clin. Physiol. Pharmaco.* **14,** 137–150.

Sakai, M., and Suga, N. (2002). Centripetal and centrifugal reorganizations of frequency map of auditory cortex in gerbils. *Proc. Natl. Acad. Sci. USA* **99,** 7108–7112.

Schaffler, T., Sonntag, J., Hartnegg, K., and Fischer, B. (2004). The effect of practice on low-level auditory discrimination, phonological skills, and spelling in dyslexia. *Dyslexia* **10,** 119–130.

Schellenberg, E. G., and Trehub, S. E. (1994). Frequency ratios and the discrimination of pure tone sequences. *Percept. Psychophys.* **56,** 472–478.

Schmid, L. M., Rosa, M. G. P., Calford, M. B., and Ambler, J. S. (1996). Visuotopic reorganization in the primary visual cortex of adult cats following monocular and binocular retinal lesions. *Cerebral Cortex* **6,** 388–405.

Schneider, P., Scherg, M., Dosch, H. G., Specht, H. J., Gutschalk, A., and Rupp, A. (2002). Morphology of Heschl's gyrus reflects enhanced activation in the auditory cortex of musicians. *Nat. Neurosci.* **5,** 688–694.

Schoups, A., Vogels, R. N. Q., and Orban, G. (2001). Practising orientation identification improves orientation coding in V1 neurons. *Nature* **412,** 549–553.

Schwaber, M. K., Garraghty, P. E., and Kaas, J. H. (1993). Neuroplasticity of the adult primate auditory cortex following cochlear hearing loss. *Am. J. Otol.* **14,** 252–258.

Shahin, A., Bosnyak, D. J., Trainor, L. J., and Roberts, L. E. (2003). Enhancement of neuroplastic P2 and N1c auditory evoked potentials in musicians. *J. Neurosci.* **23,** 5545–5552.

Snyder, R. L., and Sinex, D. G. (2002). Immediate changes in tuning of inferior colliculus neurons following acute lesions of cat spiral ganglion. *J. Neurophysiol.* **87,** 434–452.

Snyder, R. L., Sinex, D. G., McGee, J. A., and Walsh, E. W. (2000). Acute spiral ganglion lesions change the tuning and tonotopic organization of cat inferior colliculus neurons. *Hear. Res.* **147,** 200–220.

Spiegel, M. F., and Watson, C. S. (1981). Factors in the discrimination of tonal patterns. III. Frequency discrimination with components of well-learned patterns. *J. Acoust. Soc. Am.* **69,** 223–230.

Suga, N., Gao, E. Q., Zhang, Y. F., Ma, X. F., and Olsen, J. F. (2000). The corticofugal system for hearing: Recent progress. *Proc. Natl. Acad. Sci. USA* **97,** 11807–11814.

Suga, N., and Ma, X. (2003). Multiparametric corticofugal modulation and plasticity in the auditory system. *Nat. Rev. Neurosci.* **4,** 783–451.

Thai-Van, H., Micheyl, C., Moore, B. C. J., and Collet, L. (2003). Enhanced frequency discrimination near the hearing loss cut-off: A consequence of central auditory plasticity induced by cochlear damage? *Brain* **126,** 2235–2245.

Thai-Van, H., Micheyl, C., Norena, A., and Collet, L. (2002). Local improvement in auditory frequency discrimination is associated with hearing-loss slope in subjects with cochlear damage. *Brain* **125,** 524–537.

Thiel, C. M., Friston, K. J., and Dolan, R. J. (2002). Cholinergic modulation of experience-dependent plasticity in human auditory cortex. *Neuron* **35,** 567–574.

Tsukahara, N. (1981). Synaptic plasticity in the mammalian central nervous system. *Annual. Rev. Neurosci.* **4,** 351–379.

Tucker, A., Williams, P. I., and Jeffress, L. A. (1968). Effect of signal duration on detection for gated and for continuous noise. *J. Acoust. Soc. Am.* **44,** 813–816.

Turner, C. W., and Nelson, D. A. (1982). Frequency discrimination in regions of normal and impaired sensitivity. *J. Speech. Hear. Res.* **25,** 34–41.

Ward, W. (1957). Method of "single descent" in group audiometry. *J. Acoust. Soc. Am.* **29,** 371–376.

Watson, C. S. (1976). Factors in the discrimination of word-length auditory patterns. *In* "Hearing and Davis: Honoring Hallowell Davis" (D. H. E. S. K. Hirsh, I. J. Hirsh, and S. R. Silverman, Eds.), pp. 175–189. Washington University Press, St. Louis, Missouri.

Watson, C. S. (1980). Time course of auditory perceptual learning. *Ann. Otol. Rhinol. Laryngol. Suppl* **89,** 96–102.

Watson, C. S., Kelly, W. J., and Wroton, H. W. (1976). Factors in the discrimination of tonal patterns. II. Selective attention and learning under various levels of stimulus uncertainty. *J. Acoust. Soc. Am.* **60,** 1176–1186.

Webster, H. H., Hanisch, U.-K., Dykes, R. W., and Biesold, D. (1991). Basal forebrain lesions with or without reserpine injection inhibit cortical reorganization in rat hindpaw primary somatosensory cortex following sciatic nerve section. *Somatosens. Motor. Res.* **8,** 327–346.

Weinberger, N. M. (1995). Dynamic regulation of receptive fields and maps in the adult sensory cortex. *Ann. Rev. Neurosci.* **18,** 129–158.

Weinberger, N. M. (2003). The nucleus basalis and memory codes: Auditory cortical plasticity and the induction of specific, associative behavioral memory. *Neurobiol. Learning Memory* **80,** 268–284.

Weinberger, N. M. (2004a). Experience-dependent response plasticity in the auditory cortex: Issues, characteristics, mechanisms, and functions. *In* "Plasticity of the Auditory System" (T. N. Parks, E. W. Rubel, A. N. Popper, and R. R. Fay, Eds.), pp. 173–227. Springer, N. Y.

Weinberger, N. M. (2004b). Specific long-term memory traces in primary auditory cortex. *Nat. Rev. Neurosci.* **5,** 279–290.

Weinberger, N. M., Hopkins, W., and Diamond, D. M. (1984). Physiological plasticity of single neurons in auditory cortex of the cat during acquisition of the pupillary conditioned response: I. Primary field (AI). *Behav. Neurosci.* **98,** 171–188.

Weinberger, N. M., Javid, R., and Lepan, B. (1993). Long-term retention of learning-induced receptive-field plasticity in the auditory cortex. *Proc. Natl. Acad. Sci. USA* **90,** 2394–2398.

Weiss, T., Miltner, W. H. R., Dillmann, J., Meissner, W., Huonker, R., and Nowak, H. (1998). Reorganization of the somatosensory cortex after amputation of the index finger. *NeuroReport* **9,** 213–216.

Willott, J. F., Aitkin, L. M., and McFadden, S. L. (1993). Plasticity of auditory cortex associated with sensorineural hearing loss in adult C57BL/6J mice. *J. Comp. Neurol.* **329,** 402–411.

Wright, B. A. (1995). Detectability of simultaneously masked signals as a function of signal bandwidth for different signal delays. *J. Acoust. Soc. Am.* **98,** 2493–2503.

Wright, B. A. (2001). Why and how we study human learning on basic auditory tasks. *Audiol. Neurootol.* **6,** 207–210.

Yang, T. T., Gallen, C. C., Ramachandran, V. S., Cobb, S., Schwartz, B. J., and Bloom, F. E. (1994). Noninvasive detection of cerebral plasticity in adult human somatosensory cortex. *NeuroReport* **5,** 701–704.

Zwicker, E. (1965). Temporal effects in simultaneous masking by white-noise bursts. *J. Acoust. Soc. Am.* **37,** 653–663.

Zwislocki, J. (1955). Design and testing of earmuffs. *J. Acoust. Soc. Am.* **27,** 1154–1163.

Zwislocki, J., Marie, F., Feldman, A. S., and Rubin, H. (1958). On the effect of practice and motivation on the threshold of audibility. *J. Acoust. Soc. Am.* **30,** 254–262.

SPECTRAL PROCESSING IN COCHLEAR IMPLANTS

Colette M. McKay

School of Life and Health Sciences, Aston University, Birmingham B4 7ET, United Kingdom

In this chapter, the functional aspects of cochlear implants that are relevant to spectral processing are discussed. In Section II, basic information about how a cochlear implant works is given. The signal processing strategies in current clinical use are described, as well as novel signal processing strategies aimed at improving perception with an implant. In Section III, the perception of spectral and temporal information in electrical signals is discussed. Section IV describes the ability of implantees to understand complex signals such as speech and music. The coding of acoustic spectral information onto place-of-stimulation information within the cochlea has led to successful transmission of information about speech signals. However, the perception of music with cochlear implants remains problematic. Current implant systems do not provide the fine spectral or temporal information that the perception of musical pitch requires. Finally, in Section V, aspects of central plasticity that underlie the success or otherwise of implantation are discussed. The variable outcome of implantation is due largely to the variable survival of peripheral auditory neurons and the plastic changes that occur throughout the whole central auditory system in response to periods of profound deafness. There is much evidence that the central auditory system can adapt to novel peripheral inputs, particularly for tasks such as speech sound categorization.

INTERNATIONAL REVIEW OF
NEUROBIOLOGY, VOL. 70
DOI: 10.1016/S0074-7742(05)70012-X

0074-7742/05 $35.00

This is particularly true for congenitally deaf children who are implanted at an early age, but is also true, to variable extents, for older people who have a long period of auditory deprivation.

I. Introduction

Cochlear implants are now an accepted clinical device for the alleviation of severe to profound deafness. Currently more than 60,000 people are users of cochlear implants, and this number is increasing rapidly as the technology improves and criteria for candidacy widen. A multiple-electrode cochlear implant directly stimulates the remaining peripheral auditory nerve (AN) cells within the spiral ganglion of the cochlea with current delivered by an array of electrodes inserted into the scala tympani of the cochlea. As such, the array of electrodes is designed to at least partially restore the function of the cochlea as an important organ for spectral processing (see Chapter 1 by Lopez-Poveda). Because the peripheral processing of the normal external, middle, and inner ear is bypassed by a cochlear implant, it is possible to produce peripheral neural activation patterns that are different from those produced by normal hearing. Thus, cochlear implant research has the potential to provide additional insight into the normal central auditory processing of spectral information. It should be remembered, however, that severe peripheral hearing loss can be accompanied by plastic changes throughout the central auditory system (Hardie and Shepherd, 1999; Shepherd and Hardie, 2001). Therefore, the results of cochlear implant research can be applied only in a qualified way to the understanding of normal processes. On the other hand, examining how people learn to listen with a cochlear implant can provide insight into how the central auditory system responds to new patterns of stimulation.

In the following sections, a brief summary of a how a cochlear implant works is first presented. In later sections, the perception of spectral and temporal information in simple and more complex stimuli is discussed. Finally, the plasticity of the central auditory system in adapting to electrical stimulation is discussed.

II. Cochlear Implant Signal Processing

A. Hardware

A cochlear implant consists of both external and implanted components. An external microphone is worn at ear level, and the acoustic signal is conveyed to the sound processor, which is a signal-processing device either worn at ear level (like a hearing aid) or worn on the body. The processor extracts components of sounds considered to be important for speech understanding and determines (in real-time) the timing, position in the cochlea, and size of current pulses to be delivered by the electrodes within the cochlea. This information is usually encoded in radio-frequency (RF) signals that are transmitted across the skin via magnetic induction to the implanted electronics. The implanted electronics are housed in a sealed container that is positioned under the skin, in the mastoid bone behind the ear, and the electrodes are inserted into the scala tympani via the round window in the middle ear (or surgically produced opening). There are various designs of electrode array, with differing numbers of electrodes (from 6 to 22 at the present time), and different physical configuration and orientation of electrodes. The electrode array is inserted around the basal turn of the cochlea, and the electrodes are located at equal spacing around this turn. The actual insertion depth depends upon the state of the individual cochlea (whether there is bone growth in the cochlea due to pathological processes, for example), the type of electrode array, and the surgical technique used. The incoming RF signal is decoded by the internal electronics and the stipulated current pulses are produced at the appropriate electrode positions.

B. Signal Processing Strategies in Clinical Use

There are several different sound-processing strategies that are currently used in commercially available devices. However, these have several basic principles in common. First, the incoming signal is filtered by a bank of overlapping bandpass filters. The number of filters used is usually the same as the number of active intracochlear electrodes available, and the filters are assigned to the electrodes in tonotopic fashion. That is, the lowest-frequency bandpass filter is assigned to the most apical electrode position, and so on. The frequency range of the bank of filters covers what is considered to be an important range for speech perception, generally from as low as about 100 Hz to a maximum of about 8 kHz. The output of each filter is used to determine the current waveform delivered by its corresponding intracochlear electrode. The current return path can either be another intracochlear electrode (bipolar stimulation) or an extra-cochlear electrode (monopolar stimulation).

At this point it is informative to compare the spectral processing provided by the sound processor and electrode array to that provided by the cochlea in a normal ear. First, it can be seen that the function of the cochlea to spatially separate different frequencies along the length of the basilar membrane and hence to provide tonotopic information to the central auditory system has been approximated in the implant design. (For an in-depth discussion of cochlear mechanics in normal hearing, see Chapter 1 by Lopez-Poveda.) The fact that implantees can usefully understand speech and recognize environmental sounds attests to the success of this strategy of using bandpass filters to direct signal information to different parts of the cochlea on a frequency-specific basis. However, there are several differences between the spatial-based coding of frequencies in cochlear implants and normal hearing. In normal hearing, the cochlear auditory filters have continuous center frequencies, whereas in a cochlear implant, the bandpass filters are limited in number (to the number of active electrodes) and fixed in frequency. Also, the auditory filters in normal hearing are non-linear and level-dependent, increasing in width as the level increases, whereas the filters used in a sound processor are generally fixed. The bandwidth of the filters used in sound processors depends on the number of filters (or electrodes) used, and can vary from a similar width to that of a normal-ear auditory filter up to much wider than this. Thus, the ability of an implantee to resolve the different frequency components of a signal is essentially limited by the filter design in their sound processor. It should be noted that the filter bands assigned to the electrodes are at a generally lower frequency than the characteristic frequency of the electrode place (since the array is generally only inserted around the first turn of the cochlea). A second limiting factor for the implantee is the degree to which stimulation on different electrodes is perceptually distinct. As described in Section III, the ability to distinguish stimulation on different electrodes varies widely among implantees.

There are several commercial companies that manufacture cochlear implants for clinical use, and each company has several different models of implant and speech processor design. Also, the latest versions of these sound processors are capable of implementing more than one sound-processing strategy. It is beyond the scope of this chapter to describe the detailed design and implementation of all the currently available processing strategies. Instead, the basic principles underlying the most common strategies will be described. The most common type of stimulation used in implants today is a sequence of non-overlapping charge-balanced biphasic current pulses. In each stimulation cycle, the filter outputs (smoothed to extract envelope information) are scanned, and one pulse is output on each of the selected electrodes. The size of each pulse (duration and/or height) is dependent on the magnitude of the corresponding filter output. The transform between acoustic level and electrical level varies with different implementations, and in some cases can be adjusted by the clinician. Essentially, the current varies on each electrode between a minimum required to evoke a hearing sensation, and a maximum related to the highest comfortable loudness. The stimulation cycle rate varies according to the strategy and/or the choice of this parameter by the clinician, but the highest rate in any implementation is limited by the duration of the pulses themselves, how many electrodes are activated in each cycle, and the technical design of the transcutaneous information transfer.

In continuous interleaved sampling (CIS) strategies, all of a fixed set of electrodes are activated in each stimulation cycle, and the cycle rate is usually >800 Hz. In its first implementation for the (now obsolete) Ineraid device (Wilson *et al.*, 1991), the 4 to 6 available electrodes were activated via a percutaneous connector. Most more-recent versions of the CIS strategy utilize 6 to 8 fixed electrodes and stimulation rates of at least 1000 Hz. With the development of improved technology, some modern devices can output stimulus pulses at much higher rates. For example, the Hi-Res strategy (Frijns *et al.*, 2003) is a CIS-type strategy for the Clarion CII implant system, which can output up to 90,000 pulses/sec, with pulses distributed among up to 16 electrodes (cycle rates of up to about 5000 Hz).

The other class of strategies that output sequential biphasic pulses are those that, on each stimulation cycle, select a *subset* of a larger number of electrodes for activation. The filter outputs are scanned and the highest *n* of these are selected and the corresponding electrodes (assigned to those filters) are activated in that cycle. The first commercial strategy of this type, SPEAK, was implemented for the Nucleus CI22 implant following development by McDermott *et al.* (1992). Patients with this implant had up to 22 active electrodes. In general, SPEAK selected about 6 of these electrodes in each cycle, with the selection based on which filters had the highest outputs. Limitations of the electrode configuration and transmission link in this implant limited the maximum cycle rate of SPEAK to about 250 Hz. With improvements in technology, faster strategies of this type are now available. For example, the advanced combinational encoder (ACE) strategy for the Nucleus CI24 implant can implement cycle rates of up to 2400 Hz. Similarly, the *n-of-m* strategy for the Combi 40+ system (Ziese *et al.*, 2000) can produce cycle rates of more than 4000 Hz.

There are two commercial strategies implemented in the Clarion CI and CII devices that diverge from the usual output of sequential biphasic pulses (Frijns *et al.*, 2002; Zwolan *et al.*, 2001). The paired pulsatile stimulation (PPS) strategy is a variation of CIS. It outputs simultaneous pairs of pulses on distant pairs of electrodes. These pulse pairs are output sequentially in each cycle, as for CIS. This technique enables a theoretical doubling of the cycle rate compared to a normal CIS strategy. The simultaneous analog stimulation (SAS) strategy uses the (unsmoothed) scaled output waveform from the filters to produce the analog electrical waveforms for each electrode. This is a relatively new implementation of an earlier strategy (CA for compressed analog), in which the acoustic signal was first compressed and then filtered with a bank of four bandpass filters. The electrical stimulus on the four fixed electrodes was a scaled version of the output of the four filters. The CA strategy was largely abandoned in favor of the CIS strategy, after the detrimental effects of direct current summation across electrodes in the CA strategy had been demonstrated (Wilson *et al.*, 1991). The SAS strategy differs from CA in that SAS applies a mapping function from acoustic to electric level at the filter output, rather than applying compression to the acoustic input signal. The deleterious effects of across-electrode current summation found with CA are reduced in the newer SAS strategy by the use of modiolar-hugging electrode arrays (in which the electrodes are positioned closer to the spiral ganglion nerve cells that are their target), and by electrode configurations and designs that are intended to limit the lateral spread of current along the cochlea (see Section II.C).

In summary, all of the aforementioned strategies convey spectral information about the acoustic signal by transferring the information about energy in different frequency regions to stimulation at different places in the cochlea. In addition, the temporal envelope information within each frequency band is coded in the temporal envelope of the electrical stimulation at the corresponding electrode place. Both place and temporal information have limitations imposed by the hardware and software. Place information is limited by the number of electrodes used and their discriminability. Temporal envelope information is limited in sequential-pulse strategies by the smoothing of the filter outputs and by the stimulation rate (which effectively samples the smoothed temporal envelope). The SAS strategy can reproduce fine temporal information present within the acoustic signal in the electrical stimulus waveform (although it is questionable whether implantees can perceive this fine temporal information). In earlier implant systems, technical limitations led to a trade-off between temporal and place cues: CIS strategies conveyed better temporal information than SPEAK (because of the higher stimulation rates in CIS), but coarser spectral/place information (because of the small number of filters/electrodes used). However, modern versions of implant systems can output pulses at many electrode sites while maintaining a high rate of stimulation, thus simultaneously preserving fairly detailed spectral and temporal-envelope cues.

C. New Directions in Signal Processing and Hardware

One major focus of recent research has been to design electrode arrays, or stimulus waveforms, that can produce a greater number of perceptually distinct sensations. In theory, this might be achieved by placing the electrodes in closer proximity to the spiral ganglion cells. The spiral ganglion cells are contained within the modiolus, which is a bony part of the cochlea situated in the central wall. Several designs of perimodiolar (or modiolar-hugging) arrays are in current use. Unfortunately, there does not seem to be any consensus about the benefits of these new arrays. Comparisons of psychophysical, objective, and speech perception performance with different arrays is difficult because within-subject comparisons cannot be performed, and because of the large individual variability that is typical in these measures. Cohen *et al.* (2001, 2003) have suggested that electrodes closer to the modiolus (within the same subject) show less spread of excitation, at least at lower stimulus levels, and lower thresholds. In contrast Kreft *et al.* (2004) found that subjects with a perimodiolar array had, on average, similar thresholds, but smaller dynamic range, and larger intensity difference limens than those subjects with a standard array. Boex *et al.* (2003) found that subjects with a standard or perimodiolar array did not differ in the amount of channel interaction they experienced (measured psychophysically using simultaneous stimulation on two electrodes), whereas they found less channel interaction, as expected, in bipolar compared to monopolar stimulation.

Another possible way to create a more place-specific excitation pattern is to use different electrical waveforms from the standard biphasic pulse. One idea, assessed by Bonnet *et al.* (2004), is to use triphasic pulses in which the middle, cathodic pulse has four times greater current than the preceding and following anodic pulse (while adjusting the phase durations to maintain charge balance). Biphasic pulses can activate different neural populations on each phase, leading to less place-specific neural excitation, whereas the triphasic pulse will tend to activate neurons only on the large, cathodic phase. Bonnet *et al.* found a small improvement in speech perception compared to the standard CIS strategy when using this new stimulus waveform. Another way to manipulate the stimulus waveform is to use simultaneous stimulus pulses of reversed phase on electrodes adjacent to the activated electrode (tripolar stimulation), thus sharpening the current field (Jolly *et al.*, 1997). Figure 1 shows spatial tuning curves for single auditory neurons in cats for acoustic stimuli and three different electrical stimulation modes. It shows that the place specificity of electrical stimulation improves when moving from monopolar to bipolar to tripolar modes.

In summary, several different ways have been investigated of making the neural activity evoked by each electrode more place-specific, in the belief that this will lead to improved auditory perception by implantees. However, these attempts have not yet led to a major improvement in speech perception. Such manipulations will reduce the amount of direct current summation across channels in new strategies that use simultaneous or overlapping current pulses. However, the benefits of using simultaneous stimulation on multiple electrodes have not yet been convincingly demonstrated.

Research efforts are continuing in the field of signal processing. As previously described (Section II.B), all of the current commercial strategies are fairly similar in that they use the output of fixed filters to determine the electrical output at the different electrode positions, and the stimulation is in the form of sequential biphasic pulses (or analogue waveforms). New efforts have concentrated on improving the fine temporal information content of the electrical signal, or creating stimulus patterns that more closely mimic the spatial/temporal patterns elicited by normal hearing. The SpeL strategy (McDermott *et al.*, 2003a) aims to provide loudness percepts that are more closely related to those that normal-hearing

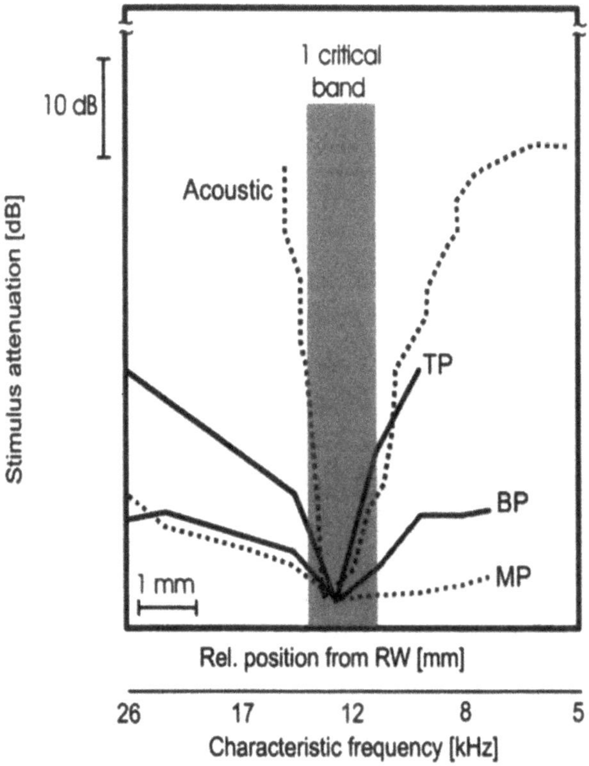

FIG. 1. Acoustic and electric spatial tuning curves measured on a single auditory nerve fiber in a cat. It can be seen that electrical stimulation exhibits place specificity. The tripolar (TP) stimulation is most place-specific followed by bipolar (BP) and monopolar (MP) stimulation. (Figure reproduced from Hartmann and Kral, 2004, with permission; data from Kral *et al.*, 1998.)

people perceive. Instead of a map between acoustic level and electrical level in each separate channel, as previously described, the SpeL strategy uses a map between normal-hearing specific loudness and electrical specific loudness. The outputs from non-linear filters that mimic the behavior of a normal ear are analyzed with a normal-hearing loudness model to obtain real-time estimates of specific loudness (the loudness contribution from each part of the cochlea). An electric loudness model (McKay *et al.*, 2003) is then used to determine the correct current to elicit the appropriate specific loudness at each electrode position. This strategy has been shown to improve the perception of the relative loudness of sounds with varying bandwidths (McDermott *et al.*, 2003a) and produces improved speech perception at soft levels compared to the SPEAK strategy, and equivalent to ACE, within several weeks of experience (McDermott *et al.*, 2003b, 2004).

Several research groups are developing strategies in which the timing of the individual pulses is manipulated in some way, either to mimic the phase delays of cochlear traveling waves, or to provide frequency-specific fine temporal information. For example, the peak-derived timing (PDT) strategy (van Hoesel and Tyler, 2003) places stimulus pulses on each electrode at times corresponding to fine-timing peaks in filter outputs. This strategy was motivated by the desire to produce bilateral stimulation (for those with two cochlear implants) in which the cues for interaural time delay (ITD) in the acoustic signal are preserved. Experiments with two bilaterally implanted subjects using the PDT strategy have shown that they experience a small bilateral masking level difference benefit (using a 500-Hz narrow band noise as masker and a 500-Hz tone as signal, in-phase and 180^0 out of phase) of 1 to 2 dB (van Hoesel, 2004).

As the average hearing performance of cochlear implantees increases with technical advances, the criteria for implant candidacy continue to widen to include people who have more residual hearing ability. Many implantees gain advantage from wearing a hearing aid on the ear opposite the implant (Ching *et al.*, 2001, 2004; Dettman *et al.*, 2004; Tyler *et al.*, 2002), and recent research has investigated the option of wearing a hearing aid in the same ear as the implant. Turner *et al.* (2004) have shown that noise consisting of multiple talkers has less effect on speech perception for implantees with a significant amount of residual hearing in the same ear, compared to those using electrical stimulation alone. These subjects had a new short electrode array that is designed to reduce insertion damage (and hence preserve hearing) in the implanted ear, thus allowing for the continued use of acoustic hearing in that ear. Much of the benefit seen when using a combination of acoustic and electric stimulation can be attributed to the improved perception of fundamental frequency information that acoustic stimulation facilitates (see Section IV.B).

Finally, recent research has investigated the use of neural growth factors (NGFs) to halt the neural degeneration that occurs after the hair cells in a severely hearing-impaired cochlea die. Studies in guinea pigs have shown that such neurotrophins can halt the neural degeneration that occurs after deafening (Kanzaki *et al.*, 2002). On the assumption that better neural survival will lead to better hearing with a cochlear implant, such a therapy in humans may lead to improved perceptual ability.

III. Psychophysical Measurements of Perception of Spatial and Temporal Information

In this section, we will discuss the perception of electrical stimuli as deduced from psychophysical experiments, starting with simple, single-electrode stimuli, and moving on to multiple-electrode stimuli. The electrical stimuli in these experiments are generated by experimental software, and are independent of the signal processing that occurs when wearing a sound processor (and thus do not correspond to a sound processor output for any real sound). It should be remembered, however, that sound processors code spectral shape information onto electrical place-of-stimulation information, and temporal envelope cues onto electrical temporal envelope information (see Section II). Thus, psychophysical studies of the perception of place and temporal cues allow us to deduce the likely limitations of the sound processor output for conveying important spectral information to the implant user. In Section IV, we will discuss the perception of complex sounds (particularly speech and music) via a sound processor.

A. Perception of Place of Stimulation

Many early experiments with implantees showed that stimulation at different places along the scala tympani produced sounds that differed in pitch or timbre (Shannon, 1983; Tong *et al.*, 1979). Figure 2 shows the results of a pitch-scaling experiment, in which the subjects were asked to provide a numerical estimate of the relative pitch produced by pulse trains on different electrodes. The data of

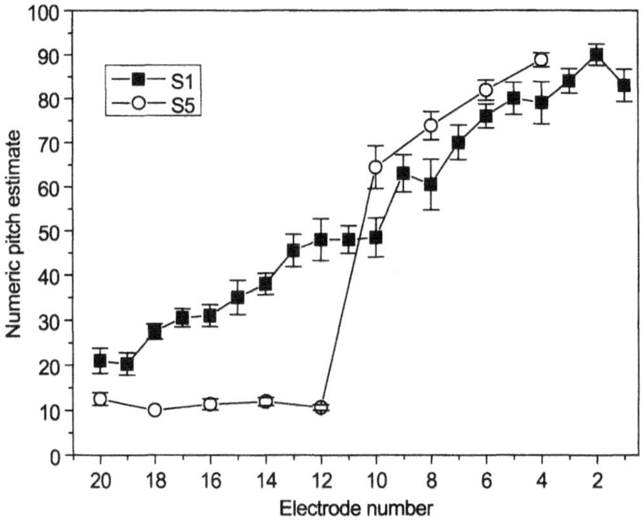

FIG. 2. Data for two subjects in a pitch-scaling experiment, in which the subjects were asked to provide a numerical value that represented the pitch. The data of S1 show the usual pitch progression from high to low as the electrode position moves from basal to apical, but the data of S5 show a broad area in the apical region where pitch does not change with electrode position. (Figure reproduced from McKay, 2004, with permission; data adapted from Cohen *et al.*, 1996b.)

subject S1 generally conform to what would be expected based upon tonotopic ordering of the neural population in the spiral ganglion. That is, as the electrode position moves from apical (high-numbered electrodes) to basal (low-numbered electrodes), the perceived pitch increases monotonically. However, it is not uncommon for implantees to experience non-monotonic pitch percepts or to have electrodes that are perceptually indistinguishable. Subject S5 in Fig. 2 is one such subject: it can be seen that there is a large apical region that elicits the same low pitch percept. Place-pitch anomalies such as these can arise from factors such as poor neural survival, pathological processes such as new bone growth, and poor placement of electrodes due to surgical procedure difficulties. There are also cases in which the percepts evoked by different electrode positions are not well-described by a single perceptual dimension (corresponding to pitch or timbre) (Henshall and McKay, 2001). These cases are probably associated with the stimulation of non-auditory neural processes by some electrodes, or a widely varying spatial extent of neural excitation at different electrode positions. The ability to distinguish stimulation on different electrodes varies greatly among implantees (Collins *et al.*, 1997; Henry *et al.*, 2000), with the smallest discriminable electrode separation varying from the smallest available separation (0.75 mm for Nucleus implants) to separations up to about 5 mm. The reasons for this variation are probably the same as those previously discussed in relation to non-monotonic pitch percepts.

The degree of overlap of neural populations excited by different electrodes can be assessed by several psychophysical and objective means. Forward-masking experiments measure the threshold shift of a probe stimulus due to the presence of a preceding stimulus (Chatterjee and Shannon, 1998; Cohen *et al.*, 1996a, 2003). The pattern of threshold shift across probe electrode positions is thought to be related to the degree of excitation overlap between the probe and masking electrode. Figure 3 shows three such forward-masking patterns. All three masking patterns are quite wide, but each shows a different place-specificity related to electrode position. Similar masking patterns can be obtained objectively by measuring the forward-masked electrically evoked compound action potential (ECAP) at different probe electrodes with a fixed-electrode masking stimulus (Cohen *et al.*, 2003). Suprathreshold measures of place-specificity can be inferred from several psychophysical experiments. For example, McKay and McDermott (1996) showed that subjects could perceptually combine the temporal patterns across two electrodes only if the electrodes were separated by less than a critical amount (about 3 to 4 mm).

The ability to discriminate different electrode positions does not appear to be highly related to the degree of overlap in excited neural populations. For example, stimulation at higher currents does not result in reduced electrode discrimination (McKay *et al.*, 1999; Pfingst *et al.*, 1999). Also, wider modes of stimulation do not seem to greatly affect electrode place discrimination. For example,

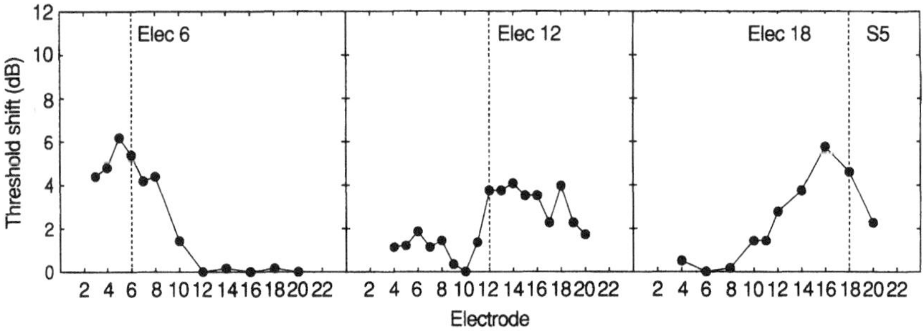

FIG. 3. Forward-masking patterns from masking stimuli on three different electrodes (6, 12, and 18). The vertical axes represent the difference between the masked and unmasked thresholds (in dB of pulse duration change) for each probe stimulus. The probe-stimulus electrode is the variable on the horizontal axis. The pattern of probe-stimulus threshold shift across the electrode array has peaks in different places for the three cases. (Figure reproduced from McKay, 2004, with permission; data adapted from Cohen *et al.*, 1996a.)

McKay *et al.* (1996) showed that wider bipolar stimulation (using electrodes for active and return current path that are more widely separated) did not produce percepts that were less discriminable than narrow bipolar stimulation (see the following text). Thus, as with discrimination of pure tones in normal hearing (see Chapter 2 by Moore), there is evidence that changes in the edge of the excitation pattern may facilitate discrimination.

The perceptual task that is potentially more related to the spread of excitation in the cochlea is the ability to resolve percepts related to different electrode positions when more than one electrode is activated. An early experiment by Tong *et al.* (1983b) showed that when stimulus pulses were interleaved between two electrode positions, the subject heard two perceptual dimensions related to the two different electrode positions. Figure 4 shows a Euclidean stimulus space for one subject, which was the result of a similar multidimensional scaling experiment performed by McKay *et al.* (1996). The letters 'a' to 'e' represent five adjacent electrode positions (separation 0.75 mm). The single letters represent stimuli at a single electrode position and the double letters represent dual-electrode stimuli, where stimulus pulses were interleaved at the two electrode positions. It can be seen that the stimuli are grouped in a two-dimensional array. The dimensions of the array can be interpreted to be an average electrode position (dimension 1) and electrode separation (dimension 2). Alternatively, if the space is rotated by 45°, the two dimensions can be interpreted to be the apical and basal electrode positions, respectively. Tong's experiment in 1983 underpinned one of the first processing strategies (called F0F1F2), in which the first and second formant frequencies of speech were estimated and interleaved stimulation applied on two corresponding electrode positions. In this strategy, the fundamental frequency of voiced speech was coded as the cycle repetition rate. An

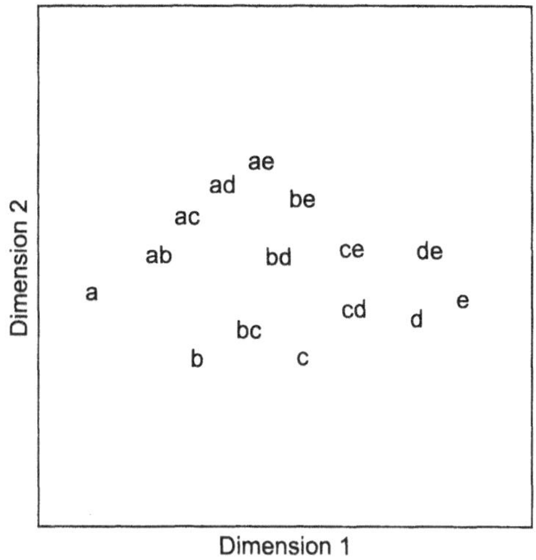

FIG. 4. The stimulus space resulting from a multidimensional scaling experiment with one subject. Each stimulus is represented by a letter or pair of letters. 'a' to 'e' are electrode positions 10 to 14. Single letters and pairs of letters represent single- and dual-electrode stimuli respectively. Stimuli were pulse trains (or pairs of interleaved pulse trains) with rate of 250 Hz. (Reprinted with permission from Mckay *et al.* © 1996, Acoustical Society of America.)

interesting finding of the McKay *et al.* (1996) study was that the stimulus space obtained with the five adjacent electrode positions still showed the clear two-dimensional structure even when a very wide bipolar mode was used (so that there was a high degree of overlap in the two excitation patterns).

The multidimensional scaling experiments using dual-electrode stimulation previously described suggest that implantees can perceptually resolve stimulation at different electrode places, even when the neural excitation patterns significantly overlap. However, the output of a sound processor has a higher degree of complexity, usually involving activity on six or more electrodes in every cycle. The spectral shape of the acoustic signal is replicated in the pattern of different current levels across electrodes, and it is intended that this spectral pattern is perceived by the listener as an 'internal spectrum.' Experiments with normal-hearing listeners have measured this internal spectrum by using a forward-masking technique (Moore and Glasberg, 1983). The threshold shifts of probe tones of various frequencies are measured when the tones are preceded by the complex masking sound. The pattern of threshold shift across probe frequency is related to an internal representation of the signal. A similar experiment has been conducted using forward masking in electrical stimulation (Stainsby, 2000). Figure 5 shows the electrical output from the SPEAK strategy for the sound of a cello (vertical bars represent the current levels on each electrode) and the

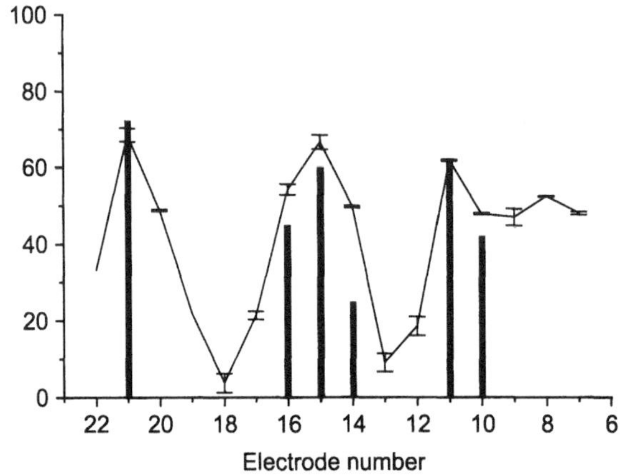

FIG. 5. The forward-masking pattern (or internal spectrum) produced by a multiple-electrode electrical stimulus. The masker is a simulation of the output of a spectra processor, using the SPEAK strategy, for an acoustic input produced by a cello. The bars represent the amount of current (in % dynamic range) on each electrode in the masking stimulus, and the line represents the amount of forward masking (in % probe dynamic range) at each electrode position. (Figure reproduced from McKay, 2004, with permission; data from Stainsby, 2000.)

corresponding internal spectrum (line graph), which is the amount of threshold shift for probe stimuli on each electrode. It can be seen that for this subject, the internal representation is fairly consistent with the electrical stimulus.

A more recent experiment was performed by Laback *et al.* (2004), who measured the 'internal spectrum' of vowel-like stimuli by using a simultaneous-masking technique. The vowel-like masking stimuli were constructed from steady-state portions of (acoustic) German vowels, and the thresholds of probe tones of various frequencies were measured, in which the probe signals were embedded in the temporal center of the masking signal. Both signals were presented acoustically to subjects who were using the Combi-40 or Combi-40+ implants and the CIS strategy (see Section II.B). The results showed that the vowel-masking patterns closely mimicked the electrode activation pattern (current amplitude versus electrode place), and that the ability to discriminate pairs of the vowel stimuli was correlated with the spectral distance of the two corresponding masking patterns.

B. Perception of Temporal Information

As with acoustic stimulation, many temporal properties of the electrical signal are retained in the temporal pattern of peripheral neural spikes. Studies on single-nerve cells in animals have shown that the phase locking of auditory neurons to

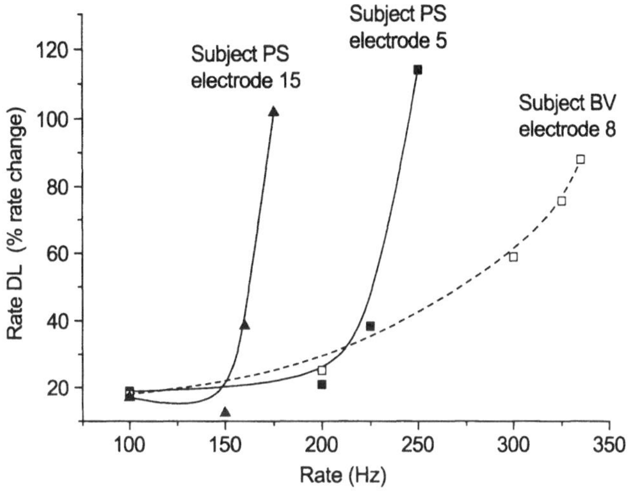

FIG. 6. Data for two subjects from our laboratory, showing how their rate discrimination ability changes with the rate of stimulation. Note that the rate at which discrimination becomes difficult varies between the two subjects and also between the two electrodes for subject PS. (Figure reproduced from McKay, 2004, with permission.)

pulsatile electrical stimuli of different rates, or sinusoidal stimulation, is enhanced over the phase locking in response to acoustic pure tones (Dynes and Delgutte, 1992; van den Honert and Stypulkowski, 1987). Implantees hear a pitch associated with the rate of stimulation. Pitch increases as rate increases until a saturation point of approximately 300 Hz. The pitch intervals for changes in rate below this limit, as assessed using musically trained subjects, appear to be closely related to those perceived by normal-hearing people listening to changes in fundamental frequency (McDermott and McKay, 1997; Pijl and Schwarz, 1995). Figure 6 shows three functions of rate difference limen (DL) versus rate. It can be seen that the saturation point can vary among different subjects as well as among different electrodes in the same subject. As the electrophysiology studies show good phase locking of ANs to more than 1000 Hz, the much lower limitation in the rate-pitch percept may be due to more-central processes of the auditory system. Nevertheless, rate-pitch appears to be more limited in cochlear implantees than in normal-hearing listeners. Carlyon and Deeks (2002) showed that normal-hearing listeners can extract temporal pitch to rates higher than 300 Hz. Several hypotheses, based on differences between electric and acoustic stimulation, that attempt to explain why implantees cannot access temporal information above about 300 Hz have not been supported by experiments. For example, McKay *et al.* (2000) showed that consistent place and temporal changes (i.e., in the same pitch direction) did not allow better discrimination of two stimuli than inconsistent changes in rate and place. Also, Baumann *et al.* (2004) showed that rate discrimination on deeply inserted electrodes (where the characteristic frequency is more nearly matched to the rate of stimulation) was no better than discrimination of rate on basal electrodes.

Implantees, in general, show a similar ability to normal-hearing people in temporal resolution tasks such as gap detection (Shannon, 1989) and amplitude modulation detection (Busby *et al.*, 1993; Shannon, 1992). However, there are individual differences, and these appear to be correlated with speech perception performance (Cazals *et al.*, 1994; Fu, 2002). As in normal hearing, temporal resolution exhibits a low-pass characteristic, with smaller gaps or higher-frequency amplitude fluctuations being harder to perceive. For example, the temporal modulation transfer function shows a cut-off frequency of between 100 and 300 Hz, above which modulations become increasingly difficult to perceive.

Amplitude-modulated pulse trains also evoke a pitch that is, in general, related to the modulation frequency. However, there are several situations in which the pitch perceived is different from that of the modulation frequency. McKay *et al.* (1994) showed that the carrier pulse rate had to be at least four times the modulation frequency (or the modulation frequency a submultiple of the carrier frequency) in order for the perceived pitch to be related to the modulation frequency. That is, implantees are sensitive to the temporal distortions that occur when the modulation pattern is sampled too sparsely by the carrier pulses. Additionally, for small modulation depths, the pitch is also influenced by the carrier frequency (McKay *et al.*, 1995) and, in some situations, both the carrier and modulation frequencies contribute independent pitch percepts (McKay and Carlyon, 1999).

The auditory mechanism for extracting the pitch of pulsatile stimuli does not appear to always rely on the modulation period or fundamental frequency. Several experiments (using stimuli that have two or more unequal pulse intervals within the stimulus period) have shown that the pitch experienced by cochlear implantees (or normal-hearing people listening to filtered pulse trains) is more closely related to the first-order intervals within the stimuli (Carlyon *et al.*, 2002; McKay and McDermott, 1996; van Wieringen *et al.*, 2003). These results are in conflict with the autocorrelation types of temporal pitch theories. Pressnitzer *et al.* (2002) have pointed out that first-order intervals are not evident in peripheral neural interspike intervals, and so a pitch mechanism that extracted first-order intervals would of necessity have to combine information across different nerve fibers.

As previously mentioned, the neural response to pulsatile electrical stimulation is more tightly phase locked to the stimulus than the response to acoustic stimulation. There are two other aspects of the peripheral neural response that differ between electrical and acoustic stimulation. First, without the influence of hair cell synapses, the spontaneous firing of acoustic nerve fibers is essentially absent in a deaf ear (Shepherd and Javel, 1997). Second, electrical stimulation is capable of synchronously activating a large number of nerve fibers. These effects mean that responses to electrical stimulation are not as stochastic as those to acoustic stimulation. Theoretically, therefore, temporal information is not transferred to the neural response as well as it would be in acoustic stimulation. For example, it has been shown that when the rate of a pulse train exceeds a certain rate, the neural refractory time interacts with the period of stimulation resulting in an alternating response in the AN (as measured with electrically evoked compound action potential) (Wilson *et al.*, 1997). For example, the first pulse in a current pulse train will elicit a large response, and all the nerves that fired on that pulse will be refractory for the second pulse, and have recovered for the third pulse, and so on. Increasing the stochastic nature of the electrical response would reduce this phenomenon and improve the capability of the neural response to accurately follow the temporal envelope of the stimulus. Several experiments have investigated ways of increasing the stochastic nature of responses to electrical stimulation. For example, the application of a sub-threshold high rate pulse train can produce changes in the threshold of detection of a low-frequency signal that are similar to what would be expected with stochastic resonance (Hong *et al.*, 2003; Litvak *et al.*, 2003; Rubinstein *et al.*, 1999).

C. Relationship of Spatial and Temporal Cues

Several researchers have investigated the perceptual relationship between temporal and spatial cues in electrical stimulation. An early experiment by Tong *et al.* (1983a) showed that the rate of stimulation and place of stimulation were perceived as independent perceptual dimensions. In that experiment, a multidimensional scaling task was used to obtain a stimulus space representing the perceptual differences among single-electrode stimuli differing in rate and electrode place. The resultant perceptual stimulus space (see Fig. 7) showed a two-dimensional solution, with each dimension related to rate and place respectively. It can be seen from this experiment that changes of rate cannot be compensated for by changes in place, as the percepts are not in the same perceptual dimension. This result was supported by a later study by McKay *et al.* (2000). In that experiment, the discrimination of combined place and rate changes was consistent with the optimal processing of independent observations. In particular, the discrimination of stimuli did not differ in cases where the place and rate changes were consistent or inconsistent, respectively. The independence of temporal and place frequency cues is reflected in the way that these cues are used in speech processors. In general, voice pitch or musical pitch is conveyed by amplitude fluctuations at the fundamental frequency, rather than as a place-of-stimulation (see Section IV.B).

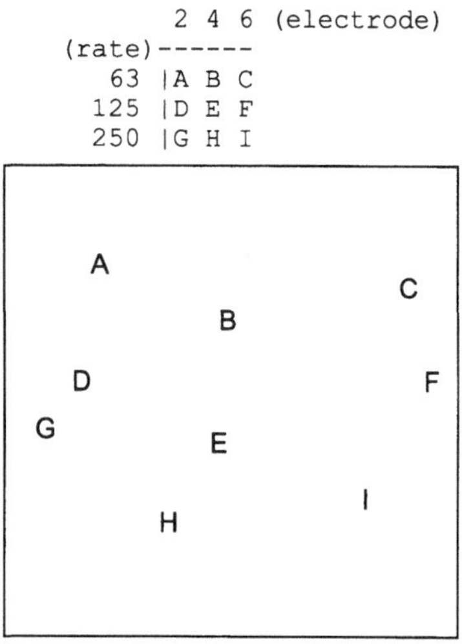

FIG. 7. The stimulus space resulting from a multidimensional scaling experiment by Tong *et al.* (1983a). It can be seen that the rate of stimulation and place of stimulation were perceived as two independent dimensions. (Reprinted with permission from Tong *et al.* © 1983, Acoustical Society of America.)

The auditory processing of temporal information in electrical stimulation (as in normal hearing) is easier when the cues are within auditory channels rather than across channels. The detection of phase differences across different electrodes is easiest when the electrodes are less than a subject-dependent critical distance apart. McKay and McDermott (1996) found that subjects could not detect phase differences when the electrode separation was >3 to 4 mm. On the other hand, Carlyon *et al.* (2000) demonstrated that subjects could discriminate stimuli in which pulses across two widely separated electrodes were nearly synchronous (interchannel delay of 0.1 ms) from stimuli for which the interchannel delay was greater. Similarly, a gap in electrical stimulation is easier to detect when the stimuli on either side of the gap are on the same, or nearby, electrodes (Hanekom and Shannon, 1998). These experiments with phase differences or temporal gaps across electrodes yield very similar results to analogous experiments in normal hearing. That is, within-channel processing of temporal information is easier than across channel processing (Nelson, 1994), although there is some evidence that cross-channel processing occurs (Yost and Sheft, 1989).

An interesting question still under investigation, and important to implant signal processing, is "does the processing of frequency (or pitch) information depend upon correct characteristic frequency representation?" As previously mentioned (Section III.A) the electrode array generally reaches only around the first turn of the cochlea, and the acoustic frequencies assigned to the electrodes are hence usually lower than the characteristic frequency at the place of stimulation. Many current theories of complex pitch perception rely solely on temporal analysis of the neural responses to sound (e.g., Meddis and O'Mard, 1997), although some also rely upon the correct tonotopic place of components of a complex sound (e.g., Shamma and Klein, 2000). An interesting recent paper by Oxenham *et al.* (2004) described an experiment addressing this question. These authors translated the temporal information in low-frequency sinusoids onto amplitude modulated high-frequency tones, thus transposing the temporal information to an 'incorrect' place in the cochlea. They found that frequency discrimination of the transposed tones was poorer than with the untransposed tones. This was not surprising in that, although both types of stimuli had similar temporal information, the sinusoids of differing frequency also produced changes in cochlear place. More importantly, they found that when transposed tones were combined to form three harmonics (3 to 5) of a complex tone, and presented to three (incorrect) cochlear places, subjects could not extract the fundamental frequency of the tones. These results suggested that correct tonotopic place is very important for complex pitch perception. The fact that implant arrays have fixed electrode positions, and the frequency-to-electrode allocation is mismatched relative to characteristic frequencies, suggests that implantees will be very unlikely to hear the pitch of complex sounds, unless the fundamental frequency is directly depicted in the temporal modulations of the speech processor electrical output (see Section IV.B).

Finally, in this section, we will discuss whether implantees can use across-channel temporal cues to separate two different sounds. This issue has been investigated by Carlyon *et al.* (2003) and Deeks and Carlyon (2004). Deeks and Carlyon (2004) used acoustic simulations of a six-channel CIS scheme, in which sentences were processed in three alternate channels and noise was processed in the three intervening channels. In each channel, the speech or noise was represented by filtered pulse trains modulated by the temporal envelope of the filtered signal. There was no consistent advantage for sentence recognition (by normal-hearing listeners) when the speech and noise channels had very different rates of stimulation (80 and 140 Hz), suggesting that implantees would be unable to exploit rate differences across electrodes to perceptually segregate two signals. Experiments with cochlear implant listeners (Carlyon *et al.*, 2003) showed that implantees could not exploit a difference in rate to 'hear out' one channel within a four-electrode stimulus (consistent with the acoustic simulation experiments), but they could use a difference in onset time to achieve this task. That is, when the stimulation on the target electrode was offset in time from that on the remaining three electrodes, this improved the ability to 'hear out' the stimulus on the target electrode.

IV. Perception of Complex Sounds by Implantees

A. Speech Perception

The major purpose of cochlear implants is to facilitate verbal communication for people with severe to profound deafness. To a large extent, this purpose has been fulfilled, with many implantees understanding speech well, at least in quiet situations. As explained in Chapter 4, speech understanding is fairly good when normal-hearing listeners are presented with the temporal envelope of the speech signal in a relatively small number of frequency bands. Similarly, studies with implantees that have examined the effect on speech perception of varying the number of analysis channels or electrodes have shown that increasing the number of channels above 7 to 10 produces ever-diminishing gains in speech understanding (Fishman *et al.*, 1997; Friesen *et al.*, 2001; McKay and Henshall, 2002; Zeng and Galvin, 1999). In those experiments the number of electrodes used in the maps was reduced, and the output of more than one analysis filter was assigned to each remaining electrode. This resulted in an inevitable and unavoidable alteration in the frequency-to-electrode allocation as the number of electrodes was reduced. As discussed more fully in the following text (Section V), such shifts result in an immediate reduction in speech perception, followed by a gradual improvement over time as the new tonotopic code is learned.

An alternative method of measuring the resolvability of spectral information by cochlear implantees, which does not necessitate altering the frequency-to-electrode allocation, was presented by Henry and Turner (2003). These authors used acoustic rippled noise stimuli, with the spectral ripples having differing spectral spacing. The task of the implantees was to discriminate two stimuli of the same spectral ripple spacing, but inverted spectral shape (i.e., peaks in the spectrum where there were valleys and vice versa). The threshold spectral spacing (for which the subjects achieved 70.7% correct) when using a 12-channel CIS processor varied greatly, with a range from 802 Hz to about 8 kHz. This resolution ability was significantly worse than that of normal-hearing listeners who listened to an acoustic simulation of the same processor with 12 channels (a range of 309 to 555 Hz spectral spacing). Furthermore, the ability of implantees to resolve the spectral ripples was significantly correlated with their ability to identify vowels (see Fig. 8).

Henry and Turner (2003) also measured the improvement in discriminable ripple spacing as the number of channels (or electrodes) was increased. They found that a small number of implantees performed similarly to normal-hearing listeners up to 8 channels (and one subject up to 12 channels), but most implantees did not show improvements in performance with more than 4 to 6 channels. These results underline the fact that, provided that the processor has up

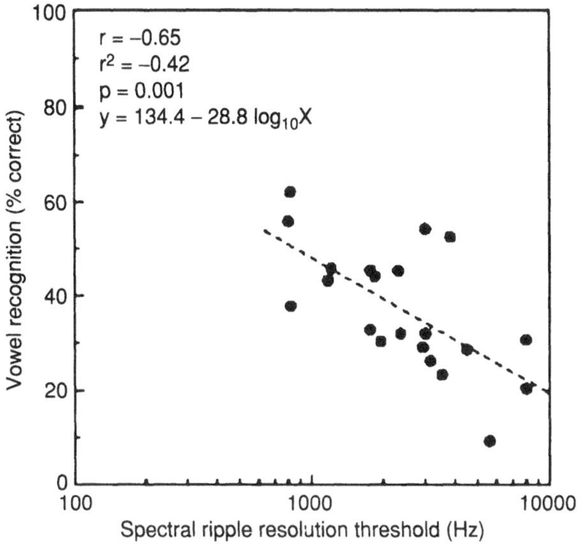

FIG. 8. The relationship between spectral ripple resolution and vowel recognition for 21 cochlear implantees using a 12-channel CIS processor. (Reprinted with permission from Henry and Turner © 2003, Acoustical Society of America.)

to 10 to 12 channels, the main factor inhibiting the ability of implantees to resolve frequencies is the individual amount of perceptual spectral smearing they experience. As previously mentioned, this smearing is probably due to differing degrees of spread of excitation resulting from differing patterns of neural survival and pathological processes within the cochlea.

The impact of spectral smearing on speech perception is greatest for low to mid-frequencies (up to 2 to 3 kHz). This is where the main energy in vowels is located, and where the resolution of vowel formant frequencies is important for vowel identification. The relative importance of resolution for these frequencies over that for higher frequencies is reflected in the differential effect of reducing the number of analysis channels on vowel and consonant identification. That is, consonant identification is affected to a lesser degree than vowel identification when the number of channels is restricted or the bandwidth of the channels is increased (e.g., Fu and Shannon, 2002). Henry *et al.* (2000) measured the relative importance of spectral information in different frequency regions by correlating the ability to distinguish adjacent electrodes when the current levels were randomly varied with the amount of speech information transmitted in the frequency region assigned to the same electrodes. A high correlation was found only for those electrodes assigned to frequencies below ~2.5 kHz. This result is consistent with better speech perception being related to better ability to resolve frequencies in the region of vowel formants, but not dependent as much on frequency resolution in higher frequency regions.

Two studies have investigated the effect of changing the pattern of how filter center frequencies are distributed over electrode place, without changing the total number of analysis filters or the total frequency range, and while providing a significant period of experience with the altered frequency-to-electrode map. Given the previous discussion, it might be predicted that using wider spatial steps between electrodes or using narrower filter bandwidths in the vowel formant region would improve the perception of vowels without a corresponding degradation of consonant information. A study by McKay and Henshall (2002) supported this idea. The same 18 analysis filters were distributed among the same subset of 10 electrodes in two different ways: in the first, the filters were distributed evenly (essentially two per electrode) across the 10 electrodes; and in the second, the lowest 9 filters (up to 2600 Hz) were assigned to the 9 most apical electrodes of the set, and the remaining 9 filters to the single most-basal electrode. It was found that vowel perception and sentence perception in noise were poorer in the evenly allocated map, and consonant perception was unchanged, compared to the map that devoted 9 of the 10 electrodes to the low frequencies. This result supported the idea that devoting more channels to the low frequencies (and correspondingly fewer to the high frequencies) would improve speech perception. A second experiment (Leigh *et al.*, 2004) further tested this hypothesis, by creating a map that retained the full set of electrodes and filters (up to 20). The low-frequency filters were reduced in bandwidth (creating 2 to 3 additional channels below 2600 Hz) and the high-frequency filters were broadened, with the total frequency range being unchanged. The results were mixed. Vowel and consonant confusion tests using tokens from a male speaker supported the hypothesis, showing better transmission of F1 information for vowels, and unchanged consonant information with the altered map. However, tests using whole words spoken by a female showed a degradation of both vowel and consonant perception with the altered map. It is unclear why there was an interaction between speaker gender and outcome, but one possibility is that, using the SPEAK strategy and narrow filters, the electrode selection in each cycle (see Section II.B) represented the lower harmonics of the signal rather than the vowel formant frequencies, leading to perceptual confusion. This interaction of the effect of using narrow filters and speaker gender on speech perception with the SPEAK strategy has been noted also by Fu *et al.* (2002).

In summary, the speech perception of cochlear implantees is limited to some extent by the spectral smearing that occurs due to poor neural survival and spatial spread of intracochlear currents. Excellent speech perception in quiet is possible (depending on the individual) with as few as 10 analysis filters covering the range of approximately 200 Hz to 8 kHz. The distribution of these analysis filters across the frequency range is important: the ability to resolve frequencies in the range up to 2600 Hz is more important than the ability to resolve higher frequencies. Research to optimize frequency-to-electrode allocation is difficult because of the co-existing detrimental effects of shifting the frequency allocation away from that to which the subjects are accustomed. Such shifts always result in an immediate degradation of speech perception, followed by a period of adjustment. The ability of implantees to adjust to shifts in frequency allocation is discussed further in Section V.

B. Complex Pitch/Music Perception

For periodic (or harmonic) complex signals, such as voiced speech or musical sounds, the *fundamental frequency* (F0) leads to the percept of voice pitch or musical pitch. F0 for these sounds generally lies between 50 Hz and at least several hundred Hz. F0 is an important feature of harmonic complex sounds. It provides information about the identity of the sound source, in particular the frequency of vibration of the sound generator. For speech, F0 equals the vibration frequency of the vocal chords. F0 information helps to identify the speaker, and its variation within an utterance provides useful prosodic information (for example, whether the utterance is a question or statement), as well as information about the emotional state of the speaker. In tonal languages, the meaning of single words may be changed when different F0 modulation patterns are applied. Perception of the fundamental frequency is crucial to the appreciation of musical sounds by providing the musical pitch. Importantly, F0 also plays a role in the ability to perceptually segregate different simultaneous sound sources.

The central auditory processing by which the fundamental frequency of a harmonic complex is perceived can be facilitated by either spectral or temporal means. When multiple harmonics fall within the same auditory filter, the output of that filter contains amplitude modulations at the F0 (as well as periodicities related to the harmonic frequencies) that may be perceived via temporal processing of the peripheral neural responses. For example, when acoustic pulse trains are filtered into a high-frequency bandpass in which separate harmonics cannot be resolved, or when a noise is amplitude modulated, a pitch related to the F0 is heard (Burns and Viemeister, 1981; Hanna, 1992; McKay and Carlyon, 1999; Shackleton and Carlyon, 1994). On the other hand, when a series of harmonics is spaced so that they can be resolved (i.e., they activate non-overlapping auditory filters), the central auditory system can extract the F0 by combining spectral or temporal information across the filters. The method that the central auditory system uses to combine information across filters to extract pitch is still a matter for research. Central template models propose that the central auditory system compares the spectral or temporal (phase locking) information across filters with a set of central pitch templates (Cohen *et al.*, 1995; Wightman, 1973). Shamma and

Klein (2000) have demonstrated that such a template can arise by a system of coincidence detection across the output of different auditory filters. An alternative class of models for how the central auditory system may use temporal information to extract the pitch of a complex sound relies on a possible neural autocorrelation mechanism (Meddis and O'Mard, 1997). Peaks in the autocorrelation function pooled across filters provide the cues to pitch in these models. The question of whether complex pitch is represented cortically by spectral or temporal features has also been studied using brain imaging techniques (see Chapter 10 by Hall). As previously mentioned, a recent study by Oxenahm *et al.* (2004) suggested that the temporal information of individual higher harmonics cannot be used to extract F0 information, if the harmonics are presented at an incorrect cochlear place. Thus, it appears that the extraction of F0 from a harmonic complex may require place-specific information.

The aforementioned research in normal hearing has implications for the transmission of F0 information via a cochlear implant, and helps to explain the difficulty that implantees have in perceiving musical pitch. The first potential issue is the inevitable and unavoidable mismatch between each analysis filter frequency and the characteristic frequency associated with the corresponding electrode place. Thus, any model of pitch perception that relies upon a spectral template model based upon characteristic frequencies would predict that implantees could not correctly identify the complex pitch. Even if 'new' central templates could be formed, in which the 'new' frequency-to-place pattern is incorporated, the signal processing applied in existing implant sound-processors (SPEAK, ACE, or CIS), in which fixed overlapping bandpass filters (up to 22) are used, does not allow the implantee to resolve many harmonic components of complex sounds or to identify the precise harmonic frequencies. In the temporal domain, the output of the analysis filters will have amplitude modulations at the F0 rate, provided that the filters are wide enough to encompass more than one harmonic, and provided that smoothing functions at the output of the filters do not attenuate this modulation. In theory, these modulations could provide an F0 pitch percept, provided that the rate of stimulation is high enough to accurately 'sample' this modulation pattern.

However, many studies have shown that implantees have great difficulty in recognizing melodies (e.g., Leal *et al.*, 2003). This is particularly true if the cues due to rhythm are removed or the melodies are played without accompanying words. Figure 9 shows the results of a pitch ranking task (McDermott, 2004) in which implantees were asked to rank the pitch of two sung vowels separated by one-half an octave (a 41% shift in F0). The average scores across subjects for 4 of the six stimulus pairs were at chance levels. Furthermore, six of the nine implantees reported a pitch change for specific notes opposite to the direction expected. Unfortunately, this situation does not seem to be greatly improved with higher-rate strategies such as CIS and ACE, which would be expected to

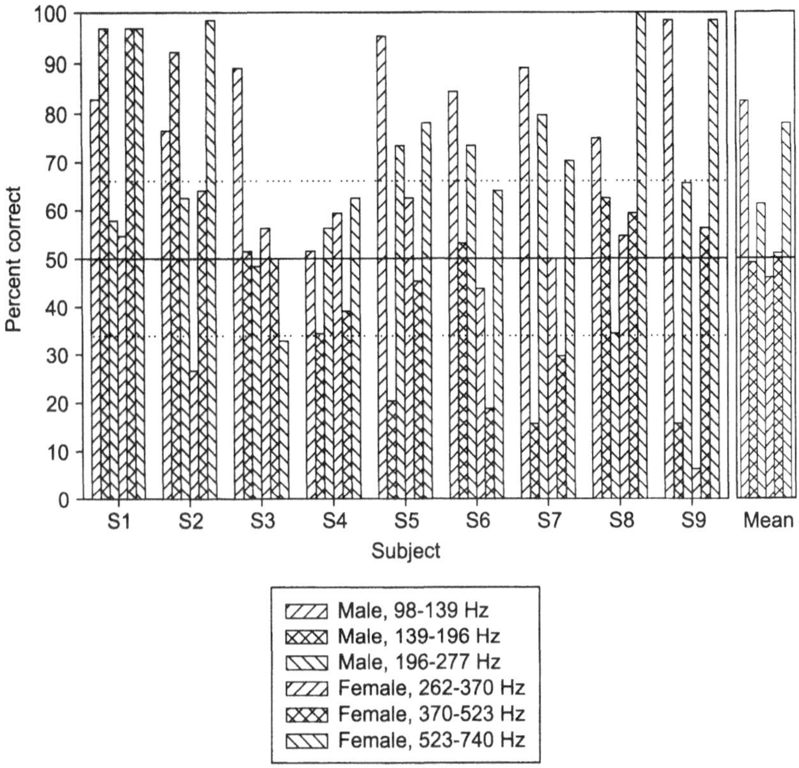

FIG. 9. Results of a pitch ranking experiment in which users of the SPEAK strategy were asked to rank the pitch of two sequential sung vowels (/a/ and /i/) differing in fundamental frequency by half an octave (McDermott, 2004). Chance score is 50% correct (solid line) and the dotted lines delimit the region where scores are not significantly different from chance at the $p < 0.01$ level. Note that the results that are below this region are pitch reversals. The six bars for each subject represent the results for male and female singers for intervals relative to three reference frequencies.

represent modulations at F0 more adequately than SPEAK. A later study (Tsang, 2004) showed that these same sung vowels were correctly ranked on average 71% of the time by users of CIS and ACE (with a similar incidence of reversed pitches). Inspection of the processor outputs for these strategies showed the presence of amplitude modulations for low F0s, however, the modulations on different electrodes appeared to have somewhat random phase relationships. It has been shown psychophysically (McKay and McDermott, 1996) that phase shifts between nearby electrodes will impair the ability of the implantee to perceive the modulation frequency, or may lead to a spurious pitch not related to F0. It was also noted by Tsang (2004) that spectral (timbral) changes were evident (as different current amplitudes across electrodes) between different sung notes, even when the same vowel was used, and that these timbral changes may have been the source of the reversed-pitch responses. It was postulated from these experiments that temporal cues to complex pitch are not represented in the processor output in a way that implantees can consistently or successfully use. In particular it was hypothesized that the out-of-phase nature of the modulations at F0 may be preventing implantees from using this temporal cue. Further support for this hypothesis has been provided by recent psychophysical experiments by Swanson *et al.* (2004), in which tones were better ranked in pitch by processing that led to in-phase periodicities in current pulse patterns across electrode positions.

In summary, complex pitch is currently very difficult for implantees to perceive. It is unlikely that implantees will be able to perceive complex pitch via the resolution of harmonics of the sound unless many more perceptual channels are provided (see Chapter 4 by Shannon), and even then it would perhaps be necessary for the analysis channels to provide temporal input to the correct tonotopic place in the cochlea (Oxenham *et al.*, 2004). The direct application of amplitude modulations at F0 should be perceivable as a pitch by implantees, but current processing schemes do not provide optimum information about F0, either because of the smoothing of filter outputs or because of the uncontrolled phase differences on nearby electrodes.

V. Learning a New Tonotopic Code

This section will examine the evidence of plasticity in the auditory system following implantation with a cochlear implant and in response to a shift in peripheral tonotopic map. As previously mentioned, the allocation of analysis frequencies to cochlear places in a cochlear implant does not reproduce the characteristic frequency pattern of normal hearing. For patients who have enjoyed normal (or near-normal) hearing before the advent of their deafness, the fitting of a cochlear implant means that the central processing of frequency information (or the interpretation of this information) must undergo a change in order for signals such as speech to be understood. Unless the onset of deafness is very sudden, and followed quickly by implantation, there is also the issue of the plastic changes that occur as a result of auditory deprivation. Finally, some implantees have a congenital profound deafness, leading to questions of whether their auditory systems have similar spectral processing functionality to that of normal-hearing people.

Several studies in animal models or in humans have examined the impact of congenital hearing loss or auditory deprivation on the central auditory system. The primary auditory cortex in congenitally deaf cats (with no prior auditory experience) can be activated via electrical stimulation and shows rudimentary cochleotopic organization (Hartmann *et al.*, 1997). However, there is evidence that the cortical auditory system remains immature and even degenerates during this period of auditory deprivation (Shepherd and Hardie, 2001). For example, Sharma *et al.* (2002) and Ponton and Eggermont (2001) used the latency of the P1 auditory-evoked response to investigate the maturation of the central auditory system in children and adults with congenital hearing loss who received an implant. The P1 latency usually decreases over the first 6 years of life, with smaller decreases up to as many as 20 years, showing maturational development of the cortical response. Both groups found that the P1 latency of congenitally deaf children who were older than 3.5 years at implantation remained at a very immature value (similar to 1- to 2-year-old normal-hearing children), even after many years of electrical stimulation. In contrast, Sharma *et al.* (2002) found that children who were implanted before the age of 3.5 years exhibited an age-appropriate latency within 6 months of implantation. The latter result suggests that there are at least some aspects of cortical plasticity that have a sensitive period after birth (3.5 years in humans). The existence of a sensitive period for plasticity is further supported by studies with congenitally deaf white cats, in which the area of cortical activation following implantation and electrical stimulation increased over time with the implant for cats who were implanted at <6 months of age, but not for cats who were implanted after 6 months of age (Kral *et al.*, 2002).

It should be noted that, although there is evidence for a sensitive developmental period, there is also evidence that cortical re-organization occurs throughout life. An example of this is provided by animal studies in which a frequency-specific cochlear lesion leads to an expansion of the cortical representation of neighboring frequencies into the area previously occupied by the missing frequencies (Irvine *et al.*, 2000). At the perceptual level, there is some evidence that congenitally deaf children who are implanted before 4 years of age develop expressive and receptive language at a faster rate than those implanted later (Dowell *et al.*, 2002). However, increasing duration of deafness is associated with poorer outcomes even for older congenitally deaf children (Sarant *et al.*, 2001) and post-linguistically deafened adults (Blamey *et al.*, 1996). In the latter two groups, the effect may be due to ongoing degeneration of the auditory system during auditory deprivation, as well as a reduced plasticity.

The largest effects of auditory deprivation are seen in the auditory association areas of the cortex. It has been shown that these areas are not as well activated in people (or animals) that have experienced a lengthy period of auditory deprivation (Giraud *et al.*, 2000, 2001b). Furthermore, Tobey and Devous (2003) found that implantees with the poorest speech perception ability were those who had the least activation of auditory association areas and very little activation of the cortex ipsilateral to the stimulation. There is evidence that some auditory areas of the brain are taken over by visual processing tasks in congenitally deaf people (Nishimura *et al.*, 1999), and conversely some visual areas are activated by (auditory-alone) speech stimuli after implantation (Giraud *et al.*, 2001a). Hartmann and Kral (2004) have proposed that there is a decoupling of the primary and secondary cortex in the congenitally deaf, leading to cross-modal reorganization, and that the degree of success with the implant depends on how well this de coupling can be overcome.

Given the previous discussion, it can be seen that there are several factors that might affect how well or how quickly an implantee can adapt to the unfamiliar electrical stimulation after a period of profound deafness. After implantation, the auditory processing functions that are important for analyzing and categorizing speech elements need to adapt to an altered (or completely new) peripheral input. Even if the central auditory processing of stimuli is quite normal (e.g., the person has only recently suffered a hearing loss and experienced normal auditory development), the processing of speech categories will still need to be adapted to the new tonotopic map in the primary auditory cortex. If the person has experienced a long period of deafness, or is congenitally deaf, many of the auditory processing functions will be degraded or undeveloped. Thus, plasticity in the auditory processing areas of the brain is crucial to the success or otherwise of implantation.

Experiments with normal-hearing listeners in which the temporal envelope of speech is applied to shifted frequency bands simulate the effects of the shifted tonotopic map that implantees experience. Such experiments always show a large and immediate reduction in ability to understand speech (e.g., Baskent and Shannon, 2003; Shannon *et al.*, 1998). However, given only limited experience listening to the shifted speech, normal-hearing listeners will show some recovery in speech perception (Fu and Galvin, 2003; Rosen *et al.*, 1999). Of course, unlike the experiments with normal-hearing listeners, implantees gain exclusive and long-term experience with their new tonotopic code, and this would be expected to maximize their adaptation. After implantation, there is usually a period of up to two years in which the speech perception ability of post-lingually deaf adults continues to improve, with most of the improvement happening in the first 3 to 6 months (Blamey *et al.*, 1996; Hamzavi *et al.*, 2003; Staller *et al.*, 1997; Waltzman *et al.*, 1999; Weston and Waltzman, 1995). Whether this improvement is due solely to adaptation to the shifted tonotopic code, or is also due to other plastic changes occurring in the cortex is unknown. It is evident, however, that the duration of deafness (or existence of significant residual hearing) does affect the rate at which speech perception improves (Blamey *et al.*, 1996; Hamzavi *et al.*, 2003; Shipp and Nedzelski, 1995).

An interesting experiment that aimed to find out whether the frequency shift experienced by implantees is an important factor affecting speech understanding was performed by Svirsky *et al.* (2004). In that experiment, four subjects were tested on the day of fitting of their implant and at various times for 2 years following implantation. Subjects were presented with a two-dimensional vowel space in which they could choose synthetic vowels of varying F1 and F2 formant frequencies. Their task was to choose stimuli that most sounded like their memory of given vowels. It was expected that, on the day of fitting, the vowel space would show a basal-ward shift relative to that of normal-hearing listeners (due to the fact that the positions of the electrodes in the cochlea did not match the analysis frequencies assigned to the electrodes) and that the vowel space would gradually shift to match that of normal hearing as the implantee learned to re-categorize the speech sounds. Surprisingly, no basal-ward shift was seen. Instead, the subjects had central ill-defined vowel spaces immediately after switch-on, and the vowel spaces gradually shifted to be the same as normal-hearing listeners over a period of time ranging from 1 month to 2 years. The three subjects who became deaf well into their adulthood showed normal vowel spaces within 3 to 4 months, whereas the one subject who became deaf prelingually (at 20 months) needed 2 years experience before the vowel space became normal.

These results suggest that memories of vowel categories may decay during periods of profound deafness and, therefore, that the shift in frequency allocation from that of normal hearing is not very relevant (except perhaps for those who receive an implant immediately after becoming deaf). For people who are likely to have better-preserved auditory processing skills (e.g., those that show better activation of auditory association areas or who have more residual hearing and have used hearing aids) the task of re-categorizing vowel spaces can take between 1 and 4 months, whereas, for people with long-term deafness, it may take significantly longer, or may never reach a 'normal' proficiency.

Given the previous discussion, it would be expected that implantees could adapt to additional changes to their frequency-to-electrode allocation given further significant time and experience. Experiments that have investigated the alteration of frequency-to-electrode allocation always show an immediate detrimental effect of such shifts away from the allocation in the subject's usual processor map (Dorman and Ketten, 2003; Fu and Shannon, 1999a,b). However, experiments in which subjects are provided with significant experience with the altered map show total or at least partial recovery of speech perception. It is difficult to test the effect of a frequency-to-electrode shift in cochlear implantees without introducing other changes that will influence the result. For example, keeping the analysis channels and number of electrodes unchanged while shifting the electrode positions requires the use of fewer active electrodes than would normally be used in the subject's clinical map. Alternatively, keeping the number and location of electrodes constant while changing the analysis filters will lead to changes in the filter bandwidths and/or the total overall analysis range. The first type of experiment was performed by Dorman and Ketten

(2003). They programmed a patient with a 6-channel Ineraid implant with three 4-channel CIS processors. The same four filters were assigned to electrodes 1–4, 2–5, or 3–6. Each electrode position change represented a basal-ward shift of characteristic frequency of ratio approximately 1.6. The subject was tested immediately following each change and one week later. The results showed an immediate drop in perception, and a recovery over one week of consonant perception but no recovery of vowel perception. Although this report was only for one subject, it suggests that more than one week is needed to re-establish vowel categorization. The second type of experiment was performed by Fu *et al.* (2002), Skinner *et al.* (1995), and McKay and Henshall (2002). McKay and Henshall showed a full recovery of word perception in quiet and sentence perception in noise after two weeks when the analysis filters were shifted upward by a ratio of 1.33. Skinner *et al.* showed an improvement in vowel perception (i.e., better than the unshifted map) after 2 or 3 weeks' experience when the filters were shifted downward by a ratio of 0.8. Fu *et al.* shifted the frequencies of the analysis filters down by one octave (ratio of 0.5). In that experiment, three users of the Nucleus-22 implant using the SPEAK strategy wore the experimental map for three months. Two of the three subjects were quickly (over 2 to 4 weeks) able to fully adjust to the frequency-allocation shift when listening to men's voices. However, for female voices, and even more strikingly for children's voices, vowel perception was severely affected over the whole 3-month period for all three subjects. This speaker-gender interaction effect may have been caused by the fact that the filter bandwidths were halved (to 100 Hz in the first 8 filters) and the analysis range was lowered by an octave. These changes, together with the use of voices with high fundamental frequency (200 to 400 Hz for females and children) would lead to the channel selection of SPEAK representing harmonics of the speech rather than formant frequencies.

In summary, it would seem that implantees can adapt to mild shifts in frequency-to-electrode allocation, given at least two weeks to accommodate to the new frequency code. In a similar way, when the implant is first fitted it may take several months (or in some cases up to two years) for the categorization of vowel sounds to be established in a way that is similar to normal hearing. In the case of initial fitting, the vowel perceptual spaces may be highly abnormal due to degradation over the period of profound deafness. Thus, the mis-match in electrode position and characteristic frequency may not be as important as the degradation of speech processing systems within the auditory association areas. It is interesting to note that it has been reported that implantees or normal-hearing people do not need a second period of adjustment to previously experienced tonotopic maps after an experiment has been completed (Fu and Galvin, 2003; Fu *et al.*, 2002). This suggests that, once established, a new perceptual vowel space is resistant to decay for a period of time, even if a second perceptual space is established in the intervening time.

VI. Conclusions

In this chapter, the functional aspects of cochlear implants that are relevant to spectral processing have been discussed. The coding of acoustic spectral information onto place-of-stimulation information within the cochlea has led to successful transmission of information about speech signals. However, the perception of music with cochlear implants remains problematic. The current implant systems do not provide the fine spectral detail or temporal information that the perception of musical pitch requires. The variable outcome of implantation is due largely to the variable survival of peripheral auditory neurons and the plastic changes that occur throughout the whole central auditory system in response to periods of profound deafness. There is much evidence that the central auditory system can adapt to novel peripheral inputs, particularly for tasks such as speech sound categorization. This is particularly true for congenitally deaf children who are implanted at an early age, but is also true, to variable extents, for people who have experienced a long period of auditory deprivation. Future research will focus on habilitation and remedial strategies that limit the degree of degeneration of the auditory system experienced by people with profound or partial deafness, as well as signal processing strategies to improve the transmission of fine temporal and spectral information.

References

Baskent, D., and Shannon, R. V. (2003). Speech recognition under conditions of frequency-place compression and expansion. *J. Acoust. Soc. Am.* **113,** 2064–2076.

Baumann, U., and Nobbe, A. (2004). Pulse rate discrimination with deeply inserted electrode arrays. *Hear. Res.* **196,** 49–57.

Blamey, P., Arndt, P., Bergeron, F., Bredberg, G., Brimacombe, J., Facer, G., Larky, J., Lindström, B., Nedzelski, J., Peterson, A., Shipp, D., Staller, S., and Whitford, L. (1996). Factors affecting auditory performance of postlinguistically deaf adults using cochlear implants. *Audiol. Neuro-Otol.* **1,** 293–306.

Boex, C., de Balthasar, C., Kos, M. I., and Pelizzone, M. (2003). Electrical field interactions in different cochlear implant systems. *J. Acoust. Soc. Am.* **114,** 2049–2057.

Bonnet, R. M., Frijns, J. H., Peeters, S., and Briaire, J. J. (2004). Speech recognition with a cochlear implant using triphasic charge-balanced pulses. *Acta Otolaryngol.* **124,** 371–375.

Burns, E. M., and Viemeister, N. F. (1981). Played-again SAM: Further observations on the pitch of amplitude-modulated noise. *J. Acoust. Soc. Am.* **70,** 1655–1660.

Busby, P. A., Tong, Y. C., and Clark, G. M. (1993). The perception of temporal modulations by cochlear implant patients. *J. Acoust. Soc. Am.* **94,** 124–131.

Carlyon, R. P., and Deeks, J. M. (2002). Limitations on rate discrimination. *J. Acoust. Soc. Am.* **112,** 1009–1025.

Carlyon, R. P., Geurts, L, and Wouters, J. (2000). Detection of small across-channel timing differences by cochlear implantees. *Hear. Res.* **141,** 140–154.

Carlyon, R. P., van Wieringen, A., Long, C. J., Deeks, J. M., and Wouters, J. (2002). Temporal pitch mechanisms in acoustic and electric hearing. *J. Acoust. Soc. Am.* **112,** 621–633.

Carlyon, R. P., Long, C. J., and Deeks, J. M. (2003). Temporal cues for concurrent sound segregation and pitch perception in electric and acoustic hearing. "Proceedings of the 2003 Conference on Implantable Auditory Prostheses," p. 36. Asilomar, California.

Cazals, Y., Pelizzone, M., Saudan, O., and Boex, C. (1994). Low-pass filtering in amplitude modulation detection associated with vowel and consonant identification in subjects with cochlear implants. *J. Acoust. Soc. Am.* **96,** 2048–2054.

Chatterjee, M., and Shannon, R. V. (1998). Forward masked excitation patterns in multielectrode electrical stimulation. *J. Acoust. Soc. Am.* **103,** 2565–2572.

Ching, T. Y., Psarros, C., Hill, M., Dillon, H., and Incerti, P. (2001). Should children who use cochlear implants wear hearing aids in the opposite ear? *Ear Hear.* **22,** 365–380.

Ching, T. Y., Incerti, P., and Hill, M. (2004). Binaural benefits for adults who use hearing aids and cochlear implants in opposite ears. *Ear Hear.* **25,** 9–21.

Cohen, L. T., Busby, P. A., and Clark, G. M. (1996a). Cochlear implant place psychophysics. 2. Comparison of forward masking and pitch estimation data. *Audiol. Neuro-Otol.* **1,** 278–292.

Cohen, L. T., Busby, P. A., Whitford, L. A., and Clark, G. M. (1996b). Cochlear implant place psychophysics. 1. Pitch estimation with deeply inserted electrodes. *Audiol. Neuro-Otol.* **1,** 265–277.

Cohen, L. T., Saunders, E., and Clark, G. M. (2001). Psychophysics of a prototype peri-modiolar cochlear implant electrode array. *Hear. Res.* **155,** 63–81.

Cohen, L. T., Richardson, L. M., Saunders, E., and Cowan, R. S. (2003). Spatial spread of neural excitation in cochlear implant recipients: Comparison of improved ECAP method and psychophysical forward masking. *Hear. Res.* **179,** 72–87.

Cohen, M. A., Grossberg, S., and Wyse, L. L. (1995). A spectral network model of pitch perception. *J. Acoust. Soc. Am.* **98,** 862–879.

Collins, L. M., Zwolan, T., and Wakefield, G. H. (1997). Comparison of electrode discrimination, pitch ranking, and pitch scaling data in postlingually deafened adult cochlear implant subjects. *J. Acoust. Soc. Am.* **101,** 440–455.

Deeks, J. M., and Carlyon, R. P. (2004). Simulations of cochlear implant hearing using filtered harmonic complexes: Implications for concurrent sound segregation. *J. Acoust. Soc. Am.* **115,** 1736–1746.

Dettman, S. J., D'Costa, W. A., Dowell, R. C., Winton, E. J., Hill, K. L., and Williams, S. S. (2004). Cochlear implants for children with significant residual hearing. *Arch. Otol. - Head Neck Surg.* 612–618.

Dorman, M. F., and Ketten, D. (2003). Adaptation by a cochlear-implant patient to upward shifts in the frequency representation of speech. *Ear Hear.* **24,** 457–460.

Dowell, R. C., Dettman, S., Blamey, P. J., Barker, E. J., and Clark, G. M. (2002). Speech perception in children using cochlear implants: Prediction of long-term outcomes. *Cochlear Implants International* **3,** 1–18.

Dynes, S. B., and Delgutte, B. (1992). Phase-locking of auditory-nerve discharges to sinusoidal electric stimulation of the cochlea. *Hear. Res.* **58,** 79–90.

Fishman, K. E., Shannon, R. V., and Slattery, W. H. (1997). Speech recognition as a function of the number of electrodes used in the SPEAK cochlear implant speech processor. *J. Speech Hear. Res.* **40,** 1201–1215.

Friesen, L. M., Shannon, R. V., Baskent, D., and Wang, X. (2001). Speech recognition in noise as a function of the number of spectral channels: Comparison of acoustic hearing and cochlear implants. *J. Acoust. Soc. Am.* **110,** 1150–1163.

Frijns, J. H., Briaire, J. J., de Laat, J. A., and Grote, J. J. (2002). Initial evaluation of the Clarion CII cochlear implant: Speech perception and neural response imaging. *Ear Hear.* **23,** 184–197.

Frijns, J. H., Klop, W. M., Bonnet, R. M., and Briaire, J. J. (2003). Optimizing the number of electrodes with high-rate stimulation of the clarion CII cochlear implant. *Acta Otolaryngol.* **123,** 138–142.

Fu, Q. J. (2002). Temporal processing and speech recognition in cochlear implant users. *Neuroreport* **13,** 1635–1639.

Fu, Q. J., and Galvin, J. J., 3rd (2003). The effects of short-term training for spectrally mismatched noise-band speech. *J. Acoust. Soc. Am.* **113,** 1065–1072.

Fu, Q. J., and Shannon, R. V. (1999a). Effects of electrode location and spacing on phoneme recognition with the Nucleus-22 cochlear implant. *Ear Hear.* **20,** 321–331.

Fu, Q. J., and Shannon, R. V. (1999b). Effects of electrode configuration and frequency allocation on vowel recognition with the Nucleus-22 cochlear implant. *Ear Hear.* **20,** 332–344.

Fu, Q. J., and Shannon, R. V. (2002). Frequency mapping in cochlear implants. *Ear Hear.* **23,** 339–348.

Fu, Q. J., Shannon, R. V., and Galvin, J. J., 3rd (2002). Perceptual learning following changes in the frequency-to-electrode assignment with the Nucleus-22 cochlear implant. *J. Acoust. Soc. Am.* **112,** 1664–1674.

Giraud, A. L., Truy, E., Frackowiak, R. S., Gregoire, M. C., Pujol, J. F., and Collet, L. (2000). Differential recruitment of the speech processing system in healthy subjects and rehabilitated cochlear implant patients. *Brain* **123,** 1391–1402.

Giraud, A. L., Price, C. J., Graham, J. M., Truy, E., and Frackowiak, R. S. (2001a). Cross-modal plasticity underpins language recovery after cochlear implantation. *Neuron* **30,** 657–663.

Giraud, A. L., Truy, E., and Frackowiak, R. (2001b). Imaging plasticity in cochlear implant patients. *Audio Neuro-Otol.* **6,** 381–393.

Hamzavi, J., Baumgartner, W. D., Pok, S. M., Franz, P., and Gstoettner, W. (2003). Variables affecting speech perception in postlingually deaf adults following cochlear implantation. *Acta Otolaryngol.* **123,** 493–498.

Hanekom, J. J., and Shannon, R. V. (1998). Gap detection as a measure of electrode interaction in cochlear implants. *J. Acoust. Soc. Am.* **104,** 2372–2384.

Hanna, T. E. (1992). Discrimination and identification of modulation rate using a noise carrier. *J. Acoust. Soc. Am.* **91,** 2122–2128.

Hardie, N. A., and Shepherd, R. K. (1999). Sensorineural hearing loss during development: Morphological and physiological response of the cochlea and auditory brainstem. *Hear. Res.* **128,** 147–165.

Hartmann, R., and Kral, A. (2004). Central responses to electrical stimulation. *In* "Cochlear Implants: Auditory Prostheses and Electric Hearing" (F. G. Zeng, A. N. Popper, and R. R. Fay, Eds.), Ch. 6, pp. 213–285. Springer, NY.

Hartmann, R., Shepherd, R. K., Heid, S., and Klinke, R. (1997). Response of the primary auditory cortex to electrical stimulation of the auditory nerve in congenitally deaf white cats. *Hear. Res.* **112,** 115–133.

Henry, B. A., and Turner, C. W. (2003). The resolution of complex spectral patterns by cochlear implant and normal-hearing listeners. *J. Acoust. Soc. Am.* **113,** 2861–2873.

Henry, B. A., McKay, C. M., McDermott, H. J., and Clark, G. M. (2000). The relationship between speech perception and electrode discrimination in cochlear implantees. *J. Acoust. Soc. Am.* **108,** 1269–1280.

Henshall, K. R., and McKay, C. M. (2001). Optimizing electrode and filter selection in cochlear implant speech processor maps. *J. Am. Acad. Audiol.* **12,** 478–489.

Hong, R. S., Rubinstein, J. T., Wehner, D., and Horn, D. (2003). Dynamic range enhancement for cochlear implants. *Otol. Neurotol.* **24,** 590–595.

Irvine, D. R., Rajan, R., and McDermott, H. J. (2000). Injury-induced reorganization in adult auditory cortex and its perceptual consequences. *Hear. Res.* **147,** 188–199.

Jolly, C. N., Clopton, B. M., Spelman, F. A., and Lineaweaver, S. K. (1997). Guinea pig auditory nerve response triggered by a high density electrode array. *Med. Prog. Technol.* **21,** 13–23.

Kanzaki, S., Stover, T., Kawamoto, K., Prieskorn, D. M., Altschuler, R. A., Miller, J. M., and Raphael, Y. (2002). Glial cell line-derived neurotrophic factor and chronic electrical stimulation prevent VIII cranial nerve degeneration following denervation. *J. Comp. Neurol.* **454,** 350–360.

Kral, A., Hartmann, R., Mortazavi, D., and Klinke, R. (1998). Spatial resolution of cochlear implants: The electrical field and excitation of auditory afferents. *Hear. Res.* **121,** 11–28.

Kral, A., Hartmann, R., Tillein, J., Heid, S., and Klinke, R. (2002). Hearing after congenital deafness: Central auditory system plasticity and sensory deprivation. *Cerebral Cortex* **12,** 797–807.

Kreft, H. A., Donaldson, G. S., and Nelson, D. A. (2004). Effects of pulse rate on threshold and dynamic range in Clarion cochlear-implant users. *J. Acoust. Soc. Am.* **115,** 1885–1888.

Laback, B., Deutsch, W. A., and Baumgartner, W. D. (2004). Coding of vowel-like signals in cochlear implant listeners. *J. Acoust. Soc. Am.* **116,** 1208–1223.

Leal, M. C., Shin, Y. J., Laborde, M. L., Calmels, M. N., Verges, S., Lugardon, S., Andrieu, S., Deguine, O., and Fraysse, B. (2003). Music perception in adult cochlear implant recipients. *Acta Otolaryngol.* **123,** 826–835.

Leigh, J. R., Henshall, K. R., and McKay, C. M. (2004). Optimizing frequency-to-electrode allocation in cochlear implants. *J. Am. Acad. Audiol.* (in press).

Litvak, L. M., Smith, Z. M., Delgutte, B., and Eddington, D. K. (2003). Desynchronization of electrically evoked auditory-nerve activity by high-frequency pulse trains of long duration. *J. Acoust. Soc. Am.* **114,** 2066–2078.

McDermott, H. J. (2004). Music perception with cochlear implants: A review. *Trends Amplif.* **8,** 49–82.

McDermott, H. J., and McKay, C. M. (1997). Musical pitch perception with electrical stimulation of the cochlea. *J. Acoust. Soc. Am.* **101,** 1622–1631.

McDermott, H. J., McKay, C. M., and Vandali, A. E. (1992). A new portable sound processor for the University of Melbourne/Nucleus Limited multielectrode cochlear implant. *J. Acoust. Soc. Am.* **91,** 3367–3371.

McDermott, H. J., McKay, C. M., Richardson, L. M., and Henshall, K. R. (2003a). Application of loudness models to sound processing for cochlear implants. *J. Acoust. Soc. Am.* **114,** 2190–2197.

McDermott, H. J., Sucher, C. S., and McKay, C. M. (2003b). Speech intelligibility with a cochlear implant sound-processing scheme based on loudness models. *In* "Proceedings of the 25th Annual International Conference of the IEEE Engineering in Medicine and Biology Society" (R. S. Leder, Ed.), pp. 2006–2009, Cancun, Mexico.

McDermott, H. J., Sucher, C. M., and McKay, C. M. (2004). Speech perception with a cochlear implant sound processor incorporating loudness models. *Acoustics Research Letters Online* **6,** 7–13.

McKay, C. M. (2004). Psychophysics and electrical stimulation. *In* "Cochlear Implants: Auditory Prostheses and Electric Hearing" (F. G. Zeng, A. N. Popper, and R. R. Fay, Eds.), Ch. 7, pp. 286–333. Springer, NY.

McKay, C. M., and Carlyon, R. P. (1999). Dual temporal pitch percepts from acoustic and electric amplitude-modulated pulse trains. *J. Acoust. Soc. Am.* **105,** 347–357.

McKay, C. M., and Henshall, K. R. (2002). Frequency-to-electrode allocation and speech perception with cochlear implants. *J. Acoust. Soc. Am.* **111,** 1036–1044.

McKay, C. M., and McDermott, H. J. (1996). The perception of temporal patterns for electrical stimulation presented at one or two intracochlear sites. *J. Acoust. Soc. Am.* **100,** 1081–1092.

McKay, C. M., McDermott, H. J., and Clark, G. M. (1994). Pitch percepts associated with amplitude-modulated current pulse trains in cochlear implantees. *J. Acoust. Soc. Am.* **96,** 2664–2673.

McKay, C. M., McDermott, H. J., and Clark, G. M. (1995). Pitch matching of amplitude-modulated current pulse trains by cochlear implantees: The effect of modulation depth. *J. Acoust. Soc. Am.* **97,** 1777–1785.

McKay, C. M., McDermott, H. J., and Clark, G. M. (1996). The perceptual dimensions of single-electrode and nonsimultaneous dual-electrode stimuli in cochlear implantees. *J. Acoust. Soc. Am.* **99,** 1079–1090.

McKay, C. M., O'Brien, A., and James, C. J. (1999). Effect of current level on electrode discrimination in electrical stimulation. *Hear. Res.* **136,** 159–164.

McKay, C., McDermott, H., and Carlyon, R. (2000). Place and temporal cues in pitch perception: Are they truly independent? *Acoustic Research Letters Online* **1,** 25–30.

McKay, C. M., Henshall, K. R., Farrell, R. J., and McDermott, H. J. (2003). A practical method of predicting the loudness of complex electrical stimuli. *J. Acoust. Soc. Am.* **113,** 2054–2063.

Meddis, R., and O'Mard, L. (1997). A unitary model of pitch perception. *J. Acoust. Soc. Am.* **102,** 1811–1820.

Moore, B. C. J., and Glasberg, B. R. (1983). Masking patterns for synthetic vowels in simultaneous and forward masking. *J. Acoust. Soc. Am.* **73,** 906–917.

Nelson, D. A. (1994). Level-dependent critical bandwidth for phase discrimination. *J. Acoust. Soc. Am.* **95,** 1514–1524.

Nishimura, H., Hashikawa, K., Doi, K., Iwaki, T., Watanabe, Y., Kusuoka, H., Nishimura, T., and Kubo, T. (1999). Sign language 'heard' in the auditory cortex. *Nature* **397,** 116.

Oxenham, A. J., Bernstein, J. G., and Penagos, H. (2004). Correct tonotopic representation is necessary for complex pitch perception. *Proc. Natl. Acad. Sci. USA* **101,** 1421–1425.

Pfingst, B. E., Holloway, L. A., Zwolan, T. A., and Collins, L. M. (1999). Effects of stimulus level on electrode-place discrimination in human subjects with cochlear implants. *Hear. Res.* **134,** 105–115.

Pijl, S., and Schwarz, D. W. (1995). Melody recognition and musical interval perception by deaf subjects stimulated with electrical pulse trains through single cochlear implant electrodes. *J. Acoust. Soc. Am.* **98,** 886–895.

Ponton, C. W., and Eggermont, J. J. (2001). Of kittens and kids: Altered cortical maturation following profound deafness and cochlear implant use. *Audiol. Neuro-Otol.* **6,** 363–380.

Pressnitzer, D., de Cheveigne, A., and Winter, I. M. (2002). Perceptual pitch shift for sounds with similar waveform autocorrelation. *Acoustics Research Letters Online* **3,** 1–6.

Rosen, S., Faulkner, A., and Wilkinson, L. (1999). Adaptation by normal listeners to upward spectral shifts of speech: Implications for cochlear implants. *J. Acoust. Soc. Am.* **106,** 3629–3636.

Rubinstein, J. T., Wilson, B. S., Finley, C. C., and Abbas, P. J. (1999). Pseudospontaneous activity: Stochastic independence of auditory nerve fibers with electrical stimulation. *Hear. Res.* **127,** 108–118.

Sarant, J. Z., Blamey, P. J., Dowell, R. C., Clark, G. M., and Gibson, W. P. (2001). Variation in speech perception scores among children with cochlear implants. *Ear Hear.* **22,** 18–28.

Shackleton, T. M., and Carlyon, R. P. (1994). The role of resolved and unresolved harmonics in pitch perception and frequency modulation discrimination. *J. Acoust. Soc. Am.* **95,** 3529–3540.

Shamma, S., and Klein, D. (2000). The case of the missing pitch templates: How harmonic templates emerge in the early auditory system. *J. Acoust. Soc. Am.* **107,** 2631–2644.

Shannon, R. V. (1983). Multichannel electrical stimulation of the auditory nerve in man. I. Basic psychophysics. *Hear. Res.* **11,** 157–189.

Shannon, R. V. (1989). Detection of gaps in sinusoids and pulse trains by patients with cochlear implants. *J. Acoust. Soc. Am.* **85,** 2587–2592.

Shannon, R. V. (1992). Temporal modulation transfer functions in patients with cochlear implants. *J. Acoust. Soc. Am.* **91,** 2156–2164.

Shannon, R. V., Zeng, F.-G., and Wygonski, J. (1998). Speech recognition with altered spectral distribution of envelope cues. *J. Acoust. Soc. Am.* **104,** 2467–2476.

Sharma, A., Dorman, M., and Spahr, A. J. (2002). A sensitive period for the development of the central auditory system in children with cochlear implants: Implications for age of implantation. *Ear Hear.* **23,** 532–539.

Shepherd, R., and Hardie, N. (2001). Deafness-induced changes in the auditory pathway: Implications for cochlear implants. *Audiol. Neuro-Otol.* **6,** 305–318.

Shepherd, R. K., and Javel, E. (1997). Electrical stimulation of the auditory nerve. I. Correlation of physiological responses with cochlear status. *Hear. Res.* **108,** 112–144.

Shipp, D. B., and Nedzelski, J. M. (1995). Prognostic indicators of speech recognition performance in adult cochlear implant users: A prospective analysis. *Ann. Otol. Rhinol. Laryngol. Suppl.* **166, 104,** 194–196.

Skinner, M. W., Holden, L. K., and Holden, T. A. (1995). Effect of frequency boundary assignment on speech recognition with the SPEAK speech-coding strategy. *Ann. Otol. Rhinol. Laryngol. Suppl.* **166, 104,** 307–311.

Stainsby, T. H. (2000). The perception of musical sounds with cochlear implants. Ph.D. Thesis, Department of Otolaryngology. The University of Melbourne.

Staller, S., Menapace, C., Domico, E., Mills, D., Dowell, R. C., Geers, A., Pijl, S., Hasenstab, S., Justus, M., Brunelli, T., Adam, A., Borton, T., and Lemay, M. (1997). Speech perception abilities of adult and pediatric Nucleus implant recipients using the Spectral Peak (SPEAK) coding strategy. *Otolaryngol. - Head Neck Surg.* **117,** 236–242.

Svirsky, M. A., Silveira, A., Neuburger, H., Teoh, S. W., and Suarez, H. (2004). Long-term auditory adaptation to a modified peripheral frequency map. *Acta Otolaryngol.* **124,** 381–386.

Swanson, B., McDermott, H., Blamey, P., and Patrick, J. (2004). Pitch perception with half-wave gating audio processing. Proceedings of the VIII International Cochlear Implant Conference, Indianapolis, May 2004, p. 120.

Tobey, E. A., and Devous, Sr., M. D. (2003). Functional brain imaging of speech perception via electrical stimulation. *In* "Proceedings of the 25th Annual International Conference of the IEEE Engineering in Medicine and Biology Society" (R. S. Leder, Ed.), pp. 1991–1992, Cancun, Mexico.

Tong, Y. C., Black, R. C., Clark, G. M., Forster, I. C., Millar, J. B., O'Loughlin, B. J., and Patrick, J. F. (1979). A preliminary report on a multiple-channel cochlear implant operation. *J. Laryngol. Otol.* **93,** 679–695.

Tong, Y. C., Blamey, P. J., Dowell, R. C., and Clark, G. M. (1983a). Psychophysical studies evaluating the feasibility of a speech processing strategy for a multiple-channel cochlear implant. *J. Acoust. Soc. Am.* **74,** 73–80.

Tong, Y. C., Dowell, R. C., Blamey, P. J., and Clark, G. M. (1983b). Two-component hearing sensations produced by two-electrode stimulation in the cochlea of a deaf patient. *Science* **219,** 993–994.

Tsang, D. (2004). Musical pitch perception with cochlear implants: A comparison of strategies. *Thesis for Advanced Medical Science degree,* The University of Melbourne.

Turner, C. W., Gantz, B. J., Vidal, C., Behrens, A., and Henry, B. A. (2004). Speech recognition in noise for cochlear implant listeners: Benefits of residual acoustic hearing. *J. Acoust. Soc. Am.* **115,** 1729–1735.

Tyler, R. S., Parkinson, A. J., Wilson, B. S., Witt, S., Preece, J. P., and Noble, W. (2002). Patients utilizing a hearing aid and a cochlear implant: Speech perception and localization. *Ear Hear.* **23,** 98–105.

van den Honert, C., and Stypulkowski, P. H. (1987). Temporal response patterns of single auditory nerve fibers elicited by periodic electrical stimuli. *Hear. Res.* **29,** 207–222.

van Hoesel, R. J. (2004). Exploring the benefits of bilateral cochlear implants. *Audiol. Neuro-Otol.* **9,** 234–246.

van Hoesel, R. J., and Tyler, R. S. (2003). Speech perception, localization, and lateralization with bilateral cochlear implants. *J. Acoust. Soc. Am.* **113,** 1617–1630.

van Wieringen, A., Carlyon, R. P., Long, C. J., and Wouters, J. (2003). Pitch of amplitude-modulated irregular-rate stimuli in acoustic and electric hearing. *J. Acoust. Soc. Am.* **114,** 1516–1528.

Waltzman, S. B., Cohen, N. L., and Roland, J. T., Jr. (1999). A comparison of the growth of open-set speech perception between the nucleus 22 and nucleus 24 cochlear implant systems. *Am. J. Otol.* **20,** 435–441.

Weston, S. C., and Waltzman, S. B. (1995). Performance as a function of time: A study of three cochlear implant devices. *Ann. Otol. Rhinol. Laryngol. Suppl.* **165,** 19–24.

Wightman, F. L. (1973). The pattern-transformation model of pitch. *J. Acoust. Soc. Am.* **54,** 407–416.

Wilson, B. S., Finley, C. C., Lawson, D. T., Wolford, R. D., Eddington, D. K., and Rabinowitz, W. M. (1991). Better speech recognition with cochlear implants. *Nature* **352,** 236–238.

Wilson, B. S., Finley, C. C., Lawson, D. T., and Zerbi, M. (1997). Temporal representations with cochlear implants. *Am. J. Otol.* **18,** S30–S34.

Yost, W. A., and Sheft, S. (1989). Across-critical-band processing of amplitude-modulated tones. *J. Acoust. Soc. Am.* **85,** 848–857.

Zeng, F. G., and Galvin, J. J., 3rd (1999). Amplitude mapping and phoneme recognition in cochlear implant listeners. *Ear Hear.* **20,** 60–74.

Ziese, M., Stutzel, A., von Specht, H., Begall, K., Freigang, B., Sroka, S., and Nopp, P. (2000). Speech understanding with the CIS and the n-of-m strategy in the MED-EL COMBI 40+ system. *J. Otorhinolaryngol. Relat. Spec.* **62,** 321–329.

Zwolan, T., Kileny, P. R., Smith, S., Mills, D., Koch, D., and Osberger, M. J. (2001). Adult cochlear implant patient performance with evolving electrode technology. *Otol. Neurotol.* **22,** 844–849.

INDEX

G

H

J

L

Q

R

S

CONTENTS OF RECENT VOLUMES

Volume 39

Volume 40

Volume 41

Volume 42

Volume 43

Volume 44

Volume 45

Volume 46

Volume 47

Volume 48

Volume 49

Volume 50

Volume 51

Volume 54

Volume 55

Volume 56

Volume 57

Volume 58

Volume 59

Volume 60

Volume 61

Volume 62

Volume 63

Volume 64

Volume 65

Volume 66

Volume 67

Volume 68

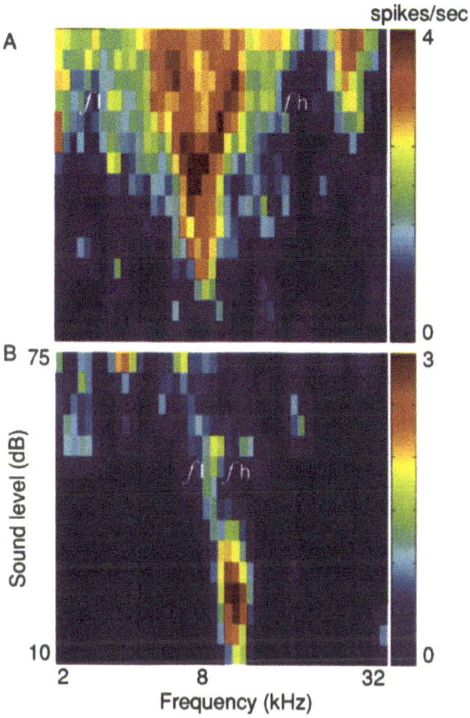

ESCABÍ AND READ, FIG. 1. Two cortical FTCs for pure tone sounds. Tone frequency and amplitude are varied over 5 octaves and 70 dB, respectively. Multi- or single-unit neuronal spike rate histograms are plotted for each frequency and level. The rainbow color spectrum shows lowest to highest spike rates (blue to red). Threshold is the lowest sound level eliciting a response above baseline. Characteristic frequency (CF) is the frequency at threshold. Linear bandwidth is the difference between the high (fh) and lower (fl) cutoff frequency at a given sound level. Examples A and B are broad bandwidth and narrow bandwidth FTCs, respectively. The rate level functions increase monotonically and non-monotonically with sound level for A and B, respectively. Neurons with non-monotonic narrow bandwidth FTCs in cortex are likely azimuth tuned (Clarey *et al.*, 1995; Barone *et al.*, 1996).

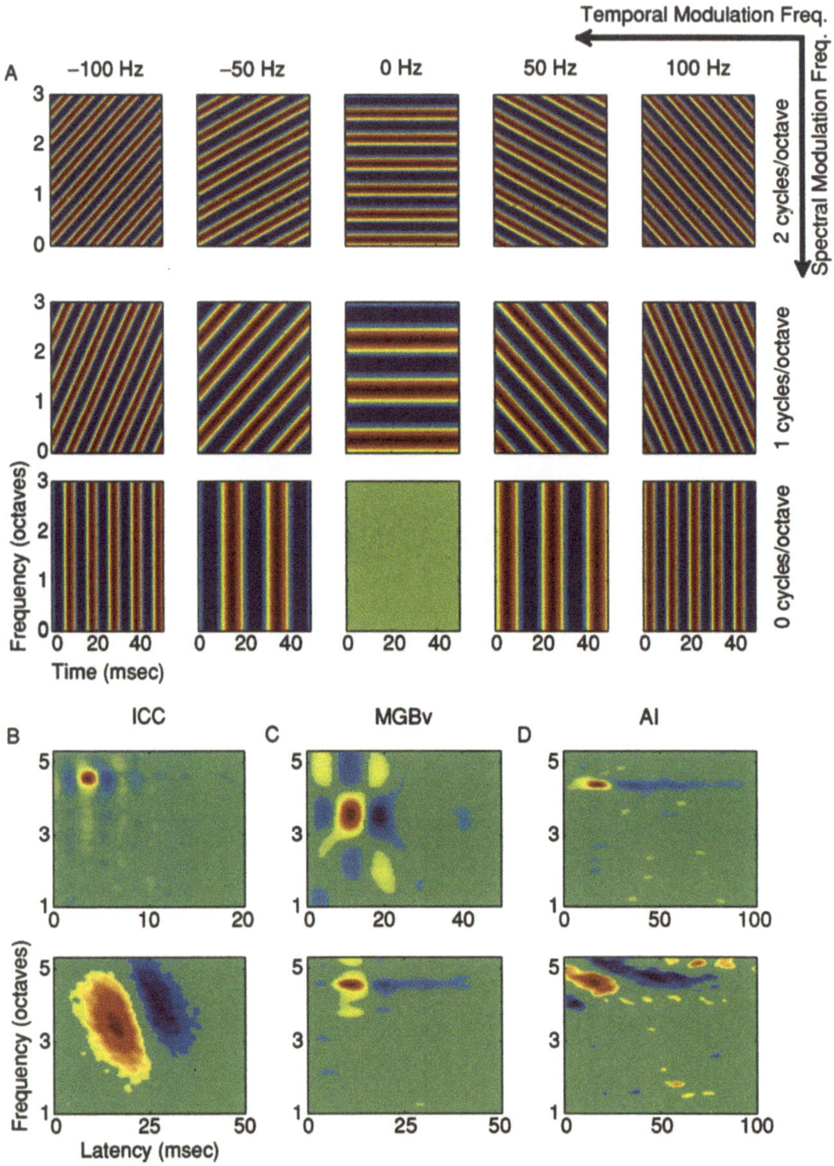

ESCABÍ AND READ, FIG. 3. (A) The spectrotemporal envelope of moving ripple sounds are illustrated at as a function of spectral and temporal resolution (SMF and TMF, respectively). Negative SMF produces ripples with upward moving frequency sweeps (top left). Alternately, a high to low frequency transition is observed for ripples with positive SMF (top right). Higher SMF and TMF correspond to finer spectral and temporal resolutions, respectively. Examples of spectrotemporal receptive fields (STRF) from the ICC (B), MGBv (C), and AI (D) were obtained with dynamic ripples sounds. A number of receptive field arrangements are possible: STRF features that are often observed include narrowband versus broadband tuning, presence of sideband inhibition, FM sweep selectivity, temporal inhibition and/or suppression. In general, a progression from fast to slow timing preferences is observed along the transition from the ICC to AI.

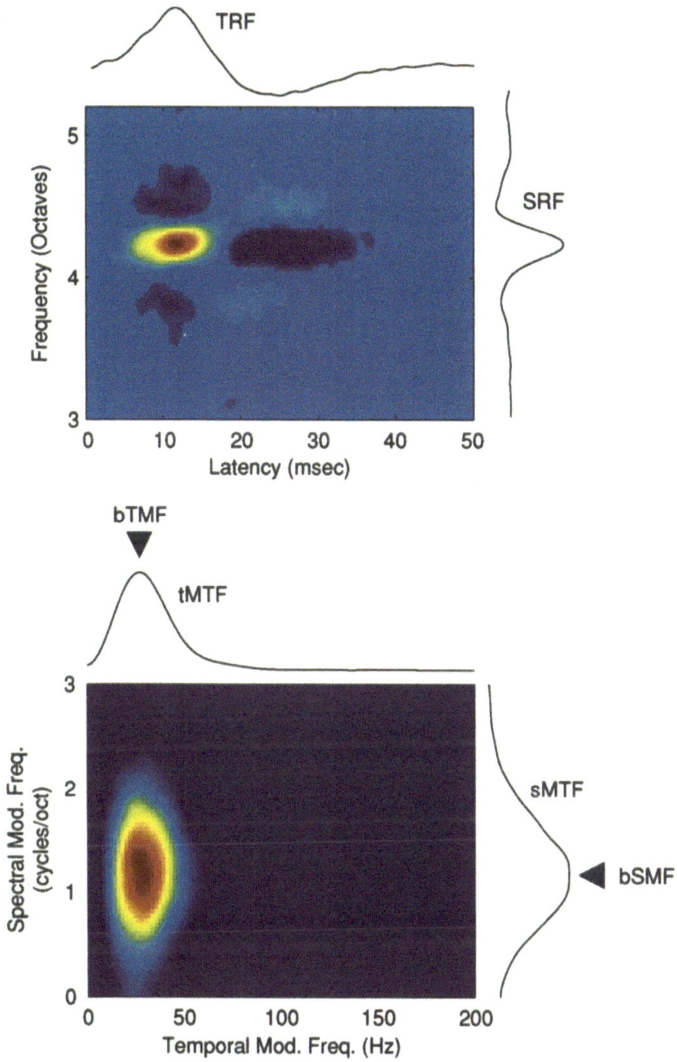

ESCABÍ AND READ, FIG. 4. Spectral and temporal modulation analysis of the STRF preference. A typical ICC receptive field is shown (left) along with its corresponding spectrotemporal modulation transfer function (stMTF) (right). The STRF can be decomposed into a temporal receptive field (TRF) and spectral receptive field (SRF) (Qiu *et al.*, 2003). This neuron's TRF is marked by an on-off temporal arrangement whereas its SRF exhibits an off-on-off response pattern. The modulation preference of this neuron can be determined by transforming the STRF to its stMTF, which exhibits a peak response at a preferred spectral and temporal resolution. The spectral and temporal cross sections of the stMTF (to the right and above the stMTF) show a peak about the neurons best temporal modulation (bTMF) and best spectral modulation (bSMF) (denoted by arrows). This neuron responds optimally to 30 Hz temporal modulations and spectral modulations about 1.25 cycles/octave.

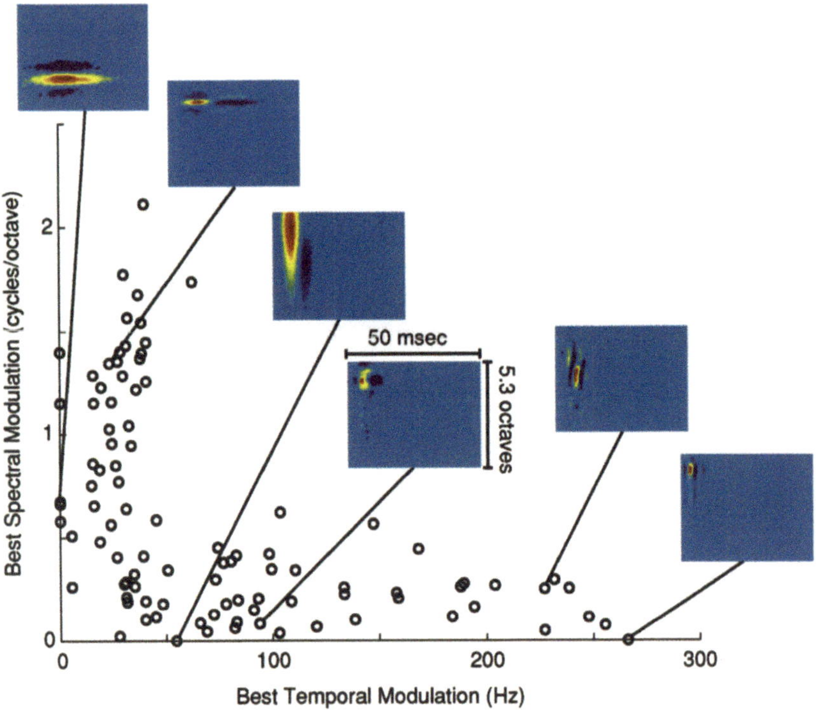

ESCABÍ AND READ, FIG. 6. The best temporal and spectral modulation parameters from ICC neurons exhibit a marked tradeoff in spectral and temporal integration resolution. A similar tradeoff is not observed in the MGBv or AI (Miller *et al.*, 2002; Qiu *et al.*, 2003). Insets show STRFs examples from the ICC at various bSMF and btMF. All STRFs are shown on the same scale (5.3 octave frequency range, ordinate; 50 msec temporal range, abscissa). At fast temporal modulation rates (>50 Hz), ICC neurons are characterized by short temporal duration, fast temporal modulation (Hz) and broad spectral bandwidths. STRFs with high (>50 Hz) modulation rates generally lack spectral sideband inhibition and respond optimally to spectral modulations below 0.5 cycle/octave. Alternately, slow temporal integration neurons (<50 Hz temporal modulation rates) are often characterized by narrow spectral tuning and significant spectral side-band inhibition in the STRF. Neurons with simultaneously high temporal (large bTMF) and spectral resolution (large bSM) are not observed.

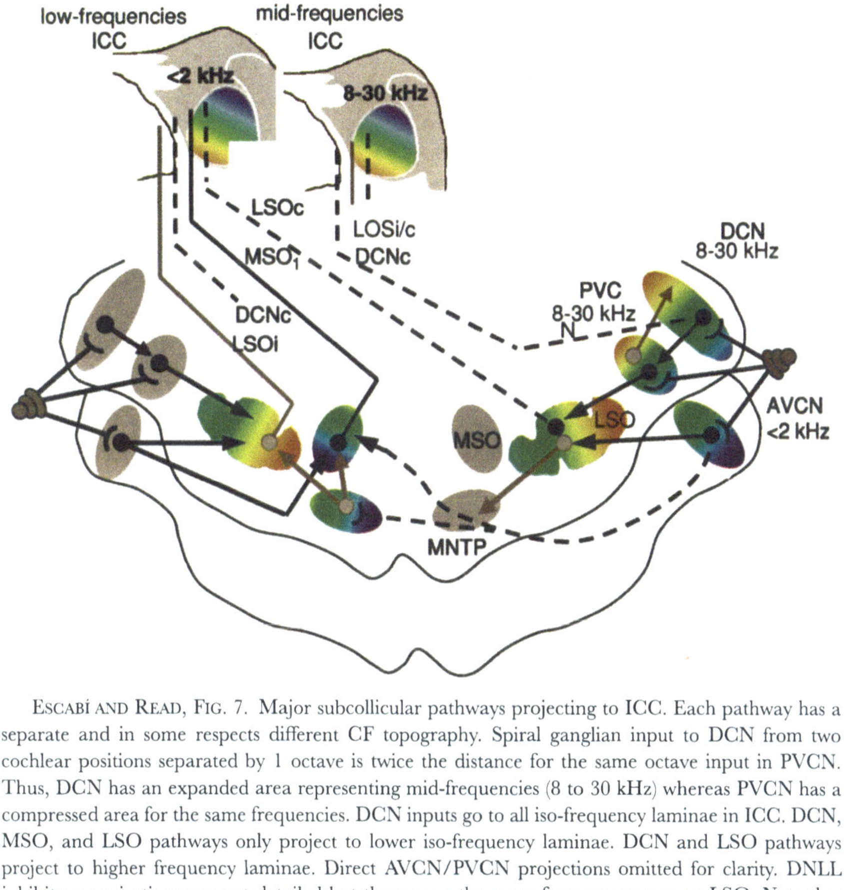

Escabí and Read, Fig. 7. Major subcollicular pathways projecting to ICC. Each pathway has a separate and in some respects different CF topography. Spiral ganglian input to DCN from two cochlear positions separated by 1 octave is twice the distance for the same octave input in PVCN. Thus, DCN has an expanded area representing mid-frequencies (8 to 30 kHz) whereas PVCN has a compressed area for the same frequencies. DCN inputs go to all iso-frequency laminae in ICC. DCN, MSO, and LSO pathways only project to lower iso-frequency laminae. DCN and LSO pathways project to higher frequency laminae. Direct AVCN/PVCN projections omitted for clarity. DNLL inhibitory projections are not detailed but they cover the same frequency range as LSO. Note that ipsi-hemispheric LSO projections to ICC are inhibitory. This figure is intended to summarize the cochleotopies feeding into ICC and many details have been omitted. Black solid and dotted lines represent excitatory input to ICC. The color scale (blue, green, red, orange) is used to represent the low to high frequency cochleotopic continuum.

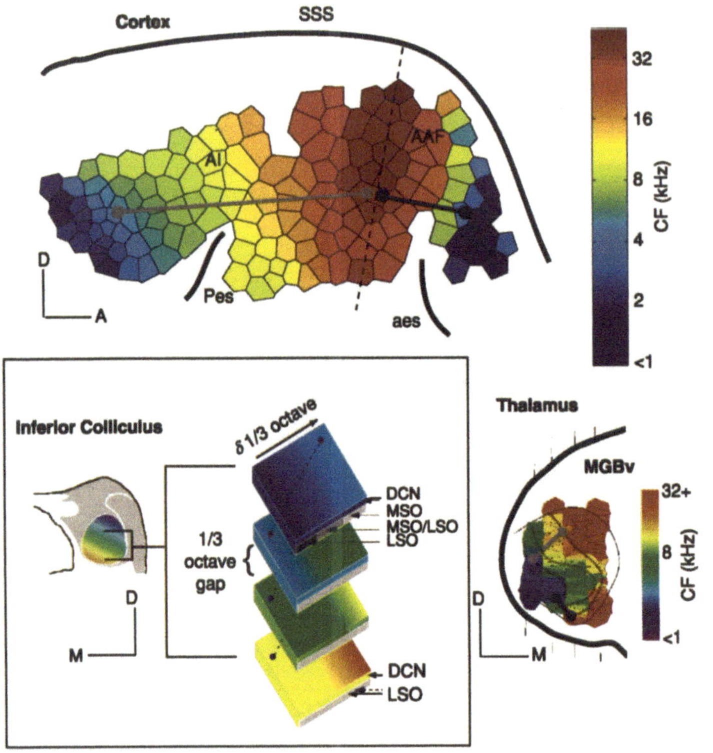

Escabí and Read, Fig. 8. Multiple cochleotopies in the lemniscal auditory pathway. Rainbow color spectrum indicates CF for each recording site. Blue and red indicate CFs ≤ 1 kHz and ≥ 32 kHz, respectively. Inset: Three-dimensional organization of CF in the IC. CF change is small for tangential shifts along the "iso-frequency laminae" (dorso-medial to ventro-lateral). CF changes step-wise with $\{1/3\}$ octave shifts along the dorso-ventral dimension. MSO/LSO and DCN all project to different sub-laminae of the same low CF iso-frequency laminae. Only DCN and LSO project to the high frequency CF (e.g., >16 kHz) laminae. Thalamus: Rainbow color spectrum indicates CF. Recording positions of four recording tracks within MGBv of the cat reconstructed from data of Imig and Morel (1985). Notice that the mm distance for one octave of CF in the inferior part of MGBv is smaller than that of the superior MGBv. Anatomical studies also show two cochleotopies in cat MGBv (Read *et al.*, 2004). Cortex: topographic map of CF showing low-to-high CF gradient for AI and a mirror reversed CF gradient for AAF (Imaizumi *et al.*, 2004). Note that the mm distance for one octave in AI is about twice that of AAF. Apparently the coarse cochleotopy in AAF is inherited from inferior MGBv.

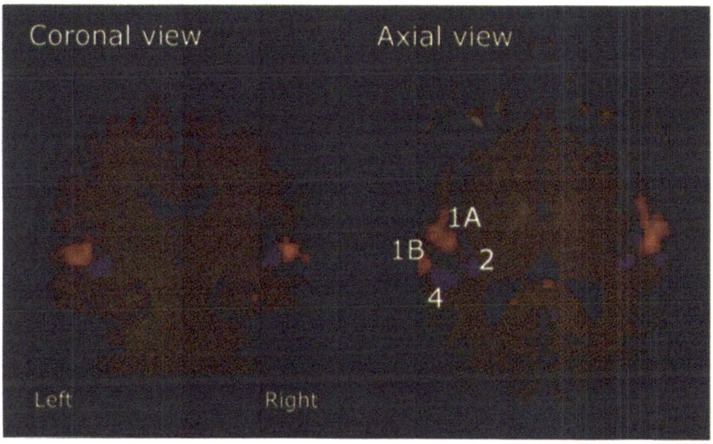

HALL, FIG. 5. Coronal and axial views of frequency-sensitivity in a single listener. Red regions represent activity that was significantly greater for stimuli with 800-Hz center frequency than for 4.5 kHz center frequency ($p < 0.001$). Blue represents the significant converse comparison. The coronal view illustrates the dominant medial-lateral frequency axis that has been described by electromagnetic studies. Numerical labels identify the multiple frequency-sensitive sites around HG as defined by Schönwiesner *et al.* (2002). The source is unpublished data from the author.

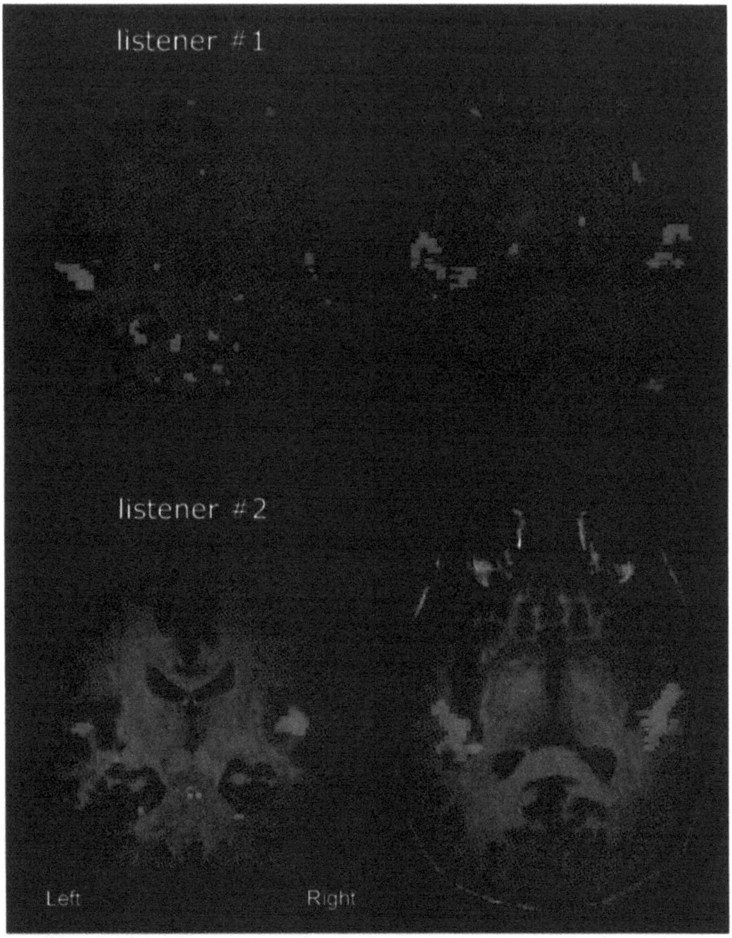

HALL, FIG. 7. A comparison between single-frequency tones and wide-spectrum noise that had a 0.75 octave bandwidth centered at the same frequency ($p < 0.001$). Coronal and axial views are shown for two listeners. Listener #1 is the same as that presented in Fig. 5. Noise evokes greater activation in medial and lateral subdivisions close to HG, although its precise distribution is not identical across listeners (based on unpublished data from the author).

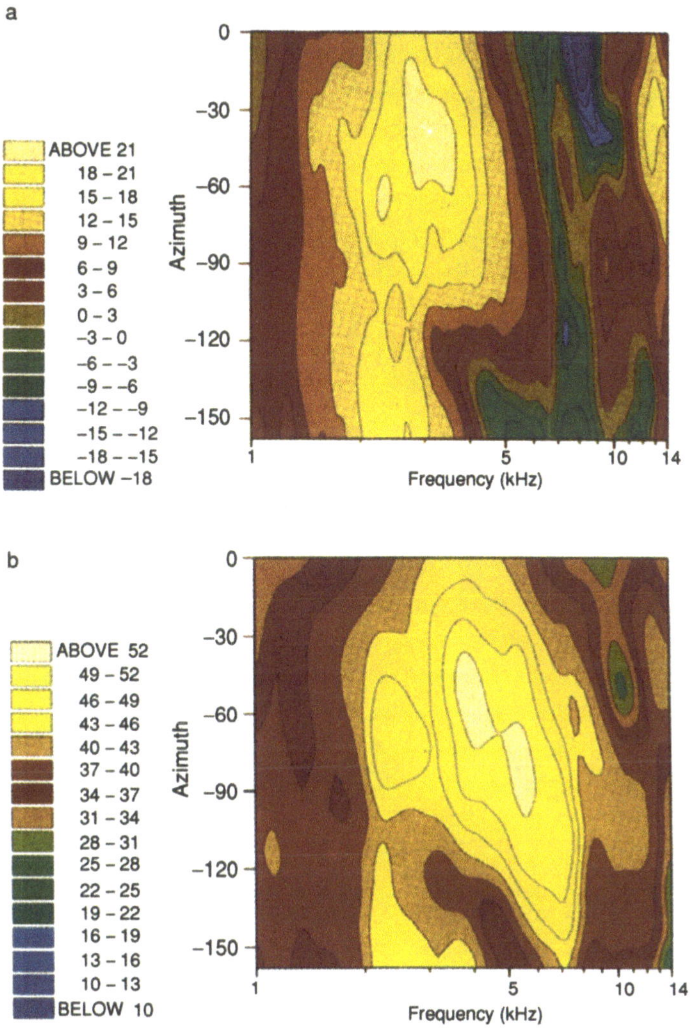

CARLILE, *ET AL.*, FIG. 1. (a) The variation in the head-related transfer function as a function of the azimuthal location of the sound source is shown for the left audio visual horizon of a subject (0° directly ahead). The transfer functions were measured using a small probe microphone located about 3 mm from the subjects ear drum. Color scale is in dB. (b) To illustrate the features of the transfer functions that are likely to be encode by the auditory system the data in (a) has been passed through a cochlear filter model (see Carlile and Pralong, 1994, for more details).

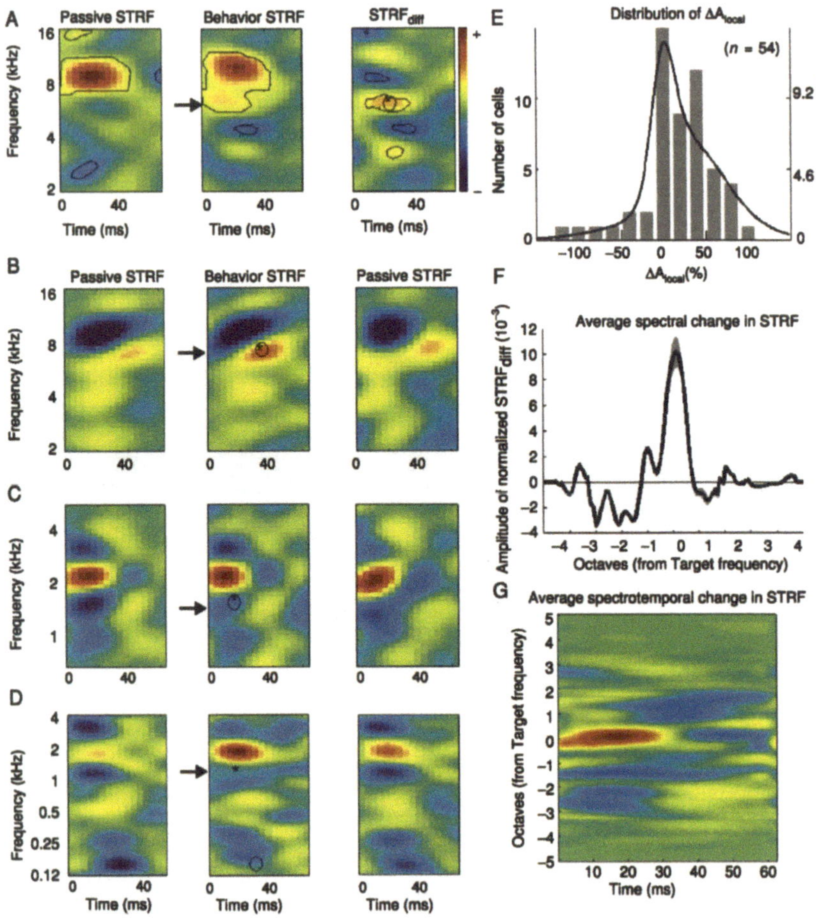

IRVINE AND WRIGHT, FIG. 2. Plasticity of spectrotemporal receptive fields (STRFs) of neurons in AI of the ferret. (a to d) STRFs of four single neurons showing changes typical of those observed during performance of the detection task. For each neuron a pre-behavior "passive STRF" is shown in the left column, and a "behavior STRF" recorded during performance of the detection task is shown in the middle column. The STRF in each panel was normalized, and all STRFs were then depicted using the same grey scale. The contours in (A) demarcate the excitatory (white) and inhibitory (black) regions with statistically significant fluctuations (± 3 SD) from the mean (see Methods in Fritz *et al.*, 2003, for details). These contours are not drawn on the lower panels to avoid cluttering the figures. The black arrows in the middle column indicate the frequency of the target tone during the detection task. (A) In the right column, the difference between the normalized passive and behavior STRFS (STRF$_{diff}$) is shown. An asterisk marks the location of maximum local change (A_{local}; 45%), and a circle marks the global change (A_{global}; 45%), defined in Fritz *et al.* (2003). In this case the local and global maximal changes were both at the target frequency, as in about one-half of all neurons. (B) Localized enhancement of an excitatory region in the STRF during behavior (left and middle). The post-behavior passive STRF (right column) reverted immediately to its original shape

(A_{local}, 37%; A_{global}, 42%). The two maxima were nearly coincident at the target frequency. (C) Local decrease or elimination of inhibitory sidebands in the behavior STRF. The inhibition recovered quickly afterwards, but the overall STRF shape was different (A_{local}, 40%; A_{global}, 40%; coincident maxima at target frequency). (D) A global weakening of inhibitory fields during behavior. Immediately following behavior the STRF recovered its pre-behavior shape (A_{local}, 64%; A_{global}, 86%). In this example the local maximum difference occurred at the target tone frequency, whereas the global maximum was located over a low-frequency inhibitory field (which was also knocked out during behavior). (E) Summary histogram and smoother distribution of local STRF changes from all STRFs in the study. The histogram (left ordinate) and distribution (right ordinate) are significantly skewed toward positive changes (overall mean, +20.2%). (F) Average spectral change in the STRF at all frequencies relative to the target frequency. This curve shows the average plastic change in STRF amplitude as a function of log-frequency distance from the target frequency (centered at 0 and marked by the dot). The shaded region around the curve represents the variance around the mean. There was facilitation for about one octave around the target (half-width of ± 0.56 octave) and asymmetric suppressive sidebands outside of this range. (G) Average spectrotemporal changes in the STRF derived from all neurons. The image was constructed by computing the $STRF_{diff}$ for each neuron, centering it around its target frequency, and then averaging over all neurons. The facilitative and suppressive changes near the target frequency, as well as the relatively rapid onset of these STRF changes, can be seen here. Reproduced with permission from Fritz *et al.* (2003).

FSC
www.fsc.org
MIX
Papier | Fördert gute Waldnutzung
FSC® C083411

Kolibri 360 GmbH
Ferdinand-Jühlke-Straße 7
99095 Erfurt, Deutschland
produktsicherheit@kolibri360.de